"So involved is Brinkley with his material that this groundbreaking history reads like a saga. Of course, the glorious natural setting and a spectacular cast of characters comprising a 'noble band of conservationist revolutionaries' have an enormous draw."
—*The New Orleans Times-Picayune*

"Let's hope that conservation historians of the future will write about our era with the same zest and gratitude Brinkley brings to his landmark book about a time when people spoke out on nature's behalf, and were heard. . . . Brinkley's avidly detailed profile of [William] Douglas is one of this altogether revelatory book's most important sections. It is here that one sees most clearly how instructive a microcosm Alaska is for today's environmental concerns."   —Donna Seaman, *The Kansas City Star*

"Even when he covers well-known controversies, Brinkley's forceful storytelling can add new insights. . . . If, like me, you enjoy reading bracing accounts of conservation battles won against great odds by impassioned activists, writers, and artists, you should find *The Quiet World* engrossing."
—*The Austin American Statesman*

"What makes the scope of his book so remarkable is that Brinkley doesn't limit it to those who wielded direct political or economic power. Painters, poets, writers, scientists, naturalists, mystics, and seemingly ordinary people—all of whom share the distinction of devoting their lives to a great cause, a lifework, no matter the cost—are given due as well."   —Marc Covert, *The Oregonian*

"Charles Sheldon, Bob Marshall, William Temple Hornaday, Olaus and Margaret 'Mardy' Murie, Lois and Herb Crisler, among others, rise in three dimensions in Brinkley's hands, getting overdue credit for their varied roles in saving wilderness in a state that was their cause célèbre and muse."
—*The Anchorage Daily News*

The Quiet World

Also by Douglas Brinkley

The Wilderness Warrior: Theodore Roosevelt and the Crusade for America

*The Great Deluge: Hurricane Katrina, New Orleans,
and the Mississippi Gulf Coast*

Windblown World: The Journals of Jack Kerouac, 1947–1954 (editor)

*Wheels for the World: Henry Ford, His Company,
and a Century of Progress, 1903–2003*

The Mississippi and the Making of a Nation
(with Stephen E. Ambrose)

American Heritage History of the United States

The Western Paradox: Bernard DeVoto Conservation Reader
(editor, with Patricia Nelson Limerick)

Rosa Parks

The Unfinished Presidency: Jimmy Carter's Journey Beyond the White House

The Majic Bus: An American Odyssey

Dean Acheson: The Cold War Years, 1953–1971

Driven Patriot: The Life and Times of James Forrestal
(with Townsend Hoopes)

— The Quiet World —

• • • • •

SAVING ALASKA'S
WILDERNESS KINGDOM,
1879–1960

• • • • •

DOUGLAS BRINKLEY

HARPER ● PERENNIAL

NEW YORK • LONDON • TORONTO • SYDNEY • NEW DELHI • AUCKLAND

HARPER ● PERENNIAL

A hardcover edition of this book was published in 2011 by HarperCollins Publishers.

THE QUIET WORLD. Copyright © 2011 by Douglas Brinkley. All rights reserved. Printed in the United States of America. No part of this book may be used or reproduced in any manner whatsoever without written permission except in the case of brief quotations embodied in critical articles and reviews. For information address HarperCollins Publishers, 10 East 53rd Street, New York, NY 10022.

HarperCollins books may be purchased for educational, business, or sales promotional use. For information please write: Special Markets Department, HarperCollins Publishers, 10 East 53rd Street, New York, NY 10022.

FIRST HARPER PERENNIAL EDITION PUBLISHED 2012.

Designed by Leah Carlson-Stanisic

The Library of Congress has catalogued the hardcover edition as follows:

Brinkley, Douglas.
    The quiet world : saving Alaska's wilderness kingdom, 1879-1960 / Douglas Brinkley.—1st ed.
      x, 576 p., [32] p. of plates : ill. (some col.), maps (some col.) ; 24 cm.
    Includes bibliographical references (p. [509]-550) and index.
    ISBN 978-0-06-200596-0
    1. Nature conservation—Alaska—History. 2. Environmental protection—Alaska—History. 3. Conservationists—Alaska—History. 4. Nature conservation—Alaska—History. I. Title.
    333.72'09798 2011
    333.78216   2011280865

ISBN: 978-0-06-200597-7 (pbk.)

    12   13   14   15   16   OV/RRD   10   9   8   7   6   5   4   3   2

To

SAM HAMILTON

Visionary at U.S. Fish and Wildlife . . . stout friend of Alaska's Arctic Refuge . . . and a true believer in the Quiet World.

&

STONE WEEKS

My twenty-three-year-old assistant at Rice University . . . killed in a trucking accident in Virginia on July 23, 2009. . . . He was an angel of pure future . . . with an intense love of wild Alaska.

&

EDWARD A. BRINKLEY

My father . . . who served in the U.S. Army as a sergeant with the 196th Regimental Combat Team during the Korean War from 1950 to 1952, based out of Fort Richardson, Alaska. . . . For telling me many great army stories about encountering grizzlies on his Alaska Range ski patrols from Haines to Fairbanks.

And I brought you into a plentiful country, to eat the fruit thereof and the goodness thereof; but when ye entered, ye defiled my land, and made mine heritage an abomination.

—Jeremiah 1:6

When roads supplant trails, the precious, unique values of God's wilderness disappear.

—William O. Douglas, My Wilderness: The Pacific West *(1960)*

Is it not likely that when the country was new and men were often alone in the fields and the forest they got a sense of bigness outside themselves that has now in some way been lost. . . . Mystery whispered in the grass, playing in the branches of trees overhead, was caught up and blown across the American line in clouds of dust at evening on the prairies. . . . I am old enough to remember tales that strengthen my belief in a deep semi-religious influence that was formally at work among our people. The flavor of it hangs over the best work of Mark Twain. . . . I can remember old fellows in my hometown speaking feelingly of an evening spent on the big empty plains. It had taken the shrillness out of them. They had learned the trick of quiet.

—Sherwood Anderson, letter to Waldo Frank (November 1917)

CONTENTS

* * * * * *

Contents

PROLOGUE:

JOHN MUIR AND THE GOSPEL OF GLACIERS

Glaciers move in tides. So do mountains. So do all things.
—JOHN MUIR

I

How sad *John Muir, founder of* the Sierra Club, would be to learn that in the first decades of the twenty-first century many of the great glaciers of Alaska were melting away at an astonishing rate. Like the Creator himself, glaciers were architects of Earth, sculpturing vast ridges, changing bays, digging out troughs, making concavities in bedrock, and creating fast-flowing rivers.[1] Global warming—the alarming increase of the Earth's near-surface air temperature exacerbated by carbon dioxide emissions from gasoline-powered vehicles and by the burning of coal—was stealing away the glacial ice fields of Alaska. Nevertheless, big oil companies such as Shell, Exxon-Mobil, and BP still put climate change and greenhouse gases in scare quotes, as if the hard science were a myth conceived by tree huggers. Fossil fuel merchants were determined to keep Americans hooked on petroleum-based products until they choked. The Swedish physical chemist Svante Arrhenius was worried, in 1896, as the automobile revolution was just taking hold, that widespread fossil fuel combustion could someday cause enhanced global warming. Arrhenius, now considered the "father" of climate change, understood that the doubling of carbon dioxide (CO_2) concentration would lead to a temperature rise of five degrees Celsius; glaciers would melt, seas would rise, and the Arctic would slowly vanish.[2]

John Muir—the naturalist whom Ralph Waldo Emerson called "more wonderful than Thoreau"—had erected a tiny observation cabin near a thirty-mile-long glacier that was one of Alaska's stunning heirlooms.[3] Born in Dunbar, Scotland, in 1838, Muir had immigrated to America in 1849, just after Mr. James K. Polk won the Mexican-American War. When Muir turned twenty-nine, following an industrial accident in Indianapolis that had caused temporary blindness, he made a far-reaching personal decision to dedicate his life to the natural world and to enduring wilderness. Although he was a talented machinist, nature was his muse. Solitary and on foot he roamed through America's wide valleys, towering mountains, pristine woodlands, sublime deserts, and flower-filled meadows, filling his voluminous notebooks with vivid descriptions of plants, animals, and trees. Recording his scientific observations along the way, the peripatetic Muir tramped through the primordial forests and smoky ridges of the Appalachian Mountains, then headed south to survey the humid swamplands of Georgia's Okefenokee and the golden beaches of Florida's Gulf Coast. Shedding the dictates of his strict Presbyterian upbringing (his father was a fundamentalist minister), in 1867 Muir scrawled his home address on a weathered journal cover as "John Muir, Earth-Planet-Universe."[4] Eventually making wild California his North Star, Muir, a pioneer ecologist, began climbing the peaks of his beloved Sierra Nevada, camping under the stars, memorizing botanical details through the timeless art of sitting still. "The more savage and chilly and storm-chafed the mountains," Muir wrote, "the finer the glow of their faces."[5]

Despite all of Muir's cross-country tramps, nothing prepared him for the sheer poetic depth of the Alaskan wilderness. Muir considered himself a student of Louis Agassiz, an internationally celebrated Harvard zoologist and geologist, whose *Études sur les glaciers* (1840) was the definitive word on glaciers in the 1870s. Agassiz had explored *live* glaciers, studying their origins in the Piedmont and Tidewater regions. Glaciers could be snow-white like typing paper or a brazen virtual blue, as gray as a gravel pit or as clear as H_2O. Some extended over twenty square miles and could be as smooth as velvet or as wrinkled as a bull walrus's neck. They had blotches, slashes, stripes, and swirls. Other cirque glacier remnants covered less than a square mile. When calving, a glacier rumbled and roared, then as the ice sank or floated a strange vibration, like wind chimes, curled the air as if a tuning fork had been bonked. Unbeknownst to most Americans of the late nineteenth century, glaciers constituted the biggest freshwater reservoir on Earth.[6]

Muir was frustrated that in Yosemite he could analyze only the *effects* glaciers had on mountains; it was all the geological past. For his professional glaciology career to advance, he needed to see the real deal—to experience glaciers themselves, in raw action. Alaska was, to Muir, the ideal laboratory for studying "frozen motion" as it flowed downhill as if icy blue lava. All glaciers were cold, solid, scalloped, and slippery. But besides those four basic features, each glacier had a distinct personality of its own. Muir, with the keen eye of a farmer inspecting his crops, was looking for fresh scientific evidence of glacial deformation, recession, and retreat. Every nuance mattered. Keys to Earth's geological history could possibly be found by studying ice fields. Alaska's umpteen glaciers were to become his field teachers. "When a portion of a berg breaks off, another line is formed, and the old one, sharply cut, may be seen rising at all angles, giving it a marked character," Muir reported. "Many of the oldest bergs are beautifully ridged by the melting out of narrow furrows strictly parallel throughout the mass, revealing the bedded structure of the ice, acquired perhaps centuries ago, on the mountain snow fountains."[7]

Muir, America's legendary naturalist, first traveled to southeast Alaska's Inside Passage from June 1879 to January 1880.[8] Throughout his seven months in the district he wrote "wilderness journalism" for the *San Francisco Daily Evening Bulletin*; one expanded article actually became a tourist booklet for the Northern Pacific Railroad.[9] In April 1879 *Scribner's Monthly* had published Witt Ball's article on Alaska, "The Stickeen River and Its Glaciers."[10] A creatively competitive Muir probably figured he could top the pedantic Ball. Seeing the live glaciers of Alaska, and writing about them factually but with gusto, would allow Muir to verify his long-held hunches on glacial action and tectonic activity. Known for his abiding love of Yosemite Valley. Muir promoted the somewhat controversial notion that the gorgeous California Valley had been carved out by glaciers (not rivers). Muir's first published work, for what was then a handsome fee of $200, was an article for the *New York Tribune*, "Yosemite Glaciers"; it appeared on December 5, 1871.[11]

Muir's journey began aboard the *Dakota*, which steamed out of San Francisco near Alcatraz Island and two days later churned past the high cliffs and tree-lined shores of Puget Sound, and then entered the waters of British Columbia. The Inside Passage, through which Muir was traveling, included all the waterways from north of Puget Sound to west of Glacier Bay. Next the *Dakota* threaded through the Alexander Archipelago islands to Sitka, Alaska. The ship, though occasionally protected by land, was terribly vulnerable to the Pacific gales. To the lean, bearded Muir, however, these 10,000 miles of southeastern Alaskan islands and fjords (long, deep arms of the ocean, carved out by a glacier) and 1,000 camel-back islands, dense with western hemlock and Sitka spruce, were "over-abundantly beautiful for description."[12] Giant cliffs billowed straight out of the seawater, rising 500, 600, 700 feet over the Pacific Ocean. A frustrated Muir kept pleading with the captain to stop and let him quickly climb a mountain, but to no avail.

As the *Dakota* ventured farther up the Inside Passage (now the longest protected marine waterway in the world), Muir—a taut man of forty, with red-brown hair and beard, always stooping over to jot notes—played the populist professor. He kindly explained to tourists aboard that the snouts of glaciers shed blocks of ice in a "calving" process. With his thick Scottish brogue, Muir, a natural raconteur, made even the most citified tourist ready to paddle into quiet coves around Baranof Island, to kayak down a cleaved river as it roared out into Sitka Sound and then out to the Pacific. So excited had Muir become by the breathtaking scenery that he fantasized about climbing mountains up to Alaska from California someday, exploring Mount Shasta, Mount Hood, and Mount Rainier. What made Muir so special, the quality in his character that had made Emerson take note, was the way the enthusiastic naturalist fully integrated scientific knowledge with romantic wildness. Nobody could resist Muir's charm.

That fall of 1879 Muir furiously scribbled astute observations about Native Alaskan people, gold seekers, lumberjacks, canneries, and cosmic natural features. Muir even developed his own "glacial gospel": that fjords and wilderness, like gentle magic, lifted the soul on a journey of self-discovery filled with an infinity of unknowns. Inner peace could be found in glaciers. Southeastern Alaska was an immortal land that would, in turn, immortalize him.[13] Picking his way through a sea of sparkling bergs, sometimes leaping across slippery, deteriorating ice floes, Muir reveled in the innate dignity of his surroundings. "A new world is opened," Muir wrote in his journal, "a world of ice with new-made mountains standing vast and solemn in the blue distance roundabout to it." [14]

It took Muir only a day to become a booster for Alaska's magnificent Glacier Bay. The land uplift rate—1 inch per year—was among the highest in the world, because the glaciers receded, thus removing their considerable weight from the land. In his wilderness journalism Muir urged Americans to journey to paradisiacal Alaska and let their jaws drop. Although Muir didn't discover Glacier Bay, his enthusiasm made the bay internationally celebrated. "Go," Muir cried, "go and see." [15] Alaska, purchased from Russia for $7.2 million only twelve years prior, had just started to be discovered by nature lovers who cruised up the southeast coast from Seattle. Muir, in a way, was the first great ecotourist of Alaska. Go to Kachemak Bay . . . Catch a halibut . . . Go pick yellow-reddish salmonberries and currants on the banks of the Chilkat River . . . Tramp the glacier ice mantle of the Coast Range . . . Go eye bald eagles nesting in Juneau . . . Go gather seashells at Calvert Island beach during low tide . . . Go spy on the white mountain goats of Howling Valley . . . Go to the boulder-bound Chugach Mountains . . . Go see the northern lights' "auroral excitement" and "bright prismatic colors" flash across the starlit night at the Yukon River . . . It was the Earth's halo . . . Didn't you know? [16]

Muir's first landfall aboard the *Dakota* was Fort Wrangell, Alaska. Here he joined thirty-year-old S. Hall Young, a Presbyterian missionary hoping to Christianize the Chilkat Tlingit. Together Muir and Young would travel all over the Inside Passage, constantly in ice range, to Sitka, the Stikine River, Fairweather Range, and, last but not least, Glacier Bay. Young later wrote a memoir—*Alaska Days with John Muir*—about their fine times together. But Fort Wrangell, crude and vulgar, devoid of even an iota of charm, was an end-of-the-line outpost where lawlessness reigned supreme. A grumbling Muir didn't cotton to the devil-may-care attitude of the Euro-Americans looking for quick mining profits in such a picturesque setting. Fort Wrangell was an ugly row of low wooden buildings (not too far as the crow flies from today's Misty Fiords National Monument Wilderness). Some of Muir's "Go . . . go . . . go to Alaska" evangelism tapered off in Fort Wrangell, where he slept on the dusty floor of a carpenter's shop. Muir described his quarters as "a rough place, the roughest I ever saw . . . oozy, angling, wrangling Wrangell." [17] Locals didn't know

what to make of Muir. "What can the fellow be up to?" one resident in-
quired. "I saw him the other day on his knees looking at a stump as if he
expected to find gold in it. He seems to have no serious object whatever."[18]
A few years earlier, Young had tried breaking colts but had ended up with
both shoulders seriously dislocated. Carrying a backpack up glaciers
was understandably challenging for him. "Muir climbed so fast that his
movements were almost like flying, legs and arms moving with perfect
precision and unfailing judgment," Young wrote. "I must keep close behind him or I would fail to see his points of advantage."[19]

Clad in a Scottish cap and long gray tweed ulster, Muir could have been
a shepherd from the island of Skye. Lured by his ethereal surroundings,
he even wandered around in a rainstorm, eager to learn what "songs" the
Alaskan trees "sing" when wet.[20] Muir wanted to map Glacier Bay—shaped
like God's horseshoe and opening out to the Gulf of Alaska, with immense glacial walls of ice tumbling out of snouts at Icy Strait—as a freelance service for the U.S. government. No cartographer had yet done the
job. Mapmakers aren't keen on moving ice. Yellowstone—America's first
national park—was only seven years old in 1879. Muir—who in 1901 would
write *Our National Parks*, perhaps the most seminal preservationist essay in American history—wanted to see many such public wonderlands
created by Congress. Perhaps Glacier Bay, he intuited upon his first visit,
would someday meet that criterion. "Muir's depiction situates Alaska as
the New World's 'new world,'" the ecocritic Susan Kollin argued in *Nature's State*, "a Last Frontier that enables the United States to once again
unmap and remap itself."[21]

Passing the coast of Admiralty Island, Muir and Young, canoeing amid
the fjords, saw a couple of brown bears, which seemed to smell their leaf
tobacco, rice, bread, and sugar. It was monumental scenery, wild beyond reach, with deep vistas and glacier-carved valleys that surpassed
the Swiss Alps or the Norwegian fjords.[22] Eventually they discovered an
amazing ice expanse, soon dubbed Muir Glacier. Its terminus was at a
maximum during the Little Ice Age around 1780 (between 1914 and 2010,
this thirty-mile glacier retreated by almost twenty miles).[23] Frequently
paddling into eddies for breaks, their arms always sore from fighting currents, Muir and Young bonded. The Chilkat Tlingit village up the Lynn
Canal, where they camped, became the village of Haines in 1884 (named

after Mrs. F. E. Haines, chairwoman of the committee that raised funds for its construction). "I know of no excursion in any part of our vast country where so much is unfolded in so short a time," Muir wrote. "Day after day, we seemed to float in a true fairyland, each succeeding view more and more beautiful. . . . Never before this had I been embosomed in scenery so hopelessly beyond description."[24]

Glacier Bay was a touchstone landscape to Muir. The Tlingit, who had lived around Glacier Bay for 8,000 years, called the region Sitakaday ("the bay where the ice was").[25] Muir had spent 1861 to 1862 at the University of Wisconsin learning about glaciers from his geology professors. Hiking around the Sierra Nevada, Muir had been able to study the effects of the glacial process. But now, in October 1879, with four Tlingit Indian guides— experts at catching all five species of Pacific salmon (sockeye, king, coho, pink, and chum)—he was experiencing the glacial ice firsthand. The geologic force of ice, he was convinced anew, shaped Alaska *and* the canyon lands and peaks of the Sierra Nevada. Glaciers, he decided, were truly the divine spirit of nature writ large, more priceless than gold, able to carry away entire mountains, "particle by particle, block by block and cast them into the sea."[26] One of the Tlingit guides complained to Young that Muir "must be a witch" to "seek knowledge" in "such a place" as Glacier Bay, especially in the "miserable weather" of a blinding snowstorm.[27]

Muir admired the prowess of the Tlingit with their handcrafted thirty-foot dugout canoes carved from cedar, which had twin sails, allowing them to stealthily cover vast distances in good time. By the campfire, he enjoyed hearing their trickster stories about ravens, known to lead bears to their prey and even to play hide-and-seek with wolves. With a keen eye for masks, paddles, and jewelry art, Muir studied Tlingit totem poles. He chuckled, however, at ancient Native American superstitions regarding glaciers as supernatural or extraterrestrial or weird natural phenomena. For all of Muir's high-octane romanticism and use of tropes about scenic wonders, he was a botanist-naturalist-glaciologist addicted to scientific fact. Tlingit folklore went only so far with him. The Tlingit, for their part, didn't care that Muir was an encyclopedia of literature about moraines (both medial and terminal). Generally speaking, First Nation people interested Muir less than the glaciers; he still saw them as "savage." In *First Summer*, for example, Muir wrote that the "uncleanliness" of Sierran Indians bothered him tremendously. If Young, the missionary, was going to help the Tlingit prosper, Muir thought hygiene had to come first.

At night while the Tlingit guides stayed at camp, the ecstatic Muir would climb up the glacial slopes to feel the full power of phantasmagoric geology at work. During the summer months it stayed light almost all night long in Alaska. This worked to Muir's favor. At a glance Muir knew if a glacier was advancing or retreating, or whether the precipitation during any given year had caused the ice to surge.[28] Like Michelangelo measuring luminosity in the Sistine Chapel, Muir studied the Inside Passage as light struck the dense glacial ice. Every shade of blue in the spectrum dominated by a wavelength of roughly 440 to 490 nanometers miraculously appeared, scattered by the crystalline ice; and the blue glow was dispersed and refracted in such a subtly distinguished array of tints that no words existed for them in *Webster's Dictionary*.[29] Unlike the Alaska Range, which lay in the district's interior, and where the glacial process was slowed by the fierce cold, the Fairweather Range and Coast Mountains, where temperatures were mild yet there was lots of compact snow, were an ideal setting for glaciers to develop. A layer of snow could transmute into glacial ice in a few decades. For the study of glaciers, the Inside Passage was like Greenland, a hypernatural landscape that seared itself forever in Muir's fervent imagination.

For Young, keeping up with Muir's glacier terminology could be frustrating. Absolute verity was essential to everything Muir did. When the professor espoused the gospel of glaciers, Young was reduced to listening. There was a glossary of Muir's terms to understand: *hanging glacier* (above a cliff or mountainside); *kettle pond* (created when a massive iceberg melted, leaving behind a water-filled hollow); *firn* (grainy ice, which is formed from snow about to become glacial ice). Before traipsing around Glacier Bay with Muir, Young hadn't realized that in 1794 the British explorer George Vancouver (British Columbia's fantastic city is named after him) had demarcated the entire Glacier Bay area as a *single* ice mountain, which then separated into the twelve smaller ones. For Young every moment with the great Muir was like being taught by Charles Darwin or Thomas Huxley. Naturally inquisitive about the Glacier Bay, Young asked his naturalist friend a lot of questions. The world's authority on glaciers—John Muir—was canoeing with him for hours at a time in Alaska, espousing the glacial gospel like a preacher at a revival meeting.[30]

Instead of being self-centered, Muir at Glacier Bay was life-centered. Feeling he belonged to wild Alaska, a child of the tidal flat, Muir understood anew that the whole Earth was a watershed, just one giant dewdrop. He thanked God for such a magnificent plan. To get around the Alexander Archipelago, Muir used a reprint of George Vancouver's old nautical charts to help him navigate.[31] At Glacier Bay he filled his journals with vibrant writing about his canoe trips, the maritime currents, and the ice features. Ice chunks drifted all around them as they canoed; they felt minuscule. Wave-sculptured pieces of ice floated by blue-green runaway rafts with a mind of their own. Alaska—whose name derived from the Aleut word *aláxsxaq*, meaning, roughly, "great land"—truly came as advertised. And glaciers spanned the entire southern perimeter of the colossal territory, from just north of the Canadian border in the southeastern region to midway along the Aleutian Islands chain. Less than 0.1 percent of the nearly 100,000 Alaskan glaciers had a name. "I stole quietly out of the camp, and climbed the mountain that stands between the two glaciers," Muir wrote from the Coast Mountains. "The ground was frozen, making the climbing difficult in the steepest places, but the views over the icy bay, sparkling beneath the stars, were enchanting. It seemed then like a sad thing that any part of so precious a night had been lost in sleep."[32]

Muir ended up publishing numerous articles in the *San Francisco Daily Evening Bulletin* about the Inside Passage, where "ice and snow and newborn rocks, dim, dreary, mysterious" had engulfed him. An outpouring of theological emotion about Alaska emanated from the great naturalist. All these Inside Passage glaciers regularly thawed and refroze as they muscled and ground downslope. Nothing lasted forever in glacier country. Using religious language, Muir declared the glaciers God's temples, the theology of ice, frozen temples. Many of the glaciers seemed to have a heavenly blue lantern light glowing from within. Even in wild weather, with "benumbed fingers," Muir had eagerly investigated the "shifting avalanche slopes and torrents." With so much weird, picturesque, sublime ice all around him, Muir could barely sleep at night. Every minute he

paddled around the Inside Passage, even with constant foggy precipitation, he felt "wet and weary and glad."[33]

Regularly, Muir shouted "God Almighty!" and "Praise God!"[34] when
confronted with a spectrum, or crazy quilt, of icy green-blue hues. The
colors of the bay were his stained-glass altar. With his narrow atten-
tiveness to every detail of glacial ice, Muir might as well have had a full-
immersion baptism in the Gulf of Alaska. In the surrounding waters Muir
continued watching humpback whales showing their flukes, barnacles
visible on their sleek backs. Nearly all of Alaska's glaciers were within six
hundred miles of the Pacific Ocean, so there was plenty of whale watching
for fun.[35] There was a glassy tranquillity to the currents of the Inside Passage that Muir hadn't expected, adding to the spiritual aura. According to
Young, Muir was a "devoted theist" at Glacier Bay, melodramatically paying homage to the "immanence of God in nature [and] His management
of all affairs of the universe."[36]

In the fall of 1879, Muir left Alaska a changed man. En route back to
California, he first traveled around the Pacific Northwest, journeying
up the Columbia River, preaching the gospel of the glaciers to anybody
who would listen. Just a few months later, he married Louise Stenzel, the
daughter of a wealthy agriculture businessman. As a wedding gift, Stenzel's father gave the Muirs a ranch house with a twenty-acre orchard—
including a lot of pear and cherry trees—in Martinez, California. Working
as a fruit farmer now, Muir nevertheless remained committed to preserving the integrity, stability, and beauty of Alaska's glacier community.
When picking fruit and filling baskets for market, Muir daydreamed
about Alaska, wishing he could slide down an ice sheet on his back, as he
had done on a toboggan during his youth in Wisconsin.

II

The following summer of 1880, Muir returned to Alaska's tidewater glacier land. The Reverend S. Hall Young, recently married to
a fellow missionary, was very excited to see his naturalist friend.
"When can you be ready?" Muir said upon greeting him in Fort Wrangell,
cutting to the chase; "get your canoe and crew and let us be off."[37] Young
hired three Tlingit guides in Fort Wrangell—the ones he had been

Christianizing—to help him get around the Inside Passage. On this trip Muir, anxious to observe the summer moods, visited by dugout canoe Sum Dum Bay and its maze of tributaries, Taku Inlet, Glacier Bay, and Taylor Bay.[38] Glaciers are particularly stunning when viewed from the water level of a canoe or kayak. And the arrogance of sightseers is likely to be squelched by the feeling of smallness that a boat's-eye view induces. Sailing through glacial fjords was the outdoors thrill of a lifetime for Muir and the others. "Every passage between the islands," Young wrote in *Alaska Days*, "was a corridor leading into a new and more enchanting room of Nature's great gallery."[39]

When hiking in Taylor Bay by himself, with only his mutt Stickeen as a companion, Muir had a hair-raising near-death experience. The higher they climbed, the less hemlock and spruce forest there was; then there was no plant life at all. Muir had brought with him only an ice ax and half a loaf of bread. Foolishly he had left his gun, rain gear, blankets, and matches back at camp. Impetuous enthusiasm had its shortcomings. A sense of doom now fell over the outing from the first. Stickeen was limping. A thunderstorm soaked them. Muir was determined to find Taylor (now Brady) Glacier, even in the rain. But then ominous darkness started to close in on man and dog. It was clearly time to head back down to camp.

Both Muir and Stickeen did a lot of fancy footwork, leaping across crevasses like Dall sheep in search of lichens. When a forty-foot crevasse manifested itself in front of him, Muir feared death. Somehow they had gotten themselves stuck in an ice maze. Muir was not a man prone to panic. But the only way out of his predicament was to cross an ice bridge eight feet below him. Muir dropped down, somehow managing not to slip—a slip would have meant instant death. The warm rain was creating a melting effect. Using his ax pick, Muir now made his way across the bridge, inch by inch. Poor Stickeen was terrified, howling and barking in fear of being left behind. Muir coaxed his dog to muster courage and follow his path. Eventually the frightened dog scaled down the glacier and somehow managed an acrobatic walk across the ice bridge. Muir and Stickeen embraced each other with a kind of shivering born-again love. "The joy of deliverance burned in us like a fire, and we ran without fatigue," Muir wrote, "every muscle with immense rebound glorying in its strength."[40]

Once back from the trip, Muir fleshed out the story to publish as an article for *Century* and eventually as an essay-length book, *Stickeen*. When it finally was published in 1909, it became a solid best seller. Besides using his journal notes, Muir had drawn on George Romanes's *Animal Intelligence*, published in 1881, to include new scientific data on the psychology of nonhumans.[41] "The spread of evolutionary thinking, animal-welfare legislation, bird-watching, and other challenges to homocentrism all gave this story of an ordinary-looking but brave little dog a deeper significance," the biographer Donald Worster explained in *A Passion for Nature*, "exactly as Muir had hoped."[42]

The Tlingit had made Muir an honorary chief during this visit in 1880; they supposedly called him "Great Ice Chief." The indomitable Muir routinely camped alone to study the calving glacier more closely.[43] Crouching to study the ice for hours at a time, he gleefully started naming landmarks around Muir Glacier as if they were boyhood friends dyed blue: Black Mountain (5,130 feet), Tree Mountain (2,700 feet), Snow Dome (3,300 feet), and Howling Valley—all part of today's Muir Glacier, which is a feature in Glacier Bay.[44] He drove stakes into the ice so that he could take measures on future trips. Young tells a comical story about what a powerful whim it was for Muir to designate nameless features. One afternoon Muir named an entire area after his Presbyterian friend. "Without consulting me, Muir named this 'Young Glacier,' and right proud I was to see that name on charts for the next ten years or more," Young recalled in *Alaska Days*. "But later maps have a different name. Some ambitious young ensign of a surveying vessel, perhaps, stole my glacier, and later charts give it the name of Dawes."[45]

Pilgrimages to Glacier Bay became Muir's Alaskan trademark. After his second trip in 1880, he returned to Alaska four more times, longing for the ethereal highs of Glacier Bay, the life-affirming crisp gray weather, the no-man's-land of wingspread mountains unfolding seemingly forever.[46] With imaginative leaps Muir's Alaskan journals sang Whitmanesque rhapsodies about the dazzling "thunders of plunging, roaring icebergs," surrounded by avalanche chutes and ice fields. And then there were frozen granite wilderness places—like Tracy Arm, Misty Fjords, and South Prince of Wales—which Muir embraced with the same love he held for Yosemite. *Travels in Alaska* was published in 1915, the year after he died. It's a valentine to Glacier Bay.

On all of his trips to Alaska, Muir sketched glaciers with pencil or ink in his journals. Some of the drawings—housed in the Holt-Atherton Special Collections at the University of the Pacific in Stockton, California, the primary depository for Muir's papers—stand alone on single sheets. Considering that many were drawn from a canoe or in the rain, they are quite remarkable.[47] Little has been written on Muir as a visual artist, but his drawings of glaciers were impressive. (By contrast, whenever he included humans in an Alaskan landscape, they looked like mere doodles, stick figures, or silhouettes.) What fun it is to study thirty-plus drawings of glaciers sketched between 1879 and 1899. There are pictures of glaciers at Kachemak Bay, Chugach National Forest, and Prince William Sound. But his most loving studies are of Muir Glacier at Glacier Bay, drawn from many different angles.[48]

After two summers in Alaska inspecting glacial motion—essentially, a study of velocity—Muir returned to northern California a changed man. The American West held a highball fascination for him, and Glacier Bay joined Yosemite as his obsession. "I am hopelessly and forever a mountaineer," he wrote to a friend. "Civilization and fever, and all the morbidness that has been hooted at me, have not dimmed my glacial eyes, and I care to live only to entice people to look at Nature's loveliness."[49] Modest, self-effacing, and with a permanent twinkle in his intense eyes, Muir was nevertheless zealous in his approach to *everything* wild. His enthusiasm for Alaska was so intelligently *real* that even his critics never tried to belittle him by calling him fanatical about glaciers. "Waking and sleeping, I have no rest," Muir wrote. "In dreams I read blurred sheets of glacial writing, or follow lines of cleavage or struggle with the difficulty of some extraordinary rock-form."[50]

Spoiled by Alaska's wild wonders, Muir had a hard time readjusting to living in Martinez, California. Domestic life had all the appeal of being chloroformed. Stuck with paying bills, operating an orchard, and answering an ever-increasing amount of correspondence, Muir constantly dreamed of Glacier Bay. He regularly complained to Young, who was doing missionary work in southeastern Alaska, about being stuck in California, and he was desperate for news about his beloved glaciers. Celebrity in America had its strains. Muir was constantly grappling with editors while trying to manage land tracts. Politically active in the saving of Yosemite, Mount Shasta,

Kings Canyon, Mount Rainier, and other treasured American landscapes, Muir missed being a wandering glaciologist, working in the glacier lands of Alaska and mastering the art of not fatally slipping. One afternoon Young, who was in the San Francisco Bay area on church business, unexpectedly dropped in on Muir. The naturalist was out in the fields, supervising cherry picking, holding a basket full of fruit. "Ah! My friend," Muir exclaimed like a wistful prisoner hoping to be freed. "I have been longing mightily for you. You have come to take me on a canoe trip to the countries beyond—to Lituya and Yakutat bays and Prince William Sound; have you not?" [51]

III

In May 1881, Muir expanded his Alaskan knowledge base by joining the USS Corwin on an expedition up the Arctic coast to search for the missing steamer Jeannette. This voyage afforded Muir the chance to explore the Bering Sea while simultaneously doing a good deed. Muir's primary goal was to study the ice on the frostbitten islands in the Bering Sea and the Bering Strait. The Jeannette had disappeared off Point Barrow when Muir had first traveled up the Inside Passage. Muir, on the Corwin, now got to expand his field studies to the Pribilof Islands (the largest fur seal rookery in North America) and Kotzebue Sound (home to polar bears and a wide variety of birds). The Lower Forty-Eight had less than 200 square miles of glaciers, in nine states: Washington, Wyoming, Oregon, California, Colorado, Idaho, Utah, Montana, and Nevada. All those glaciers, taken together, didn't equal a single large one in Alaska. Further expanding his sightseeing, Muir became one of the first humans to set foot on rocky Wrangel Island (between the Chukchi and East Siberian seas at meridian 180). This island had the highest density of polar bears in the world and was believed to be the last place on Earth inhabited by woolly mammoths. "How cold it is this morning!" Muir wrote to his wife from aboard the Corwin. "How it blows and snows!" [52]

Throughout the six-month Arctic cruise, to contribute to glacial science, Muir kept a daily record of the landscape he encountered. He also discussed the history of New England whalers, who had plied Alaskan waters since 1848. There were approximately 100,000 glaciers in Alaska; his fieldwork was endless. He wrote a handful of letters to be published in the *San Francisco Evening Bulletin*. His botanical reports on the flora found in the Arctic were elegant and pioneering. In 1883, the U.S. Treasury Department printed Muir's botanical investigation as Document No. 429. "I returned a week ago from the polar region around Wrangell Land and Herald Island," Muir wrote to the great protégé of Charles Darwin, Asa Gray, on October 31, 1881, "and brought a few plants from there which I wish you would name as soon as convenient, as I have to write a report on the flora for the expedition. I had a fine time and gathered a lot of exceedingly interesting facts concerning the formation of the Bering Sea and the Arctic Ocean, and the configuration of the shores of Siberia and Alaska. Also, concerning the forests that used to grow there, etc., which I hope some day to discuss with you."

Near Cape Thompson, Muir discovered a new species of *Erigeron*. Asa Gray was astounded. The asteraceous plant resembled a daisy and grew in clusters of three. Muir reported that it was abundant in the Arctic—confusing people who thought that the northern latitudes were a wasteland of ice. Gray classified it as *Erigeron muirii* (known to botanists as Muir's fleabane). A decade earlier, Gray had challenged Muir to discover a new flower. "Pray, find a new genus, or at least a new species, that I may have the satisfaction of embalming your name, not in glacier ice, but in spicy wild perfume."[53]

Although not published until 1917, *The Cruise of the Corwin*, Muir's account of the Arctic trip, became one of his signature books. Unlike *Travels in Alaska*, which was primarily about glaciers, this new memoir expressed Muir's deep compassion for animals. When members of the *Corwin*'s crew shot at a nearby harbor seal (*Phoca vitulina*), Muir flinched, writing that the creature had "large, prominent, human-like eyes," and therefore it was "cruel to kill it."[54] When a steamer owned by the Western Fur and Trading Company pulled up next to the *Corwin*, Muir sadly inspected the huge bundles of black and brown bearskins, marten, mink, beaver, lynx, wolf, and wolverine. "They were vividly suggestive of the far wilderness whence they came," Muir wrote, "its mountains and valleys, its broad grassy plains and far-reaching rivers, its forests and its bogs."[55] In *The Cruise of the Corwin*, Muir presented himself as an advocate of wildlife protection. Chapters were titled "Caribou and a Native Fair," "The Land of the White Bear," and "Tragedies of the Whaling Fleet."

IV

Twenty years after Muir's first visit to Alaska, the tycoon E. H. Harriman, owner of the Union Pacific Railroad, assembled a group of elite scientists and Thoreauvian naturalists for a ten-week cruise on the custom-built steamer *George W. Elder* to Glacier Bay and other Alaskan landmarks; the steamboat was, as Muir called it, "a floating university." [56] The entire party—including the ship's crew and officers, and servants—added up to 126 persons from both the Atlantic and the Pacific coasts. [57] This was Muir's seventh trip to Alaska. After boarding in Seattle, the sixty-one-year-old Muir would get to visit Victoria, Fort Wrangell, Juneau, Glacier Bay, Sitka, Prince William Sound, Cook Inlet, Unalaska, and Saint Lawrence Island—and to play the distinguished glaciologist and resident wise man on the 9,000-mile voyage. He didn't get back to Martinez, California, until late August. Never before had he seen such a variety of glaciers and ever-craggier peaks in such a short time span; the Chugach Mountains and Prince William Sound made him incredibly happy. Here was the greatest concentration of tidewater, calving glaciers in the world. [58]

The Harriman Alaska Expedition of 1899 voyaged up the Inside Passage, passing hundreds of forested islands, isolated coves, towering glaciers, and white-dipped mountains rising in waves against the mainland. The expedition—which included Muir's fellow naturalist John Burroughs, the scientist William H. Dall, the botanist William Brewer, the conservationist and ethnographer George Bird Grinnell, the artist Louis Agassiz Fuertes, and the ethnographer and photographer Edward S. Curtis—eventually crossed the Bering Sea all the way to the Chukchi Peninsula to catch a glimpse of Siberian soil before heading back to Puget Sound. They spent five days in Glacier Bay—one of the first scientific expeditions to this ecosystem—with Muir as their teacher with regard to glaciers.

What shocked members of the Harriman Expedition more than the wild beauty itself was how imprudently coastal Alaska was being stripped of its natural resources. They noted deforestation, clear-cutting, over-fishing, animal slaughter. Canneries and extraction companies were in the process of recklessly slashing many natural features. "At places," Burroughs wrote, "the country looks as if all the railroad forces in the world have been turned loose to delve and rend and pile in some mad, insane folly and debauch." [59] Most troublesome of all were the fifty-five salmon canneries along coastal Alaska, many around the Inside Passage and far west at Bristol Bay. Refusing to pay Native Alaskans fair wages, these big canneries hired cheap Chinese labor. Determined not to be federally regulated, these canneries formed the Alaska Packers' Association. [60]

In Prince William Sound the *Elder* explored the largest concentration of tidewater glaciers in Alaska. Many were actively calving. The surrounding Chugach and Kenai mountain glaciers were so powerful that they had cut more than forty fjords into the margins of the sound. The expedition spent perhaps the finest hours of the journey at College Fjord, twenty-five miles long and three miles wide. The members even discovered an unmapped inlet, dubbed Harriman Fjord as a tribute to their benefactor, containing over 100 glaciers. Muir burst with childlike excitement at seeing these glaciers. Instead of sleeping on the *Elder*, he pitched a tent along the shore to be closer to them. Grove Karl Gilbert, a glaciologist, always with binoculars in hand, likewise thrilled at seeing the Prince William Sound glaciers, taking invaluable notes on the stunning topography. "Gilbert's work on the Harriman Expedition was a major contribution to glacial geology," the historians William H. Goetzmann and Kay Sloan wrote in *Looking Far North*. "He had described the Ice Age horizons and he had outlined the physical mechanics of glaciers and glacial action." [61]

What came from the expedition was the publication of the thirteen-volume Harriman Expedition reports (usually called the *Harriman Alaska Series*). These scientific volumes, organized around information gathered on the cruise, captured the public imagination about wild Alaska as nothing had before. The fact that the northern third of Alaska (above the Arctic Circle) had yet to be properly explored or mapped excited people's imagination. Want to have a mountain named after yourself?—head to the Brooks Range or the Aleutian Range. Also, Harriman's eminent scientists brought back a wealth of data that opened up Alaska to natural history for the first time. Muir, however, was frustrated with the penchant of the expedition's members for hunting bear and catching the biggest fish. Muir also found the opulence aboard the *Elder* (the expedition's ship) off-putting; too much faux positioning went on. "Why, I am richer than Harriman," Muir bluntly declared. "I have all the money I want and he hasn't." [62]

Some fifty scientists compiled the *Harriman Alaska Series*; editorial work was done in New York; Washington, D.C.; and Berkeley, California. Harriman, as always, was generous with pay. The team modeled the scientific volumes on the old U.S. Geological Survey reports once famously issued by Clarence King and John Wesley Powell. Never before had coastal Alaska been analyzed from so many scientific perspectives. Every contributor revealed in detail what he had learned on the *Elder*. Grove Karl Gilbert wrote on glaciers; John Burroughs provided the definitive summary text; John Muir also wrote about glaciers and the harmony of nature; George Bird Grinnell wrote on the Tlingit, Aleuts, and other Native Alaskan peoples; Charles Keeler wrote on birds (with Louis Agassiz Fuertes brilliantly illustrating the descriptions of tufted puffins, harlequin ducks, and cormorants); B. E. Fernow wrote on forests. Unlike the expedition's other intellectuals, Muir wrote his reports in a lyrical tone. Upon seeing College Fjord's Western Wall in the Chugach, he wrote of the glacier group that "they came bounding down a smooth mountainside through the midst of lush flowery gardens and goat pastures, like tremendous leaping, dancing cataracts in prime of flood." [63]

What these reports accomplished was to teach Americans that Alaska was a unique, untrammeled, sui generis wilderness in need of preservation on many levels. In Henry Gannett's *General Geography*, written after Gannett participated in the Harriman Expedition, Alaska is envisioned as a future gigantic national park. "For the one Yosemite of California," he wrote, "Alaska has hundreds." Doubtful that mining gold, coal, and copper could be sustainable in the long run, Gannett prophesied that Alaska's destiny was wilderness tourism. "The Alaska coast is to become the show-place on earth, and pilgrims, not only from the United States, but from beyond the seas, will throng in endless procession to see it," Gannett wrote. "Its grandeur is more valuable than the gold or the fish or the timber, for it will never be exhausted. This value, measured by direct returns in money, received from tourists, will be enormous. Measured by health and pleasure, it will be incalculable." [64]

Muir has been called the "mentor of the conservation movement"; it's a reasonably apt accolade. Better than George Bird Grinnell, John Burroughs, or C. Hart Merriam, he understood nature's rhythmic cycles both emotionally and scientifically. While Muir has been given a lot of well-deserved credit for helping to create Yosemite National Park and starting the Sierra Club in 1892, he was also America's most enthusiastic Alaskan glaciologist prior to 1900. His teaching method wasn't merely to illuminate listeners about snouts, crowded bergs, calving, or retreating ice. Glaciers, to Muir, were great indicators of weather, climate change, and tectonic plate shifts. As a glaciologist he held his own with the brilliant Gilbert. But as a preacher of the "glacier gospel" Muir was a one-man show. Burning with enthusiasm, Muir promoted Alaska's seacoast wilderness, temperate rain forests, and green-ice glaciers as ever-changing masterpieces of creation. When Muir was on top of glaciers, he could see the ocean. Muir even dug a snow pit to study the layers within; all of Glacier Bay was his field laboratory; every inch of ice was a psalm.

By championing Alaska's Glacier Bay as a site that had to be seen to be believed, Muir helped create today's national park as surely as he had done with Yosemite. Muir had asked Americans to imagine glaciers along a stretch of mountain-hemmed sea . . . to crave calving ice . . . prehistoric forests . . . gamboling orcas . . . thousands of bald eagles . . . salmon runs . . . ice floes like bottles with messages drifting in clear waters. In southeastern Alaska, he was like a happy-go-lucky marooned seafarer, pleased to uncork the frozen essence of pressure melting when ice flowed around to the downhill side and then froze. Muir believed that a glacier had five main parts: the *face* was the front; the *terminus* was the downhill end; the *surface* was the top; the *base* was like a belly where it scraped against the valley bottom; the *source* was the area from which it flowed.[65]

The Harriman Expedition of 1899 was Muir's last visit to Alaska. Nevertheless, Muir continued to espouse the protection of the eighteen tidewater glaciers (the glaciers that reach the sea) as Glacier Bay National Park. The sheets of living ice were thousands of feet thick and a few miles wide. If lucre was the reigning force of American life, then Muir wasn't above promoting tourism to Alaska to protect the "solitude of ice and snow and newborn rocks, dim, dreary, mysterious" of the Inside Passage, Prince William Sound, and Cook Inlet.[66] Glaciers existed in the entire southern perimeter of the state from just north of the Canadian border in the southeast to the last Aleutian Islands. Glaciers bespread the Fairweather Range, in the Coast Mountains, on the peaks of the Saint Elias Mountains, and the Alaska Range. The Chugach, Kenai, and Wrangell mountains all have glaciers—though more are melting. Muir was the protector and poet for all of Alaska's more than 100,000 glaciers.

Today more than 1 million tourists a year head up the Inside Passage and Prince William Sound on cruise ships, loosely tracing Muir's routes from 1879 to 1899. What Muir—like the Harriman Expedition itself—was offering Alaskans was another revenue stream besides the extraction industries: ecotourism. The heavy cruise ship traffic in Glacier Bay and Prince William Sound, in fact, has caused the National Park Service to turn away business rather than overly disturb the harbor seals, orcas or icebergs. Few passengers study glaciation processes in detail, but Muir believed that the more people saw of Alaska's frozen wonders, the more likely they were to become conservationists. "Muir believed with evangelical passion that nature's glaciers could form men as well as mountains, and he might well have viewed the proposed trip to Alaska as a pilgrimage as much as a scientific expedition," the historians Robert Engberg and Bruce Merrell wrote. "In this way, his motivation may not have been so clearly distinct from that of the modern tourist who wishes to get away from it all by a visit to Alaskan wilderness." [67]

Alaska . . . the three syllables had a magic radiance in 1899. And its primeval tundra north of the Brooks Range had yet to be explored by a single Darwinian biologist. Serious dry-fly anglers of the Izaak Walton League sort had yet to feel the weight of the clear, cold, fast streams against their legs. Few sportsmen had ventured anywhere near Lake Clark–Lake Iliamna to hunt the free-ranging moose. (But Native Alaskan hunters were part of these ecological systems for more than 10,000 years.) Most adventurers, however, weren't interested in the glories of Mother Nature—they were after a quick fortune in mining, promised to them by recurrent come-ons: "There's gold in them thar hills." With the gold rushes of 1897 to 1899, more than 30,000 people stampeded to the Alaska and Yukon territory, most with the sole intention of extracting riches from the suddenly valuable land. Alaska, once derided as "Seward's folly," the most foolish real estate deal in American history, was suddenly a glittering boom land where gold nuggets could be panned out of any swift-moving stream. For every John Muir who came to see the grandeur of huge glaciers spilling over the rough-hewn landscape, a hundred others stood by, ready to harvest the glacier ice and sell it for a profit.

A battle was on between those who wanted to preserve Alaska's wilderness and those who wanted to extract wealth from minerals, salmon, glacier ice, timber, and, later, oil. The Nobel Prize–winning novelist Knut Hamsun, of Norway, once described Americans' obsession with get-rich-quick commerce in this way: "They never allow themselves a day of quiet. Nothing can take their minds off figures; nothing of beauty can get them to forget the export trade and market prices for a single moment."[68] His words perfectly describe the mentality behind the dozens of Alaskan gold rushes and all the Alaskan oil rushes ever since. Yet there was from the get-go a cult of determined "wilderness believers" who fought against the private sector's extraction mania in Alaska. To these nature lovers, often supported by the U.S. government, Alaska was a paradise for poets, scientists, recreationists, and tourists alike.

"In God's wildness lies the hope of the world," Muir wrote, with timeless Alaska in mind, "the great fresh unlighted, unredeemed wilderness. The galling harness of civilization drops off, and the wounds heal ere we are aware."[69]

Chapter One

. . . . . .

ODYSSEY OF THE SNOWY OWL

I

oung Theodore Roosevelt could barely believe his good fortune. Taking a long break from studying for his Harvard University entrance exams in Manhattan, he headed to Long Island for an outdoor ramble in the calming woods. A dedicated birder, the seventeen-year-old Roosevelt was hoping to add a couple of new species to his growing North American list. Suddenly, Roosevelt heard a faint barking *hoot* and looked up. Blessed with a marvelous aural ability, as if in compensation for poor eyesight, Roosevelt stopped dead in his tracks. There in front of him in the sylvan stillness was an inscrutable migrant from somewhere around the Arctic Circle, the imaginary line that runs around the globe at a latitude 66° 33' 43" north.[1] It was a snowy owl (*Bubo scandiacus*). Bright white in plumage, with velvety, fine-textured downy feathers, this huge owl had a flat humanlike face with piercing yellow eyes that glowed like railroad lanterns. The bird's insulating white plumage protected it from ambient temperatures of minus forty degrees Fahrenheit. The protective coloration of the snowy owl, much like that of the polar bear, arctic fox, or Dall sheep, was a marvel: evolutionary adaptation principles on gallant display. To Roosevelt's amazement this circumpolar Odyssean from the dim blue north was overwintering in—of all places!—Oyster Bay, New York. Instead of preying on lemmings or voles around Arctic Alaska, it was gulping down small rodents in the frozen fields of Nassau County.[2]

One by one, and with an ornithologist's care, Roosevelt checked off the owl's otherwordly anatomical features, marveling at its biological ingenuity. He was awed by the purity of its evolutionary composition. Even the owl's talons were camouflaged with white feathers and had extra-thick pads designed to endure subzero weather. They were strong enough to carry off an arctic vole or medium-size goose. Although freeze-tolerant snowy owls had reportedly been encountered as far south as the Rio Grande valley of Texas, it was a genuine aberration for Roosevelt to stumble randomly upon one in Greater New York City. For a few moments Roosevelt must have held his breath, determined not to break the tranquillity, mesmerized by this living testimony of migration. Then, without further hesitation, he raised his shotgun and killed the snowy owl. Proudly carrying the carcass back to his parents' house in Manhattan, the future president of the United States performed taxidermy on the adult male bird, using arsenic to preserve the skin, as was typical during the Victorian era.

The snowy owl—the official bird of Quebec—is still among the most coveted, by bird lovers, photographers, ornithologist-collectors, of the world's 200 owl species. It is often regarded as a talisman from the aquamarine ice lands of the North Country—along with the white morph gyrfalcon (*Falco rusticolus*) and ivory gulls (*Pagophila eburnea*). Human fascination with snowy owls is as old as recorded history. Paleolithic hieroglyphics of these owls were etched on stone walls in ancient France. In recent years the author J. K. Rowling used the snowy owl as a symbol of eternal wisdom in her *Harry Potter* books. When Roosevelt entered Harvard in September 1876, his stuffed owl was a prized possession in his apartment on Winthrop Street in Cambridge, encased by a bell jar on the mantel. Oddly, the bird's plumage became whiter as it aged.

After his encounter with the snowy owl, Roosevelt maintained a deep-seated fascination with all Arctic Circle creatures—even the Alaskan beetle (*Upis ceramboides*), which can live at temperatures as low as minus ninety degrees Fahrenheit; and the wood frog (*Rana sylvatica*), which hibernates beneath the snow and is protected by a concentration of glucose in its cells and bloodstream.

A voracious reader of literature about the Arctic Circle (or the region above the tree line), Roosevelt particularly treasured the eyewitness reports of polar bears (*Ursus maritimus*) in William Scoresby's *An Account of*

the Arctic Regions with a History and Description of the Northern Whale Fishery (1820) and James Lamont's *Yachting in the Arctic Seas* (1876). Stories
about the Hudson Bay bears also interested him. Roosevelt, however, was
skeptical of Scandinavian and Dutch reports from the Arctic seas that
polar bears regarded humans as merely "an erect variety of seal." Polar
bears, he correctly believed, were generally aloof and skittish, instinctively scattering when people appeared. "A number of my sporting friends
have killed white bears," Roosevelt wrote, "and none of them were ever
charged."[3]*

Arctic Alaska's signature species, the polar bear, is Earth's largest terrestrial carnivore. Polar bears, like the snowy owl, were isolated in the
north on an ice sheet during glaciation; in the course of adaptation to this
extreme environment, their coat became entirely white. A male polar bear
measures eight to nine feet long and weighs up to 1,500 pounds. Females
are typically around six to seven feet long and weigh around 600 pounds.
The Beaufort and Chukchi seas make up America's Arctic Ocean. (Most
Americans don't realize that Alaska has roughly 50 percent of the con-
tiguous U.S. coastline.) Blanketed primarily by sea ice, this shore habitat
along the Beaufort and Chukchi is considered one of the finest polar bear
denning areas in North America; the Harriman Expedition, however,
wasn't able to find a single one on its Alaskan voyage in 1899.[4] Every December through January a mother polar bear will give birth to one to three
cubs along these Arctic seas. The cubs accompany their mother for two
years before striking out on their own. There are also polar bears along
the Chukchi Sea between Point Hope and Point Barrow in Arctic Alaska.
Of the eight bear species currently studied, only the polar variety are exclusively carnivores. Their diet consists of one thing: meat. Unlike brown
bears, which have round faces, polar bears have a more slender head with
a pointy nose: an excellent snout for sniffing out elusive seals burrowed in
snow or ice (seals are their primary food source).[5]

Enraptured by forbidding Arctic tales, Roosevelt affectionately called

---

* Roosevelt would probably have been amazed at the report by *National Geographic*
in 2009 that DNA (genetic) testing had confirmed the existence of a grizzly–polar
bear hybrid in the Arctic.

polar bears the "northern cousin" of grizzlies.[6] Reading about polar bears by lamplight amid the comforts of Manhattan or Cambridge, however, was not comparable to exploring Arctic Circle landscapes himself. He dreamed of someday kayaking down wild Arctic rivers where the sun didn't set from May to August. Imagining himself an outback citizen in Nome, Nunivak Island, or Kotzebue—where simply to inhale fresh air in winter was to frost one's lungs—Roosevelt dreamed of someday hunting a polar bear in the unforgiving Bering, Chukchi, and Beaufort seas.[7]

In the late nineteenth century, Alaska—from southeastern rain forests to Aleutian volcanoes to barrier islands along the Arctic coast to the ice glaciers of the Inside Passage—was a never-never land of unnamed mountains, unnamed rivers, and unnamed species. For sheer spatial perception, Alaska's 591,004 square miles dwarfed the Mojave Desert, the Rocky Mountains, or the Appalachian chain. Stand on any mountain in the Brooks Range or Alaska Range, peer out over the gray granite upthrusts, and you were bound to see a hawk pass a raven in the strongest headwinds known to mankind outside Patagonia and Antarctica. How to describe Alaska's prodigious natural world in mere words, art, or photography is daunting. As Muir understood, a single Aleut word— *Alaska*—encompassed so much dramatic geographic beauty, intricately laced mountains, glaciers, valleys, and coastline that it seemed surreal; the territory encompassed four different time zones. Whether you lived in Homer, Fort Wrangell, Fairbanks, or Point Barrow, scenic wonders worthy of a national park abounded. Alaskan place-names themselves, as provocative as Ed Ruscha's minimalist word paintings, are far more evocative of Alaska's wild austerity than even the *National Geographic*'s best photos. The North Slope. Wrangells. Beaufort Lagoon. Mount McKinley. Tongass. Chugach. Kenai Peninsula. The Yukon and Tanana rivers. Mendenhall Glacier. Gates of the Arctic. Plover Glacier. Bristol Bay. Lake Clark. Nunivak Island. Izembek. The Alexander Archipelago. There was wildlife in abundance in all these varied Alaskan places— bears, caribou, wolves, whales, otters, moose, sea lions, and seals. There were Alaska's Native peoples—among them Tlingit, Haida, Athabascan, Eyak, Yupik, Inupiat, Tsimshian, and Aleut tribes. There were two major "Eskimo" peoples: the Yupik (of western Alaska from the Kuskokwim Bay area to Unalakleet northeast of the Yukon River mouth) and the Inupiat (from that point northward and eastward to Barter Island and beyond to the Beaufort Sea). There was the new breed of far north wanderers—lumberjacks, whalers, salmon merchants, hikers, oil sniffers, dogsledders, fishermen, seal hunters, missionaries, sourdoughs, prospectors, and the occasional John Muir—the wanderer in nature. All these colorful character *types* shared one undeniable reaction: amazement at the bounty of wild Alaska.

It was Alaska's abundant wildlife that first brought Asian hunters to cross the Bering Strait land bridge—which joined eastern Siberia with North America—more than 25,000 years ago. These nomads wandered from Asia, surrounded by the world's northernmost ocean, chasing such grazing mammals as the woolly mammoth, camel, mastodon, antelope, ground sloth, and bison. Following the jagged berglike pressure ridges—today's Seward Peninsula to Brooks Range to the coastal plain of the Beaufort Sea—they trekked across the Bering Sea land bridge, hundreds of miles wide, with no intention of returning to Asia. Then a cataclysm occurred. At the close of the Pleistocene ice age, the Bering Strait land bridge was swallowed up by rising seas. Most of this land bridge today lies beneath the icy waters of the Bering and Chukchi seas. (The U.S. Interior Department now oversees the Bering Land Bridge National Preserve, which contains heritage sites of prehistorical and geological interest.) Stuck along the Arctic rim, these nomadic hunters made the best of the new situation. They survived by harvesting whales, fish, caribou, and other game.[8]

Enter Vitus Bering, a Danish sea captain, 10,000 years later. Commissioned by Peter the Great in the 1720s to determine if North America and Asia were linked by land, the brave explorer set sail from eastern Siberia in a square-rigged ship for Alaska on a couple of occasions. In 1741 Bering made landfall on Kayak Island (located off Cape Suckling on the southern coast of Prince William Sound). Russia wanted to exploit these Alaskan lands in search of furs, timber, and minerals. Survivors of Bering's expedition brought back from Alaska all sorts of luxurious sealskins and sea otter pelts. Walrus were easily found in groups numbering ten to fifty. This, however, didn't bode well for the future of these great rookeries.

As a consequence of his voyage, Bering's name became famous.

Residents in twenty-first-century Alaska are regularly reminded of Vitus Bering because of the Bering Strait, the Bering Sea, Bering Island, the Bering Glacier, and the Bering land bridge. Early Russian explorers, for their part, named other geographical features after people favored by the czar: Cape Tolstoy, Belkofski, Olga Rock, Poperechnoi Island, and Wosnesenski Island are just a few.[9] In 1790 Lieutenant Salvador Fidalgo of Spain voyaged to Alaska in search of the Northwest Passage. The shortcut to Asia was never found, but the Spanish did find Prince William Sound, and named today's Valdez, Port Fidalgo, Gravina, and Cordova.[10]

Germany's most eminent naturalist-botanist, Georg Wilhelm Steller, a physician by training, was the first scientist to document the unique flora and fauna of wild Alaska. Vitus Bering, at the request of the Russian Academy of Science, had invited Steller to come along on the 1741 voyage to record wildlife sightings. Working quickly under severe time constraints, Steller took excellent notes on climate, soil, and resident flora and fauna. Allowed only ten hours on Kayak Island, principally to help collect freshwater, he nevertheless discovered Steller's jay (*Cyanocitta stelleri*), recognizing it as resembling the eastern American blue jay. "This bird," Steller wrote, "proved to me that we were really in America."[11] The same afternoon he found Steller's eider (*Polysticta stelleri*), Steller's sea eagle (*Haliaeetus pelagicus*, now endangered), and Steller's white raven (a mystery). He discovered all sorts of new fish. As the historian Corey Ford pointed out in *Where the Sea Breaks Its Back*, Steller never missed an opportunity to attach his name to an Alaskan discovery in need of instant classification. There were also Steller's greenling (*Hexagrammos stelleri*), a colorful rock trout; Steller's sea cow (*Hydrodamalis gigas*), a giant northern manatee; and Steller's sea monkey (which was never formally identified). That was a lot of naming for a single working day.[12]

Steller also stumbled on a Native encampment, where the campfire coals were still warm but nobody was to be seen. Fearful that enemies were lurking around, Steller swiped a few Indian artifacts and fled back to the ship.[13] Steller's naturalist studies were sui generis in eighteenth-century Alaska. He was a man far ahead of his time. On the return voyage to Russia many of the sailors on the Bering Expedition were sick with scurvy. Serving as a herbalist, Steller administered antiscorbutic broths that were credited with saving lives. "He was brilliant; he was arrogant; he was gifted as are few men," the former director of the Alaska Game Commission Frank Dufresne wrote of Steller. "Though he spent no more than ten hours on Alaskan soil, his accomplishments in that short day were such that his name will live on forever." [14]

Alaska's biological diversity seemed to explorers a strange remnant from the ice age. American geographers around the time of the Harriman Expedition divided the territory into five very distinct ecosystems: (1) the Arctic, (2) Western Alaska, (3) the Interior, (4) Southwestern Alaska, and (5) the Southeastern Panhandle (including the Inside Passage cities of Sitka, Skagway, Ketchikan, Wrangell, Haines, and Juneau). Depending on where you went, there were icy fjords, sedge meadows, glacial fields, volcanic ranges, and tundra regions. What the Mississippi River had been to Mark Twain's imagination, the 1,980-mile Yukon River—whose watershed comprised nearly half of Alaska—was to the new generation of fortune seekers. For a natural scientist wanting to start a career, the banks of the Yukon River were (and still are) an all-you-can-gaze-at smorgasbord of wildlife. Despite the presence of scientists on the *Elder*, mysteries such as caribou migratory routes or wolf ecology were largely propagated by unreliable oral tradition. A university-trained biologist, one who wrote well, could make a distinguished reputation seemingly overnight by trekking north from the Lower Forty-Eight and investigating the biological face of roadless Alaska. [15]

Alaska belonged to the Native tribes and wildlife while Roosevelt was growing up in New York City following the Civil War. Muir in *The Cruise of the Corwin* had deemed the Indians "the wildest animals of all." [16] Alaska was far removed even from the slow crop-growing pulse of rural American life. Farmers had yet to settle there. A few rogue gold miners made their way from British Columbia hoping to strike a vein. But wandering fur hunters from the Rockies and whalers from Russia, Great Britain, and Canada were the most prevalent new arrivals. During the summer months, whales swam the coastal waters in pods; their sheer numbers would have baffled and delighted a New Englander. Musk oxen (*Ovibos moschatus*) roamed wild, shaggy relics of the ice age. But Danish, Norwegian, and American hunters were quickly driving them toward extinction. Walrus (*Odobenus rosmarus*)—evolved from eared seals more than 20 million years ago—lived in and bred on remote Alaskan islands in the

Bering and Chukchi seas; these pinnipeds would hook their two tusks on ice floes to help haul themselves out of the water. Dall sheep (*Ovis dalli*), native to Alaska-Yukon, climbed snowcapped peaks; their curled keratin horns were coveted by trophy hunters. There were more brown bears (*Ursus arctos horribilis*) on Alaska's Admiralty Island alone than in all other U.S. states and territories combined. John Muir, as perspicacious as ever, wrote that in Alaska grizzlies wandered "as if the country had belonged to them always." [17] Today there are 31,000 brown bears in Alaska, while their populations have been drastically reduced in the Lower Forty-Eight. [18]

The Native totem poles ("story poles") of Alaska celebrated ravens, bald eagles, and halibut as the holy spirit of life incarnate. [19] Discovering these tall carved monuments, central icons of the northwestern coast region, became a rage at New York's American Museum of National History during the "gilded age." Roosevelt himself was fascinated by Tlingit, Haida, Kwakiutl, and Nootka craftspersons who honored animal life in Alaska. The totems weren't inspired by religion or sorcery. Rather, totem poles matter-of-factly told the life stories of Indian tribes. The wooden poles, sometimes fifty feet high, were, in a sense, a substitute for books. And every totem pole was different. A hawk, whale, or bear often crowned the log-post top. Feuds sometimes broke out between villages over who had the highest pole. Tribal elders perceived the totem pole as a monument to nature and to village life, an emblem of human strength and the bounty of the land and sea. To New Yorkers the poles were Indian art and were coveted for museum collections. A movement was started to help preserve them from weather, rot, and vandalism. "The carved totem-pole monuments are the most striking of the objects displayed here," Muir reported in 1879. "The simplest of them consisted of a smooth, round post fifteen or twenty feet high and about eighteen inches in diameter, with the figure of some animal on top: a bear, porpoise, eagle, or raven, about life-size or larger." [20]

The scientists of the Harriman Expedition liked Native Alaskan artifacts too much. The photographer Edward S. Curtis told how the steamer *George W. Elder* came upon a deserted Tlingit village; everybody was probably out hunting or fishing. Hurrying to shore, the Harriman crew stole everything from children's clothing to pottery to bring back to New York as museum-worthy artifacts. Muir, who refused to participate, described the incident as "robbery" in his journal. Curtis didn't record whether he participated in the raid, but he later openly criticized three scientists who stole "a ton of human bones" from a Native cemetery.[21]

Whereas American settlers saw the wilderness as an adversary, an obstacle to overcome, Alaskan Natives saw nature as something they belonged to; the totem pole was a symbol of oneness between people and animals. The heyday of Alaskan totem poles occurred between 1820 and 1890. (In 1893 twelve totem poles were displayed at the Chicago World's Fair, to great acclaim.) Carvers were ordered by tribal chiefs—who preferred using red cedar—to honor wildlife in wood effigies. The storytelling aspect of the totem pole was prioritized over its external appearance. Still, to decorate the poles, carvers made glowing paints from animal oil and blood, charcoal, salmon eggs, ocher, wildflowers, and moss. The Bella Bellas of the Kwakiutl nation of British Columbia learned astonishingly innovative ways to mix colors. Some moonlighting carvers also chiseled wooden boats to resemble killer whales. But mainly the totem poles paid respect to favored species such as the halibut, frog, and beaver.[22]

II

The *Alaska Purchase by the Andrew* Johnson administration had taken place on May 28, 1867, when Roosevelt was only eight years old and Muir had just recovered his eyesight. Through the bold initiatives of Secretary of State William Seward, the United States acquired more than 586,000 square miles of northern territory from Russia for a song—$7.2 million (less than 2 cents an acre).[23] Seward defended the purchase as the final act of western expansionism, claiming that Alaska would provide salmon runs, mineral wealth, and forest resources. (Alaska was also where Seward planned on having lines laid for the international cable being promoted by the American Telegraph Company.) At the time, anti-expansionists called the purchase "Seward's Folly," considering the region a frozen wasteland not worth a trillionth of a dollar. But expansionists, including Theodore Roosevelt, later celebrated the Alaska Purchase as a trophy of great worth. Roosevelt described Alaska as glacier-streaked territory of "infinite possibilities" that the U.S. government had wisely purchased "despite bitter opposition" of many small-minded men.[24] And

Seward himself, who visited Sitka in 1869, understood that his purchase of Alaska would someday be seen as the high-water mark of his long, distinguished career in public service.[25]

For a few decades the Russians prized Kachemak Bay as a source of lignite coal. In 1855 alone the Russian-American Company, operating out of Port Graham, employed 131 men and produced 35 tons of coal daily. The coal was shipped to San Francisco, but sold at a loss, so the company abandoned the export trade. Russia, which never claimed more than 800 settlers in the colony, was beginning to see that, given the harsh weather and the vast export distances, mining Alaskan coalfields wasn't particularly profitable. The Russian Orthodox Church, however, flourished in the Kenai Peninsula. The Old Believers split from the main church in 1666, refusing to implement reforms. In Alaska the Old Believers clung to Slavonic texts, used two fingers for the sign of the cross, and practiced triple-immersion baptism. They colonized little villages such as Nikolaevsk, Razaldna, and Kachemak Selo. They resembled the Amish of Pennsylvania in some ways, such as their old-fashioned clothing—these Russian women wore head scarves—and they represented Russian Alaska well into the twenty-first century.

Starting in October 1867 U.S. troops relieved Russian soldiers at the colonial capital, Sitka; the American flag now flew over the District of Alaska.[26] The USS *Ossipee* brought two government officials to the transfer ceremonies. The secretary of the navy publicly declared that a couple of ships were headed to Alaska to collect information on "harbors, production, fisheries, timber, and resources."[27] Rudyard Kipling once wrote discouragingly of Alaska, "Never a law of God or man/Runs north of Fifty-three."[28] Contrary to Kipling, in coming decades, spiritual pilgrims, a cult of wilderness devotees like Muir and Young, *found* God in the blue-green ice of Glacier Bay, the upper reaches of the austere Brooks Range, and the caribou-thick coastal plain of the Beaufort Sea. Early dispatches out of frontier mining and timber towns, however, proved that Kipling's assessment was spot-on. Alaska, in fact, was so underpopulated by U.S. citizens in the late nineteenth century that it had been administered in musical-chairs fashion by several government departments: Army, Treasury, Customs, and Navy.

Alaska's first census came in 1880, while Muir was on his second voyage up the Inside Passage. Of the 33,426 people residing in the territory, fewer than 500 were non-Native. At the time of the Alaska Purchase, Seward had wisely refused to offer free land to attract homesteaders. The U.S. Mining Laws of 1824 had banned freelance prospecting. This bar was amended a decade later. Alaska belonged to the federal government, and various agencies dispatched wildlife biologists, cartographers, and forest experts to write reports on what exactly Seward had acquired.[29] Anthropologists started writing about how Native nomads had crossed from Siberia to Alaska over the Bering land bridge. Reports from the *Corwin* noted that the Inupiat and Yupik were dispersed throughout the northern and western regions of Alaska. Whalers knew for certain that the Aleuts were primarily based in the island chain named for their tribes: the Aleutians. Around the Alaskan interior—near present-day Fairbanks—were the Athabascan people. Then in southeastern Alaska there were the totem pole peoples—Tlingit, Haida, and Tsimshian—who lived in a green paradise: they had rich forestland, a mild climate, and fish and game galore.

The Alaska district, a colossal subcontinent, was a relatively new and unknown addition to the United States. The naturalist Steller's old notes, in fact, were still relevant to zoologists. Another naturalist, William H. Dall—known to the scientific clique at the Cosmos Club in Washington, D.C., as "the dean of Alaska experts"—had befriended various Native Alaskan tribes including the Aleuts and Tsimshian. Besides being amazed at their arts and crafts, he considered them all great fishermen. Dall was more worried about the American drifters headed into Alaska looking for quick fortunes in salmon fishing than about the Natives. Dall, America's first serious "Alaska naturalist," wrote that from 1867 to 1897 the district was marked by a surprising amount of lawlessness and the slaughter of seal herds for market.[30] No citizen could make a legal will or own a homestead. Polygamy was widespread throughout the territory. Occasionally there was even a burning of accused witches. With no courts in the region itself, Alaskan land claims had to be defended in the courts of California, Oregon, and Washington.

Dall, who had traveled in interior Alaska with mush dogs, began lobbying the U.S. government to regulate timber and mining claims, hoping that Alaska could be sensibly developed and eventually achieve statehood. Brimming with encyclopedic knowledge about the Alaska Range and the Kenai Peninsula, Dall insisted that the U.S. government had to regulate timber and mineral claims; it was a legal imperative. Dall saw Alaska as having ecological, moral, scientific, and spiritual values that would help preserve the frontier spirit if properly managed by the federal government. A Victorian-era classifier of animals, Dall had two Alaskan species named in his honor: Dall's porpoise (*Phocoenoides dalli*) and the Dall sheep (*Ovis dalli*). He also called on the U.S. Navy to stop Japan and Russia from slaughtering the northern fur seal (*Callorhinus ursinus*) for pelage. The luxuriant dark coat of northern fur seals—males are a handsome brown and females gray-brown (dorsally) with a streak of chestnut-gray (ventrally)—was coveted by trappers for a global market. Only sea otters had a denser underfur than these seals—so dense that ocean water never touched their skin. A ringed seal pelt, with bold black stripes, as in a Franz Kline painting, was sought after by Paris and London merchants and furriers. Dall envisioned a time when the great northern fur seal herds of Alaska—like the animals of Charles Darwin's Galápagos—would attract tourists from all over the world.[31] Ignoring Dall's call, the U.S. government decided to lease "killing privileges" on Alaska's seal rookeries to private businesses, with royalties coming to the general treasury. There was a strong movement in Congress, in fact, to get back, by way of the skins of fur-bearing mammals, the $7.2 million that the Alaska Purchase had cost.[32]

Nobody captured the horror of the slaughter of Alaskan seals and otters quite like the novelist Rex Beach, of Michigan. Beach's first novel was *The Spoilers*, a 1906 best seller about government officials stealing from gold prospectors in Nome, Alaska, but he later turned to the ruthless U.S. fur industry and wrote a blistering fictional exposé, considered by some scholars a pioneering environmental work. He had zero tolerance for seal blood in tidal pools of sea grasses and kelp. "Jonathan Clark, for one, considered the wholesale destruction of harmless and bewildered creatures as a thoroughly dirty and degrading business," Beach wrote in *The World in His Arms*. "He was ready to wash his hands of it in more ways than one." Clark, the novel's hero, confronts the Alaskan territorial government in the 1870s about the need to ban the killing of marine mammals. "You probably won't believe that a man of my sort can have a respect—a reverence, I may say—for the wonders of nature," Beach wrote. "But a rogue can revere beauty or grandeur and resent their destruction. Those fur seals are miraculous; it's a sacrilege to destroy them." [33]

But as Beach made clear in *The Winds of Change*, first published in 1918, the Russian, Canadian, and Japanese pelagic hunters continued slaughtering Alaska's northern fur seals indiscriminately. Seal fur brought money. And law enforcement, as represented by federal agents in Washington, D.C., was far, far away. To these market hunters, the Pribilofs—rocks with only clusters of creeping willows and a few shrubs bearing black currants and red salmonberries—were Fort Knox; actually, truly fine pelts were worth more than gold. Disdainful of federal seal protection laws, vessels from these countries would anchor just outside the three-mile U.S. limit and slaughter the great herds. Rudyard Kipling included in his second *Jungle Book* the short story "The White Seal," a saga of the Bering Sea about nations slaughtering Pribilof fur seals and otters. Using high-powered rifles, hunters in the Aleutian Islands shot at the heads of seals and otters, hoping their bodies would wash ashore, where skinning could commence.

III

By the time Roosevelt graduated from Harvard in 1880 he had become envious of naturalists like Dall—a latter-day American version of Steller—who roamed the strange and forbidding Alaskan tundra with mush dogs, sledding past grizzly bears. The ribbon seal (*Phoco fascita*) in the Bering Sea, the giant tusked walrus, ice cascades, fierce gales, alpine tundra, root-digging grizzlies, unspoiled conifer forests, ripping tidal currents to match those of New Brunswick's Bay of Fundy—all of Alaska's extraordinary ensemble of natural wonders tugged on his psyche. Such wild grandeur was incomprehensible on the East Coast. Excitedly, Roosevelt devoured everything published about the Alaskan frontier. But the real excitement in Alaska from 1870 to 1914 was gold. From the early prospectors of the 1870s to the crazed strikes of the late 1890s in the Klondike and the stampedes in Nome to the El Dorado gold fever triggered by discoveries in Tanana, Ruby, Iditarod, and Livengood, gold ruled Alaska. Only the discoveries of oil fields in Alaska, first up the Cook Inlet and then along the Arctic Ocean coastal plains in the middle to late twentieth century, equaled the wild-eyed hunger for gold. [34]

The part of the permafrost Arctic expanse owned by the United States—northern Alaska above the Arctic Circle—was home to millions of birds from all over the world. In the air—arriving in swirls from Antarctica, Australia, Asia, South America, northern Canada, and the Lower Forty-Eight—were migratory birds that had flown thousands of miles. Every spring geese, ducks, swans, and sandhill cranes were the first to arrive, even before the ice melted and the rivers were free. Native Alaskan tribes, Roosevelt learned, had flourished for thousands of years, living hand-to-mouth off the frozen land and rough sea. According to the *U.S. Geographical Survey* in 1877, most of Alaska was an open book for any faunal naturalist willing to collect quantitative data. After reading Henry Wood Elliot's *A Report upon the Condition of Affairs in the Territory of Alaska*, Roosevelt craved the rock, snow, and ice of the territory even more. Russia had made one of the worst blunders of the nineteenth century in selling more than 586,000 square miles so cheaply. Alaska sprawled over 21 degrees of latitude and 43 degrees of longitude. As the explorer Alfred Hulse Brooks—who gave his name to the Brooks Range—noted, Alaska was truly a place of "continental magnitude."[35]

Alaska was one-fifth the size of the continental United States, larger than California, Texas, and Montana combined. If superimposed onto a U.S. map, the state would stretch all the way from California to Florida. Alaska had an astounding 33,000 miles of coastline; seventeen of America's twenty highest peaks were in the territory. There were more active volcanoes there than in Hawaii and the Lower Forty-Eight combined. This was the land of 100 Yosemites. Texans could brag all they wanted to about open space, but Alaska was well over twice as large as the Lone Star State. "In Alaska," the conservationist Paul Brooks wrote in *The Pursuit of Wilderness*, "everything from the price of eggs to the antlers of moose is more than life size."[36]

Of all the major U.S. politicians following Seward's Alaska Purchase of 1867, it was Roosevelt who initiated giving Alaska's citizens—largely Aleut, Inupiat, Tlingit, and other Native tribes—constitutional rights. A fist-pounding Roosevelt had urged Congress in 1906 to "give Alaska some person whose business it shall be to speak with authority on her behalf to Congress."[37] Roosevelt saw Alaska as a primitive wilderness, full of game, its waters teeming with fish without end, serving as a long-term salve to the inherent rottenness of industrialization. There was no dollar value to put on magnificent places like the Alaska Range, Alexander Archipelago, or Aleutian chain. To Roosevelt, all of Alaska could become a vast federal district whose natural resources would be tightly controlled from Washington, D.C. By the late nineteenth century wilderness preservation societies were sprouting up across the Lower Forty-Eight: the Appalachian Mountain Club (1876); the Sierra Club (1892); the Mazamas of Portland, Oregon (1894); and the Camp Fire Club of America (1897), to name just a few.[38] To members of these nonprofit organizations, Alaska was a great cathedral, the last place to worship the most spectacular, untrammeled wilderness in North America.

When Roosevelt left the White House in March 1909, after serving as America's twenty-sixth president, he donated his handsome snowy owl mount to Frank M. Chapman (head of ornithology at the American Museum of Natural History and an early proponent of federal bird reservations). A grateful Chapman gladly accepted the specimen, tagging it as accession No. 15600. He then proudly put the owl on public display. Instantly, it became the most popular artifact of Roosevelt in the museum's natural history collection, housed in the appropriately named Roosevelt Memorial Hall.[39] Everybody wanted to see the snowy owl. The bird was like a messenger from the far north, possibly from Alaska, that had winged its way thousands of miles from the quiet world to crowded New York. The snowy owl proved to Roosevelt and others that the Arctic was *real*—not remote or otherworldly. Although Roosevelt never visited Alaska, his conservation policies, specifically his bedrock belief that the federal government had to save Alaskan wilderness tracts en masse from despoilers, profoundly influenced how future generations thought of the district turned territory turned state.

Chapter Two

·   ·   ·   ·   ·

THEODORE ROOSEVELT'S
CONSERVATION DOCTRINE

I

Roosevelt would have given his eyeteeth to be an ornithologist on the Harriman Alaska Expedition of 1899, exploring what are today Glacier Bay National Park, Misty Fiords National Monument and Wilderness, and Chugach National Forest to observe bald eagles, whales, seals, bears, and more. But, alas, he was at that time governor of New York and couldn't get away to a far-distant sphere. Albany seemed to him like an eddy in a stream where branches float backward and accumulate in the mud, a logjam of bureaucracy—not a free, wild place like southeastern Alaska. The rhythm of natural history discoveries was Roosevelt's passion in life. Oh, to have been able to discuss the king eider (*Somateria spectabilis*) with John Burroughs and mountain goats (*Oreamnos americanus*) with John Muir! Governor Roosevelt's stars, however, did not align in 1899; he would have to wait for the future to see the Inside Passage, Prince William Sound, and the Bering Sea land bridge site for himself. But Roosevelt was gearing up—like the twenty-eight prominent Americans documenting the natural world along 9,000 miles of Alaska's coastline from the *Elder*'s deck—to make protecting the "great land" a key component of his conservationism.

An accomplished naturalist and adventurer, Roosevelt had eagerly read the scientific reports written by the faunal naturalists on the Harriman Expedition as they were periodically issued. He was especially impressed with the work of George Bird Grinnell (editor of *Forest and Stream*), Dr. C. Hart Merriam (chief of the U.S. Biological Survey), and William H. Dall (paleontologist of the U.S. Geological Survey and honorary curator of mollusks at the U.S. National Museum). Sometimes Roosevelt was envious of Dall, who had three species—the Dall porpoise, Dall sheep, and Dall limpet—named after him. By 1899 only the Olympic Range elk (*Cervus roosevelti*) had been named—by Merriam—in Roosevelt's honor. The Harriman Expedition was Dall's unprecedented fourteenth trip to wild Alaska; his first visit had been in 1865, to study the possibility of an intercontinental telegraph line. Dall had encyclopedic knowledge of all things Alaskan and was teasingly nicknamed "Inuit Dall." His book *Alaska and Its Resources* (1870) was a bible to U.S. government agents traveling in the district after Seward's purchase.[1]

Once Roosevelt started reading Dall on Alaska, he began thinking about ways to protect Alaska's species (and their habitat) in perpetuity. Now, in 1899, as New York's conservationist governor, he was setting the tone for the rest of America, including Alaska. The big-game hunter Dall DeWeese had also gone to Alaska in 1897 to shoot a trophy bull moose (*Alces alces*) on the Kenai Peninsula. He bagged an antler rack that set a record at the Boone and Crockett Club. Antlers of an adult moose in Alaska weighed around seventy-five pounds and had a seventy-two-inch spread. Most adult caribou (*Rangifer tarandus*) and moose shed their antlers by January, after the rut. Female caribou shed theirs in the springtime after calving (often not until June). Female moose have no antlers.

But when DeWeese went back the following summer, the moose population on the Kenai Peninsula had been severely diminished by market hunters working in the lowlands. Roosevelt was livid over DeWeese's report—the Alaskan moose might soon go the way of the Great Plains bison. Roosevelt, working with the New York Conservation Society and the Boone and Crockett Club (which he had cofounded), started planning to create a moose refuge in Alaska, on Fire Island (the first in the world). To Roosevelt's chagrin, the reports of the Harriman Alaska Expedition, of which he started getting advance copies in 1900, were short on moose biology. (That wouldn't have happened if *he* had been a mammalogist on the *Elder*, along with Merriam and Grinnell.)

Roosevelt believed that, like bison on the Great Plains, moose added an alluring charm to the Alaskan landscape. He wanted a tough law that Alaskans had to get special permits to hunt moose only *in* season, when the antlers were biggest. His views weren't far removed from those of the Koyukon people of Alaska, who claimed that wild animals weren't property. "Wild beasts and birds are by right not the property merely of those who are alive today," Roosevelt said, "but the property of unknown generations whose belongings we have no right to squander." [2]

Paradoxically, stories of the Harriman Expedition mooring off Kodiak Island to hunt bears fascinated Roosevelt no end. Back in the 1880s in the Bighorns of Wyoming, he had shot grizzly bears. It was an ambition of his to bag a Kodiak bear as E. H. Harriman had done in Alaska. (Harriman, in turn, had partially modeled his thirteen-volume report of the expedition on Roosevelt's three outdoors memoirs about the Dakota Territory.) There was, however, a dissenter: Muir thought that Harriman, Merriam, and others stomping around Kodiak Island to bag bear trophies in the name of "science" were a childish and pathetic spectacle. To Muir, his cruise compatriots were cruel fools, idiotically abandoning the glories of Prince William Sound to become "gun laden" actors preparing "for war." [3] The excuse Merriam offered was that he was writing the definitive study of Kodiak bears; he needed an "old bruin" to study biologically. E. H. Harriman, the railroad tycoon, was the expedition member who shot the biggest brown bear. A Russian hunter, paid as a scout, had killed a second. The expedition taxidermist, Leon J. Cole, was dispatched from the *Elder* to skin the enormous bears and bring their furs back to the ship. [4]

Roosevelt's attitude—that of a faunal naturalist—seemed to envelop the Harriman Expedition. A lot was accomplished in a short time. Roosevelt's old friend, the illustrator Robert Swain Gifford, was chosen by Harriman to sketch scenes from the two-month voyage. Gifford had ably done the illustrations for Roosevelt's book *Hunting Trips of a Ranchman*, published in 1885. Burroughs, Roosevelt's dear friend, promised to tell the governor about all the Alaskan birds when he returned to New York. Merriam had reviewed Roosevelt's first book—a pamphlet, really—titled *Summer Birds of the Adirondacks*, back in 1878; they became fast friends. Merriam headed the Biological Survey and was the person Roosevelt corresponded with most often about North American mammals and birds. Then there was

Grinnell, founder of the original Audubon Society in 1886, with whom Roosevelt had started the Boone and Crockett Club. Together with Grinnell, they saved the Lower Forty-Eight herds of bison, antelope, deer, elk, and moose from extinction. These cronies of Roosevelt, traveling together on the *Elder*, were determined now to save parts of wild Alaska just as they had done for Yosemite, Yellowstone, and the Adirondacks.

E. H. Harriman was so rich in 1899 that he didn't need a letter of introduction in Alaska. But Roosevelt was close to Governor James Brady of Alaska—they had a family connection—and saw to it that Brady rolled out the red carpet for the Harriman Expedition in Sitka, a fishing and forest town in the Alexander Archipelago. If they were going to save wild Alaska, including what birds and game weren't shot-out, Brady would be a crucial ally. At this time, Harriman was fifty-one years old and, as chairman of the Union Pacific Railroad, one of the richest and most powerful men in America. By the time of his death in 1909—when he was worth $100 million—Harriman had overseen the Union Pacific, the Southern Pacific, the Saint Joseph and Grand Island, the Illinois Central, the Central of Georgia, the Pacific Mail Steam Ship Company, and the Wells Fargo Express Company. Years afterward, however, it was his scientific expedition to Alaska that earned him his permanent place in history.[5]

When Roosevelt became governor of New York in January 1899, Alaska was very much in the news. While he formed the Rough Riders in San Antonio, Texas—the volunteer cavalry outfit—to fight in the Spanish-American War in 1898, the Klondike gold rush was on. The U.S. Army had charted the upper 500 miles of the Yukon River in Alaska, inadvertently opening up the Klondike gold fields to placer mining. Prospectors and preachers, prostitutes and poachers, thieves and roustabouts—all came tumbling into Alaska in record numbers. While only a few men made fortunes in 1890, Alaska's mineral production was estimated to be $800,000; by 1904, gold production alone had risen to $10 million.[6] A few of these boomers, too, became millionaires; but most found themselves cursing the cold, inhospitable climate.

Because of his elite upbringing as a New York Knickerbocker, Roosevelt rejected the kind of get-rich-quick schemes that the novelist Knut Hamsun had condemned. Nor did Roosevelt believe that corporate monopoly should have a role in the Alaska district. What interested Roosevelt most about Alaska—besides the moose in the Kenai Peninsula—was that whole-some, God-fearing pioneer families were starting to put down permanent roots. The principal Christian denominations in the Lower Forty-Eight, through the Federal Council of Churches, had divided up zones in which to bring New Testament principles to the Native Alaskan people. Different sects sent missionaries to different regions: Kenai Peninsula (Baptists); Point Hope (Episcopalians); Brooks Range, Anaktuvuk Pass, Barrow, Wainwright, the Alexander Archipelago (Presbyterians); and Anchorage (Baptists). At their best, the missionaries taught sanitation, medicine, and math. At their worst, they prohibited dancing and frowned upon Native arts and crafts as perverse. Both benignly and purposefully, an erosion of Native Alaskan culture was under way.

One Presbyterian missionary hoping to Christianize the Chilkat Tlingit was John Brady (who served as governor of the district of Alaska from 1897 to 1906).[7] Brady, living amid mile-long glaciers and shimmer-ing coastal waters, would do *anything* for Roosevelt—literally anything—because he owed his life to Theodore Roosevelt Sr. (the president's father). Brady was based in Sitka (on Baranof Island) with the Pacific Ocean serv-ing as his backyard, and he was properly concerned that thirty-seven salmon canneries were operating at capacity around the Inside Passage, the glorious waterways that Muir had extolled in the *San Francisco Daily Evening Bulletin*. In 1894, for the first time, packers in the Alexander Ar-chipelago had exceeded the million-case mark by late fall. Overfishing was becoming a menace. Grinnell had likewise complained in his report for the Harriman Expedition that Alaskan Natives were being swindled by the rapacious salmon industry. "For hundreds of years," he wrote, "the Indians and Aleuts had held those fisheries with an actual ownership which was acknowledged by all and was never encroached upon. . . . No Indian would fish in a stream not his own."[8]

Perhaps in repayment for having been born with a silver spoon in his mouth, Theodore Roosevelt Sr. regularly found foster parents for home-less children in New York during the late nineteenth century. John Hay, a family friend, claimed that TR Senior had a "maniacal benevolence" to help slum children. He was a founder of the Children's Aid Society; he paid support stipends for dozens of waifs through that society; and he spent every Sunday at the Newsboys Lodging House, offering counsel as a sort of father figure. One afternoon in 1854 TR Senior saw little John Brady, in ragged clothes and with disheveled hair, begging with a tin can around Chatham Square. Within a few weeks TR Senior had gotten the ragamuffin Brady, whose father was a drunken longshoreman, placed with a foster family in Indiana. He took personal pride in Brady's graduation from Yale University in 1874. Next for Brady was a scholarship to Union Seminary; he excelled at ecclesiastical scholarship and was eventually ordained as a Presbyterian minister. Then Brady studied law, with a plan to help Native Americans get ahead in society. In 1878 he moved to Sitka, Alaska, and founded the Presbyterian mission school there. (This school, which later became the Sitka Indian Industrial Training School and still later Sheldon Jackson College, was an industrial vocational institution for Native Alaskans.) The words of Jesus had arrived on the panhandle.

When Muir was touring Glacier Bay in 1879 and 1880, Brady, the only serious rose horticulturist in Alaska, had been teaching Tlingit and Haida children how to speak English, and introducing them to the scriptures. The huge glacier where Muir and the dog Stickeen had almost died would be named after Brady. On July 15, 1897, after McKinley's election as president, Brady became governor of Alaska. Brady was a protégé of Sheldon Jackson, a Presbyterian minister who had been trained at Princeton University and who was pro-temperance, pro–Native American rights, and pro-conservation. Jackson, in fact, imported reindeer from Siberia on the cutter *Bear* and released them at Amaknak Island in the Aleutian chain. Reindeer—domesticated caribou—were then introduced near Port Clarence in Northwest Alaska, and Jackson hired Laplanders to teach Native peoples how to become reindeer herders.[9] Much of Brady's work, like Jackson's, involved trying to properly educate Alaska's Native populations—the Aleut, Athabascan, Tlingit, and other peoples—and insisting that they were equal to whites. Although Brady and Jackson meant well, their promotion of Christianity and of English as a first language led to the destruction of many Native American cultural mores.

Brady's town, Sitka, once a peaceful village, was populated in the late 1890s by 40,000 non-Native argonauts trying to make a strike in the Klondike. Like the voyageurs of the fur companies in the Canadian northwest, they came hungry for wealth. Brady telegraphed the U.S. Army for immediate help; troops arrived with what might today be described as riot gear and dispersed the gold-seekers away from Sitka.[10] Roosevelt had lobbied for him. "Your father picked me up on the streets of New York, a waif and an orphan, and sent me to a Western family, paying for my transportation and early care," Brady reminded Governor Roosevelt when they met in 1900. "Years passed and I was able to repay the money which had given me my start in life, but I can never repay what he did for me, for it was through that early care and by giving me such a foster mother and father that I gradually rose in the world until I greet his son as a fellow governor of a part of our great country."[11]

When Roosevelt suddenly became America's twenty-sixth president in September 1901—after the assassination of William McKinley in Buffalo, New York—the district of Alaska hadn't yet earned U.S. territory status (until May 7, 1906, it was simply federal district property). With Brady serving as the leading light of Alaskan politics, Roosevelt called for Alaska's representation in Congress (a half measure was adopted in late 1906, when Alaska was given a voteless delegate in Congress). While Alaska had executive and judicial officers, the district was without a legislative body until 1912. This gave Brady, as district governor, political clout in Alaskan affairs during Roosevelt's presidency.

Harriman himself praised Brady's hospitality and intelligence after spending time with Brady seeing the natural wonders around Sitka. Brady had taken Harriman, Burroughs, Merriam, and others swimming in a hot spring outside Sitka (at a camp frequented by the Sons of the Northwest). The bubbling waters, the smell of sulfur, and the steam hissing out of the little pond made the hot spring like a spa—and a curative for the cabin fever that had beset the passengers on the *Elder*. But Muir, the moralist, objected to the Sitkans' hunting practices. Muir recalled that the overseer at the hot spring "murdered a mother deer and threw her over the ridge-pole of the shanty, then caught her pitiful baby fawn and tied it beneath the dead mother."[12] Such brutality to animals, including killing at point-blank range, always turned Muir's stomach.

Until the advent of Roosevelt, officials in Washington, D.C., were baffled about what to do with Alaska's soaring mountains, hidden caves, sumptuous forests, and extensive coastline. Extraction in all its many manifestations seemed wisest. The intimidating Alaskan landscape was so large that rivers just disappeared over horizons and mountain ranges unfolded in staggered rows toward a surreal blue infinity. Deeply influenced by the Harriman Expedition's reports—the publishing venture overseen by Merriam, which grew into thirteen thick, illustrated volumes—Roosevelt knew that the U.S. government had to properly manage the remarkable forestlands, fisheries, and wildlife resources. These reports became his administration's all-purpose reference points for Alaskan scientific research and management of public lands. A conservation ethos had to prevail in Alaska, to prevent the twentieth-century industrial order from turning the district into slagheaps, cesspools, and tracts of stumps. Important gold discoveries—Fairbanks (1902); Valdez Creek (1903); Kantishna (1905); Richardson, Chandalar, and Innoko (1906)—were becoming annual occurrences. Gold, however, wasn't going to save Christian souls, help First Nation people prosper, or protect the forestlands of Alaska.

Besides reading the Harriman Expedition's reports, Roosevelt received from a Seattle studio copies of images produced on the 1899 cruise. Edward Curtis's black-and-white photographs of Alaska's natural wonders floored the president. Curtis (born February 16, 1868, in Whitewater, Wisconsin) was well known to people interested in the Pacific Northwest wilderness for taking amazing landscape photos of Mount Rainier, the Olympic Mountains, and the island-dotted Puget Sound. Grinnell, an expert on Native American culture, met Curtis one afternoon at Mount Rainier. A fast friendship ensued. Curtis had just taken his first portrait of a Native American: Princess Angeline (the daughter of Chief Sealth of Seattle). Grinnell became an enthusiastic booster of the youthful thirty-year-old Curtis, who thus got the job with the Harriman Expedition.

Using a six-by-eight camera, developing his own film in the ship's darkroom along the Alaskan coastline, Curtis brilliantly documented the surreal boldness of Alaska in 1899, using the high-latitude light to produce textured prints documenting glacial action. Regularly, Curtis explored glaciers with Muir, perfecting his photographic techniques. While Muir was sketching glaciers, Curtis recorded their advances and retreats. Scientists concerned about global warming in the early twenty-first century used Curtis's prints as archival evidence of what used to be.

Historians have a sense of Alaskan glaciers, icebergs, and fjords in 1899 because of Curtis's devotion to his craft and to science.

When the expedition ended, Curtis was commissioned by Harriman to make the *Souvenir Album* of Alaska. Feverishly working overtime to get this volume ready by 1900, Curtis did a remarkable job of laying out what he saw as the *two* Alaskas—soaring nature and desolate poverty. His Native American portraits, soon to become his calling card, sadly display acculturation at work. Eskimos, seen through Curtis's honest lens, were abysmally treated by whites as lowly servants. Curtis's village images were a haunting testament to the ramshackle, dilapidated fishing communities of the Alaskan coast, which looked nothing like the happy cottages of Nova Scotia or Newfoundland. In Curtis's portfolio, sealing camps were slums where gutted whale carcasses and bloody dirt dominated the landscape. Curtis frowned on the vicious slaughter of walrus—strong animals, but defenseless against a harpoon or gun. Captains of the hunting schooners in the Bering Sea all seemed amused by a new cutthroat attitude: the idea that Alaska was like a ripe melon to be sliced and diced for profit by outsiders. Washington, D.C., was too far away to enforce anything. Curtis's bold photographic images demonstrated that if overfishing continued, treasured places like Prince William Sound would become whale cemeteries. Curtis, always studying the light, started being referred to by Native Americans as a "shadow catcher."[13]

But Curtis made nature in Alaska radiate with transcendent light and love. Prefiguring Ansel Adams's Alaskan photographs by almost half a century, Curtis's icebergs looked like marble sculptures by Henry Moore. Curtis's photographs of volcanoes in the Aleutian Range, some people said, were as majestic as Frederic Church's landscape paintings. Any connoisseur of natural wonders would be touched by Curtis's elegiac Alaskan images, such as *Muir Glacier*, *Orca Harbor*, *The Way to Nuntak*, and *Last View of the Pacific*. His most enthusiastic fan of all, it seemed, was Theodore Roosevelt.[14] Writing the introduction to Volume 1 of Curtis's magisterial twenty-volume work *The North American Indian*, Roosevelt said, "In Mr. Curtis we have both an artist and a trained observer, whose work has far more than mere accuracy, because it is truthful. . . . Because of his extraordinary success in making and using opportunities, [he] has been able to do what no other man has ever done; what, as far as we can see, no other man could do. Mr. Curtis, in publishing this book, is rendering a real great service; a service not only to our people, but to the world of scholarship everywhere."[15]

President Roosevelt wanted to help Alaska—its nature and its Natives—prosper. Henry Gannett, the chief geographer for the U.S. Geological Survey, was his well-placed ally in drawing up new maps for the territory. After spending several amazing weeks on the Harriman Expedition, hiking along ice-cloaked fjords with Muir and Burroughs, marveling at the majesty of coastal rain forests and pendant-shaped waterfalls, Gannett realized that "nature tourism" would become a major Alaskan industry. He rejected outright the notion that Alaska should be tapped for gold, copper, coal, and other sources of extractable wealth. "There is one other asset of the Territory not yet enumerated, imponderable, difficult to appraise; yet one of the chief assets of Alaska, if not the greatest," Gannett wrote. "This is the scenery. Its grandeur is more valuable than the gold or the fish or the timber, for it will never be exhausted. This value, measured by direct returns in money received from tourists, will be enormous, measured by health and pleasure it will be incalculable."[16] *National Geographic* echoed his sentiment in the coming years. "The Alaska coast is to become the showcase of the earth," the magazine predicted in 1915. "Pilgrims, not only from the United States but from far beyond the seas, will throng in endless procession to see it."[17]

Just as a Texan cattle rancher wouldn't tolerate claim jumpers overrunning his pastures, President Roosevelt objected to California boomers racing up to Alaska and timbering and gold mining on public lands. Natural resource management wasn't going to be a free-for-all under his administration. No matter how many gold discoveries were trumpeted in the *Fairbanks Daily-Miner* or the *Yukon Press*, Alaska belonged to the U.S. government, not to bonanza seekers—end of story. As landowner in chief, Roosevelt envisioned the Alaska district as a loosely knit fabric of well-run small towns surrounded by federal forest reserves and wildlife refuges.[18] Mining and timbering would be localized. Over time, civic responsibility would emerge. Alaska would be America's permanent wilderness zone. When asked by the *Wall Street Journal* if such an exercise of executive power might not hurt his popularity, Roosevelt scoffed at the idea, saying that he wasn't a "college freshman" and that therefore he always acted on behalf of the long-term "public interest."[19]

With the whole Harriman Expedition cheering him on, Roosevelt insisted that an honest court system had to be established in Alaska, one willing to take to trial a huge backlog of civil and criminal cases. The Aleutian Islands had virtually no courts, and the Yukon valley regularly delayed trials. Lawyers who could marshal pro-conservation arguments needed to be appointed and complemented by more honest judges, or else the 375 million acres of pristine Alaskan landscape would be ravaged. Disdainful of the disgraceful way the Russians had slaughtered seals on the Pribilofs, Roosevelt believed the judge's gavel was needed in the territory, as much as Paul Bunyan's ax. A better tax system had to be established. The Hamiltonian side of Roosevelt's political personality wanted strong federal regulations for the Alaska district, from Point Barrow all the way down to Ketchikan. Meanwhile, in Juneau, Brady served as Roosevelt's political watchdog over U.S. public lands that were being protected.

II

During the spring of 1903, President Roosevelt made a "great loop" tour of the American West, promoting conservation; it included stops in Yellowstone, the Grand Canyon, and Yosemite. By the time he arrived in Seattle on the steamer *Spokane*, the president was on a conservationist mission to protect Alaskan lands from overmining, overfishing, market hunting, and deforestation. Dry-hole oil speculators were starting to drift to Alaska, motivated by the automobile craze, calling petroleum the new whale blubber. Parading through thirty-five blocks of downtown Seattle, Roosevelt eventually made his way to the original grounds of the University of Washington. He gave a thumbs-up to the totem pole in Pioneer Square. More than 50,000 people listened to him deliver a stem-winder about protecting the natural resources of places such as Prince William Sound, the Alexander Archipelago, and the Tongass. Behind him was a banner that read, "Alaska Greets the President." Roosevelt was asserting the supremacy of federal law in the Alaska district. The great primeval forests of the Pacific Northwest and southeastern Alaska, he said, belonged to the American people. The highlight of Roosevelt's three days in the Puget Sound area was his address to the Arctic Brotherhood, a fraternal organization founded in 1899 on board a steamer traveling from Seattle to Skagway. Perhaps mellowed by the free-flowing booze, Captain William Connell had suggested en route that a "Brotherhood of the North" be formed. Membership in what became the Arctic Brotherhood was restricted to white males over eighteen who lived in Alaska, the Yukon Territory, the Northwest Territory, or parts of upper British Columbia—men who knew what it was to endure the severity and length of an Alaskan winter. The brotherhood's mascot was the polar bear, and trudging off for getaway hikes and hunts was what its members did best. Accepting honorary membership in the brotherhood, Roosevelt warned the other members that conservation had to be part of their mission.[20]

"Most of the people of this country are wholly in error when they think of the mines as being the sole, or even the chief, permanent cause in Alaska's future greatness," Roosevelt said in his address in Seattle. "Let me tell you just exactly how I mean it. In the case of a mine, you get the metal out of the earth. You cannot leave any metal in there to produce other metal. In the case of a fishery, a salmon fishery, if we are wise—if you are wise—you will insist upon its being carried on under conditions which will make the salmon fishery as profitable in that river thirty years hence as now. Don't take all of the salmon out and go away and leave the empty river to your children and your children's children." Then Roosevelt went after Pacific Northwestern timber companies that were already at the gates of southeastern Alaska, waiting to clear-cut vast stretches of forestlands. Limited logging was fine, Roosevelt said. But at all times, the "preservation of the forest for the settlers and the settlers' children that are to come in and inherit the land" had to be the governing ethic.[21]*

As the Arctic Brotherhood learned, when it came to protecting Alaska's

---

* In 1909 President William Howard Taft spoke to the Arctic Brotherhood on a visit to Seattle's Yukon-Pacific Exposition. He was given the title Post Grand Arctic Chief. Unlike TR, Taft agreed to wear the brotherhood's ridiculous Arctic robe with its polar bear collar. The current Arctic Brotherhood Web site mocks Taft (perhaps inadvertently) for wearing "the gayest-looking costume any president has dared to wear." In 1907 the fraternal organization Pioneers of Alaska had been formed to preserve the territory's history. Igloo No. 1 was founded in Nome.

natural wonders and wildlife resources, there was no gentleness or delicacy in Roosevelt's manner. Whether you entered Alaska by the Copper River, the Kuskokwim delta, or the Yukon territory didn't matter—the conservationist doctrine had to be followed. As a forest conservationist, fascinated by America's different physiographic conditions, Roosevelt was particularly worried about the fate of Alaska's boreal woodlands, which extended from the Kenai Peninsula to the Tanana valley near Fairbanks and stretched northward to the Brooks Range. Trees with high commercial value—white spruce, quaking aspen, paper birch, western balsam, poplar, and larch—blanketed Alaska in immense stands. Roosevelt knew that once the Klondike gold veins dried up, timber corporations such as Weyerhaeuser Lumber would be ready to speed up the logging of old-growth forests without even replanting, leaving places such as the exquisite Tongass barren. Roosevelt was determined to protect Alaska's millions of forested acres from senseless destruction.

Roosevelt's brain trust for Alaskan forestry was Gifford Pinchot. Born just after the Civil War in Simsbury, Connecticut, to a father who made a fortune in timbering and land speculating, Pinchot probably knew more about the perils of deforestation than anybody else of his generation. Over six feet tall with a handsome full mustache, Pinchot, Yale University's silvicultural prodigy, was a mighty fighter for the cause of forest protection. Pinchot, in fact, was considered the father of the modern forestry movement. He became the first chief of the U.S. Forest Service, appointed by President Roosevelt in 1905.

A virtual fixture in Washington, D.C., society from 1890 to 1946, Pinchot could often be seen walking in the streets around Dupont Circle, noticeable in an instant by his broad-brimmed, floppy brown hat. A believer in the wise use of natural resources, Pinchot also devoted his life to preserving many of America's pristine forestlands. Congress had passed legislation creating forest reserves in 1897, and this gave Pinchot his opening to make history. Working under Roosevelt, Pinchot helped save more than 150 million acres—mainly in the American West—as protected national forests between 1901 and 1909.

One problem Pinchot faced in Alaska was that surveys were scant. In 1902 he had dispatched William Langille—a Canadian-born Oregonian known in the Forest Service as Pinchot's eyes and ears—to do reconnaissance in both the Chugach and the Tongass forests of Alaska. Using dog-sleds and dugout canoes to get around, Langille immediately recognized that the Alexander Archipelago, Chugach, and Tongass were richer in trees than the entire Rockies of Colorado, Wyoming, Idaho, and Montana combined. "Special attention should be paid to species, age, and distribution of the forest bodies in different sections," Pinchot had instructed Langille, "noting relative sizes, etc. in reference to geographical, physical, and altitudinal position." [22]

According to people associated with Roosevelt and Pinchot, unless Forest Service rangers like Langille continually hiked or patrolled by canoe, timber trespassers around Puget Sound, backed by Wall Street capital, would swoop into Alaska, destroy forestlands, never replant, and leave the tiny timber communities to starve. Tree stumps, runaway unemployment, and silted rivers would be the legacy of "big timber." Instead of sustainable communities built by pioneer families, ghost towns would spread from Homer to Seward to Ketchikan. Roosevelt had seen this trend all over the Wild West. Determined Rooseveltian conservationists like Langille envisioned the coastal forests of southeastern Alaska—an extension of the rain forests of Oregon, Washington, and British Columbia—remaining as national forests. There would be no huge cities, no steel mills like those in Pittsburgh, no Tacoma cable running from hill to hill, no seal-skinning slums on the disappearing glaciers. Instead, little Alaskan fishing and lumber towns would be settled by permanent populations like those found in British Columbia, unencumbered by the gospel of greed. Langille, who kept scrupulous record books, would soon be tasked with arresting timber thieves, issuing occupancy permits, enforcing game laws, and working closely with the Biological Survey, Fish Commissions, and Geological Society.

When it came to Jack London's tales of prospecting in Alaska and the Yukon, Roosevelt was a scold. Although he admired London's heartiness, he hated *The Call of the Wild*, deeming it tacky "nature-faking." Roosevelt—a cheerleader for the Harriman Expedition—wanted Alaska studied for its caribou herds, its grizzly and polar bears, the giant walrus, its seabirds. He was not interested in phony wolf stories. The heroes of Alaska, to Roosevelt, were the U.S. Forest Service agents like Langille, the conservationist law enforcers who arrested the dirty, unkempt gold seekers who thought Alaska was a casino to gamble their lives in. Roosevelt disdained "cheechakos," avaricious miners flooding into southeastern Alaska who had no family values, no fashion, no manners, no code of ethics. To Roosevelt these gold seekers were like ice worms, digging into the land without any sense of community-building. London described prospectors on the California iron trail feeding Native Americans liquor, but this sad notion sickened Roosevelt. As a former police commissioner in New York City, who admired explorers, army officers, foresters, and biologists, he had zero tolerance for boozers, rabble-rousers, and malcontents, many so dirty that they stank, who didn't know a snow bunting (*Plectrophenax nivalis*) from a dusky shrew (*Sorex monticolus*). What he admired about cowboys was their work ethic, cleanliness, homestead lifestyle, and equestrian skills; none of these rancher qualities were apparent in the "let's blow up a mountain" ice worms.

On public lands leased from the U.S. government in Alaska, Roosevelt preferred mining coal from the ground rather than dynamiting mountains in search of coal seams. When he could—as in the cheerful cases of designating Crater Lake (Oregon) and Mesa Verde (Colorado) as national parks—he worked in tandem with Congress. But if legislators resisted him, he steamrollered over them, invoking the new Antiquities Act to protect, for example, the Grand Canyon of Arizona or Devils Tower of Wyoming. The Antiquities Act gave any U.S. president the prerogative, on behalf of scientific investigation, to prevent public land from being exploited. On eighteen occasions Roosevelt used this act to set aside national monuments by means of executive orders. In addition, Roosevelt issued executive orders creating fifty-one federal bird reservations, four game preserves, and more than 150 national forests. Roosevelt had also introduced a host of modern wildlife protection laws. Calling in 1909 for a World Conservation Congress, Roosevelt wanted to protect the world's oceans against overfishing, habitat destruction, and pollution from oil and sewage. Because Alaska had so much shoreline, he saw it as an ideal place to usher in a new marine conservationism, one enlightened by the image of Earth as a single pulsating biological entity, in peril from industrialization. (As the historian T. J. Jackson Lears has argued in *No Place of Grace*, Roosevelt was antimodern in his belief that "country life" was superior to urbanization.[23])

By the time of his address to the Arctic Brotherhood in 1903, Roosevelt had already made a first bold preservationist strike in Alaska. On August 20, 1902, he set aside the Alexander Archipelago, a 300-mile-long group of forested islands off the southeastern coast of mainland Alaska (named after a former head of a Russian fur-trading company). On that day more than 4.5 million acres of Alaska were protected in perpetuity. Muir, Merriam, Grinnell, and Burroughs had been eloquent about the Alexander Archipelago in the Harriman Expedition's reports. These islands contained huge rugged mountains, high tundra, plunging valleys, and blindingly green rain forests. The fir-timbered islands in the Alexander Archipelago were an incubator for tens of thousands of hermit thrushes (*Catharus guttatus*), pine siskins (*Spinus pinus*), harlequin ducks (*Histrionicus histrionicus*), northern goshawks (*Accipiter gentilis*), black oystercatchers (*Haematopus bachmani*), and a large number of whale species. There was no such massing of wildlife anywhere else. In his memoir *Travels in Alaska*, Muir said that the Alexander Archipelago, whose melting glaciers were powerful enough to carry mountains to the sea, was his new "home" of "pure wilderness."[24] Native tribes called the archipelago the "Great Raven's World" (the raven being considered cleverest of all animals). This lavishly diverse land was interlaced with spectacular inlets, fjords, glaciers, mountains, estuaries, thick meadows, muskegs, and high tundra.[25]

Roosevelt also saw this new Alaska reserve—modeled after the Thousand Islands reserve of upstate New York—as a preemptive strike against Great Britain and Japan, whose cold-blooded market hunters were still clubbing fur seals in American waters around the Pribilofs. "We have taken forward steps in learning that wild beasts and birds are by right not the property merely of the people alive to-day," Roosevelt said, "but the property of the unborn generations, whose belongings we have no right to squander."[26] Roosevelt also wanted to make sure the land rights of the Native Alaskan peoples—the Inupiat (in the Arctic), the Tlingit, and the Haida (in the panhandle)—would be protected. On a map, the Alaska Panhandle looked like an extension of British Columbia. But now, owing to William Seward's fine diplomacy of 1867, and to Roosevelt's conservationist convictions, the Alexander Archipelago was permanently safeguarded. Accepting advice from Grinnell, Roosevelt hoped that the Tlingit and Haida would be his watchdog rangers around the islands, making sure the luxuriant stands of tall natural timber remained. Roosevelt encouraged the residents to be whistle-blowers for the U.S. Forest Service if any illegal activities were being pursued by canneries.

Because Roosevelt's uncle Robert Barnwell Roosevelt had been the leading American ichthyologist from the 1860s to the 1910s, protecting fish populations and creating hatcheries was something of a family business. The Roosevelts were early believers in the idea of artificial fish propagation, useful in Alaska, where the salmon runs were getting thin. Roosevelt insisted that for every red salmon taken from Alaskan waters, at least four new fish had to hatch. This was a variation of his policy "Plant two trees for every one cut down." In 1903 Roosevelt moved control of Alaskan fisheries from the U.S. Treasury Department to the recently founded U.S. Department of Commerce and Labor, which was tasked with regulating the industry. Roosevelt wanted the Bureau of Fisheries to fund Alaska's growing salmon propagation program. Tension rose between Alaskan citizens and the federal government. But Roosevelt, bypassing Congress, created two huge fish hatcheries in Alaska, which were vehemently opposed by fish packers. Toughening his regulation policies even more, Roosevelt wanted Secretary of Commerce William C. Redfield to regulate the mouths of *all* Alaskan rivers, streams, and bays within a three-mile radius. A reluctant Congress grappled with Roosevelt's scheme; in the end it changed the limit to 500 yards as a compromise.[27]

III

The notion of creating an *Alexander* Archipelago National Forest first came to George T. Emmons, a former naval lieutenant. Emmons gloried in Alaska's wilderness; he supervised Alaska's display at Chicago's World Columbian Exposition.[28] At the time, Emmons was considered America's reigning authority on Tlingit and Haida totem poles, and major museums worldwide collected Native pieces acquired by him. Nobody in the East, in fact, knew more about the islands, glaciers, and waterways of southeastern Alaska than Emmons (with the possible exception of John Muir). At a glance Emmons could distinguish edible plants like nagoonberry, fiddlehead fern, and wild celery from poisonous ones. Although he was based in Princeton, New Jersey, Emmons spent long summers in Sitka, studying everything from orcas (*Orcinus orca*) to short-tailed weasels (*Mustela erminea*). A true Renaissance man, he also considered himself a novice cetacean (whale) biologist, polar authority, and climatologist.[29] When the Harriman Expedition's ship anchored in Sitka, he brought the members to a hunter who had brown bear skulls for Merriam to study properly. Emmons was a disciple of Robert Barnwell Roosevelt and considered the Gulf of Alaska one of the finest marine biological zones in the world. Serving as Roosevelt's eyes and ears concerning fishing regulation, he was determined to make sure the Alaskan coastal waters weren't overfished or degraded.

At Roosevelt's request, Emmons wrote a cut-and-dried report—"The Woodlands of Alaska"—promoting the Alexander Archipelago as a national forest and protected waterway. Emmons's knowledge of these southeastern Alaskan islands was based on personal exploration—a practice that always found favor with Roosevelt. Emmons's report noted that much of Alaska's wilderness was a patchwork of tundra, not suited to become a forest reserve. There were millions of acres of permafrost (frozen soil but *not* ice) in Arctic Alaska where thermometers regularly shattered in the severe cold. In the Arctic Ocean, there were solitary icebergs the size of Saint Louis or New Orleans. No cartographer had ever properly mapped the vast Brooks Range or coastal plain of the Beaufort Sea.[30] Anchorage wasn't even a city yet. Emmons didn't think the extraction industries could do much damage that far north. But he worried that in a generation the great primeval forests of southern Alaska would be clear-cut if the federal government did not intervene. And the offshore waters were teeming with marine life; the departments of Commerce and Labor needed to police Alaskan fishing villages like Sitka and Katalla to send a broad message: the U.S. government, not fish-packing companies, controlled the waters.

After careful scientific consideration, Emmons suggested that the Alexander Archipelago islets should remain "one immense forest of conifers," a sacred place where the coast hemlock and Sitka spruce could thrive along protected coastlines.[31] Only limited timbering would be allowed on the Alexander Archipelago islands, without the threat of the pulp industry's sawmills. Emmons, in fact, concocted a plan for how the Roosevelt administration could work around fishing camps, sawmills, and canneries, and oversee a first-rate forest reserve.[32]

With his "dream of a national forest in Alaska" accomplished by way of the Alexander Archipelago, Roosevelt set his sights on protecting other forest ecosystems in the aftermath of the Harriman Expedition.[33] Roosevelt boldly proclaimed the 4.9 million-acre Chugach National Forest—adjacent to what would become Anchorage—in 1907. (The new forest reserve absorbed the Afognak.[34]) Vast stands of Sitka spruce (*Picea sitchensis*), western hemlock (*Tsuga heterophylla*), western red cedar (*Thuja plicata*), and Alaska cedar in south-central Alaska were saved by the U.S. Forest Service from the maw of the timber and coal industries. To Muir's delight, more than 10,000 glaciers were also part of the Chugach; currents regularly broke off chunks of ice and carried them out to sea. Roosevelt was worried that the Chugach, if not managed by the federal government, would be destroyed by the pulp industry and by fish packers. Much of the federally protected ancient forest was located around Prince William Sound, which extended from the Copper River on the east to the Kenai Peninsula on the west.[35]

The Chugach National Forest—a subpolar rain forest—was perhaps Roosevelt's most ambitious move with regard to conservation. Throughout the Chugach Mountains, which provided Anchorage with an ideal natural backdrop, rivers and creeks interlaced with snowfields, salmon wove their way to the Gulf of Alaska, wolverines (*Gulo gulo*) and bears were on the prowl, and explorers could easily get lost in whipping mists and rain squalls. Geographic features were named after animals: for example, Ptarmigan Lake and Caribou Creek. (All around the Kenai Peninsula, there were natural features eventually named after former U.S. presidents, such as Harding Icefield and Grant Lake.) Overnight the Chugach became the northernmost addition to the portfolio of the U.S. Forest Service; there were many gurgling creeks and milky blue-green rivers that nobody had yet named. All along Resurrection Creek, however, could be found the scars of mining. Heaped-up cobble and gravel marred the riverbanks.[36]

Nobody in the East—with the exception of the faculty of Harriman's "floating university"—could have imagined the diversity of Chugach National Forest. Essentially there were three bioregions within its boundaries: the Copper River delta (site of the premier salmon run in the world), Prince William Sound, and the Kenai Peninsula. The mantles of glaciers—which Muir had described on the expedition of 1899—were world-class around Prince William Sound. The members of the Harriman Expedition, having a little fun, named glaciers they encountered after schools: Columbia, Yale, Bryn Mawr, Radcliffe, Dartmouth, Holyoke, Barnard, Smith, Wellesley, Amherst, Williams, and Harvard among them.[37] There were numerous tidewater glaciers calving into Prince William Sound. Other glaciers clung to mountains; Muir, poet of the Chugach wilderness, wrote about them in a journal: "The sail up this majestic fjord in the evening, sunshine, picturesquely varied glaciers coming successively to view, sweeping from high snowy foundations and discharging their thundering wave-raising icebergs, was, I think, the most exciting experience of the whole trip."[38]

The Forest Service—the largest bureau within the U.S. Department of Agriculture—had its work cut out for it in policing the sprawling Chugach. Its rangers, emboldened by Pinchot's esprit de corps, sometimes had sole responsibility for overseeing millions of acres of both land and sea. From 1903 to 1911 Langille worked like an FBI agent, hunting down timber scoundrels in the Chugach and illegal fishermen in Prince William Sound. Simultaneously, he kept pushing for more Alaskan lands to be run by the Forest Service. Locals thought the Alaskan wilderness should be clear-cut and turned over to the pulp and paper industry. Langille scoffed at the "corporate frontiersman," interested only in exploiting land for a single generation and not in preserving it for the long run. Among all the places managed by the Forest Service, only the Philippines were as isolated from the Lower Forty-Eight as the Chugach. In 1909 the boundaries of the Chugach included all of Prince William Sound along with such landmarks as Montague Island, Controller Bay, and the thousands of unnamed glaciers that Muir treasured. Sometimes it seemed to Langille that all the non-Natives in Alaska were corruptible. Even Governor Brady, Roosevelt's friend in Sitka, had been hoodwinked by a con man, Harry Reynolds, into investing in a sham home-rule railroad. Reynolds had persuaded Brady to be a member of the railroad board even though he was still governor. An embarrassed Roosevelt had Brady fired in 1906.[39]

Langille had made Yes Bay, a cannery village near Ketchikan, his Forest Service headquarters. Having grown up along the Hood River of Oregon, he was an expert skier. During the Klondike gold rush of the late 1890s he had traveled from Dawson to Nome looking for gold. For a while Jack London had been his cabin roommate in the Yukon, and he had known the dog "Buck" from *The Call of the Wild*. Rough-hewn, as fit as a lumberjack, endowed with the guts of a pioneering Arctic explorer, Langille was the kind of rugged outdoorsman the Forest Service hired to oversee Alaska's forestlands. While in Nome, prospecting for gold, Langille received a wire from Gifford Pinchot asking him to oversee federal forestry in Alaska. Langille, wearing his only suit, went to Washington, D.C., in 1902 to plot a strategy for Alaskan lands with President Roosevelt and Gifford Pinchot. His annual salary would be around $2,000.

Upon returning to Alaska, Langille went on a reconnaissance mission all over the territory. Like a French-Canadian voyageur during the old days of fur trapping, Langille canoed up the Stikine River, inspecting sawmills, canneries, and Native villages. His reconnaissance was extensive. In April 1904, he traveled from Juneau to Controller Bay all across Prince William Sound and then onward to Norton Sound. Living off the land, Langille shot rabbits and ptarmigan with his .22 rifle, reeled in grayling with homemade flies, collected wildflowers to be studied back in the East, and mapped boundaries for potential forest reserves. Langille was a one-man Corps of Discovery operating in the twentieth century, watching over Alaska. Or, perhaps more accurately, Langille was Roosevelt's top cop in the big woods.

Roosevelt, influenced by Langille, withdrew the Tongass National Forest in 1907; it consisted of nearly all of southeast Alaska from Dixon Strait (in the south) to the Yakutat forestlands (in the north). Within the forest were parcels of private land.[40] Not long after the creation of the Tongass reserve came the great consolidation. In 1908, TR combined the Alexander Archipelago reserve (which he had saved in 1902) with the new Tongass reserve to form one huge entity of 6.7 million acres. The entirety was now designated the Tongass National Forest (named for the Tongass tribe of the Tlingit people).[41] The Tongass—carved out of the public domain—stretched over 500 miles from north to south in Alaska and included more than 11,000 miles of rugged coastline (a figure equal to nearly 50 percent of the entire coastline of the Lower Forty-Eight). To look at it another way, the new national forest blanketed an area exceeding 61,000 square miles—about 80 percent of southeastern Alaska.[42] Today it is one of the world's largest surviving temperate rain forests, with a biomass comparable to tropical rain forests in Venezuela, Brazil, or Costa Rica.

While the Tongass was certainly a large landmass, it wasn't all forestland; there were many glacial waterways as well as limestone rock and varied vegetation. What mattered was keeping all the elements of the Great North Forest ecosystem intact. Roosevelt knew that soon, with the statehood movement advancing, entrepreneurs would start imagining clearcuts, smokestacks, and roads. He wanted only a few sawmills, locally run, to be licensed for timber. No huge outside companies would be allowed. "The history of the Alaska territory up to that time had been a history of exploitation," Kathie Durbin wrote in *Tongass: Pulp Politics and the Fight for the Alaska Rain Forest*. "The exploiters had come in waves, seeking sea otter pelts, gold, salmon, and crabs from the Aleutians, and ivory from the Arctic coast. These raids had depleted Alaska's bounty, producing brief booms and generating great wealth for a few."[43] Roosevelt believed that the U.S. Forest Service could protect the Tongass in perpetuity. An array of chartered fishing vessels—purse seiners, halibut boats, gill netters, and trawlers—were allowed, but there were highly regulated catch limits.[44]

With scores of high, tumbling waterfalls; five salmon species in the streams; and brown and black bears around every bend, the Tongass was a wilderness paradise, considered by many to be among America's most beautiful national forests. There were eagles and ravens everywhere. Nearly one-third of the Earth's old-growth temperate rain forest grew in the Tongass coastal ecosystem. The estuaries and coastal meadows in the Tongass have been called a "biological buffet" of marine life. Annually, anadromous wild salmon leave the ocean and return to their Tongass freshwater birth streams to perish. Every spring humpback whales (*Megaptera novaeangliae*) journey nearly 3,000 miles from Hawaii to the waters around the Tongass to breed and calve. "The Tongass is a place where people live with salmon in their streams and bears in their backyards," the naturalist Amy Gulick, granddaughter of Sierra Club's founder Edgar Wayburn, wrote. "It's a land of remarkable contrasts."[45] Like Muir, Roosevelt wanted these lands and waters to be a cherished part of the American heritage forever. To desecrate such islands of the Alexander Archipelago as Admiralty, Baranof, and Chichagof was akin to rape and unbefitting of a conservation-minded citizenry working to benefit future generations. The Forest Service would hold the line in the Tongass and the Chugach, refusing to lease land tracts to huge corporate extraction businesses.

Unfortunately, many Alaskans loathed Roosevelt's two national forests, the Tongass and the Chugach. The prevailing sentiment in Juneau-Sitka was: How *dare* Washington, D.C., tie up so much land to prevent strip-mining and clear-cutting? Before long, the Alaska boomers and sourdoughs complained, the maniacal public land conservationists would even try to save worthless muskeg bogs and thaw lakes. First- and second-generation Alaskans, particularly those in extraction and cannery businesses, turned livid at what they considered two blasphemous words: *national forest*. Mentioning Roosevelt's name at a flapjack shack in Seward or a general store in Valdez guaranteed a negative reaction. Wisely, Roosevelt kept Langille—a forester, scholar, and artist to boot—as his first federal forest officer in Alaska. Locals couldn't help liking Langille for his self-effacing, honest demeanor; he was always ready with a corny joke, though his hand was never far from his holster. Langille recommended to Roosevelt that Admiralty and Montague islands—offshore from Cordova in Prince William Sound—become protected wildlife reserves. "There is room for the frontier settler and fisherman on the shore land," Langille wrote, "but keep out the fire and wanton game destroyers."[46]

To put Roosevelt's Alaskan forest reserves into perspective, the Tongass National Forest was the size of West Virginia and the Chugach National Forest approximately equaled Massachusetts and Rhode Island combined in acreage. This was clearly not just a clump of trees (eventually 6 million acres of the Tongass would be designated a federally protected wilderness). It was a first-round knockout punch for the naturalist side of the feud over the "two Alaskas." Only California held more acres of federal reserves than Alaska.[47] Limited, regulated logging would be allowed in the Tongass and Chugach forests, as would licensed hunting and fishing. But the Roosevelt administration was opposed to wholesale clear-cutting. Alaskans could chop down trees in the Tongass and Chugach to build houses, but out-of-state corporations wouldn't be allowed to annihilate the virgin forests for quick profit. Southeastern Alaskan sawmills had to be community-based, family-owned operations. Each stage of logging on U.S. federal property in Alaska—marking, cutting, skidding, transporting, and milling—would be carefully regulated. The diverse wildlife—including deer, bears, eagles, and salmon—in these national forests would likewise be managed. Logging in Alaska, Roosevelt declared, had to be consistent with wildlife protection objectives. Simply put, Roosevelt vehemently objected to the harvesting of these national forests by timber barons or giant multinational corporations. As he had told the Arctic Brotherhood in 1903, smaller localized pulp mills should instead get the land leases. The rest of Alaska belonged to Uncle Sam.

There was in Roosevelt's Alaskan wilderness withdrawals an element of aristocratic and scientific elitism, which had arisen from the white-linen tablecloth on the luxurious *Elder*. While Roosevelt did consider the Chugach and the Tongass "people's forests," his aim was to create wildlife sanctuaries for his fellow Ivy League—educated outdoorsmen to enjoy. Roosevelt always believed that wildlife and forestry were among the noblest professions. He intuitively trusted science-minded people because they hadn't built their lives around earning fortunes. There truly were two types of Alaskans: those who saw the Chugach and Wrangell mountains as coal, gold, lands, and oil, and those who wanted to hike in magical places like Bagley Icefield and Kayak Island (where Vitus Bering had landed in 1741).

Alaska was abuzz, during Roosevelt's presidency, with the news that oil seeps had been found fifty miles southeast of Barrow near Cape Simpson (locals would cut out oil-soaked tundra to use as fuel). In the easternmost part of the Chugach National Forest, oil was discovered seeping out of the ground near Katalla. In 1896, a prospector, Thomas White, shouted *oil*, in a voice that was heard loud and clear in the nearby Prince William Sound community of Cordova. White filed the first petroleum claim in Alaska, at the time that McKinley was assassinated and Roosevelt became president. What concerned Roosevelt was that the Alaska Development Company started getting an oil flow in 1902. Well Number 1 had been successfully drilled at 366 feet. By declaring all the lands surrounding the Katalla well part of the Chugach National Forest, the U.S. government suddenly controlled the future of the area as an oil field. Alaskans saw Roosevelt's action as a federal land grab and decidedly *not* as nature preservation.

Not all the president's reasons for creating the Chugach and Tongass national forests were inspired by preservation. He also wanted to address the economic concerns of southeast Alaska. Plenty of exemptions and withdrawals were allowed for local settlers. Roosevelt also had continual concerns about Canadians raiding U.S. forestlands and smuggling timber across the international line. Rangers and fire wardens of the U.S. Forest Service would serve as border protectors. As president, Roosevelt had enforced a federal law prohibiting the export of timber boards harvested from the Chugach or Tongass to any non-American market. These trees—alive or chopped—belonged to the American people. Southeastern Alaskan businessmen, in particular, protested against what they perceived as a federal stranglehold on potential timber crops. Why should citizens of Sitka or Ketchikan be denied global free trade? Roosevelt—and his small cadre of Pinchotian forestry experts operating in Alaska—argued that the Tongass and Chugach national forests actually helped frontier enclaves become sustainable communities instead of boomtowns. The Chugach and the Tongass were an inheritance to be passed on to future generations like jewels, gifts for time immemorial.

IV

Besides the Tongass and Chugach reserves, Roosevelt created gigantic bird refuges of inestimable merit in the Alaskan territory. Far from being an empty icebox, Alaska, Roosevelt knew, was green for half the year, owing to the long days of the Alaskan summer, and swarming with great flocks of birds in every direction. The various loons—common (*Gavia immer*), yellow-billed (*G. adamsii*), red-throated (*G. stellata*), Pacific (*G. pacifica*), arctic (*G. arctica*)—were so commonplace that their haunting call was heard even in the Brooks Range, and it came to symbolize the North Slope. The significant date for Alaskan wildlife protection was February 27, 1909, during Roosevelt's last two weeks as president, while all the political pundits were focused on the appointees of the incoming Taft administration. In 1909, TR proclaimed six federal bird reservations in Alaska by means of executive orders: Tuxedni, Saint Lazaria, Yukon

Delta, Bering, Pribilof, and Bogoslof. They were far bigger in scope than all of his administration's previous bird reserves in places such as Florida, Oregon, and Louisiana. And the Alaskan reserves were unsurveyed. To Roosevelt, hunting game birds for sport or science wasn't a sin, but as a die-hard member of the Audubon Society he believed that killing an American dipper (*Cinclus mexicanus*) or black-capped chickadee (*Poecile atricapilla*) just for the hell of it *was* a crime against God.

First among Roosevelt's Alaskan bird sanctuaries was the Yukon Delta Reservation (today known as Yukon Delta National Wildlife Refuge)—eventually more than 16 million acres of flat delta stitched with rivers (mainly the Yukon and Kuskokwim) and dotted with hundreds of lakes, creeks, sloughs, and ponds. Roaming this huge lake-spattered tundra along the Bering Sea—the second largest in the United States, after the Mississippi—were caribou, lynx (*Lynx canadensis*), bears, and wolves (*Canis lupus*). Roosevelt knew, from E. W. Nelson of the Biological Survey, that in terms of density and biological diversity, the Yukon Delta terrain, like the Sacramento Valley, was an essential shorebird nesting area in the United States. In the 1870s Nelson had brought back to Washington, D.C., nests and eggs from the salmon-rich Yukon Delta, to study them more carefully than he could on the marshy tundra.[48] The reserve Roosevelt created was the size of South Carolina.[49]

Birds from six major flyways—from the Atlantic Ocean to the eastern coast of Asia—would nest on the Yukon Delta or stop to rest and feed on their way to farther-off nesting grounds. Almost the entire world populations of bristle-thighed curlews (*Numenius tahitiensis*) and black turnstones (*Arenaria melanocephala*) breed there. Sheets of white birds often blanketed the landscape, making it look like a cotton field in Dixieland. Clouds of geese and ducks regularly swept across the sky, headed for the vast marshland during the great rush of spring. (Two of the sea ducks that regularly visited the Yukon Delta—the spectacled eider (*Somateria fischeri*) and Steller's eider—are now listed as threatened and are protected under the Federal Endangered Species Act.[50]

On February 27, 1909, Roosevelt created the Tuxedni Federal Bird Reservation (consisting of Chisik Island and Duck Island) with Executive Order No. 1039. The order provided for the protection of the nesting habitat for the largest aggregation of seabirds in Cook Inlet. Pelagic birds that congregated on the gravel beaches in these reservations included black-legged kittiwakes, common murres, horned puffins, glaucous-winged gulls, double-crested cormorants, the common eider, tufted puffins, and black oystercatchers. A fair number of passerine birds and raptors also thrived on the islands. This executive order actually exceeded bird-lovers' hopes.[51] Roosevelt, like a trickster raven, was brazenly confronting the "malefactors of great wealth," claiming that a smew (*Mergellus albellus*)—the smallest Arctic sawbill—had more intrinsic value than a pulp mill.[52]

Roosevelt had first learned about the birdlife on Chisik Island from Dall, who was tasked with mapping Cook Inlet for the Coast and Geodetic Survey in 1895. Dall had published detailed field notes about Tuxedni Island for the *Bulletin of the American Geographical Society* (these notes were complemented with numerous maps). Roosevelt was enthralled by the scientific exactitude of Dall's prose pertaining to volcanoes, talus slopes, Mesozoic fossils, and glacial plains. But it was Dall's vivid description of birdlife that got Roosevelt's juices flowing. "Near the beaches the rocks are worn into cave arches and pillars," Dall wrote, "about which circle innumerable multitudes of sea birds."[53] But still, there was a dearth of reliable ornithological information, owing to a lack of local records, about the birds of Alaska until the late 1950s.

Perhaps the most enduringly fascinating of the Alaskan places that Roosevelt saved as a federal bird reservation was Saint Lazaria, a sixty-five-acre islet in the middle of the Alexander Archipelago. It supported an astonishing 500,000 seabirds. Why did Saint Lazaria attract these birds when other nearby islets didn't? Roosevelt wanted an answer from the ornithological community. He got one. First, Saint Lazaria was a good incubator because of an absence of ground predators: there were no foxes, raccoons, or wolves. Second, the soft soil was ideal for seabird burrows. A third factor was that the birds on Saint Lazaria had an endless supply of fish in the surrounding waters. Saving Saint Lazaria had been recommended to Roosevelt by Edward A. McIlhenny—founder of the company that makes Tabasco, a Louisiana hot sauce—who had donated Arctic birds (representing sixty-nine species) to be studied at the Philadelphia Academy of Natural Sciences.[54]

V

What has often been overlooked when environmental scholars list Roosevelt's conservationist accomplishments in Alaska is that they were often accompanied by a preservationist ethic. This point didn't escape the notice of the ecologist, environmentalist, and writer Aldo Leopold, a former employee of the U.S. Forest Service, whose *Game Management* (1933) is still a classic book on methods of maintaining wildlife. Leopold dutifully noted that "conservation" was a "lowly word" until Roosevelt made wildlife and forest protection his "cause." Suddenly, the game hogs, market hunters, and salmon depleters were on the run. Fire lookouts were created on top of mountains and men were trained as smoke chasers. To be selected by Roosevelt's U.S. Department of Agriculture for forest duty—as William A. Langille was in Alaska—was a high honor. "Wild life, forests, ranges, and waterpower were conceived by him to be renewable organic resources, which might last forever if they were harvested scientifically, and not faster than they reproduced."[55] According to Leopold, Roosevelt's doctrine of conservation had three primary tenets regarding game and forests, and these tenets were essential to preserving Alaska's wilderness:

1. It recognized all these "outdoor" resources as one integral whole.
2. It recognized their "conservation through wise use" as a public responsibility, and their private ownership as a public trust.
3. It recognized science as a tool for discharging that responsibility.[56]

In his classic work *A Sand County Almanac*, published posthumously in 1949, Leopold added a fourth tenet to the Roosevelt doctrine. To enjoy a wilderness, the nature lover didn't need to invade it (as a corollary, to love a species like the snowy owl didn't mean killing it for a taxidermy mount). To Leopold the blank spots on the map, places without roads or towns, needed to be left alone, untouched. There was already too much industrial-agricultural stress on the land. For example, one didn't have to travel to Arctic Alaska to hunt a Dall sheep to prove one's mettle as a sportsman or as a scientist.

"Is my share of Alaska worthless to me because I shall never go there?" Leopold asked. "Do I need a road to show me the arctic prairies, the goose pastures of the Yukon, the Kodiak bear, the sheep meadow behind Mt. McKinley?" Leopold answered his own question: *no*. As the twentieth century progressed, the new land ethic wasn't about building modern roads or hunting game in "lovely country." It was about "building receptivity" for ecosystems in human thinking, so that treasured landscapes like the Chugach and Tongass, the Yukon Delta, and Saint Lazaria could survive industrialization.[57]

o   o   o   o   o   o

THE PINCHOT-BALLINGER FEUD

I

On September 6, 1909, while on safari in the northern foothills of Mount Kenya, Theodore Roosevelt—less than six months after leaving the White House—received three cables announcing that Commander Robert E. Peary had reached the geographic north pole. Peary's telegraphed message was sent from Indian Harbor, Labrador, and read: "Stars and Stripes Nailed to the Pole." Roosevelt, on a ten-month safari to British East Africa, accompanied by his twenty-nine-year-old son, Kermit, and a retinue of eminent naturalists, was delighted for Peary. America had beaten the Russian explorers to the world's rooftop. Voyages of discovery toward the geographic north pole had always held a special fascination for Roosevelt. The ex-president considered collecting scientific information about the Arctic region a sign of national greatness (the modern-day equivalent would be NASA going to the moon). Roosevelt had followed Peary's adventures with keen interest during his eight months in British East Africa. Prophetically, the great Norwegian Arctic explorer Fridtjof Nansen had told Roosevelt at a dinner in Washington, D.C., "Peary is your best man; in fact I think he is on the whole the best of the men now trying to reach the Pole, and there is a good chance that he will be the one to succeed."[1]

Roosevelt had been riveted by Peary's heroic stories of dogsledding and building igloos between 1886 and 1891 in the Arctic light, and he considered Peary's book *Northward over the Great Ice* (1898) a valuable contribution to exploration literature. In 1905–1906 Peary, then a U.S. Navy lieutenant, had patrolled the coast of upper Greenland in the ship *Roosevelt*—another point in his favor. In 1908 President Roosevelt had boarded Peary's ship to bid him godspeed on his historic voyage to the north pole.[2] Now, with his Arctic Club expedition and his successful historic dash to the north pole, Peary had outdone himself. All his life, Peary had sought glory—the one heroic deed that would reverberate forever in history. "Too much credit cannot be given him," Roosevelt wrote to his friend William Robert Foran. "He has performed one of the great feats of the age, and all his countrymen should join in doing him honor."[3]

Barry Lopez, one of America's finest writers, meditated in his *Arctic Dreams* (which won a National Book Award) on Peary's obsession with the north pole. Lopez remarked on the hardships Peary endured for the sake of Arctic exploration: leaving his wife behind, determining to prevail over blizzards, camping on permanently frozen ground to which he clung like a dwarf plant, not seeing another human being for months at a time, facing many desperate moments, always running out of provisions, being forced to kill sled dogs for food, making astute solar observations in order to stay alive. Literature is replete with such quest stories. But Peary's undaunted attempt to find the geographic north pole transcends the imagination. Weathering temperatures of minus fifty degrees Fahrenheit (not including windchill) and combating hypothermia for days at a time, the walrus-mustached Peary had planted an American flag near the north pole, even though he had, by then, lost eight toes to frostbite.[4] Peary's entourage of twenty-three men, 133 dogs, and nineteen sleds had encountered a rich continuum of wild environments never before marked. The Arctic, an almost desert landscape, was a different realm where snow geese (*Chen caerulescens*) ice-walked and polar bears ran as fast as shooting stars. On their trek, wrapped in Arctic furs to stay warm, Peary's team shot more than 600 musk oxen to keep their camp stocked with high-protein meat. Cold weather burned away calories very quickly. By surviving the sea ice, the pressure ridges, and frostbite, Peary was, in Roosevelt's estimate, a sustainable American hero in the tradition of Zebulon Pike.

When the explorer Frederick Cook claimed to have beaten Commander Peary by reaching the north pole first, nearly a year earlier, Roosevelt was dismissive. Writing an introduction to Peary's memoir, *The North*

Pole, Roosevelt lauded Peary for his "iron will" and "unflinching courage" in overcoming physical weariness in pursuit of Arctic knowledge.[5]
He thought it idiocy that Peary was being criticized by jealous explorers
for not having a solar expert confirm his latitude. (Although Roosevelt
enjoyed reading Cook's memoir *To the Top of a Continent*—particularly
the parts about climbing the Alaskan peaks McKinley, Foraker, Russell,
and Dall—it was Peary's descriptions of the north pole that he considered
priceless.)*

But although Roosevelt was ecstatic that the American flag had been
planted at the north pole, President Taft reacted with a yawn. To Taft, the
geographical blankness of the Arctic made the region uninteresting;
he was a city man. What could America possibly do with a frozen landscape where towns like Barrow lived in darkness between Thanksgiving
and late January? Why would anybody want to explore north of the timberline where animals were lucky to survive? Roosevelt saw the Arctic
from the scientific perspective of a Smithsonian Institute expedition:
young caribou migrating 2,700 miles annually and snow geese building
up fat reserves on the tundra before flying down to Mexico. But the very
notion of twenty-four-hour light at the summer solstice, at a latitude of
66° 33' North, was disturbing to Taft; and the fact that the Yenisey and
Lena rivers flowed in the Arctic with more freshwater than the Mississippi or Nile bored him.[6] When Peary offered to put the Arctic at Taft's
disposal, the president was unable to imagine that the north pole held any
hidden biological secrets. "Thanks for your interesting and generous offer," Taft wrote to the fifty-three-year-old Peary. "I do not know exactly
what to do with it."[7]

As a student of geography, Roosevelt was perplexed that the United
States wasn't more proud of its northern lands. In Russia the Arctic was
a part of nationalistic pride. Norway considered the Arctic the core of its
economic future. Canadians viewed their Arctic real estate as a symbol

---

* The fierce debate over who first reached the north pole persists in academic
circles. Cook claimed to have reached it on April 21, 1908—a year before Peary.
Critics of Cook claim he had once faked climbing to the top of Mount McKinley
and wasn't to be trusted.

of their greatness. But in America the north pole was viewed as a frozen wasteland not worth the time it took to consider. Years later, Peary, now an admiral in the U.S. Navy, recalled how bravely Roosevelt had stood by him when envious critics claimed that he had exaggerated his exploits at the north pole. In an article published in *Natural History*, Peary wrote that old-fashioned loyalty was Roosevelt's finest quality; that was a fact. Those who actually knew Roosevelt always thought of him as a strong ally. Upon returning to American soil from the north pole, Peary sent Roosevelt the finest polar bear skin he had collected on his daring expedition. It became the drawing room rug at Sagamore Hill, Roosevelt's estate in Oyster Bay, New York. The ex-president was also given an ivory walrus tusk. Holding court over pots of black coffee, Roosevelt would talk with Peary about what a precious "heirloom" Arctic Alaska would be to future generations. "The friendship of Theodore Roosevelt was indeed a most precious possession," Peary wrote. "Whenever and wherever extended, it had the effect of a superlative to greater deeds."[8]

With boyish enthusiasm Roosevelt read everything published about the musk ox herds found in the Arctic. The last known one perished in 1909. It sickened him that the musk ox had disappeared from Alaska by the late 1800s, overhunted and weakened by blue winter cold. These hardy, stocky, oxlike bovids had roamed the permafrost valleys from Alaska to Siberia. Roosevelt had high hopes that a new subspecies of musk ox would be discovered in the north pole or Greenland. With zoological dispatches about exploration at the north pole sent to him in Africa by the Smithsonian Institution, Roosevelt started studying everything published about these weird-looking, 800-pound, curly-horned beasts that plodded Arctic Alaska's tundra. And he planned to have the musk ox (and wood bison) reintroduced in Alaska. Standing about four feet tall at the shoulder, the musk ox had a noble lineage not unlike that of the American bison. Throughout France and Germany, in fact, ivory and stone carvings 250,000 years old depicted the musk ox accurately. Musk oxen, nervous and suspicious by disposition, were coveted for their straggling *quviut* (an underfur), which was combed out and used for caps and coats.

Roosevelt's Yukon Delta Federal Bird Reservation indeed proved to be the ideal place to breed stock to help restore musk oxen to their former ranges in Alaska. By the late 1930s a small herd from Greenland was indeed shipped to Alaska and released on the sweeping tundra of Nunivak Island Refuge.[10]

As an ex-president, Roosevelt regularly wrote zoological articles for the *Outlook* (where he was on the masthead as "contributing editor"), *Scribner's*, and the *American Journal* (although these treatises seldom received much public notice). The musk ox became one of Roosevelt's favorite species to analyze from an evolutionary perspective. With meticulous precision he learned every biological fact he could about the species the Inupiat called *omingmak* ("the animal with skin like a beard"). Calling the musk ox the "last survivor of the ice age," Roosevelt marveled at how the beast used its long skirtlike guard hair—much bushier than that of bison—to stay warm.[11]

"These musk-oxen," Roosevelt wrote in the magazine *Outlook*, "which once lived in what is now Ohio and Kansas, just as they once lived in England and France, have followed the retreating glacial ice belt toward the Pole; and there, in the immense desolation of the North, they still dwell side by side with men, the Eskimos, whose culture is at the same stage of development as that of those inconceivably remote ancestors of ours who hunted the musk-ox when it was still a beast of the chase in mid-England."[12]

Roosevelt would later personally interview Peary for detailed information about the chain of life around the north pole, everything from the behavior of orcas to the patterns of wind currents. He was riveted to learn how wolves were "hangers-on" around musk ox herds, preferring ox meat to caribou. Yet the migrating musk ox had evolved to survive bitter subzero weather. To Roosevelt, the musk ox brought a certain majesty to the Arctic ecosystem. He fixed his attention on these shaggy ambassadors from ancient times, ice age relics who had once shared vast stretches of Arctic tundra with woolly mammoths and short-faced bears. "The musk-oxen [are] helpless in the presence of human hunters, much more helpless than Caribou, and can exist only in the appalling solitudes where even arctic man cannot live," Roosevelt wrote in *A Book-Lover's Holidays in the Open*, "but against wolves, its only other foes, its habits of gregarious and truculent self-defense enable it to hold its own as the Caribou cannot."[13]

II

W hen Roosevelt left the White House in March 1909, he was proud of his conservation accomplishments in Alaska. Having left Gifford Pinchot ensconced as chief of the U.S. Forest Service, ex-president Roosevelt felt confident that his Alaskan legacy of natural forests and wildlife refuges would be properly protected. All that his handpicked successor—William Howard Taft, a distinguished Yale man who had served him admirably as secretary of war—had to do was let the very able Gifford Pinchot micromanage the Forest Service and read the riveting field reports of William A. Langille. Taft, with Pinchot as a witness, had promised Roosevelt, as a quid pro quo for his support, always to put conservation first—especially in Alaska. "The way is long and cold and lone," Roosevelt's friend Hamlin Garland wrote about Alaska. "But I go . . . where pines forever moan their weight of snow." [14]

Little could Roosevelt have foreseen that Taft, not understanding the need to preserve the moaning pines, would fail him with regard to Alaskan land issues. Taft, it turned out, thought Rooseveltian conservation, while essentially a noble cause, had ventured too far in protecting forest and marine environments (Tongass National Forest), saving glaciers (Chugach National Forest), and protecting wildlife (Yukon Delta Federal Bird Reservation). The Tongass, Taft believed, should be leased off for harvesting timber and canning salmon. He also had problems with the legality of Roosevelt's executive orders on behalf of wildlife habitats. Taft, in his first year in office alone, compromised the integrity of huge expanses of wild Alaska. He gave in to syndicates involved in forest clearcutting; fur trading; salmon canning; whaling; and gold, ore, and copper mining. New sawmills in Hope, Alaska, had a cut capacity of 20,000 feet per day. In Homer, Alaska, a Philadelphia coal company had built major new facilities, complete with a shipping dock. The Chugach Mountains formed an arc 50 million years old around Prince William Sound and the Copper River delta. Many conservationists feared that mining companies were going to ruin the scenery in a decade if Taft didn't keep the brakes on.

From Roosevelt and Pinchot's perspective Taft was intent on letting the U.S. government reap lucrative revenues by exploiting the precious metals of the far north, recklessly and unregulated; such metals were usually found in the remote white wilderness, reachable only by dogsled. "Despite his promise to Roosevelt that night in the White House," Timothy Egan has explained in *The Big Burn*, "Taft believed the conservation movement had gone too far, too fast, and that too much land had been put in the public's hands." [15]

In late 1909, Roosevelt was in the Lado Enclave in the Belgian Congo, hunting white rhinoceroses, when he received a shock. A special runner sprinted into Roosevelt's holiday camp with the news that Gifford Pinchot had been fired as chief forester by President Taft ten days earlier. It was as if Roosevelt aged on the spot. He shuddered and paced about, desperate for accurate information about Pinchot's apparent contretemps with President Taft. [16] It was hard to fathom the implications. Why had America's most eminent forestry professional been dismissed? Pinchot was like a son to Roosevelt. Together they shared a historic legacy of saving more than 230 million acres of wild America by creating national parks, national monuments, national forests, and federal wildlife reserves. They had injected the concept of conservation into public discourse. A self-critical Roosevelt now wondered how he had let Taft worm his way into his good graces. Seething with contempt and feeling that he had been double-crossed, Roosevelt was dismayed that Taft didn't have the courtesy to continue his predecessor's federal forestry and wildlife protection policies in Alaska. Taft, he soon declared, was a "great pink porpoise of a man." [17] All of Roosevelt's deepest suspicions regarding Taft came to the fore. "I cannot believe it," Roosevelt immediately wrote to Pinchot on January 17, 1910. "I don't know any man in public life who has rendered quite the service you have rendered; and it seems to me absolutely impossible that there can be any truth in this statement. But of course it makes me very uneasy." [18]

Roosevelt had no way of communicating directly with Pinchot from the Belgian Congo. He was baffled by Taft's action. Was the president trying to change Roosevelt's entire approach to national forests? Was Taft seeking corporate kickbacks? Were big businessmen suddenly outmaneuvering conservationists? Or had Pinchot become an intolerable nuisance to the Taft administration? Roosevelt had a genius for understanding bureaucracy, although he loathed it, but he could not deconstruct the fact that under the Taft administration, Alaska had more than twenty separate bureaus and offices in the departments of the Interior, Agriculture, Commerce, the Navy, and War.[19] A frustrated Roosevelt sent a message to Pinchot through the American embassy in Paris, instructing Pinchot to give him a detailed report of the firing.[20]

An anxious Pinchot decided to do more than just send an "Ivy League confidential" to Roosevelt via the American embassy in Paris. Determined to discuss his firing face-to-face with Roosevelt, Pinchot left Grey Towers, his home in Milford, Pennsylvania, and bought a ticket on an ocean liner to Denmark. At the ex-president's invitation Pinchot planned to meet with his old boss that April somewhere in Europe and deliver chapter and verse on the controversy.[21] Loyally, Roosevelt wrote to Pinchot that history would vindicate him for being the "aggressive, hard-hitting leader" of "all the forces struggling for conservation."[22] Wandering around Africa with Kermit, to whom the outdoors life was an opiate, Roosevelt plotted revenge on Taft. Kermit—fluent in Spanish, French, Greek, and Romany (Gypsy); able to read Sanskrit; and with encyclopedic knowledge of animal ecology—bonded with his father in Africa as never before. They used playful nicknames for each other, encouraged by their African guides. Roosevelt was *Bwana Makuba* ("Great Master"); Kermit was *Bwana Merodadi* ("Dandy Master").[23] Peary, to both Roosevelts, was the King of the North Pole: a hero. Pinchot was . . . politics . . . politics . . . pioneering modern forestry . . . and more politics.

Pinchot's grievances against the Taft administration were many. For starters, the new president had flat-out rejected a World Conservation Congress that Roosevelt had proposed in February 1909. Roosevelt had gotten the queen of the Netherlands to join him in actually creating a United Nations for Conservation; Taft scoffed at this notion. Adding insult to injury, Taft replaced the preservation-friendly James R. Garfield (son of the assassinated twentieth president) as secretary of the interior with a Republican land dealer, Richard Ballinger of Seattle, who favored the rapid exploitation of western resources.[24] Garfield had been an excellent secretary of the interior, and had regularly gone on long hikes in Rock Creek Park and swims in the Potomac River with Roosevelt.[25] Ballinger, by contrast, had been vehemently opposed to the Roosevelt administration's creation of both the Tongass and the Chugach national forests.

An investigator for the Department of the Interior, Louis R. Glavis, had documentary evidence, which he handed to Pinchot, that Ballinger was expediting the sale of federal coalfields in Alaska's Wrangell–Saint Elias Mountains to sell to the financial titans J. P. Morgan and Solomon R. Guggenheim, sometimes called the Alaska syndicate or, more often, derided as "Morganheim." According to Pinchot, Ballinger was offering sweetheart deals to railroads, mining outfits, cattle concerns, and logging conglomerates on public lands. Ballinger insisted that the U.S. Land Office had only one job: let private concerns divvy up the public domain in orderly fashion. "Morganheim" dominated the district's economics in the early twentieth century. Starting with the Kennecott copper mine deposits, "Morganheim" wanted to form an industrial empire in Alaska. Corporation heads such as George Hazlett, Stephen Birch, and David Jarvis constantly flouted U.S. government regulations, maintaining an adversarial attitude. Luckily for America, they were thwarted by conservationist leaders of the progressive era such as Roosevelt and Pinchot, along with some muckraking journalists.[26]

Arrogant, and opposed to the very concept of forest management, Ballinger—who had strong affiliations with western land barons—even opposed fire control because it involved state, federal, and private cooperation. "A stocky, square-headed little man who believed in turning all public resources as freely and rapidly as possible over to private ownership," Pinchot complained in his diary about the new secretary of the interior.[27] With Ballinger spearheading the effort, more than 1.5 million acres that the Roosevelt administration had set aside for future federally controlled waterpower sites had been forfeited.[28] With Wall Street putting pressure on his administration to open up Alaska's storehouse of natural resources, Taft capitulated, weakening federal authority over public lands in the Chugach. "Taft's betrayal was a constant topic of conversation," Pinchot later recalled, "between TR and his intimate advisers."[29]

Ballinger, a lawyer who was a former mayor of Seattle, thought that between 1901 and 1909 President Roosevelt, Pinchot, and Garfield had withdrawn too much public land for national forests in the Pacific Northwest and Alaska. Were huge federal forest reserves such as the Tongass really necessary? Why did Roosevelt want to save the Chugach, which held one-third of all the territory's glacier-covered land, including the Bering Glacier (one of the largest glaciers in North America)?[30] Ballinger, who distrusted easterners and science, wanted the pendulum to swing back to nonregulated capitalism. He respected Puget Sound business interests. Roosevelt and Pinchot's crusade had, he said, "gone too far."[31] Not only did Ballinger consider the acreage of Mount Olympus National Monument (in Washington state)—which President Roosevelt had created with an executive order in March 1909—excessive, but he thought more Alaskan coalfields should be opened up to the private sector. The Roosevelt administration had favored leasing U.S.-owned coalfields in Alaska whereas Ballinger wanted them sold outright to the private sector.

When Garfield left Washington, D.C., Pinchot tried to stop the newly appointed Ballinger from undoing the achievement of the Roosevelt administration regarding Alaskan public lands. By the fall of 1909, Pinchot had come to the conclusion that Ballinger was two-faced, the worst swindler he'd come across in decades, a toady for "Morganheim." Pinchot met personally with President Taft in the White House, pleading with him to control Ballinger before Alaska's boreal forests became permafrost wastelands of rotted tree stumps. Pinchot was worried that in the early summer months, the towering Chugach Mountains, rising dramatically from the sea, were a tinderbox. (Ballinger and his associates believed fires didn't happen around glaciers.) A wildfire caused by an industrial accident or by a bolt of lightning could suddenly ignite the forestlands. Strong prevailing winds would spread the flames uncontrollably in all directions.[32] Pinchot strongly feared that the major development projects of the Morgan-Guggenheim syndicate—the Kennecott Copper Company and the Copper River and Northern Railway—were monopolistic in intent.

Even though Taft thought Pinchot too zealous a promoter of federal forestry, Pinchot was a Yale man of stature, with impeccable New England credentials. Strong-willed, devoted to natural resource management, and convinced that God was in forests, Pinchot would often sleep outdoors with a wood block pillow to increase his hardiness. When he went camping, he would order his valet to wake him in the mornings as bracingly as possible—by dousing him with water from an icy stream.[33] "I do regard Gifford as a good deal of a radical and a good deal of a crank," Taft wrote to his brother, "but I am glad to have him in the government." [34] But Taft also frequently mocked Pinchot for being a sycophant of Roosevelt, engaging in "sort of a rough rider fetish worship." To Taft, GP, as his troops in the Forest Service lovingly called him, was a troublemaker. "G.P. is out there again defying the lightning and storm and championing the cause, or the oppressed and downtrodden," Taft wrote his brother, "and harassing the wealthy and the greedy and the dishonest." [35]

If President Taft had an Achilles' heel, it was his stout refusal to grant interviews. This was not smart for a president trying to get traction. Ballinger called it White House "nopublicity." [36] Unsure how to cope with the Taft administration's indifference toward federal land protection and its virtual boycott of the press, Pinchot founded the National Conservation Association as a watchdog group in 1909. The outgoing president of Harvard University, Charles W. Eliot, who sympathized with Roosevelt and Pinchot's land ethic, signed on as its honorary president; Pinchot served as president until 1925. Garfield joined the executive committee; so did Henry Stimson, a leader of the Boone and Crockett Club (later to be secretary of state for Herbert Hoover and secretary of war for Franklin Roosevelt) and a rising star in the wildlife protection movement. The purpose of the new association was to enact laws to promote conservation. "This association is to be the center of a great propaganda for conservation," Pinchot wrote. "It is hoped that all organizations interested in special phases of the conservation movement will become affiliated with it." [37]

No sooner had the National Conservation Association been formed, in the summer of 1909, than Pinchot, to his great dismay, discovered that Ballinger was allowing fraud to operate in the General Land Office (GLO) in the West. Alaska—in particular the areas surrounding the Chugach—was being leased to huge timber and mining operations so that quick profits could be made. Further irritating Pinchot was the fact that President Taft had killed Roosevelt's grand notion of a Global Conservation Congress. Pinchot truly believed that Taft and Ballinger were hellbent on deprioritizing conservation, and forcing it underground, back to mere "garden club" status. The populist movement of the early twentieth century viewed big business—particularly the railroad industry—with deep suspicion. A combination of conservationists, muckrakers, and trustbusters insisted, as Pinchot did, that Alaska should remain public land saved for public use.[38]

On November 13, 1909, *Collier's* published a scathing "insider" article by Louis Glavis that linked Ballinger directly to the J. P. Morgan—Guggenheim syndicate. This New York–based syndicate had purchased the enormous Kennecott-Bonanza copper mine and monopolized the Alaskan steamship and rail transportation from Seattle; it also owned twelve of the forty canneries in the territory.[39] "Gradually," M. Nelson McGeary writes in *Gifford Pinchot: Forester-Politician*, "by highly intricate financial arrangements, this partnership extended its holdings in Alaska until by 1910 it controlled copper mines, a steamship company, and a salmon-packing concern."[40] Pinchot echoed Glavis's article, charging that greedy, oppressive trusts had subservient lawmakers in the Taft administration doing their bidding—a barely concealed smear of Ballinger. In Pinchot's mind, Roosevelt's Alaskan policy was being compromised and ignored because of Taft's complicity with "Morganheim." Pinchot declared that under President Taft, the GLO reeked as badly as sulfur water. He wanted to clean the Augean stables; he wanted Roosevelt's conservationist directive—the simple rule of always making the land better than you found it—upheld by Taft. Having the U.S. Forest Service give huge corporations and banks free rein in the Alaskan lands they leased, without federal regulation, was a recipe for disaster: long-term deforestation.

Pinchot had become a whistle-blower. Using information provided by Glavis, he declared that Ballinger was a traitor to the federal government and to the conservationist movement. If the Morgan-Guggenheim syndicate wasn't stopped, the rivers in Chugach National Forest—the Copper, Russian, and Trail—would become contaminated from copper pit runoff. The margin of life for mountain goats and Dall sheep would become narrow. The finest salmon runs in Alaska—like those in the Bristol Bay Basin—would become stinking mudholes. Backing Pinchot was James Wickersham, Alaska's lone congressional delegate in Washington, D.C., who was a rip-roaring critic of the Morgan-Guggenheim syndicate (although he preferred that the Tongass and Chugach be redesignated as state forests). An Alaskan district judge, Wickersham loved wild country. In 1903 he tried to climb Mount McKinley but aborted the attempt at 8,000 feet. Wickersham, whose memoir *Old Yukon: Tales, Trails, and*

Trials is an Alaskan classic,[41] understood that all the syndicate wanted to do was mine copper ore for its smelter in Tacoma, Washington. "The delegate approved of federal conservation policies," the historian Peter A. Coates writes, "as a restraint on outside interest that creamed off Alaskan wealth."[42]

What truly concerned Roosevelt about the Morgan-Guggenheim syndicate was that it was planning to bring hydraulic machinery to Alaska to supersede small, individual placer operations. Rooseveltian conservationists did not want *any* monopoly to get a sweetheart lease for timber, copper, or ore in Alaskan national forests. Roosevelt and Pinchot's policy was for the General Land Office to lease coalfields in Alaska, whereas Ballinger and Taft wanted outright selling of the lands—a big difference.[43] From his experience with the construction of the Panama Canal, Roosevelt knew how brutally destructive such large-scale construction projects could be to pristine landscapes. (When Roosevelt visited the Canal Zone in 1907 as president, he kept natural history records of the tropical foliage.) Pacific Northwest banks, however, were itching to clear-cut the Chugach and Tongass national forests. Because the U.S. Forest Service didn't have a team of full-time rangers, bootleggers set up distilleries on federal property, convinced that they could operate undetected in such expansive outdoors settings. Whether as president or as a private citizen, Roosevelt wasn't about to let a few New York or Seattle bankers desecrate America's great rain forests. The fact that the Morgan-Guggenheim syndicate wanted to keep its Tacoma smelter burning around the clock didn't mean Alaska should be recklessly exploited.

Roosevelt always took the high ground with regard to Alaskan affairs. But as proof that he hadn't been antidevelopment, consider this: in 1906 he had appointed Wilds Preston Richardson, a U.S. Army officer from Hunt County, Texas, who had attended West Point, to become the first chairman of the Alaska Roads Commission. During the Klondike gold rush, Richardson, in command of the Eighth Infantry (eighty men), kept law and order around Skagway. He later oversaw the construction of army posts at Rampart, Eagle, and Nome. Then in 1906 Roosevelt ordered the army to build what today is known as the Richardson Highway, the two-lane road connecting Valdez (the seaport on Prince William Sound) to Fairbanks (gateway to the Brooks Range). Clearly, Roosevelt wasn't antidevelopment. He just wanted the U.S. government, not private concerns, to control the infrastructure of Alaska.[44] Nevertheless, in 1909 the *Cordova Daily Alaskan* ran a headline that evidently spoke for the majority of district citizens: "Pinchot Is Daffy over Conservation."[45]

III

The *feud between Pinchot and Ballinger* had become a brouhaha in America throughout 1909. On August 12, the *New York Times* ran the headline "Pinchot in Danger of Losing His Place." The charge against Pinchot was insubordination. No president likes leaks from or even dissension in the ranks, let alone whistle-blowers. From Taft's perspective, Pinchot was a socialist-minded menace: arrogant, fanatical about trees, one-dimensional, and unable to understand that American politics involved the art of give-and-take. The biographer Nathan Miller wrote in *Theodore Roosevelt: A Life* that Pinchot was desperate to expose Taft's deficiencies and in doing so "courted martyrdom."[46] In truth, Pinchot was a lot more politically pragmatic than that. Under Taft, scant progress was made in pushing conservation forward. A sworn enemy of reckless corporate despoilers, Pinchot was willing to shatter the Republican Party for the sake of the western forest reserves. "Without fully intending to do so," Pinchot wrote, "I think I have probably forced Taft to take his stand openly for or against the Roosevelt policies in act as well as in word."[47]

From September to December 1909, Taft was looking for a convenient way—or any way—to fire Pinchot while TR was still collecting specimens in Africa. Pinchot stumped the West, calling citizens to fight for public land: it was their birthright as Americans. Although Pinchot had staunch allies in the establishment—for example, the agribusinessman Henry C. Wallace—leaders of the big corporations wanted the chief forester gone. Pinchot fumed that the "great oppressive trusts" existed in the United States because of "subversive law-makers."[48] In 1908 there were 770 serious placer mines in Alaska, employing 4,400 men.[49] Taft and his supporters wanted to see that number doubled, for the sake of the economy of the Pacific Northwest. They sought jobs, jobs, jobs, and quick money over long-term land management.

Taft, you might say, was complicit in the radical anticonservation movement in Alaska. He simply wouldn't enforce scrupulous federal protection of the Chugach and Tongass. With the advantage of hindsight, we can see that Taft initially ignored the issue but then became pro-development and pro–big business concerning Alaskan affairs. Clearly Taft was untouched by Thoreau's belief (shared by the Tlingit Indians) that wilderness represented the preservation of the world; money was what drove Taft forward. "We have fallen back down the hill you led us up," Pinchot wrote to Roosevelt (who was in Khartoum, in the Sudan). "There is a general belief that the special interests are once more substantially in full control of both Congress and the Administration."[50]

Feeling the pressure from being constantly in the public eye during the feud with Ballinger, Pinchot headed to Santa Catalina Island, California, in the blue Pacific, to clear his head. Armed with a fishing pole, transported by a skiff, Pinchot perhaps thought about the role of dissenters from Thomas Paine to William Lloyd Garrison to John Muir. As he was riding Pacific swells, drifting eight miles from shore, hoping to catch a few good yellowtail or albacore tuna for supper, Pinchot's rod nearly split in half from a titanic tug. Suddenly a blue marlin as large as William Howard Taft leaped from the water. "High out of the water sprang this splendid creature," Pinchot wrote, "his big eye staring as he rose, till the impression of beauty and lithe power was enough to make a man's heart sing with him. It was a moment to be remembered for a lifetime."[51]

Pinchot soon returned to Washington, D.C., ready for combat. By December, the situation concerning Ballinger had become even messier for Taft. People were always quoting Pinchot to him, and muckrakers were stepping up their attack on Ballinger as a crook. *Collier's* magazine ran an inflammatory story, "Are the Guggenheims in Charge of the Department of Interior?"[52] Meanwhile, Alaskan forests were front-page news in New York City. Should the virgin stands be federal reserves? Or should they be clear-cut for the pulp industry to help the Pacific Northwest economy? By January 1910 Taft, exhausted by the feud, knew he had to "wrestle with Pinchot," as he put it. Taft composed a stern letter charging Pinchot with disrespecting the office of the president. "By your own conduct you have destroyed your usefulness as a helpful subordinate of the government," Taft wrote, "and it therefore now becomes my duty to direct the secretary of agriculture to remove you from your office as the forester." What a bad political move by Taft! Why fire the honest protégé of TR and keep the money-grubber from Seattle? At the very least, Taft should have also asked Ballinger, who resigned in 1911 anyway, to leave simultaneously with Pinchot. Truth be told, from the White House perspective, *both* Pinchot and Ballinger were behaving badly in the public sphere.[53]

Pinchot took his dismissal like a gentleman, or so it seemed at first. But as his biographers have remarked, he was ultimately simply unable to accept it. Seeking revenge, he hatched a hidden agenda against Taft. With the help of Garfield, Pinchot composed a sixteen-page memorandum for Roosevelt to read in Africa. Written as a prosecutorial brief, the memo detailed how Roosevelt's conservation policies were being ravaged by the Taft administration, which had connections to unsavory syndicates. Taft, while not personally corrupt, was the enabler in chief. Pinchot told Roosevelt that "complete abandonment" of his Alaska policies was taking place. Furthermore, Pinchot claimed, Taft had surrounded himself with "reactionaries" from big business who were bragging about a "vicious political atmosphere" aimed at undoing Roosevelt's conservationist accomplishments. According to Pinchot, Taft had "yielded to political expediency of the lowest type."[54]

What was initiated here was the eventual breakup of the Republican Party in the early twentieth century. Ballinger represented its free enterprise, big business wing; Pinchot represented the progressive-reform wing, with the "conservation doctrine" at its core. Taft was now the leader of the corporate conservatives; Roosevelt, essentially unreachable in the African bush, was the champion of the left-leaning progressives.

While field collecting for the Smithsonian Institution along the White Nile, Roosevelt received from a runner Pinchot's sixteen-point indictment of Taft in January 1910. He pored over the bracing document with gloomy curiosity. Was this memo accurate? Or was it a distortion by Pinchot? Cleverly, Taft had appointed Henry Solon Graves as Pinchot's replacement to lead the U.S. Forest Service. Graves had been a fine director of the Yale School of Forestry from 1900 to 1910 and was a solid forester incapable of making a fuss. A graduate of Yale (in 1892), he was booksmart, and he had studied forests abroad at the University of Munich. As replacements went, Taft had chosen wisely. This did not mollify Roosevelt, however, because Graves had worked as a forester for the Cleveland-Cliffs Iron Corporation in Michigan. Graves was too much of a forest industry insider to be trusted fully as a regulator of the federal forest reserve.[55]

"The appointment in your place of a man of high character, and a noted forestry expert, in no way, not in the very least degree, lightens the blow," Roosevelt wrote to Pinchot on March 1, 1910, attempting delicately not to trumpet a rival. "For besides being the chief of the forest bureau you were the leader among all the men in public—and the aggressive hard-hitting leader—of all the forces which were struggling for conservation, which were fighting for the general interest as against special privilege."[56]

Deeply disturbed by the feud, Roosevelt asked Henry Cabot Lodge to advise him in an unbiased way. Sentimentally, Roosevelt wanted very much to see Pinchot personally. But at the same time, internal warfare in his party wearied him. His affection for and his sense of obligation toward Pinchot won out. "I'm very sorry for Pinchot," Roosevelt wrote to Lodge. "He was one of our most valuable public servants. He loved to spend his whole strength, with lavish indifference to any effect on himself in battling for a high ideal and not to keep him thus employed rendered it possible that his great energy would expend itself in fighting the men who seemed to him not to be going far enough forward."[57] Lodge, by contrast, wasn't so affectionate toward Pinchot: he warily advised Roosevelt *not* to meet with the former forestry chief in Europe. Pinchot was guilty of vicious gossip and shameless politicking and had been wrong to smear Ballinger in the press by using allegations of Alaskan fraud.[58]

Glad that Lodge had given him sound counsel, Roosevelt nevertheless wanted to hear from his forty-four-year-old protégé directly upon reaching Europe. By the time Pinchot reached Denmark in April 1910, Roosevelt was agitated about Alaskan forestlands being opened up by the Taft administration to big coal interests. But he was also cautious about publicly entering the Pinchot-Ballinger feud. Worried that his conservation legacy was deteriorating under Taft's lackadaisical custodianship, Roosevelt nevertheless stayed mum. Perhaps Roosevelt also heeded his sixteen-year-old daughter, Alice, who warned him in a letter that Pinchot was self-serving and an advocate of "practically rank socialism."[59]

By telegram, Roosevelt suggested to Pinchot that they meet in Italy in April. Together, without drama or distress, they would calmly consider how best to protect Alaska's natural resources. Word of this scheduled meeting leaked out to newspapers. "There is no question that this meeting created widespread anxiety among Republicans," Pinchot's biographer McGeary noted. "Administration stalwarts, as well as others, primarily interested in party unity, feared the political consequences of having current events presented to Roosevelt from Pinchot's point of view."[60]

When Roosevelt finally appeared in Khartoum for his first press conference after months off the beaten path in the African bush—disheveled from travel, his shirtfront wrinkled, but his face glowing with a deep tan—questions were hurled at him by anxious reporters. Why was Pinchot fired? Will you challenge Taft in 1912 for the Republican nomination? Is conservation still the most pressing issue facing America? Fearful of giving clumsy answers, and not wanting to take on Pinchot's encumbrances as his own, Roosevelt refused to discuss the controversial matter. He would talk only about his experiences in the African bush. He purposefully made many references to giant elands, but none to American politics.

When Pinchot finally met with Roosevelt in Italy on April 11, they had a lot to talk about. The dapper Pinchot looked as elegant as ever, wearing exactly the right clothes for a daytime walk through vineyards and olive groves. A breeze made it a perfect day for an outing. With regard to American politics, however, Roosevelt was between a rock and a hard place. The nasty fact was that Taft had been Roosevelt's choice as his successor. If Roosevelt attacked Taft outright, that would cause a deep rift in the Republican Party. So Roosevelt stalled. At a press conference in Porto Maurizio, he refused to talk about U.S. conservation policy until August 27, when he would deliver a major speech in Colorado.[61] And Roosevelt's stalling worked. The pack of European reporters backed off, just walking away en masse to look for a headline elsewhere. Roosevelt's tactics effectively defused Pinchot as well.[62] "One of the best and most satisfactory talks with T.R. I ever had," Pinchot wrote of their meeting in Italy. "Lasted nearly all day, and till about 10:30 at night." In *Breaking New Ground*, published after World War II, Pinchot admitted that he had put his mentor, Roosevelt, "in a very embarrassing position, but that could not be helped."[63]

That spring of 1910 Pinchot published his first book, aptly titled *The Fight for Conservation*. Capitalizing on his feud with Ballinger, Pinchot excoriated "stupidly false" businessmen who were either too greedy or ignorant to comprehend that there was no such thing as inexhaustible resources.[64] Echoing George Perkins Marsh, whose work of 1864, *Man and Nature*, remained a bible to conservationists, Pinchot warned against plagues such as wildfires, dust bowls, famines, and floods that would devastate America unless huge forest reserves were maintained. Playing Cassandra, Pinchot warned that only a fool would think America's supplies of coal, timber, petroleum, soil, forage plants, and freshwater were infinite.[65] These resources belonged to the Americans and were not to be recklessly squandered for the benefit of a single generation. Pinchot ripped into financial titans who demanded special privileges or sought a monopoly with regard to natural resources. The only person mentioned by name in the slender volume, however, was Theodore Roosevelt, who, Pinchot declared, had promoted the "rapid, virile evolution of the campaign for conservation of the nation's resources."[66]

Much of *The Fight for Conservation* reads like recycled speeches or mannerly bureaucratic white papers. After a few retrospective pages about the prescience of the founding fathers in holding American citizens responsible for "our great future," the reader could be forgiven for dozing off. There is too much dull political speechifying and schoolmarmish scolding for the volume to be truly important. Nevertheless, Pinchot built his conservationist arguments on solid underpinnings from Yale's forestry school. Ironically, as *The Fight for Conservation* celebrated its centennial in 2010, the ecosystem in the Gulf of Mexico was being destroyed by an oil spill of terrible proportions from a well owned by BP. Pinchot had always feared that corporations—if poorly regulated by the Department of the Interior—would abuse their privileges. In those hours of darkness during 2010, Pinchot seemed like an environmental sage from a distant era. Furthermore, he had envisioned the environmental movement of the 1960s when writing *The Fight for Conservationism*. Whenever U.S. natural resources were despoiled, he wrote, nature lovers would, like a "hive of bees, full of agitation," swarm down on the corporate abusers "ready to sting."[67]

IV

Never before had a former American president, not even Ulysses Grant, been sought after by the press corps as ardently as Theodore Roosevelt was in April to June 1910. Everybody in Europe wanted to read about his exploits in the wild African bush. Even the sophisticates of London, Rome, Copenhagen, and Berlin were awed by his gloriously strange articles for *Scribner's*, accompanied by bizarre grayish photographs of an ex-president attired almost like a scarecrow. A beaming Roosevelt, proud of his trophies, had made Africa accessible to all. He was irresistible. As Roosevelt traveled around Europe sightseeing, he was peppered with questions about the Panama Canal, Africa, the Great White Fleet, the Grand Canyon, and Arctic exploration. And his conservation policy had been embraced by many European intellectuals. For example, Paul Sarasin, a celebrated Swiss zoologist, promoted the Rooseveltian notion of global conservation in speeches, articles, and books.

Besides being the toast of the Sorbonne in Paris, Roosevelt was greeted in Vienna and Budapest by throngs of admirers who saw him as a representative American in Ben Franklin's tradition. Admired for his African exploits, Roosevelt was also called the "king of America"! Nobody believed he was a "former" anything. Crowds waved big sticks and rawhide thongs in his honor, stamping their feet enthusiastically. A successful new cigarette in Scandinavia was marketed as "Teddies."[68] On May 5, in Oslo, Norway, Roosevelt finally delivered his Nobel laureate's speech—he had won the Peace Prize for ending the Russo-Japanese War of 1905. He made headline news when he proposed a "League of Peace" to stop war forever; he also suggested that international disputes be mediated at The Hague.[69]

Following his travels in Europe, Roosevelt went to Great Britain to serve as the U.S. special ambassador for the funeral of King Edward VII, who had died unexpectedly. For a few days, Roosevelt stepped into the world of the British royals, regaling them with tales of wildebeests, monkeys, and swarms of bugs. He and his son made a visit to Rowland Ward Ltd., in Piccadilly, to get some trophies mounted. Elephant feet were turned into ashtrays for the Roosevelt family to hand out as souvenirs. So much for science! So much for wildlife protection! And, as prearranged, Lord Curzon, the chancellor of the University of Oxford, had Roosevelt deliver the prestigious Romanes Lectures there. George John Romanes had been an intimate of Charles Darwin and the custodian of Darwin's notebooks on animal behavior. He enraptured Roosevelt with vivid stories of the great naturalist. An impressed Roosevelt wrote to Henry Cabot Lodge that Romanes was "right in my line." [70]

Although Roosevelt's Romanes Lectures were well received, he felt that the students at Oxford were too subdued. Was there anything worse than a know-it-all twenty-year-old devoid of humor? But he fell in love with Cambridge University, which was less formal and more garden-like. He went there to receive an honorary doctorate and had a grand time, as if he were at the Hasty Pudding Club. "On my arrival [the students] had formed in two long ranks leaving a pathway for me to walk between them, and at the final turn in the pathway they had a Teddy Bear seated on the pavement, with outstretched paw to greet me," Roosevelt wrote to a friend, "and when I was given my degree in a chapel the students had rigged a kind of pulley arrangement by which they tried to let down a very large Teddy Bear upon me as I took the degree—I was told that when Kitchener was given his degree they let down a Mahdi upon him and a monkey on Darwin under similar circumstances." [71]

While Roosevelt was in London, the British foreign secretary, Sir Edward Grey (later Viscount Grey of Fallodon), a fanatical bird-watcher, escorted him around the soggy woodlands of England to hear songbirds. Grey was flabbergasted at Roosevelt's precise knowledge of avian species. If bird-watching were a trade, Roosevelt assuredly would have been a guild master. In his memoirs, Grey noted that their hike in the Itchen River valley, southwest of London, was an especially remarkable experience. Roosevelt had lectured Grey, saying that the English countryside should remain undefiled by industrialization. Bird reserves were necessary. "Though I know something of British birds, I should have been lost and confused among American birds, of which unhappily I know little or nothing," Grey wrote. "Colonel Roosevelt not only knew more about American birds than I did about British birds, but he knew more about British birds also." [72]

What especially captivated Roosevelt about ornithology in 1910 was the growing bird-banding movement. John James Audubon had long been hailed in ornithological circles as the "father of bird-banding" (in 1804 he had attached silver wire rings to the toes of phoebe hatchlings).[73] For more than eighty years, he owned the franchise. Beginning in 1899, however, Denmark started banding birds by attaching aluminum strips on the legs of white storks and starlings. It was the sort of breakthrough, Roosevelt believed, for which Nobel Prizes should be given. Denmark owned all of Greenland and was properly studying its abundant wildlife. Roosevelt hoped that at last the migratory patterns of Arctic birdlife could be scientifically understood. As U.S. president, Roosevelt had encouraged the Smithsonian Institution to follow Denmark's lead and band more than 100 black-crowned night herons (*Nycticorax nycticorax*) with the inscription "Return to the Smithsonian Institution." From 1909 to 1923, the ornithologist Paul Bartsch personally banded at least 20,000 Canada geese. Other bird enthusiasts did the same for Arctic Alaskan birdlife such as the tundra swan (*Cygnus columbianus*) and long-tailed duck (*Clangula hyemalis*).

While in Africa, Roosevelt, in fact, had praised thirty members of the American Ornithological Union (AOU) for creating the American Bird Banding Association of New York City on December 8, 1909.[74] Drumming up scientific support for the experimental monitoring technique, ornithological journals such as *Auk* and *Bird Lore* freely distributed bands to birders from Alaska to Florida. Fascinated by the migratory patterns of Arctic birds, about which virtually nothing was known, Roosevelt recognized banding as a way to monitor not only bird populations but also their migrations at the same time. The Department of Agriculture (USDA) also began issuing bulletins to farmers about how the stomach of an average mountain plover contained forty-five locusts, and the message was clear: birds would help the farmers combat pests, making the land more productive. When it came to nongame birds, Roosevelt was for leaving the bullet boxes at home. Roosevelt was also proud that the National Association of Audubon Societies had been formed by thirty-six state groups. The Audubon Movement, for which Roosevelt had signed up in 1887, was going to be around for the ages.[75]

What worried ex-president Roosevelt most in Alaska was that "fair chase" hunters were a dying breed; market syndicates were wiping out all the wildlife. Taft seemed to have the U.S. government back in the seal slaughtering business in the Bering Sea. With improved rifles and ammunition becoming easy to obtain, Roosevelt feared the age of the slob hunter was arriving. Word had it that George Bird Grinnell, longtime editor of *Forest and Stream*, the most popular conservationist periodical in America, was about to lose his job. After thirty-five years as editor Grinnell was, indeed, retired. When Grinnell was at the helm of *Forest and Stream*, Alaskan wildlife had remained front and center. No longer. Roosevelt tried to rectify the situation by telling the "governing board" that this important periodical must continue to crusade for wildlife conservation. The new owners of *Forest and Stream* placated Roosevelt somewhat, allowing Grinnell's and Merriam's names on the masthead. But in reality the new editor was catering to a new market, and its readers were uninterested in the life expectancy of Dall sheep around Mount McKinley or the need to save Medicine Lake in North Dakota as a wildlife refuge.[76] By 1915 the once irreplaceable *Forest and Stream* went from being a weekly to being a monthly. And by 1930 the magazine was defunct (although its subscription list was sold to today's magazine *Field and Stream*).[77]

What Roosevelt was experiencing in 1910 and later was a backlash against the U.S. Forest Service and U.S. Biological Survey. Leading Democrats in Congress went so far as to demand that *all* national forest lands should be turned over to the states. The "two frothing horsemen" of anti-conservationism—representatives William Humphrey of Washington and A. W. Lafferty of Oregon—deemed Roosevelt and Pinchot zealots. These westerners pushed for congressional bills to cut off all funding for the U.S. Forest Service. But Roosevelt and Pinchot had two Republican allies in the Senate who belong in any conservation hall of fame: Miles Poindexter of Washington (soon to be a Bull Moose) and later Charles L. McNary of Oregon.[78] Most important, from 1910 to 1920, the Supreme Court continually validated virtually every facet of the Roosevelt administration's conservation policies from federal bird reservations to national monuments.[79]

Also riding to the rescue of Rooseveltian conservation was the fine novelist and memoirist Hamlin Garland. When Garland was thirty-one years old, in 1891, he received wide acclaim for *Main-Traveled Roads*, a collection of short stories inspired by his days in Wisconsin as a farm boy.

Turning to the American West for material, Garland headed to the Yukon in 1899 to cover the Klondike gold rush. He ended up writing *The Trail of the Gold Seekers* in 1899, but something more important happened to him in northern Canada and Alaska: he became an ardent conservationist. The northern wilderness had him transfixed. Building on the success of Owen Wister's best seller *The Virginian*, in 1910, Garland published *Cavanaugh: Forest Ranger*, a sophisticated western dime novel in which the protagonist is a brave U.S. Forest Service officer who rides the Great Plains on his horse along a "solitary trail" protecting federal lands. Garland's realistic prose about the prairie was controlled and elegant, never purple. He described little fly-bitten cow towns like Bear Valley (paradise) and Sulphur City (grimly provincial) with marvelous exactitude.

Unfortunately, Garland's dialogue seems artificial; and what really prevents *Cavanaugh* from being first-rate literature is the hokey, cartoonish way he described American women, as damsels in need of male protectors. Nonetheless, from a modern-day perspective on environmental history, Cavanaugh—a Rooseveltian conservationist foot soldier—is a welcome new type of western hero, determined to save treasured landscapes for future generations. Like a hardwood birch, Cavanaugh was straight-grained, with few knots. Take, for example, the following dramatic exchange between Cavanaugh, anxious to explain his federal oath to protect the western reserves from clear-cutting, and his love interest, the beautiful Lee Virginia:

> She perceived in the ranger the man of the new order, and with this in her mind she said: "You don't belong here? You're not a Western man?"
>
> "Not in the sense of having been born here," he replied. "I am, in fact, a native of England, though I've lived nearly twenty years of my life in the States."
>
> She glanced at his badge. "How did you come to be a ranger—what does it mean? It's all new to me."
>
> "It is new to the West," he answered, smilingly, glad of a chance to turn her thought from her own personal griefs. "It has all come about since you went East. Uncle Sam has at last become provident, and is now 'conserving his resources.' I am one of his representatives with stewardship over some ninety thousand acres of territory—mostly forest."

She looked at him with eyes of changing light. "You don't talk like an Englishman, and yet you are not like the men out here."

"I shouldn't care to be like some of them," he answered. "My being here is quite logical. I went into the cattle business like many another, and I went broke. I served under Colonel Roosevelt in the Cuban War, and after my term was out, naturally drifted back. I love the wilderness and have some natural taste for forestry, and I can ride and pack a horse as well as most cowboys, hence my uniform. I'm not the best forest ranger in the service, I'll admit, but I fancy I'm a fair average."

"And that is your badge—the pine-tree?"

"Yes, and I am proud of it. Some of the fellows are not, but so far as I am concerned I am glad to be known as a defender of the forest. A tree means much to me. I never mark one for felling without a sense of responsibility to the future." [80]

Adorned with an introduction from Gifford Pinchot, *Cavanaugh* succeeded in showcasing Forest Service rangers as defenders of nature, honest protectors ready to arrest poachers and market hunters who disobeyed federal laws in the West. Garland, by writing the novel, had rendered America a genuine creative public service. He was trying to inform people that the forest rangers—who "represented the future"—were noble guardians of the gorgeous western landscape, protecting it from plunder by black-hatted rogues. [81] Pinchot applauded *Cavanaugh* for explaining the historic transformation of the old West (buffalo hunting) to the new West (forest conservation). "The establishment of the new order in some places was not child's play," Pinchot wrote to Garland in March 1910. "But there is a strain of fairness among the western people which you can always count on in such a fight as the Forest Service has made and won." [82]

What infuriated Roosevelt about Taft's people—including the chief forester, Graves—was the notion of running all of Alaska's natural resources under a so-called Alaskan Commission (big business before conservation). Led by Alaska's congressional delegate James Wickersham, western corporations denounced Roosevelt and Pinchot's "broad arrow" policies (i.e., locking up natural resources that rightfully belonged to miners, fishermen, and farmers). War against the Tongass and Chugach was under way.

By 1910, Roosevelt, Garland, and Pinchot were concerned that the United States had very little wilderness left. With western expansion petering out, at least from an explorer's perspective, Alaska became the last frontier. They were determined to see that its natural resources would never be exhausted. Jack London described Alaska in his adventure novels as a "vast silent" wilderness region that demanded heroism. Susan Kollin, in *Nature's State: Imagining Alaska as the Last Frontier*, describes London's and Roosevelt's obsession with the far north as a means of "reinvigorating U.S. men" to "test their strength and endurance against the challenges of wilderness." Kollin, a modern environmentalist influenced by the 1960s ecology movement, approved of Rooseveltian conservation, which allowed wild places like the Tongass and Chugach to be saved. But Kollin insisted that the motivation for men like Roosevelt and London was to save a "new frontier where Anglo Saxon males could reenact conquest and reclaim their manliness." [83]

Although London has been considered the "Kipling of the Arctic" for his stories of American expansionist fortune-seeking in Alaska and the Yukon, the novelist James Oliver Curwood brought an environmentalist perspective to his brutal tales of the far north. Curwood, a die-hard Rooseveltian conservationist, was the lead lobbyist promoting legislation to create Superior National Forest in Michigan. During the early twentieth century, more than 4 million hardcover copies of his books were sold in the United States. His novels, such as *The Alaskan* and *Son of the Forests*, were translated into twelve languages. Curwood wrote about reindeer farms, Eskimo culture, and grizzly bears.[84] In *The Alaskan* his heroine, Mary Standish, bursts out with patriotic rhetoric about the wonders of Mount McKinley, the Pribilofs, and the Tongass: "I am an American. I love America! I think I love it more than anything else in the world—more than my religion even. . . . I love to think that I first came ashore in the *Mayflower*. That is why my name is Standish. And I just want to remind you that Alaska *is* America." [85]

Curwood did a fine job of injecting conservation into his novels. Worried that Alaskan waters were overfished, Curwood lamented that the "destruction of the salmon shows what will happen to us if the bars are let down all at once to the financial bandit." More of a weekend recreationist than a wilderness cultist like Muir, Curwood championed proper game and land management ethics in Alaska. The Alaskan Native Brotherhood was founded in 1912 and promoted the same conservationist principles. "Roosevelt's far-sightedness had kept the body-snatchers at bay, and because he had foreseen what money-power and greed would do, Alaska was not entirely stripped today, but lay ready to serve with all her mighty resources the mother who had neglected her for a generation," Curwood wrote. "But it was going to be a struggle, this opening up a great land. It must be done resourcefully and with intelligence."[86]

Although Rooseveltian conservationists of the progressive era such as Garland and Curwood were often perceived as a united front, always promoting forestry and wildlife science, there was at least one fault line among them. This was as menacing as the San Andreas Fault, and it had to do with whether to dam the Tuolumne River in Yosemite National Park. Following the San Francisco earthquake of 1906, when widespread fire had caused catastrophic damage downtown, the city applied to the Department of the Interior for a water rights lease to Hetch Hetchy, a breathtakingly beautiful valley in Yosemite National Park. A vicious fight ensued between those who wanted the O'Shaughnessy Dam built and those who wanted the glacial valley protected. Ironically, Pinchot, who was working against big mining interests in Alaska, sided with San Franciscans in the controversy over Hetch Hetchy because the dam, in his mind, represented "the greatest good for the greatest number of people."[87] Pinchot objected to the views of his naturalist friends—Muir, in particular—in California, who were always ready to cut a rancher's fence or torch a sheepherder's wagon to protect the Sierras from development. "When I became Forester and denied the right to exclude sheep and cows from the Sierras, Mr. Muir thought I made a great mistake, because I allowed the use by an acquired right of a large number of people to interfere with what would have been the utmost beauty of the forest," Pinchot testified before the U.S. Congress Committee on Public Lands. "In this case I think he has unduly given way to beauty as against use."[88]

From 1910 to 1913 the fight over Hetch Hetchy, which many scholars believe was the birth of the modern environmental movement, reached epic proportions.[89] The newspapers built the drama into a feud between two types of conservationists: Gifford Pinchot, a utilitarian conservationist, who was in favor of damming Hetch Hetchy; and John Muir, the wilderness prophet of the Sierra Club, who resembled Saint Francis of Assisi and was vehemently opposed to the dam. The fracas made for good theater. Uncharacteristically, Roosevelt—who on December 8, 1908, had declared Yosemite a "great national playground" where "all wild things should be protected and the scenery kept totally unmarred"—sat on the sidelines of the controversy.[90] Defending his Alaskan forest reserves was an easy decision for Roosevelt. They were largely remote and isolated from large population settlements. But San Franciscans, still recovering from the earthquake of 1906 and needing a water reservoir, were a different matter to him. It was Muir, working on *Travels in Alaska*, who held the moral high ground; his righteous fury on behalf of Yosemite echoed all the way from the snowcapped Sierra Nevada peaks to Alaska's Brooks Range up to the coastal plain of the Beaufort Sea.

"Dam Hetch Hetchy!" a furious Muir declared. "As well dam for water-tanks the people's cathedrals and churches, for no holier temple has ever been consecrated by the heart of man."[91]

Chapter Four

BULL MOOSE CRUSADE

I

W hen *Roosevelt returned from Africa in* June 1910, one of the first
public events he spoke at was a luncheon of the Camp Fire Club
of America (CFCA) held on the roof of the Waldorf-Astoria Hotel on Park Avenue in New York City. The *New York Times* treated the stag
luncheon as a glitzy convention of the conservation movement, minus
Gifford Pinchot. In getting from Oyster Bay to Manhattan, the always
competitive Roosevelt decided to race the Long Island train in his Ford
car; he beat it by five minutes. Blind in one eye and with blurred vision in
the other, Roosevelt was reckless at the steering wheel and heavy-footed
on the accelerator—in short, a menace on the road. After talking with
reporters in his *Outlook* office, Roosevelt headed to the Waldorf-Astoria
roof, which had been decorated like a rustic camp. There were a lot of
pinecones and picnic tables. Large heraldic shields honored the heroes
of the CFCA and the conservation movement: Boone, Crockett, Carson,
Pike, Frémont, Audubon, Lewis and Clark. Roosevelt arrived with his son
Kermit and his publisher, Arthur H. Scribner. Everybody wanted to hear
Roosevelt's African tales. He delivered stories about lions, zebras, and
gazelles. And he took "nature fakers" like Jack London to task. According
to the *New York Times*, when he was done with his hourlong talk, the CFCA
members "fired their revolvers to punctuate their enthusiasm." [1]

The CFCA was the inspiration of the zoologist William Temple Hornaday. Disgruntled with the Boone and Crockett Club's ethos of trophy

hunting, refusing to count dead elk or moose antler points, Hornaday broke ranks with the hunters.* In 1897 he created the CFCA, with an emphasis on sportsmen committed to the preservation of wildlife habitats, the primitive arts of the outdoors life, and the wise use of natural resources. Based in Chappaqua, New York, the CFCA included Ernest Thompson Seton among its early founding members. One of its primary objectives was to keep the Adirondacks *forever wild*. The CFCA, in fact, had challenged New York state to immediately set aside more than 1 million new acres of forestlands. Entire Adirondack watersheds needed immediate protecting. The club also wanted New York railroads not to use coal and timber companies to stop the destructive practice of clear-cutting.[2]

Studying the map of the United States—particularly in the territories of Arizona, New Mexico, Oklahoma, and Alaska—the CFCA members wanted to create more huge federal reserves like the Yukon Delta (known as the Roosevelt Bird Reserve) in Alaska. Later that year, in December, Hornaday held a dinner for about 350 people honoring Colonel C. J. "Buffalo" Jones for helping save bison in Arizona.[3] Hornaday, who was largely responsible for the effort at the Bronx Zoo to bring about buffalo repopulation in Oklahoma and Montana, urged the CFCA to fight to protect wildlife habitats in the far west and Alaska. "Roosevelt's idea of science as a tool for conservation seems a truism to us now," Aldo Leopold wrote in *Game Management* (1933), his manifesto on wildlife protection, "but it was new in 1910."[4]

A guiding principle of the CFCA was *privacy*; no reporters have ever attended an annual meeting. The event of June 22, 1910, at the Waldorf with Roosevelt was no different; there are no transcripts of his remarks. Evidently, however, a beautiful American rose was held over Roosevelt's head, representing "campers' freedom" to speak their minds candidly and off the record. Hornaday, who had the great honor of introducing Roosevelt, called him the premier outdoorsman of the era. The club's gold medal was then handed to Roosevelt, and he received a standing ovation. On its reverse side of the medal was engraved: "For his work in the protection of wildlife and forests and for his contributions to zoology."[5]

_____

* Nevertheless, Hornaday stayed a member of the Boone and Crockett Club.

That July, Hornaday also teamed up with Pinchot to further the Adirondack Park "forever wild" program. Pinchot saw Taft's departments
of the Interior and Agriculture as a joke. Back from visiting Roosevelt in
Italy, he started investigating corporate abuses in the Adirondacks. He
spent time around Mount Marcy with Overton Price, editor of *Conservation*. The CFCA had achieved a victory in New York with a bill forbidding
the sale of wild game. Now, with Pinchot as point man, they were urging
a bill to forbid the sale of timber in the Adirondack Park. Two attorneys—
A. S. Houghton and Marshall McLean—were drafting a lawsuit. Hornaday
wanted the CFCA to sue "big timber" for wasteful clear-cutting of forests
that belonged to the people of New York.[6]

Just a few days after the CFCA dinner, Hornaday attacked the Taft administration harder than Pinchot had ever dared. Hornaday, as the *New
York Times* reported, accused the head of Taft's Fur Seal Board—Walter I.
Lembkey—of personally profiting from the killing of Alaskan seals and
otters. The CFCA—seemingly with Roosevelt's support—declared Lembkey "manifestly unfit" for his position. According to Hornaday, the Fur
Seal Board should be purged of such members. President Taft and his secretary of commerce and labor were complicit in the slaughter of Pribilof
Island seals, whose number had shrunk dramatically.[7] It sickened Hornaday to contemplate that his government was complicit in the harvesting of Pribilof seals—even pups—for their pelts of thickly packed hairs
(300,000 per square inch). As far as Hornaday was concerned, the Fur
Seal Board was nothing more than a band of pirates. If Taft wanted war
over protecting Alaskan seals, then Hornaday was glad to confront him.

Throughout the summer of 1910 Hornaday tore into Taft for running a
Fur Seal Board that, instead of "watch-dogging" the Pribilofs, was allowing cash-and-carry profiteers and businessmen cronies to profit while
the northern fur seals' numbers diminished. Out of all the pinnipeds—
that is, mammals with flippers—the northern fur seals intrigued Hornaday the most. For one thing, their migratory journey from the Bering Sea
to the central California coast was exceeded in length only by the migrations of harp seals of Newfoundland and some whales. From a biological
perspective, the northern fur seal had the most pronounced sexual di-
morphism of any mammal species. And these seals were tough defenders
of territory. "It is not safe to enter a rookery in breeding season, but bulls

normally will not pursue intruders beyond the edge of their own territory and much of their angry display is bluff—though not to other bulls,"
Briton Cooper Busch wrote in *The War Against the Seals*. "The northern fur
seal is fully capable of driving off an interloping Steller Sea Lion three
times its size."[8]

By the time Roosevelt arrived in Denver that August to deliver an important speech on conservation, speculation was rampant that he would
run for president in 1912. He was driven by malice against Taft and against
the "lawless man of great wealth" who was skinning public lands in New
York, Alaska, and elsewhere.[9] Every syllable Roosevelt uttered from the
podium was infused with vehemence and urgency. He called for inheritance taxes on large estates as a fair mechanism to fund more and better big government. Demanding obedience to the sportsman's code, he
pushed his conservation agenda forward, "dee-lighted" to demonize investment bankers as "debauchers" of the American landscape. Positioning himself as the arbiter of economic justice and a countervailing force
to Wall Street, Roosevelt thundered that the lamentable antinational tide
had to be reversed. Alaska's coastal waters, for example, needed to be protected by a powerful Bureau of Fisheries, or else the salmon runs would
end. Regulation of huge cannery operations, Roosevelt said, would occur
only in the form of vigorous federal regulation of Alaska's waterways. And
the Pribilofs—those rugged breeding grounds for seals, walrus, and
otters—needed to remain *fully* in the portfolio of the U.S. Biological
Survey, not in the Commerce and Labor Department. The five-year ban
against sealing, in Roosevelt's eyes, needed to become permanent.

Arriving in Osawatomie, Kansas, on August 31, 1910, Roosevelt art-
fully preached his "new nationalism," which included vigorous conservation. With a discernible intensity, Roosevelt expressed his conviction
that the U.S. government was a far better steward of the land than the self-
interested House of Morgan and similar types who populated Wall Street.
He was anxious to bring corporate power to heel. Conservation, he said,
was *the* great moral issue of the day. Roosevelt claimed that President Taft
had unnecessarily created the Bureau of Mines with U.S. Forest Service
funds. Why not more fully fund the Bureau of Fisheries, which he had
created in 1903? Government regulatory powers, Roosevelt insisted, had
to be increased dramatically to impede human degradation of wild Amer-

ica. Spontaneous enthusiasm and reverberating cheers greeted every line of conservationist populism that Roosevelt shouted out.

"I believe that the natural resources must be used for the benefit of all our people, and not monopolized for the benefit of the few, and here again is another case in which I am accused of taking a revolutionary attitude," Roosevelt said in Kansas. "People forget now that one hundred years ago there were public men of good character who advocated the nation selling its public lands in great quantities, so that the nation could get the most money out of it, and giving it to the men who could cultivate it for their own uses. We took the proper democratic ground that the land should be granted in small sections to the men who were actually to till it and live on it. Now, with the water-power, with the forests, with the mines, we are brought face to face with the fact that there are many people who will go with us in conserving the resources only if they are to be allowed to exploit them for their benefit. That is one of the fundamental reasons why the special interests should be driven out of politics." [10]

Besides preaching for conservation in the Midwest, Roosevelt was captivating audiences across the country with his riveting tales about British East Africa. The publication of his *African Game Trails* was a huge event throughout America in the fall of 1910, and the memoir became a best seller. The farther west Roosevelt traveled, the denser the crowds became. People lined up for miles just for a chance to touch the Colonel's sleeve, and they would let out a collective yell at the sight of his famous toothy smile. Knickknack booths, refreshment tents, and toy stands were set up at many appearances. Roosevelt delivered short, impromptu speeches at book signings, denouncing plutocrats and financiers but also sharing stirring adventure tales about chasing lions, sleeping in the jungle, and inventorying the Kenyan forest belt for conservation purposes. Working for the Smithsonian Institution, the Roosevelt party had collected 8,463 vertebrates, 550 large and 3,379 small mammals, and 2,784 birds.[11] Some wildlife biologists thought it was a slaughter. In city after city, Roosevelt met with conservationists, offering his support in local fights against rapacious land developers. He spoke of the need for a Global Conservation Congress—the multinational organization the Taft administration had nixed. "Conservation means development as much as it does protection," Roosevelt told a crowd of farmers. "I recognize the right and duty of this generation to develop and use the natural resources of our land; but I do not recognize the right to waste them." [12]

African Game Trails became a popular boys' book, selling more than 1 million copies. [13] Everywhere Roosevelt went that autumn, huge groups of adolescents paraded after him, hungering for stories of the wilderness and adventure. Never one to disappoint children, the ex-president regaled them with tales of Mount Kenyan fantail warblers, giraffes eating out of his hand, and the honeyguide birds that always led to trees of sweets. As if foreshadowing the New Deal, he urged young people to form a youth army to protect wilderness areas from vandals. "There are no words that can tell the hidden spirit of the wilderness, that can reveal its mystery, its melancholy, and its charm," Roosevelt wrote; "swamps where the slime oozes and bubbles and festers in steaming heat; lakes like seas; skies that burn above deserts . . . mighty rivers rushing out of the heart of the continent through the sadness of endless marshes; forests of gorgeous beauty, where death broods in the dark and silent depths." [14]

When Roosevelt stopped in Oak Park, Illinois, the ten-year-old Ernest Hemingway, awestruck, dressed in a khaki safari suit, stood with his grandfather in a receiving line to shake hands with his hero. Young Ernest had just received his first gun (a 20-gauge shotgun) from his grandfather, and he had been playing Teddy Roosevelt instead of cowboys and Indians. Hemingway also joined the Agassiz Naturalist Club, learned taxidermy, and pleaded to go on his own safari to collect specimens. The green hills of Africa were calling him. As a young adult Hemingway—aspiring to qualify for the CFCA—would retrace Roosevelt's safari to British East Africa and would befriend one of the men who had been the ex-president's guides in 1909. [15] "More than any other individual in history, Roosevelt opened the African frontier to the imagination of America's youths," Sean Hemingway, grandson of Ernest, wrote in a helpful introduction to *Hemingway on Hunting*. "The fresh scent of a new frontier and the thrill of the hunt, both with their overwhelming sense of valor and excitement, would captivate Hemingway for the rest of his life." [16]

During Roosevelt's absence in Africa, President Taft had tried to garner a little of the "teddy bear" magic for himself. At a dinner in Atlanta, Georgia, Taft had been served a southern dish, barbecued possum. Imitating Roosevelt, Taft swore it was a "dee-licious" meal. Cartoonists jumped on the anecdote, calling Taft "Billy Possums." A few cartoons ran in syndicated newspapers, and although these cartoons lacked pizzazz, Billy Possums cookouts became a brief fad in the Deep South. Also, enterprising entrepreneurs in New York quickly manufactured a new stuffed toy, Billy Possum. The sales were dismal, however. "A dealer—one of the biggest in the country—got a telegram on the night of the dinner," the *New York Evening Post* reported. "He immediately went to a manufacturer. They put their heads together and possum skins were obtained. But the genuine skin, stuffed, looked like a gigantic rat." [17]

The possum toy sank without a bubble. Nobody was going to get excited over a novelty associated with William Howard Taft. "Before long," the biographer Kathleen Dalton noted, "cartoonists parodied Taft as a lost boy searching for his Teddy Bear." [18] By contrast, everything associated with TR, from stuffed toys to bobble-head dolls, boomed after his African adventure. Abercrombie and Fitch advertised a khaki "Roosevelt Tent," completely waterproof. It was Taft's misfortune to follow such a charismatic force of nature as Roosevelt into the White House. Nobody could connect with the average American youth like the old Rough Rider. Colonel William Selig of Selig Polyscope made a nickelodeon movie of Roosevelt on a studio lot, renting tame lions to simulate a safari. The film, *Hunting Game in Africa*, featured a bad actor as Roosevelt, always in "bully" mode. It was a disappointment at the box office but it inspired the trademark roaring lion at the opening of MGM movies. [19]

In June 1910—owing in part to Roosevelt's outdoors philosophy and his African safari—the Boy Scouts of America was founded in New York City by Robert Baden-Powell; it would soon become the biggest youth organization in the United States. [20] The front porch of the CFCA headquarters in Chappaqua, New York, surrounded by beautiful wilderness, was the site where this founding had first been thought of. Young boys needed to learn how to survive in the wild, how to tell a poisonous plant from an edible one. According to Daniel Beard, a founder of the Boy Scouts, Roosevelt's promotion of faunal naturalism was the main impetus for creating an outdoors-oriented youth organization. Beard had been concerned that young boys had admired antiheroes like Blackbeard, Laffite, and Billy the Kid, so he tried to promote the likes of Theodore Roosevelt and Robert Peary. He believed that boys needed to develop honor, as well as outdoor skills such as knowing how to build campfires, tie knots, fly-fish, and use a jackknife, if they were to develop into first-class men. Only when boys understood that a bird's egg was the most perfect thing in the world would their character be strong enough to resist the lurid carnival of American decadence. Shortly after the Boy Scouts was created, Beard had a private audience with Roosevelt. There was a direct lineage from the Boone and Crockett Club to the CFCA to the Boy Scouts; Roosevelt linked all three. "The Colonel," Beard later boasted in *Outlook*, "gave me the authority to use his own name." [21]

By September 1910, Roosevelt was praising the Boy Scouts and the CFCA on his book tour. American boyhood, Roosevelt often said, should be oriented toward the outdoors and woodcraft, and away from the open-hearth furnaces of Cleveland, Pittsburgh, and Buffalo. Youngsters needed to be able to identify a common rock wren, appreciate the beauty of the tall-grass prairie, and smell fir boughs beside a campfire at night. Being in touch with nature and honoring all humans and wild creatures would help develop high moral character. Instead of becoming apathetic brats whining about money and profits, youngsters would develop into citizen conservationists of the highest order. [22] "I believe in the Boy Scouts movement with all my heart," Roosevelt said. "The excessive development of city life in modern industrial civilization which has seen its climax here in our own country, is accompanied by a very unhealthy atrophying of some of the essential virtues, which must be embodied in any man who is to be a good soldier, and which, especially, ought to be embodied in every man to be really a good citizen in time of peace." [23] Roosevelt regularly touted Alaska, the Rockies, and the Pacific Northwest as great places for a young man to climb mountains, camp, and hike—wilderness zones where the young man could test his mettle against nature. By 1914, in part owing to Roosevelt's plea, there were five Boy Scout troops in Alaska, with four scoutmasters and thirty scouts. [24]

As the Boy Scouts developed into a nationwide idea, Rooseveltian conservation became one of the organization's central tenets. The new generation of American boys needed to be both citizen-naturalists and citizen-scientists. The original Boy Scouts *Handbook* sold 7 million copies in three decades, a number second only to the Bible. [25] By 1914, the Boy Scouts had awarded its first William Temple Hornaday Gold Medal for "conservation excellence" and the Gifford Pinchot Award for "notable work in extinguishing forest fires." And Roosevelt became an honorary vice president of the Boy Scouts of America. Urging that all Boy Scouts follow the "golden rules," Roosevelt said the real qualities that made a boy a man were unselfishness, gentleness, strength, bravery, and protection of the wilderness. "One of the prime teachings among the Boy Scouts will be teaching against vandalism," Roosevelt wrote. "Let it be a point of honor to protect birds, trees, and flowers, and so make our country more beautiful." [26]

II

Throughout the summer of 1910, Roosevelt worked hard to get Pinchot to contain his anger at President Taft. After all, it was a midterm election year, and Roosevelt didn't want to be blamed for causing the Republicans to lose congressional seats and governorships. Slowing down a conservationist hothead like Pinchot, however, wasn't an easy matter. Recognizing that Taft's political power was ebbing, Roosevelt took a paternal approach toward Pinchot, never saying that Pinchot was wrong, always showing affection and concern, but always signaling, *Knock it off*. In a fatherly way, Roosevelt told Pinchot to "husband" his influence, to "speak with the utmost caution" and not to "say anything that can even be twisted into something in the nature of a factional attack." [27] Secretly, Roosevelt admired Pinchot's progressive-minded "Insurgents" movement and was pleased that Lincoln-Roosevelt clubs were being formed across the country as a Republican bulwark against Taft and "Morganism." Outwardly, however, he continued to feign uninterest in seeking the White House again. [28]

Nevertheless, Roosevelt *did* object that it was unacceptable for private concerns to despoil Alaska of its natural resources for the purposes of big mining, big timber, and big railroads. Pinchot cheered the Colonel on. Because Alaska was geographically huge, transportation was always going to be a contentious issue there. The territory had no reliable network of roads for moving cargo. In 1912 there were four practical ways to get around: walking, dogsled, horse, or steamboat (Alaska had more than 4,000 miles of navigable waterways, of which approximately 2,700 were in the Yukon watershed). The Yukon River, flowing bow-shaped for

2,300 miles, was the great artery for freight, effectively dividing Alaska east-west into two halves. Only three North American rivers were longer than the Yukon: the Mississippi, the Missouri, and the Mackenzie. Roosevelt was in favor of internal improvements in Alaska, such as roads, canals, and railroads, but only if the U.S. government was in charge of construction on leased public lands.

To Roosevelt, who had lobbied against the railroad industry's segregating Yellowstone National Park in the 1870s, too many Alaskan roads would mean too much Alaskan development. Places like the coastal panhandle of southeastern Alaska, an ecosystem of thousands of islands equalizing the size of Florida where huge schools of humpback whales, orcas, and sea lions swam along the forested shorelines of the Alexander Archipelago and the Tongass National Forest, should be treasured, not exploited. The Tongass had the world's highest density of grizzlies, black bears, and bald eagles. Their habitat should be left alone. The real value of Alaska, to Roosevelt, resided in managing its wilderness better than land skinners had managed that of the Lower Forty-Eight.

Everywhere Roosevelt looked there were scoundrels wanting to make quick dollars on dubious transportation or reclamation projects in Alaska. The Morgan-Guggenheim syndicate had finagled financing to construct a 1,550-foot steel-truss bridge on behalf of the Copper River and Northwestern Railway, to transport copper from the mines to the seaport wharf in Cordova. Dubbed the "million-dollar bridge" (it actually cost $1.4 million), the construction project smacked of a boondoggle from day one. To Rooseveltians, the bridge was an expensive ploy to eventually open up the Chugach National Forest to increased private-sector copper mining. The ribbon-cutting ceremony for the "million-dollar bridge" took place in 1910, with officials of the Taft administration smiling alongside Kennecott copper miners. Boomers in towns such as Seward and Cordova celebrated the bridge. Alaska was on the rise! But Rooseveltians were prescient about the foolishness of Alaska's first "bridge to nowhere." By 1930, the Copper River and Northwestern Railway had gone bankrupt. Few folks used the expensive train tracks.

In 1910, every Alaskan mining town wanted a road built for its district. Likewise, a priority list was established by a territorial commission to deliver mail more efficiently. The U.S. Signal Corps led the way by connecting Valdez (then the most northerly open port in North America) to Fairbanks (the practical head of navigation on the Tanana River). A 385-mile road linking Valdez to Fairbanks allowed Alaska to become an economy based on exporting natural resources. When Roosevelt left the White House, there were about 770 productive placer mines in Alaska, employing about 4,400 men. Only a few years later, owing to transportation innovations, these numbers had grown dramatically. Coal deposits could be found throughout 12,600 square miles of the territory.

At Copper Mountain, a 250-ton smelter was polluting the air, and long tramways had been built at Niblack, Skowl Arm, Karta Bay, and Hetta Inlet to transport the most valuable ores. On the Seward Peninsula auriferous lode mining was taking place along the Solomon River. The brownish-black coal on the peninsula was lignite, frozen solid. Like peat, it cracked and crumbled on exposure to sun. However, this coal, lowest-ranked in terms of energy, burned readily, leaving chalky ash billowing upward from factory smokestacks. Carbon dioxide emitted from plants using lignite coal was more toxic than that from comparable factories using black coal. Lignite was so combustible that railroad companies, fearing industrial accidents, didn't like to transport it for long distances.[29]

Clearly, Alaska wasn't a worthless icebox, even though its nicknames, according to the *Philadelphia Inquirer*, were "Walrussia," "Icebergia," and "Frigidia."[30] It was the next West Virginia: a source of coal, a storehouse of limitless rock fuel ready to be extracted for an economic bonanza. (And probably at the cost of human lives. In 1907 alone 3,242 West Virginian miners perished in mining accidents.) The Alaska-Yukon-Pacific Exposition in Seattle had promoted this notion about coal in the "Great Land" to more than 3.5 million visitors in 1909–1910. Conservationists circa 1910, by contrast, saw Alaska as John Muir had seen it—as "nature's own reservation"[31] where "nothing dollarable is safe."[32] Huge dams or copper and coal mines, these wilderness advocates believed, would kill rivers and destroy the breeding areas of migratory birds. "Conservationists and boosters were united in admiration for the frontier and in agreement on its importance as an ingredient in American culture and history," the historian Peter A. Coates wrote. "However, they differed, often diametrically, in the ways they expressed affection and how they formulated the best means to ensure the survival of their revered frontier."[33]

Writing to his twenty-two-year-old son, Ted, Theodore Roosevelt pined for the Alaska Range, longing to be thrust into a territorial
wilderness with ospreys and eagles overhead.[34] The distance from Point
Hope, Alaska (a spit of land jutting into the Chukchi Sea), to Washington,
D.C., was greater, in miles, than that from New York to Senegal. The Colonel loved this kind of remoteness.[35] He wanted to be *anywhere* outdoors
in Alaska where there wasn't a book to sign or a hand to shake. Roosevelt
had shipped his sixteen-year-old son, Archie, off to the Black Hills under the watchful eye of Seth Bulloch, a sheriff and forest ranger who was
by nature a scoutmaster and who knew how to toughen up boys. Heading
out to the University of California—Berkeley, Roosevelt wrote to its president, Benjamin Wheeler, about the difficulties of being overbooked as
both father and speaker. Proudly preaching the "new nationalism," Roosevelt made it abundantly clear that he wanted the Republican Party to
prosper in the midterm election come November. He would hold his nose
and vote for Taft. "I have a much larger following west of the Alleghenies
than east of them," Roosevelt wrote to his son Ted, "and have my own difficulties here in New York simply because New York is of course the center
of big business, of the big lawyers who guide the big business men, and
of the multitude of small business men and small lawyers who take their
care from the men at the top of their respective professions."[36]

That November the Democrats gained fifty-seven seats in the House
and ten in the Senate. The party of William Jennings Bryan now had outright control of the House (and working control of the Senate in combination with a smattering of progressive Republicans). The Democrats were
pulling down the shade on the Republican Party for the first time since
Grover Cleveland had worked his electoral magic in 1892. But Roosevelt
didn't feel paralyzed. The midterm defeats suffered by the Republicans
turned his attention more toward his conservationism. Briefly swearing
off politics, Roosevelt returned to wildlife biology, his lifetime passion,
swapping information with professional peers. The entomologist Willis Stanley Blatchley, for example, had sent Roosevelt a book on beetles.
Roosevelt knew that Darwin, just a few weeks before dying, had written
about a water beetle that attached itself to a clam in a pond in the English
Midlands. Feeling diffident about his own knowledge of beetles, Roosevelt was glad to study Blatchley's fine new research. "There was one beetle

found on Lake Victoria Nyanza that almost came in the category of big game," Roosevelt wrote to Blatchley that Christmas, using a kind of insider's shorthand. "It was considerably larger than a mouse. You of course know all about it, it is called the galia beetle." [37]

The Christmas season of 1910 also found Roosevelt defending the immense national forests and federal bird reserves in Alaska that had been created during his presidency and were now, in some quarters, targets of cynicism. To Roosevelt (prodded by Pinchot), protecting the Tongass and Chugach national forests became a high priority. The Democrats' victories in 1910 caused a wave of resource development advocacy aimed at undoing Roosevelt and Pinchot's forestland initiatives in Alaska, Washington state, and Oregon. Acting as a lobbyist, Roosevelt fired off sharp letters to new members of Congress, explaining why federal protection of timberlands in Alaska and the Pacific Northwest was imperative. On behalf of Pinchot's new, nonprofit National Conservation Association (the forerunner of today's Natural Resources Defense Council, NRDC), [38] Roosevelt urged legislators to stop desecrating mountaintops and slopes across the country. "At this very moment we are endeavoring to get the United States Government to take over from the Eastern states the Appalachian and White Mountain reserves, just because the states have not done as well as the Nation is doing or can do," Roosevelt wrote to one recently elected congressman, Abraham Walter Lafferty, a Republican from Oregon. "There are two reasons why the National Forests in Oregon, for example, should not be turned over in trust to the state. The first and most important one is that the forest in question is necessarily, through its connection with the rivers and in other ways, an interstate question, and the National Forests can be handled far better for the general welfare by the Federal Government than by the State." [39]

By January 1911, Gifford Pinchot was suggesting either that a progressive Republican (Roosevelt) should challenge Taft for the Republican nomination or (a less attractive possibility) that Roosevelt should bolt and create a third party. Certainly, Roosevelt paid close attention to all this political maneuvering. He had toured America enough, talked with enough farmers and laborers, and answered enough sacks full of mail, to believe that Taft was bad for the Republican Party. Deeply embarrassed for having chosen "Willy-Boy" Taft as his successor in the first place, he wanted to turn back the clock to March 1909 and send Taft back to Ohio. Taft's firing of Pinchot had stuck in Roosevelt's craw; also, Roosevelt couldn't believe that Taft had supported the Payne-Aldrich Act, which continued high tariff rates. Demonizing the incumbent president now became a sport for Rooseveltians. Preparing to challenge Taft for the Republican nomination, Roosevelt simply didn't want to admit that the president did anything right pertaining to conservation. (Taft's record actually wasn't all that bad. He had, for example, saved the Oregon Caves in Oregon, Rainbow Bridge in Utah, and Devil's Potspile in California by declaring them national monuments.[40] Taft had also created the first national monument in Alaska: Sitka, a lush, temperate rain forest containing more Northwest Indian totem poles than anywhere else.[41])

For the first six months of 1911, Roosevelt avoided the warfare within the Republican Party, although there was an element of burlesque in his disclaimers. Instead, he worked hard throughout the spring to get the National Museum to properly prepare the skins, fur, and skulls from his African expedition for presentation to the scientific community. He was also hoping to arrange for Charles Sheldon to publicly display his specimens from Alaskan offshore islands in a coastal diorama. It wasn't enough, Roosevelt wrote to Charles Wolcott, to merely "collect"—full reports from both Sheldon and himself should be furnished to the public at large. He didn't want the Roosevelt Collection to go unattended, shut away in closets like Carl Akeley's ape specimens at the Field Museum of Chicago. And Roosevelt wanted to keep Edward Heller—who had been on his safari in British East Africa as the Smithsonian's leading naturalist—in Washington, D.C., until all of his specimens were stuffed by taxidermists and ready for public viewing.[42]

While Roosevelt was preparing his African mounts that April, a report was published in the *New York Times* that the last bull moose in New York state had been killed. These ungulates, with their huge racks, were once plentiful in the Adirondacks, but lumberjacks and hunters had slaughtered them. The Algonquin, a New York tribe, had called them *moose* ("twig-eaters"). The last moose had weighed 1,200 pounds, had immense antlers, and was shot by a poacher and left to rot in the snow. Roosevelt's love of these generally solitary herbivores was bone-deep. Hearing their low mooing in the forest was one of the most moving experiences in the

North American wilderness. In the following years, Pinchot and Garfield saw the "last bull moose" as standing for Theodore Roosevelt himself. "A curious thing about the bull moose," the *Independent* would write the following year about what became the symbol of Roosevelt's Progressive Party, "at such moments of emotional excitement [it] readily answers a *call* and comes headlong to meet it." [43]

As president, TR had already created the Fire Island National Game Reservation (Executive Order No. 1038) on February 27, 1909, using the Antiquities Act of 1906. Fire Island—located near the head of Cook Inlet offshore from Anchorage—was the most important federally run breeding ground for moose in the United States. [44] Roosevelt nurtured in his mind the notion of having many similar moose sanctuaries in Alaska, Minnesota, Vermont, New York, and Maine. According to Rooseveltian conservationists, Alaskan miners were overkilling moose for meat for use in their placer camps. Every camp had a moose specialty: moose hash, moose tenderloin, and crown roast of moose, among other recipes. Conservationists recommended canned hams or imported beef as better alternatives. [45]

III

Because Roosevelt had initiated the protection of bull moose in Alaska, his name was mud in the mining camps of the Kenai Peninsula, where there was a tornado of sentiment against his ethos of wildlife protection. A coalition known as the Coal Party, for example, was created by Alaska boomers hoping to recover the acreage of Roosevelt's national forests and federal bird reserves. In May 1911, the former mayor of Cordova, Alaska, accompanied by an angry group of "territory rights" activists and debt-ridden people from the chambers of commerce, engaged in an act of civil disobedience reminiscent of the Boston Tea Party. They raided the wharves where the Copper River and Northwestern Railway was storing imported anthracite, split open crates, and dumped tons of imported Canadian coal into Controller Bay. [46] "The Cordovans were striking back at a distant colonial government," the historian Char Miller explained in his biography of Pinchot. "Then they put the torch to their own King George III, burning an effigy of Gifford Pinchot, denounced by *The Alaska-Yukon Magazine*, as a man who 'thinks more of trees than people.' " [47]

Well-organized protests against Roosevelt and Pinchot erupted throughout Alaska that year. When Pinchot voyaged to Alaska on a fact-finding mission in September, the predominant complaint in the territory was that federal laws were stunting the economic growth of Alaska.[48] Such legislation was called conservation colonialism. All the major Alaskan newspapers thought that Richard Ballinger—a former mayor of Seattle and the current U.S. secretary of the interior—was a hero and Pinchot a scoundrel. Covetous boomers intuited that they had a once-in-a-lifetime opportunity to drive a wedge between Roosevelt's conservationism and Taft's pro-development philosophy; and prospectors who had missed out on the Klondike gold rush believed that coal mining would give them a second chance at wealth. Alaskan town hall meetings resounded with antigovernment rants and calls for direct action. Angry protest was everywhere. In the towns of Seward (on the Kenai Peninsula) and Valdez (on the eastern side of Prince William Sound), for example, President Roosevelt's national forest orders of 1908 limiting corporate mining in the Tongass and Chugach were posted and defaced with an angry X. In the timber town of Katalla, unhappy loggers and miners burned Roosevelt's order of 1908 in a public display of defiance. In another Alaskan town, a threatening placard that looked like a "Most Wanted" notice was posted:

PINCHOT, MY POLICY
No patents to coal! All timber to forest reserves!
Bottle up Alaska! Put Alaska in forest reserves!
Save Alaska for all time to come![49]

Wherever Pinchot traveled in Alaska, he defended conservation in front of audiences full of skeptics. Pinchot argued that Ballinger had been ousted in a necessary effort to "prevent men who were trying to plunder and monopolize Alaska from carrying out their plan." [50] Town hall meetings turned volatile if Pinchot's name was even mentioned. Many citizens in Valdez, situated on the lip of the Chugach, thought him hopelessly wrongheaded for locking up the forestlands. The *Cleveland Press*, for example, reported that a satirical anti-Pinchot banner had been hung in

Seward, Alaska: "Conservation prices . . . British Columbia coal, $17 per ton . . . Wood, $7 a cord . . . But you must not mine your own coal, nor cut down your own wood . . . All reserved for future generation . . . signed Pinchot . . . 'Pinhead.'"[51]

While a cabal of defiant Alaskans were up in arms over the Tongass and Chugach national forests, which they saw as having been grabbed by the U.S. government, Roosevelt was entering a nasty (if erudite) public argument with the naturalist Abbot H. Thayer of New Hampshire, who was a theoretician. The disagreement centered on theories of concealing coloration. Roosevelt first challenged Thayer, at some length, in Appendix E of *African Game Trails*. He also inveighed against Thayer in the introduction to *Life History of African Game Animals* (a magnificent two-volume zoology reference book whose coauthors were Roosevelt and Heller). Then—in the August 23, 1911, issue of the *Bulletin of the American Museum of Natural History*—in a 40,000-word monograph titled "Revealing and Concealing Coloration in Birds and Animals," Roosevelt intensified his thesis with new field data from Africa. He particularly objected to Thayer's claim that the stripes and spots of mammals had protective value against predators. Roosevelt himself argued, correctly, that these markings attracted mates. Rattling off the names of species in which coloration was clearly not protective, Roosevelt floated the theory of advertising.[52]

During August 1911, Gifford Pinchot, James Garfield, William Kent, and other conservationists were doggedly urging Roosevelt to campaign for the Republican presidential nomination against Taft in the coming year. They argued that his candidacy was an imperative if the conservation movement was to survive. Roosevelt thought the three men were becoming too self-righteous—they had forgotten to smile. "Come, come!" Roosevelt wrote to Kent, who in 1908 had given an old-growth redwood grove, Muir Woods near San Francisco, to the U.S. government to become a national monument. "You and Gifford are altogether crazy about Taft. I have been very much disappointed in him, of course, but you use language about him that is not justified."[53]

Roosevelt believed that if there was a cardinal sin in public life, it was becoming a "dull pointless bore."[54] A political convention wasn't a corporate board meeting; it was a roller-coaster ride at Coney Island, a fiesta in San Antonio, a horse race in Kentucky, a confetti-filled celebration in

Times Square. Perhaps he would take on Taft over conservation issues. But he wouldn't do it out of anger or for revenge. "We must not preach all the time or we will stop doing any good," Roosevelt wrote to a friend who urged him to challenge Taft. "Life is a campaign, and at best we are merely under-officers or subalterns in it." [55]

For self-given Christmas presents in 1911, Roosevelt read Charles Sheldon's *The Wilderness of the Upper Yukon*, enthralled by the naturalist's field reports of fast-ebbing currents, V-shaped flocks of geese, and previously unstudied mountain ranges north of Skagway.[56] Sheldon, a young naturalist, had sent Roosevelt chapters of a proposed new book, *Wilderness of the North Pacific Coast Islands*, to proofread; it was published in 1912. In Roosevelt's mind, Sheldon was the real deal—an outdoorsman who had become the Thoreau of the Yukon River basin, a hunter who understood that unlike land (which could be bought and sold), *wild country* had a personality distinctly its own.[57] There was a touch of the old-fashioned faunal naturalist in Sheldon—a love of peace, solitude, wild things, and serenity—that Roosevelt stoutly admired.

At Sagamore Hill that Christmas, Roosevelt had a lot more to reflect on than Alaskan moose reserves, debates over bird coloration, and wilderness outings. In January, moderate Republicans split from their party and formed the National Progressive Republican League. The Progressives, championing Roosevelt, advocated reforming the political system to give control to the people, rather than to party hacks who had no ethics, no decency, and no commitment to the long-term interests of the American people. The Progressives supported the direct election of senators, presidential primaries, and the use of volunteer initiatives such as referendum and recall. They also called on Roosevelt and other leaders to challenge the anticonservationist "milquetoast mannequin"—that would be William Howard Taft—for the Republican nomination in 1912.

While Pinchot started plotting a Progressive campaign strategy for 1912, Roosevelt went back to the occupation he had preferred since leaving the White House: being a Darwinian naturalist. Roosevelt had discovered a new ornithologist with the potential to be another William Finley (of Oregon) or Herbert K. Job (of Connecticut). His name was Francis Hobart Herrick. Considered America's authority on eagles, Herrick was named a professor of biology at Western Reserve University in Ohio. In 1901, he wrote *The Home Life of Wild Birds*, and by 1917 he had published a fine two-volume biography of John James Audubon.[58]

What really caught Roosevelt's eye, however, was a pamphlet Herrick had written on nest building. Roosevelt and Herrick exchanged thoughtful letters discussing their ideas on modern biology. "Darwin and the great scientific men of his day forced science to take an enormous stride in advance in the decades succeeding the publication of *On the Origin of Species*, but for nearly fifty years now we have tended to make the same mistake that the schoolmen of the Middle Ages made about Aristotle," Roosevelt wrote to Herrick. "The rediscovery of the works of Aristotle produced an immense forward movement in knowledge. Then there came a period of fossilization, when everybody accepted Aristotle as having summed up all possible knowledge, and when in consequence he became a positive obstacle to advance. It has been somewhat so with Darwin and the Darwinians."[59]

Roosevelt was worried that a sense of complacency had engulfed university biology departments, whose members were willing to accept—without conducting new research, collecting species, or doing field studies—everything Darwin had proved about evolution. Where was the sense of excitement about the Alaskan outdoor laboratory? Who would be the new Gregor Mendel, describing the nature of inheritance? Weren't the Aleutians the new Galápagos? Why was eighty-nine-year-old Alfred Russel Wallace still clinging to a theory of natural selection that he first articulated as early as 1858? Where were the neo-Darwinians who could offer the world something more than half-baked theories of protective coloration and nesting habits? Roosevelt hoped Herrick would become one of the new bright lights. In 1866, Ernst Haeckel had promised a practical application of the theory of evolution and had achieved dramatic results. Likewise, August Weismann in 1882 denied that a species could pass on acquired characteristics to its offspring through germplasm. What had happened to these people since then? Were they resting on their laurels?[60] "I doubt if we have ever seen anything less scientific than the extreme dogmation of men like Haeckel," Roosevelt complained, "and the solemn acceptance as facts of Weismann's extreme theories."[61]

Because the universities were slow to make discoveries about the natural world, Roosevelt placed more faith in the National Geographic Society (NGS). On June 9, 1912, for example, the Novarupta volcano erupted in Alaska's Katmai district. Since the ice age there had been seven major eruptions in the Katmai volcanic cluster, and this was the worst. A preceding series of earthquakes had been followed by enormous ejections of red-hot pumice and ash over an area the size of Maine. More than forty square miles of verdant Alaskan forest were literally scorched, buried under a thick blanket of volcanic soot, in some places up to ten feet deep. To put Novarupta into a historical perspective, the blast was ten times more devastating than the eruption of Mount Saint Helens in 1980. Only the eruption on Santorini in Greece in 1500 B.C. produced more volcanic matter than Novarupta. The cracked Alaskan earth shot up steam vents more than 500 feet high at more than a thousand holes in the Katmai district. Strange gas clouds formed and emanated from Earth. Dr. Robert Griggs of Ohio State University, a botanist who worked closely with the NGS, led a scientific expedition to the Katmai in the fall of 1912 and called the weird, smoking landscape the "valley of ten thousand smokes." But Griggs optimistically understood that within a few years the ash-laden hillsides would become alive with "verdure."[62]

The fissure floor of the Katmai—at the head of the Alaska Peninsula—was declared a geologic wonderland. Roosevelt thought that Novarupta, even more than Lassen Volcanic National Park in California,* could equal Yellowstone National Park as a tourist attraction. Nowhere else could volcanism and tectonic events be better understood by schoolchildren. Because Alaska was so sparsely settled, not a single person died in the natural event at Katmai. From 1912 to 1918, scientists traveled there to study waterfalls and lava flows. "It was as though all the steam engines in the world, assembled together," Griggs wrote, "had popped their safety valves at once and were letting off surplus steam in concert."[63]

On September 24, 1918, President Woodrow Wilson declared the "Valley of Ten Thousand Smokes" the Katmai National Monument. The boundaries of Katmai National Monument, which originally encompassed forty

---

* When he was in northern California, Roosevelt liked to stay at the ranch of the former secretary of state William Seward, near Lassen Volcanic National Park, in order to study the volcanoes.

square miles of the Mount Katmai pyroclastic flow, were expanded in 1931, 1942, 1969, and 1978.[64] Then, in one of the crowning achievements of the entire post-1960s environmental movement, the Alaska National Interest Lands Conservation Act of 1980 put millions of additional acres surrounding the "Valley of Ten Thousand Smokes" under federal protection, enlarging the total area to more than 4 million acres. It was redesignated the Katmai National Park and Preserve on December 2, 1980.[65]

IV

If Taft hadn't tried to undermine Roosevelt's national forestry agenda in Alaska, it's doubtful that the ex-president would have challenged his successor for the Republican nomination in 1912. Even though, as president, Taft had prosecuted the Standard Oil and American Tobacco trusts, Roosevelt nevertheless painted him as a lackey of big business. Roosevelt was partially wrong. President Taft did enjoy automobiles more than bird-watching, but he had a decent record on conservation. Still, perception matters in politics. No matter whether they had sided with Pinchot or Ballinger in the notorious feud, journalists believed that there was a curious ambivalence about conservation issues in Taft's White House. Taft, it seemed, had an old-fashioned Abrahamic concept of land, finding no real value in wilderness. Favoring the Department of Commerce and Labor, he seemed to enjoy rejecting expansions of forestland proposed by the departments of the Interior and Agriculture.

In February, after weighing the pros and cons, Roosevelt announced that he would indeed run for president again. His declaration was welcomed by a press corps eager for a riveting news story. On every major issue of the day, TR vowed to act uncompromisingly. Taft, in a foolish, backward-thinking way, had mocked Roosevelt's Alaskan conservationism. If Taft wanted to sell off Alaskan lands to "big coal" instead of leasing them to locals, then Roosevelt would confront the president in the public arena. To Roosevelt's everlasting fury, Taft had indeed followed through on his pledge to side with the dictatorial Speaker of the House—Joseph Cannon of Illinois—and Senator Nelson Wilmarth Aldrich of Rhode Island, both in the pocket of special interests.[66]

While the Republican old guard clung to Taft, Roosevelt beat out Robert La Follette of Wisconsin as the progressives' favorite son. Before long, Roosevelt, in an ad hominem attack, cold and merciless, was calling President Taft an old-maidish "fathead" and "puzzle wit"—and many Americans loved hearing such insults. Taft shot back that Roosevelt was a "dangerous egoist" and "demagogue." What Roosevelt understood was that rural Americans weren't instinctively fond of lawyer-politicians like Taft, who supposed that the rifle was a toy for grown-ups and that dinner came from a grocery store, not from a farm or a duck blind. Although Roosevelt outperformed Taft throughout the spring of 1912 and arrived at the Republican convention in Chicago only a few delegates short of having the nomination locked up, the conservative old guard managed to stop the wilderness warrior. Roosevelt had even secured thirty-seven of Ohio's forty delegates (and Ohio was Taft's home state)—but these were of no avail in locking up the nomination. The Republicans handed Taft the nomination, by a slim margin.[67]

But Chicago hadn't seen the last of Theodore Roosevelt. The ex-president joined a third party: the Progressive, or Bull Moose, Party. On August 7, 1912, Roosevelt delivered the most impassioned speech of his political career at its convention. Surrounding him like bodyguards were a number of Rough Riders from the Spanish-American War, in full army uniform, who had served under the Colonel in Cuba in 1898. Roaring about the rights of working Americans over business conglomerates, Roosevelt laid out the Progressive platform, emphasizing conservation and promoting the principles of public domain lands, women's suffrage, regulation of corporations, roadside beautification, federal assistance to the poor, better schools, and so on. He set forth a liberal domestic agenda for the twentieth century that Franklin Roosevelt, Truman, Kennedy, Johnson, and Obama would build on. Roosevelt lambasted mechanization and human abuse of the environment. If financial titans thought the conservation movement was over, they were doomed to disappointment. Botany, biology, geology, soil science, entomology, and forestry offered clues to humans' relationship with Earth. To Roosevelt, it was impractical to discuss land policy without placing people's concerns first and foremost. However, it was blasphemous to rape and loot the landscape for profit, as the placer miners had done along East Creek near Fairbanks. If Taft was going to be anticonservation, Roosevelt would sink him like the *Titanic*—a disaster that was still fresh in people's minds. "There can be no greater issue than that of conservation in the country," Roosevelt declared. "Just as we must conserve our men, women, and children, so we must conserve the resources of the land on which they live." [68] As Pinchot pointed out, for Roosevelt and the Bull Moose Party, conservation was a "moral issue." [69]

Throughout the 1912 campaign, Alaskan fishermen went on strike against Roosevelt and Taft's policies regulating fishing. They claimed the right to use salmon traps. Big canneries likewise insisted that the traps were a necessity. In *Pacific Fisherman*, their trade journal, packers called the huge clamlike traps, which were designed to funnel migrating salmon, "the best and only friend the canners have in Alaska." [70] A tender such as *Little Tom* would take the salmon—hundreds brailed from a fish trap—all day long. The goal was to "fish out" a place, then move somewhere else. Roosevelt wanted to shut the packers down for illegal fishing methods; otherwise, these "big fish" companies would deplete Alaska of salmon.

Most sourdoughs (or old-timers) in Alaska despised everything about the Bull Moose Party. Although Roosevelt was respected as a big-game hunter, his federal land grabs between 1902 and 1909 on behalf of wildlife and forests infuriated them. None of them, however, voted in the 1912 elections, so they didn't matter. But coal and timber corporations in Washington and Oregon used the sourdoughs' antifederal attitude to arouse contempt for all regulation of the extraction industries.

By September 1912, President Taft, his popularity diminished, was looking irrelevant. Every rally for Taft was lackluster, sweltering hot, and newsless. Roosevelt, the youngest of the presidential candidates, relished attacking the "husks" of the Democratic and Republican parties, which had nominated "boss-ridden" men of weak moral fiber. [71] Roosevelt called the president a "flubdub with a streak of the common and the second rate in him." [72] The combustible campaign centered on the Democratic nominee Woodrow Wilson's "new freedom" versus Theodore Roosevelt's "new nationalism." Wilson, a former president of Princeton University and popular reformist governor of New Jersey, pushed his attacks on corporate abuses even farther left than Roosevelt. Wilson, in some cases, claimed to support federal control of companies. But Roosevelt still held the progressive high ground when it came to the environment. Roosevelt considered Wilson nothing more than a "sham reformer," embracing dull precedent because that was politically expedient.[73]

To Roosevelt and his supporters, Wilson was also hostage to the laissez-faire doctrine, an ignorant, outdated philosophy for the twentieth century. Only money-grubbers would put laissez-faire over the collective good of the American people. "Now the governmental power rests with the people, and the kings who enjoy privilege are the kings of the financial and industrial world," Roosevelt said at rallies, promoting progressive democracy. "And what they clamor for is the limitation of government power, and what the people sorely need is the extension of governmental power."[74]

On the campaign trail, Roosevelt was brilliantly successful at inspiring young, conservation-minded outdoors enthusiasts to join the Bull Moose cause. In Chicago, for example, Harold L. Ickes, an attorney deeply interested in reform politics, quit the Republican Party and signed up with Roosevelt. Reporting for the *Chicago Record* Ickes—whose clients included Jane Addams of Hull House, a leader in the social work movement—for the first time became informed about federal forest reserves, national parks, and wildlife protection. Quirky and combative, with an impish smile that often beamed forth from his thin lips, Ickes didn't look like an outdoorsman. But looks are often deceiving. The acerbic Ickes was a dyed-in-the-wool Pennsylvanian conservationist, a proud native son of Altoona, whose great love in life was the Appalachian mountain range.[75] Clear, fast-moving western Pennsylvanian rivers like the Little Juniata, and secret places like Horseshoe Cave and Blue Knob, were indelible images in his memory. Ickes, a self-proclaimed lone wolf, was a paradox: an urban wheeler-dealer who thought America's salvation was in the backcountry. "I love nature," Ickes declared. "I love it in practically every form—flowers, birds, wild animals, running streams, gem-like lakes, and towering snow-clad mountains."[76]

As Ickes noted in his diary for 1912, nobody could claim that Roosevelt wasn't striking a nerve in the body politic with his fiery Bull Moose rhetoric. At his rallies, huge crowds hung on his words as he attacked Wall Street, overcome more by emotion than by insight. Because the Socialist Party had nominated the labor leader Eugene Debs for president, Roo-

sevelt was facing an able challenger in campaigning for economic justice for the laboring class. Budding conservationists like Ickes, however,
chanted, "The Bull Moose has left the wooded hill/His call rings through
the land/It's a summons to the young and strong/To join with willing
hand." [77] Outdoors enthusiasts had long before developed a firm affection
for Roosevelt's high-purposed stagecraft; they voted for him without hesitation in 1912. As a performer, Roosevelt was raw and visceral, brimming
with defiance, insisting that he was an unshakable one-man squad for
American betterment. His words seemed to glow in the air. But stump-
ing from coast to coast was banal compared with the outdoors life. "I am
hoarse and dirty and filled with a bored loathing of myself," he wrote to
Kermit. "I often think with real longing of the hot, moonlit nights on our
giant eland hunt, or in the white rhino camp, with the faithful gun-boys
talking or listening to the strumming of the funny little native harp." [78]

It was in this circus atmosphere that John Schrank, a Bavarian immigrant from New York, arrived in Milwaukee for a Bull Moose rally
with murder on his mind. On October 14, Schrank approached Roosevelt and shot him at close range with a .38-caliber pistol. Two spectators
restrained the psychotic shooter as Roosevelt tried not to faint. Schrank,
it turned out, was angry because Roosevelt was behind laws that closed
saloons on Sunday. Luckily, a thickly folded fifty-page copy of a speech
and a metal eyeglass case (used for bird-watching) inside Roosevelt's coat
pocket stopped the bullet from piercing his heart. A scuffle ensued and
Schrank was apprehended. Roosevelt refused medical attention and went
on to speak for ninety minutes before being rushed first to Emergency
Hospital in Milwaukee and then to Mercy Hospital in Chicago. The bullet,
lodged close to his lungs, was never removed. "Friends, I shall ask you to
be as quick as possible," Roosevelt had said from the stage, his chin thrust
high. "I don't know whether you fully understand that I have just been
shot; but it takes more than that to kill a Bull Moose." [79]

When Pinchot heard the news from Milwaukee, he was at first dis-
believing. But when he learned that Roosevelt had continued to give the
speech while bleeding profusely, he knew it was God's honest truth. "It
may seem like a queer thing to say," Pinchot wrote to Roosevelt, "but your
being shot has been one of the finest things that has ever come into my life

on account of the way you have handled the whole situation."[80] Roosevelt reassured Pinchot that he remained determined to win the presidential election, refusing to give up his effort.[81]

Americans were spellbound by the unfolding drama. Was Roosevelt still on the march? Or was his campaign now over? No matter how many scholars insist that Davy Crockett died of disease at the Alamo or that Abraham Lincoln was really a bigot, the general public refuses to abandon the orthodox view of these heroes. The larger public view—whether accurate or not—is that Crockett fought for the independence of Texas, and Lincoln emancipated the slaves. After Milwaukee it no longer mattered whether Roosevelt won or lost the 1912 presidential election. By the time he arrived at Madison Square Garden on October 24, and received a forty-five-minute standing ovation, the bullet still lodged in his rib cage, he had become an enduring American icon.[82]

On November 5, however, Wilson swept the election with 435 electoral votes to Roosevelt's disappointing 88. "I won't pretend," Ickes later recalled in *Autobiography of a Curmudgeon*, "that we didn't awake the day after the election with a bad headache."[83] Roosevelt consoled himself with the fact that Taft had won only eight electoral votes and Debs—who, surprisingly, won 6 percent of the popular vote, the most ever by a socialist candidate—nevertheless failed to receive a single electoral vote. The Bull Moose Party succeeded in winning 27.4 percent of the vote and electing thirteen new members to Congress. Even more impressively, the Bull Moose Party brought more than 230 state legislators into office. To offset his own loss, Roosevelt boasted that he had fulfilled his pledge to make Taft a one-term president. But no genuine whoop of victory was conveyed by the Colonel's reasoning. "Well," he had written to Kermit on election night, "we have gone down in a smashing defeat; whether it is a Waterloo or a Bull Run, only time will tell."[84]

Chapter Five

* * * * * *

CHARLES SHELDON'S FIERCE FIGHT

I

All of Alaska brought a bounce to Charles Sheldon's gait. Like a protagonist in a novel by James Oliver Curwood, he decided that every inch of the territory was Edenic, though with a lethal component. But it was 20,320-foot Mount McKinley, its peak blanketed in deep perpetual snow, that left Sheldon in awe. Just looking at McKinley—which he first saw in mid-July 1906 from a hilltop near Wonder Lake—seemed to lower Sheldon's blood pressure and heart rate. Time stood still within a fifty-mile circumference around the base. Even in summer, the temperature on the mountain, wrapped with storm clouds and mist, frequently dropped below zero Fahrenheit. Gold prospectors had named the towering peak in 1896 to honor President William McKinley. The name stuck. To the Athabascan Indians, however, the peak was Denali ("The Great One"). Sheldon used the Indian name (although he sometimes simply said "The Mountain"). The south peak was the highest point in North America. To Sheldon the whole area around Mount McKinley—the huge glaciers, the trough-like gorges, the miles of tundra stretching out to meet other mountains on the blue horizon—was his beloved "Denali wilderness." The Alaska Range made the Colorado Rockies seem like foothills. Furthermore, in terms of its sheer rise from base to summit Denali was the tallest mountain in the world.

Traveling around Mount McKinley, Sheldon was like a cowboy riding through a well-stocked cattle ranch in Texas and eyeing his herd, except that Sheldon's cattle were migratory caribou. From halfway up the mountain the caribou looked like ants. In his field journals he waxed eloquent about caribou herds and told of risking his life to study grizzlies. Unlike the slopes in the Lower Forty-Eight, the Alaska Range—home to 161 species of birds and thirty-seven of mammals—was not heavily forested; it was primarily blanketed by snow and ice.[1] Besides protecting wildlife, Sheldon also wanted to ensure that the large quantities of hemlock, birch, poplar, alder, and willow surrounding McKinley didn't become cordwood. Alaska had more than 450 types of plants that botanists believed might be potential medicines. He feared that the Alaska Railroad line connecting Fairbanks to Seward—completed in 1914—would forever ruin the Denali wilderness. Yet he recognized that because McKinley was between the two cities, the railroad would make the national park a convenient stopover. "To America's fledgling conservationists, railroads were synonymous with wildfire, destruction," the historian Tom Walker wrote in *McKinley Station*. "Enter the railroad—gone the wildlife; gone the frontier."[2]

In the aftermath of the Harriman Expedition, it was the search for a Roosevelt elk (*Cervus roosevelti*), in 1904 on Victoria Island in Canada, that first injected Charles Sheldon into the drama of saving Alaska's wilderness.[3] The Biological Survey was looking for this subspecies of wapiti—which had survived in small herds on the Olympic Peninsula (in Washington state) and Vancouver Island (in British Columbia)—to analyze in Washington, D.C., and Sheldon volunteered to bring back mounts.[4] (There were no native elk species in Alaska.*) Although he was essentially a big-game hunter, Sheldon had become a legend in old Mexico for climbing sheer cliffs to spy on bighorn sheep. Financially comfortable and politically astute, he was a well-rounded sportsman who was difficult to ignore. His sharp features conveyed willpower, an impression confirmed by his deep, almost gruff voice. Looking like a Canadian Mountie, the pinched-lipped, muscular Sheldon stood at about five feet ten inches.

---

* Although elk aren't native to Alaska, they have been reintroduced to Afognak Island, Etolin Island, and Raspberry Island. Elk had lived there during the Pleistocene but became extinct before Euroamericans arrived.

Demure, respectful, and bookish, he was a clean-shaven embodiment of roughing it like Jim Bridger, an old-time mountain man of days past, easily able to backpack 100 pounds or so across rugged terrain. As a big-game hunter for the Biological Survey, Sheldon was always ready with rifle, field glasses, and camera. To Sheldon wildlife conservation wasn't an optional policy; it was a life force, the necessary corrective to manifest destiny and to the industrial revolution.

Reading Sheldon's faithfully kept faunal journals isn't for everybody. Much of his prose smacks of *Forest and Field* and the old-style campfire yarns of the nineteenth century (a genre in which "half-breed" Indians were backwoods scouts and educated white men made historic "discoveries" in bioregions where Native American tribes had lived for thousands of years). Sometimes, however, when he describes blunders he made on the trail, the reader can almost hear silent laughter. Anecdotes of deer carcasses hanging from clotheslines, endless winter nights, and salmon impelled to swim upstream because of "tooth and claw" mandates reveal Sheldon as a Darwinian naturalist—Theodore Roosevelt without TR's elegant, dramatic turn of phrase. When Sheldon vividly described the flora and fauna of Admiralty Island (then part of the Tongass National Forest and since 1978 a national monument) and Montague Island (at the entrance to Prince William Sound)—both around 300 square miles with sheer-sided and thickly wooded coastlines—he was superb.

Throughout the first decades of the twentieth century, Sheldon set up U.S. Biological Survey camps on sheltered beaches on Alaskan islands. Unusually for a hunter-explorer in the early twentieth century, Sheldon brought Louisa Walker Gulliver—his wife—with him as a partner on his expeditions. Sheldon, a proud family man, also regularly brought his four children on his remarkable outdoor educational adventures. Together they would study brown bears, which were thick on Admiralty and Montague islands. These enormous bears weighed from 600 to 1,700 pounds when they were fat from feeding on the spawning salmon found throughout coastal Alaska. Bears that ate a steady diet of these weighty Alaskan fish tended to put on pounds themselves. On three other Alaskan islands—Kodiak, Afognak, and Shuyak—the largest subspecies of brown bear roamed freely: Kodiak bears (*Ursus arctos middendorffi*). No fewer than five species of mammals have been named in Sheldon's honor. These include two Alaskan discoveries: *Ursus arctos horribilis* (a bear) and *Marmota caligata sheldoni* (a hoary marmot).[5]

To Sheldon, brown bears—which ranged from Wyoming to Alaska—were the most awesome creatures in North America. All bear cubs, Sheldon noted, followed three rules of survival: obey mother, trail mother, and have fun.[6] His field notes described variations in coloration from dark brown pelage to very pale or gray-brown. Whereas most hunters would either shoot a bear or climb a tree for safety, Sheldon tried to study their muscle humps to estimate size. All over Alaska bear rugs were a centerpiece of living rooms; the thick underfur and guard hairs were warmly comforting. What Sheldon found particularly fascinating about bears was that their front paws were easily twice the size of the back paws. Many a day while working in Alaska, Sheldon saw a brown bear click its teeth, froth at the mouth, put its ears back, and then just walk away. Long before Sheldon, naturalists had written about bears; but he pioneered in dispelling the myth that bears were prowling killers of humans.

There was a new environmental awareness shining forth in Sheldon's Alaskan field journals. When he hunted in places like Admiralty Island, his big-gun mentality evaporated into a more modern ecological attitude. What interested Sheldon were topics like the range of grizzlies. On Admiralty Island, unlike vast Denali in the interior, the coastal rain forests produced large berry crops and salmon ran deep, so grizzlies seldom wandered beyond a forty-mile limit. Sheldon hoped to someday conduct quantitative research on the polar bear; he was tired of hearing only tall tales and yarns spread by the trophy seekers. Alaskans were still imbued with the wasteful frontier mentality, the idea that there were no limits to America's wildlife resources. Temperamentally unsuited for city life, although New York City was his principal home, Sheldon believed that the American frontier was alive and well in the blueberry backcountry of Alaska. Some of Sheldon's colorful journal entries had the descriptive power of paintings by Hartley or Marin.

And Sheldon, while lacking the evocative flair of Muir or Thoreau, developed a smooth, marvelously controlled prose style when describing how Alaskan mammals struggled to survive human intrusions. "It was the mystic hour of evening when our work was finished, and, the clouds having lifted, the rain suddenly stopped," he wrote on September 20,

1909, after a couple of stormbound days on Admiralty Island. "It was calm, and a peaceful silence brooded over woods and waters. Mrs. Sheldon and I walked far out on a point of reefs. Everywhere ducks were lazily floating on the surface of the water, which reflected the large trees towering near the shores as well as the high, snow-crested mountains behind them. Huge reefs were scattered all about, snow-white with the thousands of gulls which flocked on them to pass the night. Little islands, covered by groves of lofty trees, were numerous, and on one of these, in the top of a gigantic dead spruce, a fine bald eagle and its mate now perched facing each other, each one calling at short intervals in a series of shrill screams which echoed about the irregular shores."[7]

For Sheldon the best places in the western hemisphere were those where the wind velocity had rip-roaring power. The Denali valleys, he said, were like swells in the ocean, boundless and breathtaking. Those who heard his appeal were instantly ready to purchase a one-way ticket to Alaska. Private gentlemen's clubs—the most elite in America, such as the Cosmos and the Century—wanted Sheldon as a lifetime member. Poised and always adaptable, comfortable both at the Metropolitan Opera and curing fish with Native Alaskans in the Kenai Peninsula, Sheldon added both hardiness and élan to any conversation, luncheon, or campfire. Alaska, he avidly declared, was *the* escapist tonic for any urban dweller sick and tired of the rat race. The hummocks, tangled streams, and forested rivers allowed a somnambulistic urbanite a chance to follow his inner compass. Self-possessed when writing about Alaska, full of perspicacity, Sheldon charmingly made first-person declarations about "my wilderness," "my river," and "my country" to express his deep love for the sprawling territory. Back in New York, he would always tell friends that he'd left his heart in Alaska. "For Sheldon the Alaskan wilderness was not a tooth-and-claw setting for the defiance of death as it had been to Jack London and Robert Service," the historian Roderick Frazier Nash wrote in *Wilderness and the American Mind*. "He saw it as a frontier, but especially in regard to big game habitat, a perishable frontier that needed protection."[8]

II

Born in *1867 (the year the* Andrew Johnson administration purchased Alaska from Russia) to a Vermont marble-quarrying family, Sheldon grew up with the beautiful Green Mountains as his backyard. All he remembered about his childhood was the beatitude of sunny summer days and the high drama of magnificent snowstorms. Kinship with nature was an inherent part of growing up. Vermont had a number of peaks over 3,000 feet high—Mount Mansfield, Mount Ellen, and Camels Hump among them—that the teenage Sheldon climbed. Canoeing in Vermont's rivers—the Winooski and Lamoille in particular—also was a skill that he learned growing up along the hogbacks of the Front Range. When he was an adolescent, his hikes in Otter Creek Valley, where the brook trout were thick, turned him into an ardent outdoor sportsman. Nobody else in Rutland, Vermont, learned how to use an ax as skillfully as Sheldon. And because the family business was marble quarrying, Sheldon was also skilled with a chisel and hammer.

Exceedingly bright, Sheldon attended Phillips Academy in Andover, Massachusetts, reading sportsmen's literature by Izaak Walton, Robert Barnwell Roosevelt, and Frank Forester. After prep school he went to Yale University. Sheldon quickly became a leader of the class of 1890. He was a lover of American poetry, particularly Longfellow and Lowell; and he joined the Elizabethan Club. Nobody, however, remembered his performances in plays. The words that his classmates used over and over again to describe him were "rugged" and "no nonsense." One afternoon a salesman came to Yale, banging on students' doors, offering boxes of Cuban cigars. Not long after the salesman's visit, Sheldon noticed that his flute had been stolen from his quarters. Immediately he turned detective. For a long day he visited all of New Haven's and New York's pawnshops, hoping to find his flute. His determination paid off. At one of the Manhattan shops, Sheldon stumbled on the petty thief, the flute sticking out of his suit coat pocket. Without hesitation Sheldon, like a linebacker, tackled him to the ground. He then made a citizen's arrest. The salesman went to jail and Sheldon returned to Yale with his treasured instrument.[9]

Sheldon's first job after college was working for the Lake Shore and

Michigan Southern Railroad. From young manhood onward, he was transfixed by the most forlorn reaches of North America. Bouncing around Mexico for a decade, he made a risky investment in Potosi, a Chihuahuan silver and lead mine, which paid out huge dividends. Endowed with a certain charisma, Sheldon became friends with the family of Don Luis Terazas, powerful landowners in Mexico, whose haciendas were over 8 million acres in size. Everything, it seemed, was going his way financially.

Independently wealthy at age thirty-five, in 1903, Sheldon abruptly retired from business. Wide-browed, with neatly parted dark brown hair, Sheldon didn't want to become a gent in a blue blazer holding court at the Polo Lounge or the Newport races. He wanted to be a ruddy-cheeked foot soldier in Roosevelt's conservationist revolution. Hoping to model himself on TR—the jaunty naturalist who happened to live in the White House—Sheldon contacted Dr. C. Hart Merriam and Dr. Edward W. Nelson at the Biological Survey in early 1904, offering his services collecting wildlife. They were immediately taken with his handsome appearance, his gentlemanly ways, and his abiding interest in the Yukon and Alaska. His whole demeanor was that of the young TR, a sophisticate who could associate easily with packers, wranglers, and backcountry iconoclasts. So the U.S. Biological Survey tapped Sheldon to collect mammal skins on Vancouver Island, in the Yukon, and in Alaska from July to October 1904. That's when he went looking for biological data on the Roosevelt elk.

At the Biological Survey headquarters on Thirteenth and B Street (later renamed Independence Avenue) in Washington, D.C., Dr. Nelson was known as "Mr. Alaska," and for good reason. During the 1870s, decades before the Klondike gold rush, Nelson, with old-time WASP ingenuity, had traveled all over Alaska for four years, serving as a weatherman for the U.S. Army Signal Corps in the Bering Sea. Besides monitoring blizzards and wind velocity from primitive weather stations, Nelson collected wildlife specimens and Eskimo artifacts for the Smithsonian Institution (known then as the U.S. National Museum). His most astounding biological discovery was collecting field data about the all-white Dall sheep with gorgeous curled horns that populated Alaska and northern Canada. He actually purchased a couple of these sheep from backcountry hunters to conduct scientific experiments on. The sheep were carefully studied by

the Smithsonian biologists, intrigued by theories about animal coloration. Dutifully Nelson wrote a zoological treatise on rams in 1884, based
largely on Sheldon's taxonomic principles. Nelson even named a new species of sheep: *Ovis dalli* (an homage to William Dall, the great Alaskan
naturalist-explorer).[10]

Merriam, Nelson, and Roosevelt welcomed Sheldon into their small
clique of biological-minded outdoorsmen. In Mexico's Sierra Madre and
the Yukon's subarctic mountains, Sheldon had studied sheep's maneuvers
on high cliffs, seeing their fancy footwork as poetry in motion. Roosevelt had written biologically accurate essays about bighorns in *The Wilderness Hunter* (1893), and Sheldon would do the same for Dall sheep and
Stone sheep (*Ovis dalli stonei*) in his own book, based on his 1904 and 1905
northern Canada–Alaska expeditions. The journals from these high-
altitude outings were eventually published in 1911 as *The Wilderness of the
Upper Yukon*. In New York's zoological circles, Sheldon was anointed the
great pathfinder of the early twentieth century, a new Deerslayer or Natty
Bumppo. Bursting with enthusiasm for everything Alaskan, Sheldon
wanted to make national parks, wildlife refuges, and wilderness zones
out of his favorite campsites in the Alaska Range and north Pacific coast
islands.[11]

Sheldon's backwoods style enthralled Roosevelt, who saw him as a
spiritual heir. Roosevelt, in fact, reviewed *The Wilderness of the Upper Yukon* in the *Outlook*, declaring his young protégé the new TR. "Mr. Charles
Sheldon is a . . . wilderness wanderer, who to the hardihood and prowess
of the old-time hunter adds the capacity of a first-class field naturalist,
and, also, what is just as important, the power of literary expression,"
Roosevelt wrote. "Such a man can do for the lives of the wild creatures of
the wooded and mountainous wilderness what John Muir had done for the
physical features of the wilderness. . . . His experiences of Alaska, and
indeed the entire Northwest, are such as no other man has had; and no
other writer on the subject has ever possessed both his power of observation and his power of recording vividly and accurately what he has seen."[12]

Imbued with a visionary streak, Sheldon wasn't trying to present the
wilderness in Alaska as a souvenir of the closed frontier. His importance to the history of conservation lay in his belief that the days of Kit
Carson had passed, but that if the primitive arts were learned, a vibrant

wilderness adventure could still be had. Much like the Camp Fire Club of America, which was created in 1897, Sheldon recognized that wildlife would survive the onslaught of civilization only if huge tracts of habitat were saved for certain species—an approach Roosevelt had pioneered with bison near Fort Sill, Oklahoma.[13]

Sheldon, having completed his apprenticeship in Mexico and Alaska, soon became a transformational leader in the conservationist movement of the progressive era. He was elected an officer in the Boone and Crockett Club, National Parks Association, and American Forestry Association, among numerous other preservationist-minded organizations. From the outset, Merriam respected Sheldon for treating the natural world with humility and restraint. Roosevelt, in fact, saw Sheldon, whom he deemed "a capital representative of the best hunter-naturalist type today," as almost a member of his extended family.[14] Roosevelt often turned to Sheldon to serve on various wildlife committees of the Boone and Crockett Club.[15] Because Dr. Merriam had failed to finish his magnum opus *North American Mammals*, Roosevelt started hinting that perhaps Sheldon should step up and fill the void.[16] While Sheldon never produced such a comprehensive study, he led the movement for America to adopt progressive game laws.[17]

Interior Alaska was an unforgiving land in 1904, when Sheldon first went to study Dall sheep in earnest. None of the territory's 30,000 residents suffered from being too gentle. Sheldon felt like a voyageur, an intrepid explorer following animal tracks all over the Alaska Range. There was only one rule of dress: stay warm. Clothed in heavy wool garments, determined to survive, Sheldon proved his mettle as a true explorer. Every day his clothes got wet and his bedroll clammy. But he didn't complain. His rifle of choice in 1906 was a Mannlicher .256 caliber. Unlike Andrew Berg—a Finnish immigrant who became the first licensed hunt guide in the Kenai Peninsula and moonlighted as a fur trapper—Sheldon carried field glasses as his favorite tool. Berg's hunt notes, however, proved to be a monument to phonetic misspellings: "at home doctoring," "no suckuss above freezing all day . . . weathre warm."[18]

In *The Wilderness of the Upper Yukon*—published in 1911—Sheldon enthusiastically described the plans and goals of the U.S. Biological Survey's Yukon-Alaska expedition: to study the golden-horned, all-white Dall sheep foraging on grasses, sedges, forbs, and dwarf willows. Drawing on his diaries to give the book a real-time structure, Sheldon analyzed all species of wild sheep of North America. He divided the family Bovidae into two species subgroups: thinhorn and bighorn. Dr. Nelson had previously accumulated valuable information about sheep's hooves and horns, but it was based on conjecture and limited biological proof. Sheldon, filling the vacuum, provided authoritative Darwinian analysis of wild sheep's range in the Yukon and Alaska. "Indeed, so little was known about the variation, habits, and distribution of the wild sheep of the far northern wilderness, that my imagination was impressed by the possibilities of the results of studying them in their native land," he wrote. "There was, besides, the chance of penetrating new regions, of adding the exhilaration of exploration to that of hunting, and of bringing back information of value to zoologists, and geographers, and of interest to sportsmen and lovers of natural history." [19]

Awed to be working with the great Dr. Nelson, Sheldon now made Alaskan mammals and birds his area of zoological expertise. For the next decade, he commuted between New York and Fairbanks, where all the roads abruptly ended. Trails in the Alaska Range during Roosevelt's presidency had been built exclusively for the mining and timber companies. After outfitting himself in Dawson and hiring Jack Haydon as a guide, Sheldon developed the daily pace of a man on the march. Living out of a backpack and duffel bag, he was prepared for extreme camping at all times. Sheldon had clearly not come to Alaska for recreation. From dawn to dusk he worked, collecting wildlife data. No matter how grueling the outdoor experience became, he never let it affect his appearance. A proper wardrobe was the sign of a Yale man, even in the outback. He refused to look scruffy, like a muskrat trapper. Venturing down the Yukon River, he declared all the nature around him worthy of a thousand Thomas Cole paintings. The forest animals he encountered in the Denali wilderness—deer, wolves, and ground squirrels—had variations of color and size he had not anticipated. Sheldon dutifully recorded the precise numbers of the migrating caribou he encountered. Eskimos claimed that up in the Arctic great herds roamed the tundra above the timberline along the coastal plain of the Beaufort Sea. He hoped to visit the Arctic someday. To the Gwich'in people (known as the Caribou people), these great herds represented the primary source of cultural and economic sustenance. As herbivores, the caribou ate willow, dwarf birch, lichens, moss, and even dried sedges in winter. In turn the Gwich'in people ate the caribou.

Sheldon returned to the Alaska Range wilderness (which formed the southern border of the Yukon basin) on August 1, 1907, and stayed through June 11, 1908. His guide for this expedition was Harry Karstens, who wanted much of wild Alaska saved from commercial exploitation. Sheldon snowshoed through fresh powder in the Alaska Range with Karstens, forging new trails at high altitudes. The saw-toothed Alaska Range had a distinct crest line, with peaks from 8,000 to 10,000 feet high (unbroken for a breathtaking 200 miles). The range south of Fairbanks was imposing, wild, forlorn, and stern. Many peaks—Foraker, Russell, Hunter, Hayes, Silverthrone, and McKinley—were over 10,000 feet high. A trifle nervous about winter in the Alaska Range, Sheldon and Karstens stayed in the snug valleys, moving their base camp according to the weather. In the winter they holed up in a cabin on the upper Toklat River. They marveled at how, in springtime, everything came alive in Denali. The sky was filled with flocks of geese that sprawled over the long fields—Canada geese (*Branta canadensis*) and white-fronted geese (*Anser albifrons*). Surrounding Sheldon and Karstens in swirling columns were arctic terns (*Sterna paradisaea*) and sandhill cranes (*Grus canadensis*), sometimes in pairs, but often in cloudlike clusters. To Sheldon there was a sense of ancient history about these great bird flocks. He was annoyed that there were virtually no U.S. laws to permanently protect them.

To Sheldon's surprise, the truly irreplaceable link in Alaska's food chain was the willow ptarmigan (*Lagopus lagopus*). These birds mated in May and their eggs hatched in June. They were particularly thick in Roosevelt's Yukon Delta.[20] Once their white downy feathers were lost, the ptarmigan's plumage turned as brown as the willow, dark birch, and spruce boughs. They were well camouflaged, clinging to the ground to feast on wild berries and willow buds. When flushed, the ptarmigan, which are plump, seem to struggle with flight, forced to keep a low trajectory only ten to fifteen feet above the ground. The naturalist Margaret E. Murie, in *Two in the Far North*, joked that the ptarmigan was a comical creature, seemingly saying, "Come here, come here, come here—go back, go back, go back."[21] Every predator in Alaska considered the willow ptarmigan fine dining. The red foxes and ground squirrels, in fact, seemingly identified the brown-speckled ptarmigan eggs as the finest delicacy on the tundra. Other birds—golden eagles (*Aquila chrysaetos*), gyrfalcons, short-eared owls, and goshawks among them—swooped down to lift away willow ptarmigan in their claws for dinner.[22] For wolves and wolverines, the willow ptarmigan was a veritable Thanksgiving dinner in waiting. Once ptarmigan were devoured, the tree sparrows (*Spizella arborea*) and white-crowned sparrows (*Zonotrichia leucophrys*) collected the feathers to use as nest lining.[23] "The ptarmigan," Sheldon reported in his field journal, "flying from rock to rock above, kept sounding their croaking chatter."[24]

The primordial bird populations in Alaska included 100 million seabirds, 70 million shorebirds, and 12 million waterfowl. (Unfortunately, there were even more mosquitoes that swarmed up out of the marshlands.) Charles Sheldon—considered by both Grinnell of the Boone and Crockett Club and Merriam of the Biological Survey as *the* best young American naturalist-hunter—ended up inventorying the birdlife around the north base of snowcapped Mount McKinley with scientific exactitude. He published his findings in the January 1909 edition of the *Auk*. More than 400 bird species inhabited or migrated through Alaska, including many Asian species. Sandhill cranes and golden eagles were found throughout the territory, feeding along mirror-still lakes. The Aleutian chain was full of colonies of raucous seabirds, which Sheldon never got to inventory.[25]

One bird that Sheldon admired during his Alaskan wanderings was the trumpeter swan (*Cygnus buccinator*). With its bright white plumage, long periscope-like neck, and black bill, the trumpeter swan—the largest waterfowl species on earth—was magnificent. Besides being the unfortunate victims of fashion—women wanted their feathers for bonnets—these swans were sensitive to contaminants. A small colony of trumpeters lived in the lower Copper River system and the Kenai Peninsula. They bred in northwestern British Columbia and in the Saint Elias Range backcountry; those populations, however, were not faring well.[26] Determined to help the trumpeter swans survive in their core range in Alaska, Sheldon worked with the Boone and Crockett Club, the New York Conservation Society, and the Camp Fire Club of America to help protect the swans. In 1968 the nonprofit Trumpeter Swan Society assumed the full-time duty of advocating the protection of these regal birds. E. B. White's *The Trumpet of the Swan*, published in 1970, memorably introduced these beautiful birds (which can stand on one leg for more than half an hour) to many children.[27]

Market hunters were the bane of Sheldon's days in Alaska. One afternoon while Sheldon and Karstens were tracking Dall sheep, they came upon a couple of hunters with sixteen dogs around a campfire. They were gorging themselves. They had slaughtered a herd of Dall sheep, including the ewes and lambs—the whole family—which Sheldon had been inventorying. "Naturally," he wrote, "I was deeply disappointed to hear that my sheep, which I had been so carefully observing, were to be disturbed by vigorous market hunting, but could do nothing to prevent it." [28] He vowed to fight for laws to protect Alaskan game against "slob hunters" who didn't even know what conservation meant. And he mistakenly warned Natives not to trust Hudson Stuck, a Presbyterian minister whose *Ten Thousand Miles with a Dog Sled*, filled with tales of cold winter journeys on behalf of Christ, became a best seller.[29] Before long, however, Stuck became Sheldon's ally. In 1913, Stuck led an expedition up the summit of McKinley. His memoir of the climb was titled *The Ascent of Denali*.[30] With almost evangelical vigor, Stuck insisted that the name of Mount McKinley was an affront to the mountain and Native people and should be changed back to Denali. "There is, to the author's mind," Stuck wrote, "a certain ruthless arrogance that grows more offensive to him as years pass by, in the temper that comes to a 'new' land and contemptuously ignores the Native names of conspicuous natural objects." [31]

III

The 1907–1908 *year on the upper* Toklat River (in an area that became part of Mount McKinley National Park) was brilliantly described in Sheldon's memoir *The Wilderness of Denali*. Nowhere in the world, Sheldon proclaimed, were there mountains as majestic in winter as the Alaska Range. He felt privileged to have walked among such towering manifestations of the ice age. The Alaska Range, filled with swollen rivers in springtime, divided the Alaska territory not only into districts but also into distinctive climates. For an experienced mountaineer like Sheldon, tramping around the crags, clefts, waterfalls, and marshlands of interior Alaska was far better than climbing the comparatively dull Matterhorn in Switzerland. Besides the Alaska Range, there was the Rocky Mountains extension that slashed across northern Alaska as the Endicott Range (about 200 miles from the Arctic Ocean). The Coast Range, which John Muir also loved, consisted of the Fairweather and Saint Elias mountains, with peaks over 10,000 feet high (here were the blankets of glacial fields). The Wrangell Mountains were a string of unsymmetrical lava cones with peaks—Blackburn, Castle, Drum, Jarvis, Regal, Sanford, Wrangell, and Zanetti—all over 10,000 feet high.

On his Alaska Expedition from 1905 to 1908, Sheldon collected specimens of the caribou (*Rangifer tarandus*), Alaska moose (*Alces alces*), white sheep (*Ovis dalli*), snowshoe hare (*Lepus americanus*), and lemming (*Lemmus trimucronatus yukonensis*) for the U.S. Biological Survey and U.S. National Museum (Smithsonian Institution), both in Washington, D.C. Camping among the dwarf fireweed along the Savage River in Denali, he performed taxidermy on the skins and preserved the skulls of four different subspecies of meadow mouse, catching the rodents with little homemade traps. A wonderfully precise sketcher, Sheldon also drew vivid illustrations of the wildlife he procured, a time-honored tradition of the Boone and Crockett Club. Regularly Dr. Merriam wrote Sheldon glowing letters about the value of his adventures in the far north to the world of biological conservation. "While his personal interest centered chiefly in the larger game animals, Sheldon nevertheless appreciated the importance of collecting the smaller mammals and took the trouble to trap, prepare, and label large numbers of mice, lemmings, shrews, and other small species, all of which he presented to the Biological Survey for permanent deposit in our National Museum," a grateful Dr. Merriam recalled of Sheldon in an introduction to *The Wilderness of Denali*. "These specimens have been of inestimable help to naturalists engaged in defining and mapping the ranges of the smaller mammals and besides have brought to light a number of species previously unknown. And it should be borne in mind that while the major part of his field work was done in Alaska and Yukon Territory, he also made important collections and field notes in British Columbia, Arizona, and northern Mexico."[32]

Perhaps Sheldon's greatest pieces of writing, in hindsight, were his flawless essays on Hinchinbrook and Montague islands (published as chapters in *The Wilderness of the North Pacific Coast Islands*). Both large barrier islands are located between the Gulf of Alaska and Prince William Sound. They are what Martha's Vineyard and Nantucket Island are to Cape Cod, but thicker with wildlife. What interested Sheldon about these islets were the vast families of brown bears. (Along Alaska's coastal waters they were called brown bears; those living in the interior were grizzlies.) As the Japanese Zen poet Bashō had written, "To learn about the pine, go to the pine. To learn about bamboo, go to the bamboo." To learn about brown bears, Sheldon regularly visited these offshore islands of Prince William Sound. There was a profusion of bears on the islands around every puddle and bend. Sheldon was determined to accurately count their distribution numbers; accurate data would be the first step toward saving the bears.[33]

"No sight in the American Wilderness is so suggestive of its wild charms than that of the huge bear meandering on the mountain-side, or walking on the river-bank, or threading the deep forest," Sheldon wrote. "He who still retains his love for wild nature, though accustomed to the sight of wild animals, and surfeited in some degree with the killing of them, feels a lack in the wilderness—perhaps the loss of its very essence—when, tramping about in it, he knows that the bear, that former denizen of its depths, is there no more—exterminated forever."[34]

What amazed Sheldon most about the Alaskan brown bear was its massive head, which, incongruously, had inconspicuous ears and tiny eyes. In the East, zoologists thought of Alaska's bears as having fur of a single color. But Sheldon found that these bears varied from pale tan, sandy, gold, silver, and cinnamon to all shades of dark brown and black. Because of their color variations, *Ursus arctos* demanded a lot of careful biological scrutiny. Owing to Sheldon's research, for example, the Biological Survey learned that the bears from coastal Alaska were much darker and more uniformly colored than those found in the interior (which had pale-tipped guard hairs). Furthermore, Sheldon inventoried everything these brown bears ate, including grasses, tender shoots, wildflowers, tree roots, tubers, mosses, willows, and especially berries. While these solitary (except during breeding season) bears occasionally grubbed for insects, larvae, and eggs, they were unable to digest coarse forage very well.[35] By cutting out the molar of a shot grizzly, then counting the annual rings of cementum, Sheldon could tell the age of a bear. Whenever he ran into a successful bear hunter, he asked for a molar.

Both Sheldon and Merriam were determined to oversee the most comprehensive inventory of Alaska's brown bear population ever undertaken. They were convinced that zoologists hadn't properly identified subspecies of brown bears like Kodiaks. Sheldon was Merriam's "bear man" in northern Canada and Alaska, trying to understand the range of brown bears, in particular. On Montague and Admiralty islands, he slept in an open canvas shelter with his wife, staying warm under reindeer skin robes. "I note that you give *Ursus horribilis* as mainly Rocky Mountains," Sheldon wrote to Merriam. "These scarcely touch Yukon Territory, except north of the Pelly River and they extend well up the Mackenzie. *Ursus horribilis* is confined to the territory East of the Lewis and South of Steward. But west of the Lewis toward the Alsek River and the White River, close to the coastal ranges I have always thought that the grizzly there resembled those of the Alaska range (*alascensis*). There is one skin from that district in your collection and two female skulls. Therefore I had Yukon Territory divided up into three regions possibly. *Urses phaeonix, ogilves* region north of latitude 64—*horribilis* in district southeast of Yukon Rivers and *alascensis* inside St. Elias and other coast ranges north."[36]

For all his skills as a naturalist, Sheldon was unusual because he enjoyed lonely reveries amid the spruce, often not seeing anyone else for weeks (that is, when his wife wasn't along). Sheldon shot, with his .22 rifle, a wide range of coneys, marmots, shrews, and ground squirrels for the Biological Survey to analyze. On one Alaska-Yukon trip, Sheldon coincidentally bumped into the world-famous British hunter Frederick Selous, who was collecting for the British Museum. For six weeks, the pair trekked through the far north together, discussing South African game and Alaskan bears. Selous was sixteen years older than Sheldon but, like everybody else, was charmed by Sheldon's show. They became lifetime friends. When Selous died in 1917, Sheldon wrote to Roosevelt about creating an impromptu memorial to the great British conservationist. "Will you bring the matter up," Roosevelt wrote to Sheldon, "before the Boone and Crockett Club?"[37]

But the outdoors life that Sheldon led had drawbacks. In Alaska, for example, the plague was mosquitoes. There are more than 3,000 species of these insects, and at times it seemed that all of them had decided to hold a revival meeting or a roundup in Alaska. When Sheldon studied paleontology as a boy, he learned that mosquitoes had been around to buzz and bite the dinosaurs. Scientists would in coming years find mosquitoes trapped in amber (petrified sap) in a tree fossil more than 38 million years old. During the summer months, swarms of mosquitoes attacked Arctic Alaska's four caribou herds, forcing the Porcupine herd to migrate 700 miles to escape them. If an Eskimo hunter killed a caribou in summer with a spear or arrow, the carcass would be blanketed by mosquitoes within seconds. There were other true flies in Alaska—crane flies, midges, and gnats—but it was the mosquito, wings beating between 250 and 600 times a minute, that became the bane of outdoorsmen, considered a hazard as menacing as wind, sleet, and snow.

IV

What haunted Sheldon, *making him seethe* with anger, was the gradual diminution of the larger mammals such as Dall sheep, moose, and deer as a result of market hunting across the tundra-covered valley. Sometimes, even when he was hungry and miserable, Sheldon nevertheless counted and collected for the Biological Survey. Only the thick swarms of biting flies and insomnia during the summer solstice really hindered him. Driven by his love of the outdoors, Sheldon, when the creeks were down and the trails melted out, kept biological diaries of his pioneering wildlife observational research on the northern slopes of the Alaska Range. "Complete enjoyment of the wilderness," Sheldon wrote, "needs periods of solitude."[38] Being alone at a high altitude gives a person plenty of free time to think. Sheldon began dreaming of the Denali wilderness as a national park—the largest in the system, millions of protected acres. Karstens's journal entry of January 12, 1908, recorded Sheldon's first hope that the U.S. government would maintain Denali National Park as a quasi-wilderness area (i.e., roadless).[39]

The McKinley River was the longest and widest of hundreds of glacier-fed rivers, streams, and creeks. Everywhere a visitor looked, there were braided brooks gurgling across the wet tundra. More than twenty ridges were involved in the drainage of the McKinley River. With his loyal packer Harry Karstens (nicknamed "the Seventy-Mile Kid" because he had once mined a claim on Seventy Miles Creek outside Dawson City), Sheldon built a weatherproof cabin along the Toklat River, located opposite the mouth of present-day Sheldon Creek (named in his honor). From the start, they split plenty of firewood to prepare for the subzero winter. At a trading post they acquired roof shakes and a small keg of nails. Using their lean-to shack as a base camp, Sheldon began wandering around the Denali wilderness. Head down, he struck out into the high-velocity wind, with gun, pad, and pencil. He was a man in his element. It sickened Sheldon that residents of the Kantishna region north of McKinley were mass-butchering game while building the Alaska Railroad.[40] It also sickened him to see market hunters butchering Dall sheep to put meat in the pots of mining camps in the Savage, Teklanika, Toklat, and Sanctuary river valleys. Much of this meat was fed to the sled dogs. Much like annual tree rings, the indented lines on a ram's horns, some spreading as much as three feet across, indicated the Dall sheep's age. Sheldon feared the species was headed toward extinction. Even the U.S. Army infantrymen stationed in Alaska at Fort Gibbon at Tanana and Fort Liscum near Valdez considered the sheep butchery repugnant.

Throughout 1907–1908 Sheldon, like a new John Muir, had shared campouts with the Chilkat (in southeastern Alaska). His journals also indicate encounters with the Minchumina, Nenana, and Tanana.[41] The Ivy Leaguer looked as if he had been born and raised amid Alaska's varied habitats—glaciers, mountains, tundra, grasslands, wetlands, lakes, woodlands, and rivers. He wore rawhide moose snowshoes and traveled in forty-foot-long bark canoes. In his log cabin, whose roof was scarcely higher than his head, he scribbled furiously about the great round moon, silvery waterfalls, icy fjords, and torrential rains. Despite all the precipitation, Sheldon worried constantly about brushfires. Ever since the U.S. Forest Service was created in 1905, men had been paid decent wages as fire lookouts. Sheldon hoped to raise funds in New York for hiring more lookouts for Alaska. "Alone in an unknown wilderness hundreds of miles from civilization and high on one of the world's most imposing mountains, I was deeply moved by the stupendous mass of the great upheaval, the vast exterior of the wild areas below," Sheldon wrote, "the chaos of the unfinished surfaces still in process of molding, and by the crash and roar of the mighty avalanches."[42]

As reflected in Karstens's remembrances, Sheldon was determined to see Mount McKinley saved as a kind of Grand Canyon of the north— a protected American wonder, a true wilderness area untouched by axes or construction crews where a citizen could go and get lost. To his mind only one two-lane road should be allowed to cut through the park. Mount McKinley, he said, was an inheritance for his grandchildren.[43]

When Sheldon returned to New York before Christmas 1908, invigorated by the stinging snows of Denali, he almost single-handedly launched a campaign to create a national park around Mount McKinley. He was the best cheerleader wild Alaska ever had. The bird flocks in the area, he said, were loud enough to throw an orchestra out of tune. The salmon-rich rivers had the cleanest, purest water that ever rushed over rocks. To see a double rainbow over the Teklanika River at summer twilight was proof that the world had a Creator. Painting word pictures, Sheldon told his audiences about seeing Mount McKinley free of clouds, lording it over the adjacent snow-clad summits, as grizzly bears patrolled the base. The great Muldrow Glacier falling down the eastern side from the snowfield between the two domes, he claimed, was one of the great sights in nature. What worried Sheldon was that hunters were slaughtering more and more game to feed mountain-ringed towns such as Nenana, Kantishna, and more distant Fairbanks. As a purist with regard to nature reserves, he disdained the filthy backwoods stump mills, placer operations, and forest "units" earmarked for cutting. Once the railroad came, connecting Seward to Fairbanks, additional market-hunting syndicates would patrol the Denali wilderness and kill everything that moved.[44]

It had taken George Bird Grinnell a full nineteen years to see Glacier National Park become a reality. But Sheldon, who always believed luck was on his side, was determined to obtain the designation within a decade. Recognizing that securing congressional approval was tough sledding, Sheldon began intensely lobbying the heavyset James Wickersham, the Alaska territory's only delegate on Capitol Hill, a quasi-Rooseveltian conservationist. Wickersham, a pioneer judge originally from Illinois, was

Alaska's voice in Washington, D.C., from 1909 to 1921. He favored both the Alaska Railroad from Seward to Fairbanks and the establishment of Mount McKinley National Park.[45] Working alongside Sheldon in lobbying were Nelson, Grinnell, and the Camp Fire Club of America. Together they vowed to have Congress vote in favor of the national park within the decade. One crucial fact was that the Alaska Railroad was being built from the southern coast of Alaska to Fairbanks. Tracks were being laid across Broad Pass, so the eastern limit of Mount McKinley National Park would be accessible by train, a plus for tourists wanting an excursion from Anchorage.[46] Wickersham thought Mount McKinley would make an ideal railroad stopover. He imagined a getaway village, built around a string of hotels, which would attract tourists from all over the world.

Something about Sheldon's fervor for protecting Alaska's wildlife heritage was very appealing in the age of Model T's, telephone wires, catchpenny devices, skyscrapers, soap bubbles, and the Wright brothers. What could be more American than a huge brown bear feeding on salmon in a fast-moving stream or a bull moose bedding down under a pine?

At meetings of the Boone and Crockett Club, Sheldon planned with friends exactly how to create a vast national park reserve the size of his home state, Vermont—a park to be run by the U.S. Department of the Interior. They got Stephen Mather, the director of the National Park Service, to sign on, with huge enthusiasm. As an inducement, Sheldon would talk about Denali as the last frontier. The 1909 edition of *Webster's New International Dictionary of the English Language* had an interesting definition of *frontier*: "the border or advance region of settlement and civilization, as, the Alaskan *frontier, chiefly* U.S."[47]

The historian Richard Slotkin, in *The Fatal Environment: The Myth of the Frontier, 1776–1890*, described the concept of the "frontier" as the "longest lived of American myths" and a "powerful continuing presence."[48] The Denali wilderness would now be the frontier. The national park would encompass broad river valleys, wildflower tundra, massive glaciers, and a portion of the lofty Alaska Range, including the unsurpassed Mount McKinley, North America's highest summit. While many people celebrated the defeat of wilderness as progress on the march, Sheldon saw it as a loss of something essential to democracy. Sheldon realized that conservation had too many misleading labels. A crusade to eradicate bark beetles wasn't Sheldon's idea of either Muirian or Rooseveltian conservationism. To Sheldon the heart of conservation was saving wild landscapes. He saw himself, in the end, as a pioneering advocate of wilderness reserves, building on the legacy of TR's federal bird reservations. *Conservation* was a term of compromise whereas *wilderness* was preservation at its purest.

In 1912, the publisher Charles Scribner's Sons brought out Sheldon's memoir *The Wilderness of the North Pacific Coast Islands: A Hunter's Experiences While Searching for Wapiti, Bears, and Caribou on the Larger Coast Islands of British Columbia and Alaska*. In it, Sheldon wrote about Admiralty Island, particularly the brown bear populations, in biological terms. There are beautifully written anecdotes about brown bears digging up wild parsnips, stalking prey, and fleeing after whiffing the scent of man. There was nothing purposefully sentimental about Sheldon's encounters with bears, which included measuring the size of tapeworms in their dung. "It is a wonderful sight to see the huge bear suddenly appear on the bank of a creek swiftly flowing through the great forest, while the salmon fight and splash and the gulls scream in the plaintive voices as they hover about the pools," Sheldon wrote. "To see a bear leap into the rapids, sweep out a salmon with its paw, and return silently into the wood to make its feast must be a stirring experience and one that would give a wonderful glimpse of wildlife in the forest of the wilderness. It is, however, a field for the photographer, not the sportsman."[49]

Sheldon did a convincing job of presenting the Denali area to the Department of the Interior as a teeming and impressive land. Helping the lobbying, and arriving at just the right time, was a memoir by the mountaineer Belmore Browne, *The Conquest of Mount McKinley* (1913), complete with anecdotes about mushing behind a team of dogs over high mountain passes.[50] Ever since the nomadic Yupik and Inupiat brought dogsleds from Siberia to Alaska, mushing had become a preferred and practical mode of transportation across the wilderness territory. Declawed, their incisors pulled, sometimes even castrated, Eskimo dogs (or malamutes) had an inbred sense of direction and made winter travel feasible in Alaska. Arctic explorers such as Leopold McClintock and Fridtjof Nansen had popularized these dogs in their adventure sagas. Jack London transformed them into symbols of the far north in *The Call of the Wild*. In 1908

Nome inaugurated the All-Alaskan Sweepstakes, a sporting event that eventually led to the Iditarod race. And now Browne, in *The Conquest of Mount McKinley*, presented these dogs as heroic mountain climbers, thus helping Sheldon's proposed national park get extra newspaper coverage in the Atlantic coast states.

Browne's unanticipated assistance convinced Sheldon of a political truth: if you stuck to your guns long enough in America, *right* would eventually prevail. Sensing an opportune moment, Sheldon wrote to Nelson at the Biological Survey that the time had come to push the legislation for Denali National Park through Congress—the letter was dated October 10, 1915.[51] This document was the opening salvo of a fierce legislative tussle. Sheldon's journals about Denali, in fact, were now carefully studied by U.S. congressmen as clear-eyed dispatches from "The roof of the continent." Every page, it was quickly understood, constituted a first-rate argument for the wilderness and wildlife preservation rather than logging in the Denali region.

V

Sheldon *finally achieved his goal in* 1917. After a flurry of last-minute negotiations about railroad entry and hunting laws, and after crucial lobbying by the Boone and Crockett Club and the Camp Fire Club of America, Congress presented Sheldon with an approved bill. Immediately, document in hand, Sheldon hurried to the White House, hoping to speed up the signing process. On February 26, President Wilson at last approved the legislation to create Mount McKinley National Park. He invited the jubilant Sheldon to attend the official signing ceremony at the White House. Sheldon's arduous treks across the Alaska Range over glaring snowfields in icy gales, counting caribou and Dall sheep, had paid off for America and the world. The U.S. government had finally recognized his vision of Mount McKinley—and the beautiful raw-bone foothills of the Alaska Range—as belonging to every citizen. Laws associated with the new national park complemented Sheldon's vision: no market hunters, no gold prospectors, and no oil-field geologists would be allowed in the 2 million-acre wilderness.

But there were some problems. For one thing, Congress rejected the name Denali in favor of Mount McKinley National Park. Sheldon and others were annoyed. Congress also refused to appropriate new money to protect Mount McKinley from the poaching of wildlife and timber. All President Wilson and Congress had really agreed to was a template for protection. With no funds set aside for the long-term preservation of Mount McKinley, Sheldon knew, the Denali wilderness wouldn't last long. Conservationist activism was a constant experience of tribulations. Disappointed, Sheldon tapped the Boone and Crockett Club for $8,000 so that the Department of the Interior could hire a superintendent for Mount McKinley.[52]

Because of Sheldon's public promotion of Mount McKinley, tourists started trickling in—very slowly—to see it. Only seven visitors came to see the new national park in 1922.[53] In 1923 the Curry Hotel opened in time for the park's formal dedication. A scenic viewpoint—the "Regal Vista"—was established so that tourists could snap photographs of McKinley without an arduous hike.[54]

The only newspaper that seemed to care about the new national park was the *Brooklyn Daily Eagle*. Not until 1924, when roads and concessions were built, did the number of visitors increase. Stephen Mather lobbied aggressively for congressional allocations to help the park develop infrastructure. A log structure (looking rather like a strip mall) became the tourist gateway of McKinley Station; it comprised a roadhouse, a general store, a post office, a public garden, and little log motel cabins to rent. The Alaska Railroad, working closely with the National Park Service, printed up attractive brochures and extolled the run from Seward to Fairbanks as the "Mount McKinley Route."[55] As the historian Alfred Runte noted in *National Parks: The American Experience*, the new park met the major preservationist criterion of the era: "monumentalism."[56]

Sheldon also achieved, like Muir before him at Glacier Bay, the promotion of discovery and recreation in Alaska for tourists of tomorrow. A later cult of wilderness enthusiasts wanted to explore Alaska's boundless forests and glaciers. Colonel A. J. "Sandy" Macnab and Frederick K. Vreeland—of the Camp Fire Club of America—represented the new breed of outdoors enthusiasts, eager to make a permanent mark as conservationists in Alaska. After World War I, Americans, aglow with victory, discovered Alaskan mountains and rivers as a leisure-time destination.

Aviation now made "doing Alaska" feasible for rich people from the East Coast. Colonel Macnab, who served under General Pershing in France, had supervised a rifle school there, outside Le Mans. Under his leadership more than 200,000 U.S. soldiers a week learned how to use a Springfield .30-06 rifle. After the war Macnab, based in Camp Benning, Georgia, dreamed of Lake Clark, Alaska—where the Dall sheep, caribou, and bears reportedly were abundant. Vreeland—an electrical engineer, photographer, and wilderness enthusiast based in New York—had a different motivation. He wanted to photograph the region and test his survival skills.

Macnab and Vreeland, as noted above, were both members of the Camp Fire Club of America (CFCA), which had been started in 1897 by the zoologist Hornaday. The club's name sounded rather bland (as if it were for aging Boy Scouts who wanted to toast marshmallows); but in truth, its elite membership consisted of approximately 100 physically fit survivalists and wilderness devotees. (Later, its membership increased to about 480, and many members came from the Westchester County area.) Deeply secretive about their club's history—it was a kind of Skull and Bones for outdoors endurance—the members often climbed the highest peaks, went down category 4 or 5 white-water rapids in kayaks, and tested themselves against hurricanes, blizzards, avalanches, and torrential rains.[57] In his book *The Forest* (1903), Stewart Edward White described the CFCA as brave men of "essential pluck and resourcefulness pitting themselves against the forces of nature."[58] Perseverance, toughness, ardor—every club member was a Theodore Roosevelt or a Charles Sheldon in the making.

It wasn't until 1910 that the CFCA became an active group for the protection of wildlife habitats in America. Wisely, under the leadership of Ernest Thompson Seton, the club purchased two heavily wooded farms in Westchester County as a retreat—the total area was 161 acres.[59] There were two lakes on the property. More land was added in 1917. The main lodge was built from local cedar logs. (Boy Scouts of America was founded on its porch.) No electricity was allowed, but gas lamps were permitted. Pistol shooting was encouraged on the range. Every spring the club had outdoor outings. To be a member, an applicant had to feel claustrophobic about big-city life—and pass twenty-one survival tests.[60]

One afternoon at the CFCA compound Macnab and Vreeland launched
a plan to hike all around the Lake Clark–Iliamna area of southwestern
Alaska; it would be, in their estimate, the next Mount McKinley (or Denali). There was a paucity of maps of the Lake Clark area in the 1910s. The
whole mountainous area was a jumble of unnamed streams and lakes essential to Bristol Bay (the preeminent salmon fishery in the world) and
Cook Inlet (the shipping route to Anchorage). What Macnab and Vreeland understood was that Lake Clark was the big-hearted country of
Alaska. If that sounded like balderdash, consider this geographic fact:
Lake Clark was the junction of Alaska's three great mountain ranges:
the Alaska Range (from the north), the Aleutian Range (from the south),
and the region's own Chigmit Mountains. There were two active volcanoes soaring over Cook Inlet—Iliamna (10,018 feet) and Redoubt (10,197
feet)—within the lands considered worthy of becoming a preserve. Going straight west, across a vast stretch of tundra, brought a traveler to
Roosevelt's Yukon Delta Federal Bird Reservation. "Glorious views of
Kachemak Bay," Muir had written of the area east of Lake Clark in his
journal for 1899, "many glaciers; bright weather. Fine views of Iliamna,
Redoubt, and other volcanoes, the former smoking and steaming distinctly at times; surrounded by sharp lower peaks and peaklets—the most
beautiful, icy, and interesting of all the mountains of the Alaska Peninsula." [61]

Seeing the sky-piercing volcanoes around Lake Clark from the luxury
of Harriman's yacht, however, was far different from hiking near their
lava base snapping photographs, as Macnab and Vreeland aimed to do.
Traversing miles of moss and muskeg, with swarms of hard-biting flies
as companions, and camping among the swamp willows and alders was no
picnic. The Lake Clark plateau resembled the Arctic terrain, with caribou
herds wandering the permafrost tundra. It was hard going for even a nature photographer like Vreeland (after whom a Canadian glacier had been
named) and a crack marksman like Macnab (who was just back from the
Great War). Besides a few Euroamericans, the main populations around
Lake Clark were the Dena'ina Athabascans (on Iliamna) and the Yupik
Aleuts (at the mouth of the Newhalen River and the southwest portion of
Iliamna Lake). These tribes were good stewards of the land. But as in-

dustrialization increased—with overpopulation becoming a new plague—Lake Clark was bound to attract the extraction industries.

There was about Vreeland a touch of the naturalist Muir. Vreeland had written a number of excellent articles in *Field and Stream* about the preservation of nature in New England. His "Passing of the Maine Wilderness," in the April 1912 issue, was credited with saving Mount Katahdin (the favorite peak of Thoreau and, later, Roosevelt) from clear-cutting. Although Vreeland failed to get the North Woods of Maine designated as a national park, his indefatigable advocacy contributed to the creation of Baxter State Park (one of the largest in America).[62] The sacred Appalachian wilderness where Thoreau had written *The Maine Woods*, published posthumously in 1909, was secured.

A few years later, in May 1916, Vreeland testified before the House Subcommittee on Public Lands for the establishment of Mount McKinley National Park. An excellent skier and a leader in the Boy Scout movement, Vreeland lectured about the need for American wildlife and for gorgeous wilderness landscapes like Denali to be handed down to future generations to enjoy. Along with Stephen Mather (National Park Service) and Robert Marshall (U.S. Forest Service), Vreeland was the most effective conservationist to testify that afternoon on Capitol Hill. Passionately defending Sheldon's field research on the Denali wilderness, Vreeland helped convince U.S. congressmen that Mount McKinley was irreplaceable.[63]

Vreeland—in his forties, always meticulously dressed with not a wrinkle in his clothes—considered himself more of a "camera naturalist" than a hunter or an angler. Growing up, he had hunted in Maine and Quebec. Like Hornaday, however, he recoiled from trophy hunting as he matured. One of his closest friends was Daniel Beard, a founder of the Boy Scouts of America; they frequently challenged each other in learning all the birds and trees of the Adirondacks. After graduating from Stevens Institute of Technology in 1895 and Columbia University in 1909, Vreeland made a fortune inventing and patenting dozens of electrical devices, including the sine-wave oscillator, a radio band selector, and the Vreeland spectroscope. He earned further renown for photographing the grizzlies of Yellowstone in their lairs. Roosevelt considered him the best wildlife photographer around, the best landscape and portrait photographer being Edward Curtis. An expert cartographer, Vreeland also mapped the mountains between the Peace and Fraser rivers in British Columbia and Alberta. In 1915 Vreeland Glacier was named in his honor by the Canadian government.[64]

Macnab and Vreeland shared at least two ideas: the CFCA's belief that wilderness defined the American character; and the certainty that market hunting, overfishing, and poaching were reprehensible acts of debauched scoundrels. Committed to the outdoors life, they saw the Lake Clark region along Cook Inlet as a first-rate locale where hardy sportsmen of the CFCA could go in the summer to camp, hike, run rivers, fish, and maybe shoot a few ducks for dinner. The *real* Alaskan fishermen—both Euro-American and Native Alaskans—were good marine stewards of nearby places like Bristol Bay, Kachemak Bay, and the Shelikof Strait. The CFCA thought the resident fisherman should have a self-imposed limit of two to five halibut a day. And any fish over 100 pounds, unless they were trying to win a contest, had to be released; it was obviously a female full of eggs. So the fishermen of Seldovia, reels down, would bring the halibut and salmon to the dock, clean up, and go home. Fair fishing made sense to most of them. But the CFCA rejected the Seattle and San Francisco fishing companies that depleted the salmon waters around Bristol Bay for a single season's profits. Such "fake fishermen" were bad actors, anticonservationists, greedy money-grubbers. It was an uphill battle because Alaskan politicians cared only about lining their own pockets with fast money.

That Roosevelt, Muir, and Sheldon had inspired men of high character such as Macnab and Vreeland to join the wilderness movement was heartening. Conservation was proving to be more than a mere fad or an obsession with the outdoors. A U.S. Army colonel (hunter) and a famous inventor (photographer), modeling their advocacy on the campaign to preserve Mount McKinley, had set their eyes on exploring Lake Clark. Once again, Merriam and Nelson of the Biological Survey were offering wise counsel on what flora and fauna Vreeland needed to collect for the National Museum.[65]

Vreeland and Macnab plotted their Lake Clark–Iliamna adventure like military logisticians, determined to open up the Cook Inlet region to hunters, hikers, and recreationists who just wanted to experience the wild (or to vacationers who liked the idea of seeing treasured Alaskan landscapes). As CFCA survivalists straight from upstate New York, they were determined to reach the headwaters of the Mulchatna River.[66] They were "extreme sportsmen" long before the phrase came into vogue.

Chapter Six

. . . . . .

OUR VANISHING WILDLIFE

I

L osing the presidential election made Roosevelt an even more revolutionary conservationist. In January 1913, he wrote a book review in the progressive opinion journal the *Outlook* that condemned Americans' indifference to wildlife protection and habitat preservation. The review, which served as Roosevelt's own manifesto on behalf of endangered species, was of the zoologist William Temple Hornaday's *Our Vanishing Wild Life*, a scientific consideration of the "appalling rapidity" of global species destruction. What Upton Sinclair's *The Jungle* had been for reform of meatpacking, *Our Vanishing Wild Life* was to the defense of disappearing creatures such as the prairie chicken (*Tympanuchus cupido*), whooping crane (*Grus americana*), and roseate spoonbill (*Platalea ajaja*). The devastation of marine mammals in Alaskan waters was particularly disturbing to Hornaday, director of the Bronx Zoo for thirty years. In his requiem, Hornaday, who had also served as president of the Permanent Wildlife Protective Association, surveyed 100 years of reckless exploitation of American wildlife. The book included a drawing of a tombstone, listing eleven North American bird species that had been "exterminated by civilized man" between 1840 and 1910; among these were the great auk (*Pinguinus impennis*), passenger pigeon (*Ectopistes migratorius*), and Eskimo curlew (*Numenius borealis*). Dedicated to William Dutcher, president of the National Association of Audubon Societies, *Our Vanishing Wild Life* was a mournful alarm intended to educate the public about a continent, if not a world, in biological peril.[1]

Intended to shake up the status quo, *Our Vanishing Wild Life* was published in the unsparing tradition of the investigative journalists Lincoln Steffens (urban politics), Ida Tarbell (Standard Oil), and Ray Stannard Baker (coal miners' union)—a take-no-prisoners assault aimed at saving buffalo, river otters (*Lontra canadensis*), flamingos, and hundreds of other creatures from further diminution. Every page was laden with punctilious zoological facts. Every page was a harassment, a humane cry to abolish coyote wagons, steel traps, and slob hunting. Biological reports, for example, had taught Hornaday that the Bering Sea had once been populated by Steller's sea cow (*Hydrodamalis gigas*), a marine mammal twenty-five feet long and weighing eight to ten tons. By 1768, however, these sea cows, sluggish vegetarians that fed on the great kelp pastures of the Aleutian Islands, were extinct.[2] They had been wiped out by irresponsible Russian market hunters.

Aroused by Hornaday's alarm bell, the Boone and Crockett Club appointed a committee for the protection of Alaska's walrus, fur seals, sea otters (*Enhydra lutris*), and other marine mammals. The Pribilof Islands, the club members believed, should remain a wildlife reserve without the threat of market slaughter of seals, otters, and blue foxes (*Alopex lagopus*).[3] Hornaday conceived *Our Vanishing Wild Life* (published in conjunction with the New York Zoological Society and endorsed by Roosevelt) as a plea to Americans to stop their reckless treatment of their most cherished animal sanctuaries. The book was full of grave assertions, and Hornaday had scores to settle with the American industrial order. Building on court battles fought on behalf of animal rights groups such as the Society for the Prevention of Cruelty to Animals (SPCA) and the National Audubon Society, Hornaday led the way toward the Endangered Species Acts that were finally enacted in 1966, 1969, and 1973. "We are weary," he wrote, "of witnessing the greed, selfishness, and cruelty of 'civilized' man toward the wild creatures of the earth. We are sick of tales of slaughter and pictures of carnage. It is time for a sweeping Reformation; and that is precisely what we now demand."[4]

Hornaday—who was born in Avon, Indiana, on December 1, 1854—did more to save wild creatures from extinction than anyone else of his era. He had been raised on Mayne Reid's adventure stories, such as *Osceola* and *The Plant Hunter*, and he had spent his formative years in Iowa (like Aldo Leopold). He developed a sense of awe for the mysteries of creation. A skilled taxidermist, husbandryman, and animal handler, Hornaday set off around the world; he was hired by museums to collect wildlife specimens in the West Indies, Cuba, Florida, Asia, and South America. Deeply eccentric and stubborn, never flinching from a fight, Hornaday believed there were two types of individuals: those who adored animals and those who didn't. A prolific author—he wrote more than twenty books—Hornaday became the greatest popular zoologist of the late nineteenth century. In 1904 Hornaday's *The American Natural History*, beautifully illustrated textbook, was a huge best seller, educating the lay public about our native wildlife.[5]

As an advocate of animal protection, Hornaday was both unrelenting and potent. His monograph *The Extermination of the American Bison* (1889), for example, was widely credited with finally stopping the random slaughter of bison on the Great Plains. When Roosevelt formed the idea of founding the Bronx Zoo in New York during the 1890s, he chose Hornaday as his chief zoologist. Bold, ornery, and fiercely argumentative, Hornaday, with a closely cropped beard like Robert E. Lee's, was a wizard at describing animal traits with scientific certitude. He worked in tandem with Roosevelt on numerous wildlife protection projects. Together they cofounded the American Bison Society, lobbied for federal laws against the selling of wild game, and endorsed the Weeks-McLean Bill of 1912, which further protected migratory birds against states' rights legislators in the Deep South and the West. Joining forces with Roosevelt and Hornaday was the automaker Henry Ford. "Birds," Ford wrote in a letter asking his dealers to back the Weeks-McLean Bill, "are the best companions."[6]

Roosevelt and Hornaday collaborated shrewdly in protecting the northern fur seal of the Pribilofs and other Alaska rookeries. Unafraid of strenuous language, they said that Taft's U.S. Bureau of Fisheries and his Fur Commission Board were full of "pelagic pirates"—employees essentially in the pockets of the Alaska Commercial Company (which later became the National Commercial Company)—and they forced the U.S. Congress to ban the slaughter of seals. Roosevelt and Hornaday were leaders of the Camp Fire Club of America (CFCA), whose members were disgusted that American women, rejecting farm-bred mink, made seal fur coats the fashion. How grotesquely Russian of them!

What stirred Hornaday and Roosevelt to battle even more was the complicity of the Taft administration in the "murders" of Alaskan seals. At the congressional hearings in 1911 and 1912, the CFCA scored a victory. The Seal Treaty of 1911 was signed by the United States, Britain, Russia, and Japan. It probably saved Alaska's northern fur seal from extinction, and helped save other mammals as well.[7] "The treaty produced a significant dividend: almost as an afterthought it prohibited the killing of sea otters," the historian Frank Graham Jr. noted in *Man's Dominion*. "At that time they were considered extinct or nearly so on our shores. Under protection, that delightful little animal has reappeared, to the nation's aesthetic profit, in some numbers off the coast of California."[8] And there was a healthy, noisy colony of sea otters on Amchitka Island in the Aleutians. The Natives had fled Amchitka, worried about a volcano; this gave the sea otters undisturbed waters, kelp, and shellfish. Amchitka was the greatest sea otter sanctuary left in the world—a fact that Hornaday cited with nationalistic pride.

Following the victory in Congress, Roosevelt and Hornaday delivered the knockout punch—concluding a twenty-eight-year conservationist fight started by Henry Wood Elliot in the 1870s. Congress agreed to ban the slaughter of all seals and otters in all American waters. The *New York Times* declared the victory in 1912 a triumph for the CFCA. "This battle against animal murder for profit was won," the *Times* said. "Congress ordered that no man should kill a seal on American territory for five years. The friends of the seal wanted a ten-year closed season, but they were pretty well satisfied with what they got, for the reason that now the seal-slaughterers are on the run it will not be hard in 1917 to get Congress to give a five-year extension."[9]

Although Hornaday—who liked to be called "Doctor"—had been a hunter all his life, the Alaskan seal slaughter caused him to drop his gun. Nobody in the CFCA—which was filled with sportsmen—held it against him, although there were murmurs that Hornaday had turned soft on animal rights. In Hornaday's mind, it was unseemly for rifle companies such as Winchester and Springfield to donate money to wildlife protection groups. Putting the seal butchers out of business encouraged Hornaday to try to save Dall sheep and caribou in Alaska. "All large hoofed animals have a weak hold on life," Hornaday wrote. "This is because it is so difficult for them to hide, and so very easy for man to creep up within the killing range of modern, high-power, long-range rifles. Is it not pitiful to think of animals like the caribou, moose, white sheep and bear trying to survive on the naked ridges and bald mountains of Yukon Territory and Alaska! With a modern rifle, the greatest duffer on earth can creep up within killing distance of any of the big game of the North." [10] Hornaday went on to say, "I have been a sportsman myself, but times have changed, and we must change also." [11]

Enter Roosevelt again. After siding with Hornaday on the fight of 1911–1912 over protecting seals, Roosevelt now favorably reviewed *Our Vanishing Wild Life* in *Outlook*. This praise, coming from the very popular ex-president, created quite a stir, and put wildlife protection in the forefront of the progressive movement alongside civil rights, women's suffrage, and public education. The Colonel blamed the American people—yes, the people themselves—for the deplorable fact that such birds as the Carolina parakeet (*Conuropsis carolinensis*), passenger pigeon, great auk, Labrador duck (*Camptorhynchus labradorius*), and sandhill crane (*Grus canadensis*) were nearing (or had reached) extinction. An incensed Roosevelt challenged citizens to change their outdated mind-set, to more fully comprehend the farmyard fact that songbirds gobbled up noxious insects and that raptors devoured rodents. As a member of the CFCA, Roosevelt was duty-bound to rid Alaska of overfishing. He wanted the traditional salmon grounds of the Haida and Tlingit of Alaska-Canada protected from corporate canneries. As Darwin taught an entire generation, there was an intricate biological order on Earth that humans barely understood. Both Roosevelt and Hornaday, however, recognized that U.S. wildlife was part of a continental biota; therefore, Mexico and Canada had to be included in all studies. As Robert B. Roosevelt, TR's conservationist uncle, had maintained in the 1870s, dams and barricades and nets erected across rivers had to be stopped to protect the fish runs. "The United States at this moment occupies a lamentable position as being perhaps the chief offender among civilized nations in permitting the destruction and pollution of nature," TR wrote in *Outlook* (for which he was a contributing editor): "Our whole modern civilization is at fault in the matter. But we in America are probably most at fault."[12]

Roosevelt's book review revealed a maturation (or heightening) of his wilderness philosophy. Writing from his command center at the United Charities Building in Manhattan, where his eight-by-ten-foot mahogany desk was considered the Grand Central Terminal of the progressive movement, Roosevelt saw bloodstained evidence that his old enemies (the market hunters, slash-and-burn developers, corporate trusts, anti-conservationists, free marketers, the predatory rich, and corporate despoilers interested only in making money) were behind the rapid decline of wildlife in places like the Alaska Range, the Kenai Peninsula, and the Pribilofs. Insisting that the U.S. conservation movement was largely about the preservation of "noble and beautiful forms of wildlife," Roosevelt wrote that it was "wickedness" to allow companies to "destroy" animals and birds indiscriminately.

One example was the plight of the Alaskan walrus, highly gregarious pinnipeds whose extremely thick hide was coveted by Eskimos, Japanese, and Russians alike in the regions around the north pole. They had breeding grounds in the northern Bering Sea and the Chukchi Sea (including Wrangell Island). During the spring months, walrus were found on pack ice. But come summer some males had hauled themselves onto the shore to molt, becoming easy targets for market hunters. (The females and young, however, remained on the offshore ice. Only during the first decade of the twenty-first century did the female walrus come ashore because, as a result of global warming, there was no ice remaining during the summer in parts of the Chukchi Sea.[13]) With the introduction of semiautomatic weapons, hunters in pursuit of hides and blubber were now slaughtering walrus herds throughout the year, including during breeding season in the offshore Islands. Because walrus were both colonial and highly social, they liked congregating rather than seeking their own space. A male walrus tusk averaged between twenty-five and thirty inches long; these tusks were prized all around the world for their smooth beauty. Market hunting of walrus intensified when whales were overharvested in the Bering and Chukchi seas during the late 1800s: whalers, desperate to recoup lost income, trained their harpoons on walrus. Roosevelt believed that "drastic action" was needed to prevent the extinction of Alaska's walrus, recommending an "absolute prohibition of killing at all."[14]

It was now time for Alaska to make permanent advances in protecting its mammals. The moose season in Alaska, a famous October event, lured scores of hunters from the Lower Forty-Eight, and as winter drew closer, they'd stomp across the autumnal reaches and slay lumbering giants, all in the name of sport. Roosevelt and Hornaday wanted the bag limit on moose immediately reduced by 50 percent. They even wanted the Tlingit, Tanana, and Ahtna to reform their ancient ways of hunting. "The indolent and often extortionate Indians of Alaska—who now demand 'big money' for every service they perform—are not so valuable as citizens that they should be permitted to feed riotously upon *moose, and cow moose at that*," Hornaday fumed, "until that species is exterminated."[15]

To members of the Sierra Club, CFCA, New York Zoological Society, and National Audubon Society, Roosevelt's critique of American indifference toward wild animals was a heady wine. John Muir—who had escorted William Howard Taft around Yosemite in October 1909—might have danced a jig when he read Roosevelt's words in the *Outlook*, telling citizens to "wake up" to the "damage done by the migratory sheep bands" that were permitted to "pasture on, and to destroy the public domain." (Muir, even though he had once been a shepherd, famously called domesticated sheep "hoofed locusts.") Pinchot was pleased that the Colonel was still going after thugs. Hornaday's prescient book, in fact, had given Roosevelt an array of devastating statistics for making his conservationist case. But to the Republican regulars, still bitter that Roosevelt had wreaked havoc on the party in 1912, the review was another indication that TR had become a wild man. "Crazy Teddy" was more interested in the ability of sea otters to raid oyster beds in the Alexander Archipelago than in the ability of hardworking Cordova coal miners to earn a living for their families.[16] Roosevelt shot back defiantly that at least he wasn't "guilty of a crime against our children," the handing down of a "wasted heritage."[17]

Roosevelt applauded the isolated efforts of some states to protect wildlife populations, such as Montana's attempts to save bison and Alaska's efforts to protect fur seals. According to Roosevelt, Vermont—the home state of the conservationists George Perkins Marsh and Charles Sheldon—had been heroic in managing its white-tailed deer population. But as a whole, the United States had a woeful record with regard to big-game preservation. To Roosevelt's dismay, Territorial Governor Walter Eli Clark of Alaska, a trigger-happy boomer without a conservationist bone in his body, was trying to abolish the law on Kodiak Island protecting brown bears. Clark's attempt was for the supposed benefit of settlers, but if bears were a problem, *control* them, Rooseveltians insisted; don't market-slaughter them for profit. The Boone and Crockett Club formed yet another committee—led by Charles Sheldon—to save Alaska's brown bears, to treat them as game to be managed, not predators to be slaughtered. "The brown bears are the greatest attraction to visiting sportsmen in Alaska, and as living animals, are worth infinitely more to natives and the white population of Alaska," Madison Grant, of the Hunt Club, wrote to Dr. E. Lester Jones, of the Department of Commerce, in 1915, regarding a bill then before Congress proposing to transfer the care of Alaska's brown bears from the Biological Survey (which offered protection) to the Bureau of Fisheries (which would eliminate them as a nuisance to the industry).[18]

Alaskans who loved the outdoors life needed to undertake a relentless war against air polluters, land degraders, and market hunters. There should be no cowering or compromising with regard to species extinction. "The wild antelope and the prairie chicken are on the point of following the wild bison and the passenger pigeon into memory," Roosevelt said. "Our rich men should realize that to import a Rembrandt or Raphael into the country is in no shape or way such a service at this moment as to spend the money which such a picture costs in helping either the missionary movement as a whole, or else parts of it, such as the preservation of the prongbuck [pronghorn antelope, *Antilocapra americana*] or the activities of the Audubon Society on behalf of gulls and terns."[19]

Championing individual species in peril, Roosevelt lamented the declining populations of the whooping crane, bald eagle, and California condor (*Gymnogyps californianus*). Taking a step in the right direction, New York had recently passed the Audubon Plumage Law of 1910, banning the sale of plumes of all native birds for the millinery trade.[20] Roosevelt was nevertheless concerned that the Atlantic puffins (*Fratercula arctica*) off the coast of Maine, which had distinctive black-and-white plumage and a colorful, almost clownlike beak, had been extirpated. American citizens, he argued, shouldn't have to travel all the way to Newfoundland or Labrador to see a puffin breeding ground. Roosevelt called for "international agreements" among all the nations of the western hemisphere to "put down the iniquitous feather trade." This was a direct jab at ex-president Taft for having canceled the World Conservation Congress. As Roosevelt said in *Outlook*, it was "inconceivable" that "civilized people should permit [this feather trade] to exist."[21] To Roosevelt, the "bird cities" in the 1,200-mile Aleutian chain, where three species of cormorants existed, along with colonies of murres, auklets, kittiwakes, and glaucous-winged gulls (*Larus glaucescens*), constituted one of God's great spectacles.[22]

Roosevelt was struck by a chapter in *Our Vanishing Wild Life* called "The Guerrillas of Destruction." In military contexts, a guerrilla fighter is one who refuses to recognize civilized rules of engagement. In Hornaday's mind (and in Roosevelt's), hunters and plumers who ignored the sportsman's ethos were like guerrillas. Oology, the collecting of bird eggs by the thousands, had to be banned. Hornaday did an impressive job of describing the culprits, identifying many by name. He aimed an entire chapter at Italian immigrants who had brought an Old World practice of market slaughter to the New World. Hornaday itemized what types of dead birds a consumer could purchase in a Venetian or Florentine market, and the chapter made for grim reading. According to Hornaday, the American South was also willfully ignoring game laws. Robins were being systematically shot and eaten in Mississippi, North Carolina, South Carolina, Tennessee, Maryland, Texas, and Florida by the hundreds of thousands. In Dallas, Texas, a man named F. L. Crow led torchlight bird hunts along the Trinity River; on one occasion, his group killed 10,517 birds in slightly over two hours—just for the hell of it. Roosevelt's friend Edward A. McIlhenny, owner of the company that made Tabasco sauce, complained that on Avery Island, Louisiana, 10,000 robins a day were slaughtered to be sold at roadside stands in the nearby town of New Iberia for ten cents apiece. "We must stop all the holes in the barrel," Hornaday fumed, "or eventually lose all the water. No group of bird-slaughterers is entitled to immunity."[23]

Hornaday offered gruesome capsule biographies of the "guerrillas," whom he identified by name. With Roosevelt's strong approval, in fact, Hornaday issued an Eleventh Commandment, an "inexorable law" that every generation of American conservationists needed to absorb: *No wild species of birds, mammals, reptile, or fish can withstand exploitation for commercial purposes.*[24] In Alaska this meant that the harvesting of northern fur seals and sea otters had to be curtailed. The Aleutian Islands Reservation was established for that purpose in 1913, to end the "exhaustion" of wildlife resources. Furthermore, the U.S. Bureau of Biological Survey (which would become the U.S. Fish and Wildlife Service in 1940)—following the model established by TR at Pelican Island, Florida, in 1903—created a reindeer reserve on Alaska's Unalaska and Umnak islands.[25]

Another virtue of *Our Vanishing Wild Life*, from TR's perspective, was that Hornaday described the pioneering accomplishments of the Roosevelt administration in species protection. (Gifford Pinchot, by contrast, was focused on forestry and had failed to recount these federal bird reservations in his memoir, *Breaking New Ground*.)* Hornaday explicitly praised Roosevelt for saving Wind Cave, the Grand Canyon, Crater Lake, and Mesa Verde (among other American wonders), and detailed how the activist warrior of the Antiquities Act of 1906 had fought to save such treasures as Jewel Cave, Montezuma Castle, Tumacacori, El Morro, Chaco Canyon, the Gila Cliff Dwellings, Muir Woods, Pinnacles, Cinder Cone, and Lassen Peak. Roosevelt had established the national monument designation as a sort of way station to protect areas he hoped would eventually become national parks. Two of Alaska's most spectacular national parks—Katmai and Glacier Bay—were monuments first. Other impressive national parks, such as Washington's Olympic, Arizona's Grand Canyon, and California's Death Valley, also began as national monuments.

Hornaday also included a chart of the fifty-one federal bird reservations created by Roosevelt from 1903 to 1909, and credited the ex-president with developing the U.S. government's wildlife protection ethos by way of the Boone and Crockett Club and the National Association of Audubon Societies. In Alaska alone, Roosevelt's bird sanctuaries—Tuxedni (Chisik

---

* Pinchot had ghostwritten some of the chapter on conservation in Roosevelt's *An Autobiography*, published with great fanfare in 1913. He self-servingly focused the book on the more than 150 national forests he and Roosevelt had founded together from 1901 to 1909. Pinchot viewed Hornaday as a bomb-thrower, constitutionally incapable of moderation or calm bureaucratic infighting.

and Duck islands in Cook Inlet), Saint Lazaria Island, Bering Sea (Saint Matthew Island Group), Pribilof (Walrus and Otter islands), Bogoslof, and the vast marshlike Yukon Delta—would eventually become parts of two national wildlife refuges: the Alaska Maritime NWR and Yukon Delta NWR. "These reservations," Hornaday wrote, "are of immense value to bird life, and their creation represents the highest possible wisdom in utilizing otherwise valueless portions of the national domain." [26]

The Alaskan wilderness was, unquestionably, still an Eden-like paradise in 1913, what a future U.S. Fish and Wildlife director, Ira N. Gabrielson, would call a "living zoological museum." [27] But that positive assessment didn't take account of the seal, otter, and walrus rookeries, which were under assault by market hunters. Using statistical graphs, Hornaday made vividly clear in *Our Vanishing Wild Life* the high percentages of walrus and seal populations in jeopardy. The prognosis for species survival was unfavorable. In Hornaday's mind (as in the minds of Roosevelt, Sheldon, and other conservationists), there were "fatal defects" in Alaskan game laws circa 1913. For example, as part of a reparations strategy, First Nation tribes enjoyed an exemption from bag limits in Alaska. Tribes were legally allowed to shoot anything that moved. Hornaday recounted the experience of the conservationist and hunter Frank Kleinschmidt at Sand Point on the Kenai Peninsula: he saw eighty-two caribou tongues piled up in a Native Alaskan's canoe, brought to market to sell for fifty cents apiece. He was aghast at this casual carnage. "The carcasses were left where they fell, to poison the air of Alaska," Hornaday wrote of the market hunters. In contrast, he praised the outcome of regulated sports hunting: "Thanks to the game law, and five wardens, the number of big game animals killed last year in Alaska by sportsmen was reasonably small—just as it should have been." [28]

Both Hornaday and Roosevelt were adamant that Sitka deer (*Odocoileus hemonius sitkensis*), which lived in southeastern Alaska, be allowed to roam thousands of miles on protected U.S. government land unmolested by market hunters. They were part of what Roosevelt called America's "deer family." The U.S. Department of the Interior had an obligation, they believed, to allow only a very limited hunting season for Sitka deer in the Tongass and Chugach national forests. Such a position was not viewed favorably by Alaska's residents, many of whom believed the federal government had no right telling a citizen of the territory what he could or couldn't shoot. Game management seemed to them like something conceived by Karl Marx. An ex-governor of Alaska, in fact, explicitly protested that Rooseveltian conservation with regard to Sitka deer and moose was socialistic. In a rugged territory like Alaska, the argument went, a man had a right, under the Second Amendment, to follow a buck and pull the trigger. "The preservation of the game of Alaska should be left to the *people* of Alaska," a territorial ex-governor argued. "It is their game; and they will preserve it all right!"

In *Our Vanishing Wild Life*, Hornaday outlined the flaws he saw in that stance against the federal government:

1. The game of Alaska does *not* belong to the people who live in Alaska—with the intent to get out tomorrow!
2. The preservation of the Alaskan fauna on the public domain should not be left unreservedly to the people of Alaska because . . .
3. As sure as shooting, they will *not* preserve it! [29]

Hornaday wanted the sale of *all* game to be prohibited in Alaska: even an Arctic prairie billy (a Euroamerican subsistence settler) or an Eskimo should be allowed to shoot only what he or she would personally eat. This was a very extreme, uncompromising stance. Hornaday and Roosevelt believed that market hunters, such as those who were killing off Bering Sea walrus for ivory and hides, should be arrested. Roosevelt also wanted to quadruple the number of wildlife wardens in Alaska. To protect seal rookeries, the Rooseveltian conservationists wanted taxpayers to provide the Biological Survey with two state-of-the-art vessels to patrol the 34,000 miles of Alaskan coastline. Poachers should be arrested, tried, convicted, and imprisoned. Congress, these conservationists argued, should *immediately* appropriate $50,000 for increased law enforcement to protect Alaskan wildlife. The sportsman's code was coming to Alaska. "It is no longer right nor just for Indians, miners, and prospectors to be permitted by law to kill all the big game they please," Hornaday wrote, "whenever they please." [30]

Alaska's declining bear population was also worrisome. There were no biological underpinnings to Alaska's policies for controlling predators; just shoot what moved. Once statehood was achieved in 1959, Alaska's bear population was appropriately managed by the Alaska Department of Fish and Game (ADFG) and the Division of Wildlife Conservation established by the Board of Game (BOG). But in 1913, it was open season all 365 days of the year for the rancher-prospectors whose tools were rope, harness, sheep dip, branding iron, nail kegs, sledgehammers, and hunting rifles. Although the smaller black bear (*Ursus americanus*) still wandered across coastal and interior Alaska, intriguing subspecies such as the blue bear (*Ursus arctos pruinosus*) of the Saint Elias Mountains were in decline. The coastal ranges were thick with brown bear subspecies, with variations depending on geography: Kodiak bears (on Kodiak), Kidder bears (on Alaska Peninsula), the Admiralty bear (on Admiralty Island), and the Sitka bear (on Baranof Island). Mammalogists were working around the clock trying to create a brown bear sanctuary on Admiralty Island in Alaska to help these mammals survive market hunting.[31]

"I think that the attention of the Game Committee of the Boone and Crockett Club should be called to the very dangerous situation as regards bears of Alaska," Charles Sheldon wrote to George Bird Grinnell in 1918, "which, at any time, may be threatened with extermination in the coast region."[32]

While Sheldon was the point man for protecting Alaska's bear populations, Grinnell had become the established voice on properly managing the territory's salmon. The problems were many. To Grinnell's utter horror, Alaskan fishermen would shoot any bear they encountered along a stream or shoreline because the bruins were competing with their commercial nets, lines, and traps. Grinnell told how, adding insult to injury, Alaskan fishermen used only about 20 percent of the salmon they caught, keeping only the choice belly meat and discarding the rest. To Grinnell, a veteran of the conflicts of 1880 to 1909 over protecting bison, the Alaskans' professed belief—mistaken and possibly disingenuous—that salmon were abundant was all too familiar.[33] According to Grinnell, if Alaskan fisheries weren't managed properly, the salmon—sockeye, chinook, coho, pink, and chum—would die out.

What really set Roosevelt's teeth on edge wasn't just the vanishing bear and salmon populations. It was also President Woodrow Wilson's cavalier attitude toward the Tongass and Chugach national forests; it suggested cowardice (like Taft's) masquerading as blissful superiority. Wilson, a bespectacled Princetonian indoorsman, had the temerity to dismiss better-informed outdoorsmen who argued that the federal government should save vast swaths of wild Alaska for future generations. Roosevelt— who thought most Alaskan lands should be federally owned—seethed when Wilson delivered his first state of the union address in December 1913, sounding like a pitchman for Morganheim. "Alaska as a storehouse, should be unlocked," Wilson announced. "We must use the resources of the country, not lock them up."[34]

Chapter Seven

THE LAKE CLARK PACT

I

In Albuquerque, New Mexico, twenty-six-year-old Aldo Leopold—whose philosophy was the antithesis of Wilson's "unlock the storehouse" approach to natural resource management—felt liberated by *Our Vanishing Wild Life*. It had the same galvanizing effect on him that *Uncle Tom's Cabin* had on William Lloyd Garrison and the other abolitionists of the pre–Civil War generation. A whole new way of considering wildlife rights infused Leopold. No longer would details of policy or a political balance swamp his conservationist principles. Smoking his omnipresent pipe, carefully reading every line of Hornaday, he thought about all the animals he had seen slaughtered in the Flint Hills of Kansas, along the Mississippi River near Davenport, by market hunters. He thought of how the Midwest lowlands he so loved had been skinned by one-crop agriculture. Determined to make Carson National Forest of New Mexico his Walden Pond, Leopold was evolving into a combination of Thoreau (preservationist), Pinchot (forester), and Hornaday (advocate of wildlife protection). "The book galvanized Aldo's conviction," Leopold's biographer Curt Meine wrote. "Never before had the case for game protection been so alarmingly stated. Never before had the argument been made so strongly that man bore a moral responsibility for the preservation and perpetuation of threatened game species." [1]

Later, in the early 1930s, when Leopold was writing *Game Management*, inspired by *Our Vanishing Wild Life*, he explained how "the crusader"

William Temple Hornaday had affected his thinking: "He insisted that our conquest of nature carried with it a moral responsibility for the perpetuation of the threatened forms of Wildlife. This avowal was a forward step of inestimable import. In fact, to anyone for whom wild things are something more than a pleasant diversion, it constitutes one of the milestones in moral evolution."[2]

That same spring of 1913, when *Our Vanishing Wild Life* was published, Theodore Roosevelt left Oyster Bay by train to explore the Southwest. He first spent time in southern New Mexico. The Roosevelt party then moved into El Tovar Hotel on the south edge of the Grand Canyon. Roosevelt's reasons for coming to Arizona were many. One was that he hoped Grand Canyon National Monument—which he had saved during his presidency, by an exective order in 1908—could be upgraded to a national park. The whole Kaibab Plateau was a wildlife paradise. Charles Sheldon, in fact, had spent much of 1912 studying the habits of bighorn sheep (*Ovis canadensis*) in the inner gorge of the Grand Canyon for the U.S. Biological Survey. "The sheep here act exactly like all the northern sheep I have ever seen—very watchful and alert," Sheldon wrote in his Havasupais field journal on November 24, 1912. "Sheep (at least a few) probably go up the rim when the snow melts to get green food which may not grow down in the canyon until later. I have only seen two lambs. There are no enemies of sheep here, except golden eagles. The bobcats are so scarce as to be negligible."[3]

With Roosevelt at the Grand Canyon were his two youngest sons, Archie and Quentin. The guide, cook, and horse wrangler was Jesse Cummings of Mesa, Arizona. In the days to come, the bristly-bearded Cummings, a native of Kentucky, would repeatedly impress the party, and Roosevelt in particular, with his expertise in this terrain. He had traveled from the Alleghenies to the western prairies and had never gotten lost. Cummings skillfully shepherded the Roosevelts toward a bank of the serpentine Colorado River where white-water rapids had cut gorges through rock for aeons. He continually pointed out colorful bird species such as mountain bluebirds, juncos, and chickadees—and homely ones, too. And Cummings, it turned out, could procure anything in the way of supplies; he was like an army quartermaster with the Midas touch.[4]

True to form, Roosevelt slept outside his tent more often than inside it.

The riparian coyote willow, arrow weed, seep willow, and western honey mesquite were like tonics. Although Roosevelt wrote about coyotes (*Canis latrans*) and cougars (*Puma concolor*) during this Grand Canyon journey, and wanted the boys to hunt these predators, his own eyes seemed more attracted to the wildflowers and birds. He was eager to share his own counts of Grand Canyon wildlife with Sheldon, proud that he was adding to the U.S. Biological Survey's cataloging of the Southwest. "Although we reached the plateau in mid-July, the spring was just coming to an end," Roosevelt wrote. "Silver-voiced Rocky Mountain hermit-thrushes [*Catharus guttatus*] chanted divinely from the deep woods. There were multitudes of flowers, of which, alas! I know only a very few, and these by their vernacular names, for as yet there is no such handbook for the flowers of the southern Rocky Mountains as, thanks to Mrs. Frances Dana, we have for those of the Eastern United States, and, thanks to Miss Mary Elizabeth Parsons, for those of California."[5]

Roosevelt's prose from the Grand Canyon in the *Outlook*, later collected in *A Book-Lover's Holidays in the Open*, was unusual for its ease and impressionistic quality.[6] His tone had tempered and softened considerably since he wrote his Dakota trilogy of the 1880s, and certainly since he wrote the gory *African Game Trails*. He now conveyed a feeling of tranquillity and harmony. Portraits and photographs from the southwestern trip, in fact, seem to confirm this alteration, capturing a less strident-looking Roosevelt—the hard lines of his famous grimace are somewhat softened by traces of a smile. The hats he wore were more floppy, no longer crisp and uncreased. He was playing the father and uncle. Roosevelt had always been a child of nature: this new Roosevelt seemed to verge on beatific pastoralism. The reader of his essay on the Grand Canyon, which appeared in *A Book-Lover's Holidays in the Open*, is almost relieved when Roosevelt finally betrays a familiar ferocity, snapping at the despoilers of nature like a provoked grizzly bear: "Continual efforts are made by demagogues and by unscrupulous agitators to excite hostility to the forest policy of the government, and needy men who are short-sighted and unscrupulous join in the cry, and play into the hands of the corrupt politicians who do the bidding of the big and selfish exploiters of the public domain. One device of these politicians is through their representations in Congress to cut down the appropriation for the forest service."[7]

One national forest Roosevelt surely had in mind in 1913 was Alaska's Chugach. In the coming months a bill was introduced in Congress to dissolve the Chugach National Forest. According to two U.S. senators—Wesley Jones of Washington and Thomas Walsh of Montana—the Forest Service was thwarting the economic development of Alaska. Likewise, the territorial government issued a report declaring that the Chugach was an example of abuse by the federal government. The commercial timber industries, these politicians argued, should be given free rein in the Chugach. Backing this campaign to abolish the Chugach was Secretary of the Interior Walter Fisher, who wanted an Alaska commission created to lease out the land for timbering. Luckily, the U.S. Forest Service still had a lot of conservationists willing to wage an all-out war over the Chugach.[8]

Roosevelt dutifully dispatched notes from the Grand Canyon, the Petrified Forest, and Utah's Rainbow Bridge for the *Outlook*, and Leopold was riveted by TR's words, amazed that the ex-president had spent time in Deming, New Mexico, an afternoon's drive from Carson National Forest.[9] When Hornaday came west to Albuquerque in 1915 on a book tour, orating with holy-roller fervor, Leopold was in the audience cheering his every word. A mesmerizing showman, full of the indignant rage of a true believer, Hornaday showed horrific slides of seals being slaughtered, clubbed, and skinned alive. The images were so gruesome that even New Mexican sportsmen in the audience, accustomed to blood and guts, winced. A cowboy hat was passed around to collect money for Hornaday's Wildlife Protection Fund (used to pay legal fees in his successful battle against the U.S. Department of Commerce and Labor for using unethical practices to hunt marine mammals in Alaska). Leopold asked Hornaday to inscribe both *Our Vanishing Wild Life* and a copy of his newest book, published by Yale University Press, *Wild Life Conservation Theory and Practice*.[10] "To Mr. Aldo Leopold," Hornaday wrote in the latter book: "On the firing line in New Mexico and Arizona."[11]

But Roosevelt, Pinchot, and Leopold's style of "wise use" conservationism was on the firing line in California. John Muir had expended all his vitality, futilely, in trying to save Hetch Hetchy, at Yosemite National Park, from being destroyed by a dam. It perplexed Muir why the people who espoused the "Roosevelt doctrine" couldn't see that Hetch Hetchy was one of the priceless Rembrandts or Raphaels the ex-president had

written about in *Outlook*—a national treasure to be protected and preserved. Throughout 1913, congressional hearings had considered the
pros and cons of building O'Shaughnessy Dam and thereby flooding the
Hetch Hetchy Valley to create a reservoir. Because Hetch Hetchy was part
of Yosemite National Park, an act of Congress would be required to build
a dam. Unfortunately, President Wilson had selected a former San Francisco city attorney, Franklin Lane—an advocate of the dam—as secretary
of the interior. Lane was actually a conservationist-minded lover of national parks. But he was no good on Hetch Hetchy. Muir used eloquent
language about Hetch Hetchy: he said it was a "mountain temple" under
attack by "despoiling gainseekers" and "mischief-makers of every degree
from Satan to supervisors, lumbermen, cattlemen, farmers, etc., eagerly
trying to make everything dollarable." This was powerful stuff. Also, U.S.
senators received bags of mail, echoing Muir, urging them not to destroy
the lovely Hetch Hetchy.[12]

But by the end of 1913 Congress, after intense debate and deliberation,
passed the Raker Bill, which approved the flooding of the Hetch Hetchy
Valley. President Wilson signed the bill on December 19. Disappointed by
the death warrant for his beloved Tuolumne Yosemite, an exhausted Muir
hoped that "some sort of compensation must surely come out of this dark
damn-dam-damnation."[13] The following year Muir hiked in the Hetch
Hetchy Valley for the last time before the huge, groaning construction vehicles entered the national park. On Christmas Eve 1914, Muir died. Many
of his loyal supporters claimed that his tireless work to protect Hetch
Hetchy had impaired his immune system and thus lowered his resistance
to disease. The Sierra Club, his lasting institutional legacy, attempted
to obtain legal injunctions, but construction of the O'Shaughnessy Dam
nevertheless commenced. In 1923, at the cost of billions of dollars and
the loss of sixty-eight lives, the dam was completed. Muir, before his
death, had felt defeated by the "despoiling gainseekers" intent on taking
"pocket-filling plunder" from his beloved Sierra Nevada.[14]

The death of Muir was like a body blow to Americans who loved the
great outdoors. Muir's lungs and legs were strong until the end; so to his
wide circle of friends his demise from pneumonia was a surprise. He had
seemed uncollapsable, imperishable, as if his enthusiasm would spill
over mountaintops forever. But although the corporeal Muir was gone,

his exaltation of the wilderness remained timeless, influencing every environmentalist for decades to come. His legacy—the Sierra Club—was stronger than ever. What had worried Muir most was that America, his hallowed land, was being recklessly destroyed by developers. "Even the sky," Muir noted, "is not safe from scathe."[15]

Muir's concern wasn't just for preservation of the land, but also for the people who were victimized by oil drillers and strip miners. Large investment banks, such as Barnette's Washington-Alaska Bank (with headquarters in Seattle), were starting to ship heavy dredging equipment to the territory. There was an array of new players, Alaska Petroleum and Coal, Clarence Cunningham, Amalgamated Development, Saint Elias Oil, and Alaska Coal Oil among them. As a rule, Muir used to say, wherever an extraction company owned a town, the long-term future of the community was bleak. In Nevada, not far from where Mark Twain saw the "celebrated jumping frog of Calaveras County," Muir once encountered a mining boomtown that had, seemingly overnight, turned into a ghost town. According to Muir, only one man remained: "a lone bachelor with one suspender."

But Muir, a true believer, never touched by pessimism or despondency, was fearless about passing from the Earth. All over California, friends of Muir wept because they would never again see him picking berries or leaning on a walking stick. The following year the John Muir Trail was established to honor the Sage of the Sierras, running 200 miles at high altitude from Yosemite Valley to Mount Whitney.[16] "Ordinarily," Roosevelt wrote in *Outlook*, "the man who loves the woods and the mountains, the trees, the flowers, and the wild things, has in him some indefinable quality of charm which appeals even to those sons of civilization who care for little outside of paved streets and brick walls. John Muir was a fine illustration of this rule. He was by birth a Scotchman—a tall and spare man, with the poise and ease natural to him who has lived much alone under conditions of labor and hazard. He was a dauntless soul, and also one brimming over with friendliness and kindliness."[17]

The words *Hetch Hetchy* became important to conservationists in Alaska. The name was a rallying cry like "Remember the Alamo!"—a call to protect Alaska's legacies, such as Glacier Bay and Lake Clark, from meeting a similar fate. If Yosemite National Park wasn't safe from desecration, then neither was Mount McKinley National Park or Tongass National Forest or Yukon Delta Federal Bird Reservation. Federal protection was a sham, and *permanency* was an elastic or slippery term that depended on the whim of Congress and the White House. The wilderness movement seemed to be losing momentum, at least in America. In November 1913, only a few months after Hornaday's *Our Vanishing Wild Life* was published, an international conference for the protection of wild places was convened in Basel, Switzerland. Sixteen nations discussed issues of global conservation and wildlife protection; but the Wilson administration had, inexplicably, refused to participate. A worldwide movement was under way to start protecting special places in every nation as something akin to the present World Heritage sites. The United States was no longer leading the world in the conservation revolution that Muir, Burroughs, and Roosevelt had popularized.[18]

Posthumously, however, Muir gave the Alaskan wilderness movement—and Rooseveltian conservation in general—a powerful boost in the age of automobiles. In 1915, Houghton Mifflin published Muir's memoir *Travels in Alaska*, modeled on Thoreau's *Cape Cod* and *Maine Woods*. It began with Muir steaming out of San Francisco on the *Dakota* in 1879, then up past glorious Seattle all the way to Sitka. Muir vividly recounted his adventures along the Alexander Archipelago and beyond. Voyaging northward, he wrote of whales ("broad back like glaciated bosses of granite heaving a lot in near view, spouting lustily, drawing a long breath, and plunging down home in colossal health and comfort") and porpoises ("a square mile of them, suddenly appear, tossing themselves into the air in abounding strength and hilarity, adding foam to the waves and making all the wilderness wilder").[19]

The Grand Canyon and the Great Smoky Mountains have never found their bard, but Muir delivered for Glacier Bay in *Travels in Alaska*. Suddenly, in 1915, the glacier rambler of 1879 was very much alive; his enthusiasm gushed forth from *Travels in Alaska* with the force of Niagara Falls. In the memoir Muir's wise take on Glacier Bay—both landscape and wildlife—stands as a high point of American travel literature: "To the lover of pure wilderness Alaska is one of the most wonderful countries in the world. No excursion that I know of may be made into any other American wilderness where so marvelous an abundance of noble, new born scenery is so charmingly brought to view as on the trip through the Alexander Archipelago to Fort Wrangell and Sitka."[20]

Muir's approach to nature was that of the "wandering eye." Calculations were made, in *Travels in Alaska*, of the discharge of glaciers, gravel deposits, and the search for wild mutton. The gray mundane flashed with the same cerebral insight as garden spots lit with the bright colors of epilobium, saxifrage, and sedges. Place-names like Sam Dum Bay, Taylor Bay Glacier, Mount Fairweather, and Island of the Standing Stone were given prominence. Religious imagery was offered, but in the subtlest ways. "A pure-white iceberg," Muir wrote, "weathered to the form of a cross, stood amid drifts of kelp and the black rocks of the wave-beaten shore in sign of safety and welcome."[21]

The Presbyterian minister S. Hall Young was among those who couldn't accept the fact Muir had died. To Young, the gray-bearded naturalist was eternal, a sequoia tree destined never to topple. At age sixty Muir was still climbing mountains, undertaking dangerous journeys through the wild lands of California. Instead of slowing down at seventy, Muir took extended voyages to South America and Africa. All his books—*Mountains in California*, *Our National Parks*, and *The Yosemite* among them—radiated youthfulness. Wanting to eulogize Muir, as ministers are apt to do, Young published his reminiscences about their days together going up the Inside Passage, titled *Alaska Days with John Muir*, later that year.

"I cannot think of John Muir as dead, or as much changed from the man with whom I canoed and camped," Young wrote. "He was too much a part of nature—too natural—to be separated from the mountains, trees, and glaciers. Somewhere I am sure, he is making other explorations, solving other natural problems, using that brilliant, inventive genius to good effect; and sometime again I shall hear him unfold anew, with still clearer insight and more eloquent words, fresh secrets of his Mountains of God."[22]

II

Charles Sheldon had initially been considered the next Rooseveltian leader, but it was Aldo Leopold who eventually led the conservationist movement—in his low-key, deeply honest, visionary, academic way—after John Muir died. In 1917 Leopold was thirty and good to look at, with a deep wrinkle between his eyes and a high forehead. He was in good trim and balding. Every day Leopold's conservationist convictions grew stronger and his controlled writing style more lyrical. Leopold never wrote a florid line in his life. Energized by Hornaday's book and by Roosevelt's dispatches to the *Outlook* from the Southwest, Leopold spearheaded the New Mexico Game Protection Association (NMGPA)—an unusual step, considering that he was an employee of the U.S. Forest Service. Sick of politicians' blather, Leopold demanded that New Mexico's game law *always* be enforced the same way. If you poached a white-tail in the Carson National Forest, for example, jail time should be imposed, no matter who was governor in Santa Fe. Inspired by Roosevelt's effort as governor of New York in 1899–1900, Leopold now claimed that a head game warden should be appointed in New Mexico, an overseer independent of political parties. Using *The Pine Cone*, a newsletter, as his megaphone, Leopold also called for new federal wildlife refuges, known as the Hornaday plan. However, unlike Hornaday, who saw refuges as places where hunting was illegal, Leopold hoped these federal reserves would be places that produced wild game *for* sportsmen. Regardless of this difference, the two men were brothers in arms for the cause: wildlife protection.[23] The Hornaday plan failed to pass Congress, but a step had been taken toward the Wilderness Act of 1964.

Conservationist circles in America during World War I were like an underground railroad, with an inexhaustible spirit. The members passed along circulars, newsletters, and correspondence, much as the Y2K generation would later do on the Internet. Nature was wounded in forestlands and waterways, and conservationists were vigilant in starting the healing process. A ranger in the Tongass knew intimately what a game warden in Okefenokee Swamp was up to. It was much more than gossip, or a grapevine. Facts about birds, insects, mammals, and trees were traded. The bourgeois were belittled for never turning down a dollar, for their predictable greed, avarice, and overconsumption. The conservationists praised the legacy of both Muir and Pinchot. There was a growing post-Darwinian belief that the natural world held the key to unlocking the mysteries of man. Among the U.S. Forest Service publications that were being privately printed across the country, Leopold's *The Pine Cone* was the most audacious. It became mandatory reading for all those in the outdoors world, including Theodore Roosevelt.

A letter that Roosevelt sent to Leopold in 1917 has, over the decades, become the connective tissue between his and Leopold's generations of conservationists. Leopold received it courtesy of the U.S. Postal Service in his mailbox at Albuquerque, and it was as unexpected as the snowy owl Roosevelt had shot in Long Island many years earlier. It was neatly typed and quite brief. But to Leopold it was a stamp of approval for his career, as when Thomas Edison told the young Henry Ford at the Oriental Hotel on Long Island that the gasoline-run internal combustion engine, not the electric car, represented the future.

> My Dear Mr. Leopold,
>
> Through you, I wish to congratulate the Albuquerque Game Protection Association on what it is doing. I have just read the Pine Cone. I think your platform simply capital, and I earnestly hope that you will get the right type of game warden. It seems to me that your association in New Mexico is setting an example to the whole country.
>
> Sincerely Yours,
> Theodore Roosevelt.[24]

Roosevelt was in his late fifties when he praised *The Pine Cone*. His health was declining. After losing the 1912 election he had several high points—such as hiking in the Grand Canyon with his family and exploring a hitherto undiscovered river in Brazil's Amazon (named Rio Teodoro in his honor) with Kermit, who had saved his father's life in the jungle. After practicing the strenuous life for so long, Roosevelt was burned out, exhausted to the point of depletion. Jack London died in 1916. Buffalo Bill died the following year, and was buried in a tomb on top of Lookout Mountain in Colorado.[25] The whole Rough Rider generation, it seemed, was going . . . going . . . gone.

Most of Roosevelt's characteristic vitality had disappeared by 1916. He was blind in one eye; a bullet was still lodged in his chest; he occasionally experienced bouts of malarial shivers and fever lingering from the arduous trip to the Amazon in 1913–1914; some minuscule parasite still lived in his body, eating away at his energy; his digestive system was a wreck. Unable to tap into his physical reserves, Roosevelt retired his gun and took up philosophizing. Instead of telling bear yarns, he spoke of nature, the universe, the planet Earth, hardship, existence, and destiny. At home at Sagamore Hill, forgetful of his bearings, looking out the window to the west and thinking for a second he might see Old Faithful or Pikes Peak, somber in its blue snow at sunset, Roosevelt grew melancholic. After he wrote *Through the Brazilian Wilderness*—a memoir of hunting and camping with Kermit in the Amazon jungle—his prose was understandably less action-packed and aimed more at the horizon, toward distant buttes, calving glaciers, and shore mud. He turned once again to the vast expanses of Alaska.

Roosevelt's infatuation with Alaska was notable in *A Book-Lover's Holidays in the Open* (1916), his elegant celebration of the world's cragsmen, explorers, scientists, and faunal naturalists. Trumpeting his own conservationist record, he called for a revolutionary ethos of game management like the one Leopold was promoting in *The Pine Cone*. He wanted Americans to take seriously the dire Biological Survey reports by Edward W. Nelson about the danger Alaska's caribou herds were in when the long winters shut down food supplies. "The man should have youth and strength who seeks adventure in the wild, waste spaces of the earth, in the marshes, and among the vast mountain masses, in the rotten forests, amid the streaming jungles of the tropics, or on the desert, or sand or snow," Roosevelt wrote. "He must long greatly for the lonely winds that blow across the wilderness and for sunrise and sunset over the rim of the empty world." [26]

As an appendix to *A Book-Lover's Holiday in the Open*, Roosevelt wrote an individual paragraph about all the federal bird reservations he had created by means of executive orders during his presidency between 1903 and 1909. They were his secular shrines. Many of them were in Alaska. Roosevelt continued to work his magic by lobbying legislators on Capitol Hill as a voice of the National Conservation Association. Although he had lost the 1912 presidential election, it was largely through his strong influence that the Morgan-Guggenheim syndicate had been thwarted in its repeated efforts to purchase mines around the Tongass and Chugach national forests. This was a policy victory for Roosevelt despite his defeat as a third-party candidate. In 1911, Roosevelt had successfully championed the Weeks Law to purchase lands for national forests in the White Mountains and Appalachian Mountains (where there was no public land). Further, in 1914, Congress passed landmark bills regarding coal and oil leasing, and these acts were in accordance with Roosevelt and Pinchot's philosophy of keeping huge corporations out of public domain lands. Roosevelt was also a powerful advocate of the Federal Water Power Act to provide for development by private enterprise (under federal ownership and control) of waterpower for the public domain and navigable streams. It even seems possible that Roosevelt's staunch conservationist agenda had influenced his former antagonist William Howard Taft. Before leaving the White House in 1913, Taft, as if in a face-saving gesture, had signed executive orders saving Alaskan bird-breeding areas on Forrester Island, Wolf Rock, and the Hazy Islands.[27]

Through lobbying, the Rooseveltian conservationists won numerous battles in Alaska, one at a time. Roosevelt's "Terminator" was Hornaday, the genius zoologist, who never pulled a punch. The Camp Fire Club of America (CFCA) had ceremoniously placed the head of a Montana bison—one that had died on the Flathead Reservation, a federal game reserve that Roosevelt and Hornaday had founded in 1908—over the fireplace at its Chappaqua lodge in New York. This head was a present to the club from Hornaday; emphasis was placed on the fact that the bison died of natural causes. Toward the end of his life, Hornaday—who remained active in the campaign to protect Alaska's northern seals until his death in 1937—crusaded against allowing motorized vehicles into national parks; these vehicles disrupted wildlife sanctuaries. "As everyone knows," Hornaday growled, "the automobile has become a fearful scourge to the game of our land, by enabling at least 2,000,000 men of the annual army of hunters to cover about four times as much hunting territory as they formerly could comb with their guns."[28]

Hornaday also sympathized with Native Alaskans who lived just outside McKinley National Park, the Tongass and Chugach national forests, and the huge bird refuges in Alaska.[29] Furious at the shoddy way Natives were behaving as stewards of the land—they were too readily bribed by timber and coal interests—he lambasted leaders of the Aleuts, Tlingit, Athabascan, and Inuit. Hornaday didn't believe that Natives should have special eminent-domain rights to shoot caribou or sell out a habitat to despoilers. From Hornaday's perspective, the parks and refuges existed for wildlife, not for people. Global overpopulation was forcing a more rigid policy for saving wilderness. Humans—including Alaskan Natives—who ignored conservation laws were, like locusts landing on a crop, a plague.

World War I also caused worries in wildlife protection circles. Roosevelt was a leading proponent of war against Germany, believing that the United States could not sit in comfort and allow the Hun to wreak havoc in Europe. But he lost his temper when game market syndicates used the war as a pretext to abolish hunting restrictions in Alaska and elsewhere. Only money-grubbing thugs, Roosevelt said, would wipe out animal species "for all time" to "gratify the greed of the moment."[30] The premise of the market hunter syndicate was that meat—including deer, elk, antelope, moose, and caribou—was needed to feed U.S. Army troops in training. Hornaday—who had a peak named after him in the Absaroka Range in Yellowstone National Park—was unleashed by the CFCA to be an attack dog. This time, however, Roosevelt was even more vehement and threatening. "To the profiteering proposal of the Pseudo-patriots, the patriots for revenue only, that protection of wildlife in wartime be relaxed, the united hosts of conservation reply," Roosevelt said, "You Shall Not Pass."[31]

A local Alaskan conservation society—the Tanana Valley Sportsmen's Association, based in Fairbanks—backed the pugnacious Roosevelt. The association was founded in 1916, and its headquarters eventually were located alongside the lovely Chena River; it was made up of hunter-anglers from interior Alaska. The members oversaw the transporting of almost thirty bison from Montana to Delta Junction, Alaska. In coming decades they also backed the repopulation of Alaska with musk ox; supported the protection of bears;[32] and helped protect the Mulchatna caribou herd, which as a result of their efforts grew into one of the largest in Alaska. All around Twin and Turquoise lakes, in what became Lake Clark National Park, these Alaskan hunter-conservationists helped the Dall sheep survive, too.

One area where Roosevelt seemingly wanted to say "You shall not pass" was the Arctic; scientists, naturalists, and explorers—not extraction industries—were needed at the pole. Roosevelt wrote a fine article for the *Outlook*, "Is Polar Exploration Worth While?" Looking at the bright side of exploration, Roosevelt said there was a need for more Pearys, Amund-sens, Stefanssons, and Shackletons. The natural history of Antarctica was an opportunity for someone hoping to make a name for himself as a mammalogist or zoologist. "The leopard seal is as fierce as the great spot-ted cat of the tropics from which it takes its name; and there are other seals, fat, good-humored, helpless, who, unless cruelly undeceived, treat men merely as friendly strangers, objects of mild curiosity only," Roose-velt wrote. "The penguins never touch dry land and never know warmth. They pass their whole lives upon the ice and in the icy water. The em-peror penguin, standing erect on its two flippers, is almost as tall as a short man." [33]

But it was the abundant wildlife of Arctic Alaska that most intrigued Roosevelt. The Inupiat (or Eskimos) actually lived above the Arctic Divide. (By contrast, there was no permanent human habitation in Antarctica.) With no hard-packed trails to follow, they traveled by dogsled over frozen creeks and shorelines along the Beaufort Sea. The whole North Slope was a tide of caribou in migration. During the fall months, hundreds of thou-sands of lesser snow geese landed like a blizzard on the coastal tundra; some observers claimed it was the greatest avian spectacle on American soil.[34] "There is an abundant life stretching very far towards the Pole, and probably there are some representatives of this life which occasionally stray to the North Pole," Roosevelt wrote. "Both in the water, and on the ice when it is solid over the water, and on the land, in the brief Arctic summer when the sun never sets, the Arctic regions teem with life as do few other portions of the globe. Save where killed out by men, whales, seals, wal-ruses, innumerable fish literally swarm in the waters; myriads not only of water birds but of land birds fairly darken the air in their flights; and there are many strange mammals, some of which abound with a plenty which one would associate rather with the tropics." [35]

Under Roosevelt's leadership, the Boone and Crockett Club started amassing data on the inequitable treatment of Alaska's game animals. Madison Grant, a cofounder of the Bronx Zoo who had a spotty reputa-tion as a eugenicist, wrote for the club a Darwinian-style essay on why the wolves, bear, moose, and deer were all *bigger* in Alaska than elsewhere. The club called for more game reserves in Alaska—like the moose reserve on Fire Island—where no hunting, trapping, or sled dogs would be allowed.

And, most significantly, the Boone and Crockett Club, America's most prestigious hunt club, was calling for a roadless *wilderness*. The members were inspired by examples such as Afognak Island, near Kodiak, Alaska, which Benjamin Harrison had put under protection during his presidency and which was now teeming with elk. The Boone and Crockett Club also took note of strips of land along the Uganda Railway in British East Africa that had been preserved as game ranges. As Madison Grant noted, those carved-out ranges were "absolutely swarming with game." [36]

Sportsmen's clubs viewed Alaska as an opportunity to *preserve and protect* a land rather than just try to restore it. For ardent outdoorsmen Alaska was the "last chance to do it right." [37] The Arctic was a unique resource, a vast land of extremes: long winter darkness and around-the-clock summer daylight; mountain ranges and permafrost prairie; snowy deserts and tundra wildflowers as far as the eye could see. Some parts of Alaska were ice fields year-round. Aldo Leopold noted that a *wilderness* like the Arctic was a unique geographic resource, which could shrink but never expand. "Invasions can be arrested or modified in a manner to keep an area usable either for recreation, or for science, or for wildlife," he wrote, "but the creation of new wilderness in the full sense of the word is impossible." [38]

Because Alaska was a ward of the federal government, the teeming caribou herds of the Arctic that migrated thousands of miles annually could be saved in a game reserve. And it was the U.S. Congress, not the residents of Alaska, that sportsmen's clubs of the Lower Forty-Eight turned to for the enactment and enforcement of suitable wildlife-protection laws. Federal control of Alaskan land was essential, they believed, if wildlife was to thrive. Some conservationists wanted to see the U.S. Army get back into the effort to protect nature, as it had done in Yellowstone from 1872 to 1917. As Madison Grant wrote in *Hunting at High Altitudes* (copublished by the Boone and Crockett Club and the Camp Fire Club of America), "The men who live in Alaska constitute a floating population—for the most part of miners who have no permanent interest in the country in the sense that farmers are attached to the soil. . . . The stable elements of the population are chiefly the keepers of local saloons or roadhouses. Miners are accustomed to live off the country, with little care for its future. It would be extreme folly to entrust to such a population the formulation and

enforcement of complicated game laws, which require a thorough knowledge of the habits of animals."[39]

The late John Muir was still, through his published works, beckon-
ing naturalists to explore and preserve underreported areas of Alaska.
Muir's literary executor, William F. Bade, skillfully put together the great
naturalist's scientific articles and unpublished journals about the Arctic
as *The Cruise of the Corwin*; it was published in 1917. Presented as a seafar-
ing adventure story, Muir's book described the Arctic Ocean as a boundless nursery for bird flocks and marine mammals. The farther north the
Corwin went, the less heat the sun provided, and the richer Muir's prose
became. "This is the region," Muir declared on his 1881 trip, "of greatest
glacial abundance on the continent."[40]

III

Frederick Vreeland had admired Muir's memoir *The Cruise of the Corwin* because it had opened up an unknown Alaskan ecosystem—the
Bering Sea—to the general public. The Lake Clark region—named
after a trader of the Alaska Commercial Company (ACC), John C. W. Clark
(1846–1896)—was one of the least explored areas in the territory.[41] Lake
Clark is the sixth-largest lake in Alaska; it covers 110 square miles and
is at least 900 feet deep. Clark first came to Russian Alaska in 1866 (with
the Western Union Telegraph Company's Russo-American Expedition). Among his first customers at ACC were Yupik Eskimos (who lived
along the Bering Sea coast), and Dena'ina Athabascans (situated in the
Iliamna–Lake Clark region of the 55,000-square-mile Bristol Bay basin).

Recognizing that Bristol Bay was the greatest wild salmon area in the
world, Clark ran the ACC post very profitably from 1879 until his death in
1896. His most lasting achievement was his pivotal role in the creation of
the shore-based commercial salmon industry in Bristol Bay. Clark put up
thousands of barrels of salt salmon for the ACC to feed their Aleut employees, who were living in the Pribilof Islands and killing northern fur
seals for the company. Clark produced salt salmon (sold by the wooden
barrel) at his trading post. And he founded the Clark's Point cannery
and was the leading investor for the Nushagak Canning Company (which
owned the Clark's Point cannery in 1887). Clark, as a representative of

ACC, also traded furs with Alaskan Natives throughout the Bristol Bay region.[42]

The ACC valued Clark for his entrepreneurial attitude. The German-Jewish businessmen, Louis Sloss and Louis Gerstle, who owned ACC, did close business deals with Clark at Nushagak Canning and trusted him with thousands of dollars. There wasn't much Clark didn't sell. He mass-marketed red fox furs, walrus ivory, caribou hides, and beaver pelts. Clark, in fact, knew of the existence of what would soon be named Lake Clark because he served customers from the remote village of Kijik. These Natives shopped at his Nushagak trading post in the 1880s and 1890s, loading up on staples such as tea, sugar, pots and pans, tobacco, pilot bread, traps, guns, knives, and axes as provisions for the winter.

Although Clark never wrote a word himself, he successfully led the *Frank Leslie's Illustrated Newspaper* Expedition of 1891 to look for the northern source of Iliamna Lake. The members of the expedition were going to do a census around Lake Clark and Iliamna Lake. A writer for *Frank Leslie's Illustrated Newspaper*, after an expedition to the Bristol Bay basin, named the big lake after John W. Clark. Even though Clark knew that the Dena'ina called the lake Kijik, as had the Russians, Lake Clark stuck; it was the white explorers' prerogative.[43]

Now, in June 1921, with Colonel A. J. Macnab as a trail mate, Vreeland held his own scaled-down expedition to Lake Clark. Instead of weathering winter conditions, the two men were equipped for the best weeks of summer. What worried Vreeland and Macnab more than grizzlies or hailstorms was disease. In 1902, a lethal combination of measles and flu had devastated the village of Kijik on Lake Clark. Between 1902 and 1909, the epidemic's survivors relocated their village to Old Nondalton (located twenty-five miles southwest on Sixmile Lake). In 1909, the explorers G. C. Martin and F. J. Katz of the U.S. Geological Survey visited the Iliamna–Lake Clark country on a reconnaissance mission. Their 1912 map was used by Vreeland and Macnab to get around the region (unfortunately, it was an incomplete map north and east of Tanalian Point on Lake Clark). They also relied heavily on a work by Wilfred Osgood of the U.S. Biological Survey: *A Biological Reconnaissance of the Base of the Alaska Peninsula*.

Voyaging down the Cook Inlet coast, Vreeland and Macnab were amazed by how underpopulated the landscape was south of Anchorage.

They hadn't expected to see *nobody*. There were enough huge ice fields and braided streams, however, to uplift any outdoorsman's spirit. A cannery boat first took these advance agents of the Camp Fire Club of America (CFCA) to Iliamna Bay. Their adventure at Lake Clark commenced in earnest after a wonderful night's sleep at the Iliamna Pass. In the morning, they laced their heavy boots and put light wood on the fire to make coffee. The wind was roaring. The Chigmit Mountains engulfed them, calling out to their romantic yearnings. Vreeland and Macnab hiked over the twelve-mile Iliamna portage, where they met a local man, Fred Phillips. With relative ease they then crossed the seventy-mile Iliamna Lake in a canoe (towed behind Hans Seversen's gas boat). They soon reached the lower Newhalen River. They put their canoe in the upper part of the river, which was twenty-two miles long, and paddled hard to Sixmile Lake and on to Lake Clark. Having become friends through CFCA, they were ready to explore Lake Clark. They kept diary notes tracing their route from Cook Inlet to Lake Clark: Iliamna Bay, Iliamna Portage, Old Iliamna Village (on the Iliamna River), Iliamna Lake, Seversen's Roadhouse, the upper Newhalen River, Sixmile River, Lake Clark itself, and at last Tanalian Point on Lake Clark.

What most startled Vreeland about the Lake Clark–Lake Iliamna region was the shortage of visible wildlife. "This has every evidence of having been once a good game country," Vreeland reported in a long letter to E. A. Preble of the Biological Survey. "But at present every native is armed with a high-powered rifle and kills everything that he sees with the result that there is very little game left except in inaccessible places. Grouse and Ptarmigan however quite plentiful."[44]

The Lake Clark country had some of the prettiest meadows in America, soft grass sloping up hillsides, fringed by forest; the whole panorama was one of stunning mountains and creeks, with no settlements in sight. Sometimes a sudden, absolute silence made it seem sacrilegious to talk on the trail. The rivers in the region—Beluga, Chakachatna, McArthur, Drift, Tuxedni, and Big—created the tidal flats of Cook Inlet. These rivers, as Muir knew, had been born from the huge glaciers located northwest on the Chigmit and Tordrillo mountains.[45] The largest tributary of the Cook Inlet—the Susitna River—was here. Iliamna Lake—the largest body of freshwater in Alaska—was enormous, covering 1,000 square miles;

it was seventy-eight miles long, twenty-two miles wide, and as much as 1,192 feet deep.[46]

Unfortunately, the moose had largely been shot out of the vast region. The wildlife deficit didn't prevent Vreeland and Macnab from enjoying the sight of lordly Redoubt Mountain in the distance. But their overriding opinion that summer, as they investigated the region with naturalists' eyes, was that the CFCA had to start campaigning on Capitol Hill to protect the lowland forests of spruce and balsam from the timber industry. Around the gravel bars, where the bears ate, there were scattered clusters of cottonwood. Much of their hiking was on damp moss—sphagnum, mainly. The mountain country was ideal for hiking. Most peaks were about 3,000 feet high; some were broken by tributary canyons. Macnab's diary was vague about color, unusual for that of an outdoorsman. "From this region," Vreeland wrote, "can be seen to the northeast a dense tangle of rugged mountains of the Alaska Range, as far as the eye can reach."

The Vreeland report to the Biological Survey was gloomy about the depletion of game around Lake Clark. Vreeland had never seen such abuse of land in his life. Salmon were being overfished by the Bristol Bay canneries at tidewater. The Natives at Iliamna and Lake Clark—Yupik and Dena'ina Athabascans—had always depended on subsistence fishing to survive. However, their catch was a very small fraction of what the commercial canneries were hauling in at tidewater. "The native name means 'salmon go up,'" Vreeland wrote to the Survey. "The salmon in this region have been depleted to a very alarming extent by many canneries on Bristol Bay, and unless prompt action is taken their early extinction is threatened. The Fisheries Bureau has adopted the policy which I feel is very unfortunately endeavoring to exterminate the trout in the lakes because of their habit of eating salmon spawn. . . . It is a great pity to destroy wantonly these splendid fish, especially as their destruction can have only at best a slight mitigating effect on the terrible depletion of the salmon by the canneries."

While Vreeland was sounding like Cassandra in 1921, writing a long, painstaking memorandum that would start Lake Clark on the long road to designation as a national park (it would take until December 2, 1980), Macnab, who loathed café society, was having a fine time hunting and canoeing. His field diaries revealed none of Vreeland's anxiety about endangered species. Often, Macnab—whom the humorist Will Rogers called "the greatest fellow you ever saw"—wrote arch critiques of Alaskans encountered on the way to Lake Clark–Lake Iliamna. "Visited the village and called on the U.S. Commissioner named Phillips," Macnab wrote on August 5. "He is a Holy Roller—holier than thou S.O.B., a human fish."[47] Macnab's diaries are full of military terminology such as "we make a reconnaissance on foot" and "we hang our clothes on a tree and lean our other impedimenta against it."[48] He grumbles about salmon drying on racks at Iliamna Lake and about constantly having to cope with rain, rain, rain. Yet Macnab clearly loved the primitive country, putting memories of the Great War behind him, writing straightforwardly about hunting, a genre in which the first rule was to be direct about death. A typical entry read: "Killed a ptarmigan from door cabin."[49] Macnab—who in 1938 would go to East Africa for the American Museum of Natural History and write the book *The White Giraffe*—marveled at the salmon-rich waters of the Bristol Bay region (Iliamna Lake and Lake Clark were the two largest salmon producers in the bay system). As an outdoorsman extraordinaire, Macnab was flabbergasted that five—*five*—main rivers drained into Bristol Bay, thereby making it the world's richest salmon fishery. Nowhere else could boast of having five species of Pacific salmon, or of offering the financial returns of the Bristol Bay canneries.

Reading Vreeland's fact-filled field reports from Lake Clark and Macnab's diaries in close succession is somewhat jarring. Their literary styles are diametrically opposite. Vreeland, like Muir before him, wrote with something of F. Scott Fitzgerald's narrative verve. Macnab, by contrast, wrote in a clipped, Hemingwayesque style. It would seem that these two men didn't belong on the same trail together. But that was the genius of the CFCA in the 1910s, 1920s, and 1930s. What brought these two high-profile, successful men together was the great Alaskan outdoors. There was no reason for either Vreeland or Macnab to gloss over the grim reality of the time: the big country was being bought, sold, subdivided, carved, and developed. If the CFCA acted fast, this stretch of Alaska could be forever wild. Vreeland wanted Lake Clark saved for photographers, hikers, and bird-watchers—a people's wilderness in the Chigmit Mountains far away from the crunch of New York City. Colonel Macnab—a fervent gun lover who would serve on the board of the National Rifle Association

(NRA) between 1925 and 1935—firmly believed that hunting and fishing turned men into soldiers. A day in the Lake Clark–Lake Iliamna region was far better basic training, he believed, than all the push-ups demanded by a drill sergeant at Camp Benning, Georgia.

The friendship formed between Vreeland and Macnab at Lake Clark became a model for how the conservation movement could stay politically potent in the age of Harding, Coolidge, and Hoover. Without a charismatic Roosevelt—neither Theodore nor Franklin—as a leader in prioritizing wildlife protection (particularly the saving of habitats), the "hook and bullet" recreational types (like Macnab) and the wilderness preservationists (like Vreeland) intuited that public policy would be enacted by Congress as law *only* when these factions collaborated in harmony. Each faction on its own, for example, was politically too weak to wage a sustained sixty-year campaign to permanently preserve the 4 million acres of the Lake Clark region. They were, after all, going against the "Big Three": commercial fishing, mining conglomerates, and oil exploration. (There was no commercial logging in the Bristol Bay region in the 1920s.) If the hunters and anglers, however, could learn to tolerate the Vreelands, and if, in turn, the preservationists could accept the robust hunters like Macnab, then pristine landscapes and big-game species in Alaska could be saved.

Chapter Eight

∘ ∘ ∘ ∘ ∘

RESURRECTION BAY OF ROCKWELL KENT

I

During the summer of 1918, Alaska finally received the literary treatment it had long deserved—and in a far more effervescent style than what Jack London had delivered in Burning Daylight four years prior. The impoverished painter, illustrator, commercial designer, and printmaker Rockwell Kent moved with his nine-year-old son, Rockwell Jr. (nicknamed Rocky), to an abandoned cabin on picturesque Fox Island, across Resurrection Bay twelve miles from the little fishing town of Seward, Alaska. For seven months, the Kents abandoned the fast pace of New York City, turning their backs on the "beaten, crowded way" for the glories of raw nature on this largely uninhibited island owned by the federal government.[1] A book-learned transcendentalist and wandering mystic, Kent recorded his impressions of Alaska from August 1918 to March 1919 in his journals (published in 1920 as Wilderness: A Journal of Quiet Adventure in Alaska). Kent was an original voice, and his musings were cosmic meditations on the inherent liberty found in wilderness settings: sea, sky, islands, and icy fjords. Deeply misanthropic, Kent—a socialist, pacifist, misfit, and activist who was sickened by the Great War—found stark relief in the dark, lonely waters around Fox Island. "There are the times in life when nothing happens," he wrote, "but in quietness the soul expands."[2]

Holed up on a lonely island in south-central Alaska, taking a respite from city life, staring at distant mountain panoramas of the Aialik Peninsula from his beach cove, Kent found solace and exhilaration in the utter remoteness just outside his cabin door. Here was an asylum for the troubled human soul, a soul anxious to abandon the "confusing intricacy of modern societies."[3] If you felt that city life was an unendurable torment, then Alaska was a logical prescription. In truth, Kent, with his son Rocky at his side, never felt alone. On Fox Island, waters abounding with river otters, Dall porpoises, harbor seals (*Phoca vitulina*), and Steller's sea lions (*Eumetopias jubatus*) surrounded them. Humpback whales cavorted in the bay outside their cabin door, and scores of ducks flew overhead. Sometimes orcas maneuvered into shore to rub their huge bellies on the pebble beach. When the air was mild, the seabirds turned silent, poking for crabs in the soft blue dusk. But Kent knew that Resurrection Bay was a shipwreck zone where "frightful currents and winds" would have daunted even the bravest New England fishermen.[4] Every Alaskan town built a little memorial to honor the seafarers who never returned home with their catch.

From the air, Fox Island would look like a backward number 3. There were two coves on the west side, and on the eastern part of the dollop savage cliffs seemed as forbidding as Alcatraz. Occasionally a bear would swim out to Fox Island from the mainland, and then leave, disappointed, after a fruitless look around for food. Kent treated with reverence every bird nest and seal rock he found around Resurrection Bay. To keep his son company on Fox Island, Kent adopted a little porcupine (*Erethizan dorsatum*) as a family pet. The largely nocturnal herbivore ate twigs, leaves, and plants, nourishment for growing the quills needed for its self-defense system. When easels were set up outside the cabin, the little porcupine also developed a liking for licking wet paint. A typical painter might not find this humorous. But Kent, who had a truant disposition, was different. Surrounded by wilderness and ocean, he preferred stormy weather to a safe anchorage.

Kent found the kaleidoscopic radiance of wild Alaska, and even the inconveniences associated with frontier conditions, exhilarating. The color scheme of the Kenai Peninsula landscape was dramatically different from that of Maine or the Adirondacks. In 1918 the most famous painter who had tried to capture Alaskan landscapes on canvas was Albert Bierstadt; his *Wreck of the Ancon, Loring Bay*, done in 1889, became a symbol of man's inability to defeat nature. It took Kent all of a minute to realize that

Bierstadt's so-called realist paintings falsified the true outdoor nuances of color, darkness, and shadows in the far north. Surrounded by the primordial landscape, Kent used a broad range of electric blues and bleached whites. The overcast aura of Fox Island also called for the gunmetal grays on his palette. Each place Kent went in Alaska that impressed him inspired its own refined painting. Yet the lifestyles of the hardened, blistering Seward fishermen, trappers, and prospectors, all physically battered by the inhospitable elements, caused Kent to also paint crudely, like a folk artist. "Alaska is a fairyland in the magic beauty of its mountains and water," Kent wrote. "The Virgin freshness of this wilderness and its utter isolation are a constant source of inspiration. Remote and free from contact with man, our life is simplicity itself." [5]

Finding Kent's proper place in American art history has proved to be difficult. His paintings of Bear Glacier (today part of Kenai Fjords National Park), sketched from the south end of his beach on Fox Island, stand out as major works of American art. Only one other American painter has ever come close to capturing Alaska's landscape with the illuminating and haunting halo effect that Kent created. In 1904 Sydney Mortimer Laurence arrived in Alaska from Brooklyn hoping to find gold. Unable to do that, Laurence, an amateur oil painter, decided to earn money by capturing the iridescent glow of Mount McKinley on canvas. Both Muir and Sheldon had done a pretty good job of starting the tourist business in the territory. Diligently, Laurence painted the tallest summit from at least a dozen different angles. His first large painting—*Top of the Continent*—was exhibited by the Smithsonian Institution's National Collection of Fine Arts. Some art critics wrote condescendingly that Laurence the marketeer owned Mount McKinley; it was his only subject. They had a point. Laurence opened a photography shop in Anchorage, eager to sell his *Visions of Denali* to tourists. "I was attracted by the same thing that attracted all the other suckers: gold," Laurence said. "I didn't find any appreciable quantity of the yellow metal and then, like a lot of other fellows, I was broke and couldn't get away. So I resumed my painting. I found enough material to keep me busy the rest of my life, and I have stayed in Alaska ever since." [6]

If Alaskan landscape painting were an Olympic event, then Kent would surely have won the gold medal (with Laurence getting the silver). Whereas Laurence stayed stationary around the McKinley station at Anchorage, Kent was intrepid. Sometimes, in good weather, or even with mixed seas, Kent hopped into his skiff and traveled twelve miles across Resurrection Bay to get a close look at Bear Glacier (if the motor failed, he rowed). Kent's paintings of that glacier have dazzling blue shades—azure, cobalt, and sapphire—as intense as those van Gogh painted in Arles; you can almost feel the solid ice sprawling over hundreds of miles under the ultramarine sky. Enduring the "seething" squall of the bay, the sea spray "whipped into vapor," Kent would get within fifty yards of places like Frozen Falls, Caines Head, or Hive Island. Sometimes he painted the same outdoors scene in different seasons: for example, *Alaska Winter* and later *Indian Summer, Alaska*. This approach allowed him to show his uncanny ability to use different biting yellows and ice blues with sunshiny radiance.

What was the secret of Kent's success as a painter in Alaska? Why was he more proficient than Laurence at painting the rich blue colors? Talent is hard to measure. But going the extra mile uphill for your art or craft isn't. Kent actually got close to the inside of a glacier crevasse. While lake ice and river ice are clear or white, the glacier crevasse he inspected blazed a blue unknown in Maine or Newfoundland. When light strikes an object, some of the colors of the spectrum are absorbed and others are reflected; it all depends on the matter that makes up the object. Glacier ice, when fairly thick, absorbs red and yellow, reflecting only pure blue light for humans to see. Kent became a connoisseur of that blue tint, which looks electric or like a glowing gas flame.

The blue glacier wonderland where the Kents stayed was essentially today's gateway to Kenai Fjords National Park (established on December 2, 1980, during the last days of Jimmy Carter's presidency). As kayakers know, ice floats deceptively calm near outflowing glaciers and coastal fjords. Here, as nowhere else in Alaska, mountains, ice sheets, rockfall, and ocean are intertwined with dancing rivers. This is the edge of the North Pacific Ocean, with stair-step glaciers clustered together. Anywhere from 400 to 800 miles of snow accumulates annually in the mountain knuckles here. Kent knew this fjord country was outstandingly *wild*—nature didn't get any more beautiful. Determined to capture the glory of the landscape, and riveted by the icy outburst of Bear Glacier, Kent set up easels. He captured Bear Glacier, in all its austere elegance, on canvas better than any other photographer or painter. All the fjords and glaciers, surrounded by open water and shoreside mud pools, some hidden in the wave-carved grottoes, became Kent's secret sanctuary, his escape.[7]

II

K ent was born in Tarrytown, New York, in 1882. When he was only five years old, his well-to-do father died, leaving him a silver flute as a token remembrance. This little musical instrument became a good-luck charm for Kent. (He kept it with him when he went to the far reaches of Arctic Alaska, Newfoundland, Tierra del Fuego, and Greenland.) Early on, Kent became smitten with the solitude of northern seas. On a trip to Oregon, he was particularly taken with the works of the German painter Caspar David Friedrich. A naturally talented graphic artist, Kent took classes in New York with both the impressionist William Merritt Chase and the innovative Robert Henri of the Ashcan school. Kent, it seemed, wanted to paint with the religiosity and spirituality of William Blake, while exploring the world's places in all their geologic forms.[8]

After a well-received show at Knoedler Galleries, primarily of his Atlantic coast paintings, Kent was considered a rising star. George Bellows, the famous painter of *Both Members of This Club*, a dramatic boxing scene, saw him as a genuine rival, the most intriguing of the up-and-coming modernists. Kent's colors glowed on the canvas. Because of his representational symbols—heavily influenced by Nietzsche—his works resembled those of the German painter Franz Marc. Early in Kent's career the wild Adirondacks served as an inspiration for his intense nature studies.[9] Eventually Kent moved to Boothbay Harbor, Maine, hoping to soothe his troubled mind, painting the enchanting cliffs and fish houses of Monhegan Island. Even though Kent painted like a man possessed, he managed to read the collected works of Emerson, Thoreau, and Wordsworth for inspiration. While working as a lobsterman, struggling to find a commercial audience for his art, he developed into a Spartan survivalist, a singular craftsman comfortable living in genteel poverty. All Kent needed to be happy was a floor to sleep on, a bedtime vodka, and mediocre food. "If minds can become magnetized, mine was: its compass pointed north," he wrote. "I set out for the golden North, for Newfoundland, to prospect for a homeland."[10] For a while he lived in Newfoundland, finding comfort walking over the steep hillsides and rock outcroppings that dropped dramatically into the Atlantic Ocean.[11]

Kent was also a pugilist, and his arms were muscular from boxing. With dark hazel eyes, his head prematurely balding, his hands fidgeting with whatever object was nearest, he was an intimidating adversary, able to quote Nietzsche verbatim and to eat halibut raw. Practical jokes were an important part of his everyday life, which, to Kent, was a dutiful exercise in carpe diem. Local mariners saw him as a peculiar piece of work, a human clock that ran backward. "Do you want my life, in a nutshell?" Kent once wrote. "It's this: that I have only one life and I'm going to live it as nearly as possible as I want to live it."[12] Sleeping, however, didn't come easily to the hyperactive Kent, who wrote in *It's Me, O Lord*, "Insomnia isn't nice."[13] Physical exertion outdoors, what Theodore Roosevelt called the *strenuous life*, appealed mightily to Kent, who felt that it directed his inner compass and uplifted him. His favorite verse of Blake's was "Great things are done when Men and Mountains meet./This is not done by jostling in the Street." Kent lived by this creed.[14] *Going to Alaska* wasn't a random impulse; it was an imperative—the three words stood for everything free, unspoiled, and democratic.

Life for Kent was always hand-to-mouth, and an ordeal. By the time he was thirty-six years old, he was stone broke, rudderless, and furious at Woodrow Wilson for not keeping America neutral during the European war. Making matters worse, Kent was often estranged from his wife, and he had a string of affairs that proved corrosive to his family. His temper was volatile, his willpower unnerving, his attire indecorous. Deeply self-centered—his friends floated into and out of his life—the fatalist Kent knew that in the end only *he* would be around for the curtain call. Disappearing to south-central Alaska, living a Thoreauvian life in a little shack on Fox Island, and using the inhospitable remoteness of Resurrection Bay to bond with Rocky made perfect sense to Kent. A bundle of energy, always an escapist, Kent had very little to lose by going to Fox Island. "I crave snow-topped mountains, dreary wastes, and the cruel Northern Sea with its hard horizons at the edge of the world where infinite space begins," he said of Alaska. "Here skies are clearer and deeper and, for the greater wonders they reveal, a thousand times more eloquent of the eternal mystery than those of softer lands. I love this Northern Nature, and what I love I must possess."[15]

III

To get to Alaska, the Kents traveled across America by passenger train. From their windows they saw the rolling prairies of the Great Plains that Washington Irving had once written about so memorably. The Kents now understood for the first time Walt Whitman's rapture in Leaves of Grass, where he had written of "Pioneers! O Pioneers!" and "peaks gigantic" and "high plateaus." Sometimes the Kents stayed at old, rickety railroad depot inns, lured to their meals by wooden boards out in front advertising specials: fish stew, meat loaf, and beef tenderloin. In Alaska it would be halibut steak, salmon jerky, and a shot of vodka. Father and son felt like tenderfeet entering the storied Colorado Rockies in search of the northern paradise of Alaska and rumbling across Montana. Westerners, the Kents learned, had a language all their own: draws were "dells" and buffalo were "grazing cattle." Domesticity had created no flower beds in this stark, rugged country. The Kents studied horse towns, outposts, and raw forestlands from their wooden passenger seats until their train finally arrived in Seattle. The temperature was well above fifty degrees Fahrenheit all around Puget Sound. Rusted Russian ships at dockside had long unpronounceable words painted on their sides in Cyrillic script.

Following a day's rest in Seattle, the Kents traveled up the Inside Passage to Alaska on the SS Admiral Schley. Their ship felt its way past a stunning succession of fjords, bays, straits, sounds, and promontories. Boisterous in praise of this picturesqueness, the Kents passed from Yakutat Bay (an eighteen-mile area, rich in fish, that extended southwest between Disenchantment Bay and the Gulf of Alaska) to Prince William Sound. For five days the Kents lived at a Swift and Company salmon cannery surrounded by glacier-carved mountains looming over the open ocean. Somehow Kent had procured a "letter of introduction" to stay there. They were in earthquake country (a quake of 8.0 on the Richter scale had happened as recently as 1899). Local folklore held that the explorer Vitus Bering of Russia had visited the bay on his expedition of 1741.

The Kents shared an upper bunk and ate in the mess hall along with the weather-beaten crab trappers. "What meals they were!" Kent said in his autobiography, *It's Me, O Lord.* "And how those hungry fellows wolfed them! A free for all, it was, and no holds barred. Never had either of us tasted better food or seen so much. And it disgusted us to watch our opposite at table—say at breakfast—flood his huge soup-plate full of oatmeal with undiluted evaporated milk, heap on six tablespoons of sugar; follow this with two vast stacks of six-inch flapjacks, with butter and corn syrup to match; then eat four eggs with bacon and drink a quart of coffee; and all the while goddamn the company for starving him." [16]

Kent had originally hoped to begin his spiritual rebirth on the Kenai Peninsula along Kachemak Bay. But a mail clerk, working on the steamer *Dora*, told Kent about remote islands clustered offshore from the Resurrection Bay port town of Seward. Off they went. Seward was the southern terminus for the Alaska Railroad, which had been built by the U.S. government and always seemed to be behind schedule. It was a larger city in 1918 than Fairbanks, Juneau, Sitka, or Kodiak. All around were villages, fish shacks, open mines, and quarries, but the glorious wilderness remained undiminished. Alaska's Second Organic Act of 1912—which had officially established Alaska as a territory with an elected legislature—meant that Alaskans no longer had to endure colonial status, although it wouldn't get true congressional representation until 1959, with statehood.

The Kents loaded up on provisions such as beans, rice, flour, barley, and other foodstuffs. A deal was made with Thomas Hawkins, a local landowner, to let them live on Fox Island in a lean-to cabin or goat shed that needed refurbishment. For all his machismo, Kent's diaries are quite honest about his lack of hardiness and lack of stamina. Like any father, he feared for his young son's welfare. Rain gear was (and still is) mandatory in this part of Alaska. Because Kent had once built a few houses on Monhegan Island, including a small one for himself and another for his mother, he felt confident that he could remodel the hovel on Fox Island in exchange for living there rent-free.

At last, on September 24, 1918, the Kents packed a tiny dory with their essentials, including a stove and box of wood panels, and prepared to go by motorboat from Seward to Fox Island. Because they were weighed down, the three-mile voyage out to the island wasn't for the weak-hearted. A dangerous problem, potentially a lethal one, manifested itself. The engine of the Kents' 3.5-horsepower Evinrude, after 100 yanks, wouldn't turn over. The dory, weighed down with about 1,000 pounds of cargo (including the Kents' own body weight), almost capsized. Remaining undaunted, refusing to consider retreat, Kent started rowing toward his destination, using a pail to bail water out. Without modern navigational devices or survival suits, it was an act of foolhardy recklessness.

Only by the grace of God did they somehow manage to traverse or perhaps navigate Resurrection Bay safely. An intense pressure system always hovered over the Gulf of Alaska in the North Pacific, like a perpetual category 1 hurricane, regularly blanketing the vast area with heavy winds, thick fog, and whipping rain; for a sailor the region was among the most challenging on Earth to navigate. The everlasting, unpredictable waves seemed to carry a Norse wallop (as the salt wind seemed to carry an Oriental scent) and had, over the decades, gulped down and sunk British dreadnoughts and Russian vessels. "Over the water the wind blew in furious squalls," Kent wrote, "raising a surge of white caps and a dangerous chop." [17] The Kents finally moored along the northwest harbor of Fox Island, glad to be alive and able to chuckle at their own foolishness.

After settling into the cabin that evening, father and son hugged each other. They had a new lease on life. The fierce easterly winds that had been howling down from above at fifty to sixty knots dissipated into small sighing gusts. The next day the Kents roamed the woods and thickets of Fox Island and watched river otters friskily playing along the El Dorado Narrows. Although Kent had little money—he wore the same wrinkled work shirt almost daily, and couldn't afford even the blue plate special at the Seward Grill or the Sexton Hotel—he was an able carpenter, caulker, and workbench tinkerer. Like any survivalist, he knew how to live off the bounty of the sea and land. He practiced ahimsa—the Hindu and Buddhist belief that all living creatures deserve respect.[18] On Fox Island, he converted Hawkins's goat shed into a livable rustic cabin. He took a farmer's approach to the clock. There were Angora goats to milk and chicken eggs to collect. Kent's groundskeeper, who came with the house, was a seventy-one-year-old Swede, Lars Matt Olson, a retired trapper and sea dog with a pocked face and rope-burned arms indicating endurance. Olson became an adopted uncle to the Kents. It was Olson who told them stories about earless seals, tidewater glaciers, sledding black bears, and how to scratch a kid goat. His minimalist philosophy of life boiled down to: "Very little matters, and little matters a lot."

For extra money Kent painted a portrait of Hawkins's absent daughter, Virginia. (The itinerant John James Audubon had earned his keep likewise in Louisiana during the 1820s.) Hawkins also donated lumber and hardware to the remodeling of the goat shed. Occasionally Kent would row into town for drinks. Wandering around the mud streets of Seward, reading Goethe and Schopenhauer aloud, playing his battered flute for tips, rowing out to Bear Glacier at a speed of one knot, Kent was like an offbeat Adirondacks hermit in exile. Townspeople had high hopes that Kent, a well-connected New York artist, would promote the virtues of Seward over Anchorage: the two towns were vying to be *the* tourist hub of Alaska. The painter Henry Culmer had recently done a fine job of painting the interior region for the Alaska Steamship Company, and people in Seward thought that Kent might follow Culmer's example. Kent, however, let the city fathers down in this regard. The people of Seward were too focused on self-promotion, too phlegmatic, and too eager for tourism to interest him or to stoke his artistic imagination.

During the 1920s, homesteading had increased in the coastal regions of Alaska. Along the beaches, log cabins with spectacular views from the front porches were being built. The pioneers who lived in these cabins gathered coal and seafood on the shore. With remarkable ingenuity, they made their own furniture by hand. In their gardens, because of the rich glacial till and the long summer days, cabbages grew to the size of pumpkins, though the homesteaders did miss having fresh fruit. Many families carved out a decent living but often dreamed about moving to a warmer climate. A favorite sourdough joke in Homer, on the tip of the Kenai Peninsula, was that homesteaders grew "sour on the country" but "didn't have enough dough to get out."[19]

Kent, who was intrigued by ethnography, also befriended a number of Aleuts he encountered in Seward. He venerated Native Alaskan groups and thought that the Aleuts, like all maritime peoples, were riveting storytellers. As a modernist, Kent preferred Aleut primitive art rather than that of the Hudson River Valley school. The intrepid Aleuts were similar to Kent himself in caring little about social structure or about laying down permanent roots. A large Aleut village would have makeshift dwellings, and would usually be situated on an island in the Bering Sea where the fishing was good. Aleuts, to Kent's surprise, were sexually permissive. Kent marveled at how they used animal parts for tools. Clams, mollusks, and sea urchins were part of their regular diet. Excellent hunters, they used *atlatl* (a throwing stick) to bring down ducks, geese, and loons in flight. Wild berries grew abundantly on Fox Island, and the Alutiiq instructed Kent on which ones were edible.

Kent's series of abrupt drawings and rhapsodic paintings of Resurrection Bay are, arguably, the finest landscapes ever done on Alaskan soil. They were influenced by Aleut art. Because Fox Island was often foggy, Kent thought of sunny days as a benediction; sunshine was good for painting the brotherhood of man and nature. "The wonder of wilderness was its tranquility," Kent wrote. "It seemed that there both men and the wild beasts pursued their own paths freely and, as if conscious of the freedom of their world, molested one another not at all." [20]

Many of Kent's brush-and-ink drawings and engravings, free from presuppositions, accompanied the prose of *Wilderness: A Journal of Quiet Adventure in Alaska* in perfect harmony. The cold far north appealed to his love of forlornness. Even the rotten ice—called *aunniq* by the Inupiat—had its charms to a symbolist painter. [21] "It's a fine life," Kent wrote to a friend, "and more and more I realize that for me such isolation as this . . . is the only right life for me." [22]

Some of Kent's Alaskan work is reminiscent of the intricate illustrations by the poet Vachel Lindsay, who tramped around America promoting the "gospel of beauty." Kent drew Resurrection Bay in a biblical, folklorish way, the style common in the "outsider" art movement of the 1980s. Celebrating the spectacle of life, his paintings and drawings still defy easy categorization. Taken with totemic symbols, Kent populated his Alaskan paintings with Norse gods in a semimodernist style, almost like socialist realism. Some of his images of laborers—*The Whittler* (1918), *The Snow Queen* (c. 1919), *Lone Man* (1918)—have a touch of Dürer; they're paeans to heroic hardworking Alaskans who understood the power of biomass, forlornness, and self-reliance. In another context, they might be considered proletarian art. (Later, Kent would draw recruitment posters for the Industrial Workers of the World—the IWW—though he refused to join the Communist Party.)

Kent's bizarre *Mad Hermit* series—included at the end of *Wilderness*—celebrated the age of voyages. His Alaskan sun was a Cyclopean eye. The legend of the Viking Leif Eriksson—possibly the first European to land in North America, almost 500 years before Christopher Columbus—was suggested in many of his lyrical paintings of the 1920s and 1930s. It was as if Scandinavia came into his every brushstroke; his painter's fascination with light was piqued by the ever-shifting scenes created by the northern lights.

Never before had such a gifted poet-philosopher-painter contemplated Alaska's subzero climate, long winter nights, and rainy landscapes with such imaginative flair. The broad glare of winter afternoons had a bracing effect on him. Life opened up every morning in the most amazing ways, and he was there to document the pageantry. "Cold?" he once said: "We had come to love it. The snow lay deep. The sun at noon now rose above the mountain, flooding our clearing with its golden light. The north wind raged and swept up clouds of vapor from the steaming sea." Who knew that getting drenched could be fun? Kent's attitude anticipated the back-to-nature movement of the 1960s: Scott and Helen Nearing; *The Whole Earth Catalog*; organic gardeners; and the rejection of plastic, chemicals, and prepackaged food.

While he was on Fox Island, Kent would sometimes write a newspaper column either for amusement or for a little extra money. He could, it must be said, be abrasive and self-righteous at times. Locals discovered that he was a man of great humor but also was very difficult. No matter what the discussion topic or issue was, he refused to be a shrinking violet; he preferred a stance of competitive firmness. Sometimes he literally threw paint at a canvas and then ran around naked in the snow. But he was not insane (although bipolar disorder is a possibility). A brouhaha occurred in Seward when Kent's son Rocky was asked by a teacher which of several flags shown in a book was his favorite. While the other children went with Old Glory, Rocky chose the German flag because it had an eagle at its center. The angry schoolteacher thought Rocky was being treasonous and expressing support for the kaiser. The Great War had ended, but anti-

German sentiment was still strong. Kent nobly defended his son's honor to the teacher; he also challenged people in Seward to fisticuffs. Upon leaving Alaska, Kent wrote a frank, open letter in the Seward *Gateway*, denouncing local busybodies but also proclaiming that Alaska was "the only land that I have ever known to which I wanted to return."[23] The lines and colors and illuminations of Resurrection Bay, he said, spoke to him like a hymnal. "As graduates in wisdom," he wrote, paraphrasing Muir, "we return from the university of the wilderness."

What Kent philosophically promoted in *Wilderness* was the power of solitude and ahimsa. It didn't matter that a "heartless ocean" eliciting a "terror of emptiness" surrounded Fox Island. Neither the five-foot chops in the ocean nor a steady "miserable drizzling rain"—about 300 inches annually—could deflate him. For Kent had a rare gift of optimism wherever he traveled, even in Alaska's "luminous abyss," as long as his paint kit was at hand. Kent convinced himself that the desolation of Fox Island, where winds raced in swirls, was a bracing cure for the neurotic anxiety associated with the modern condition. Solitude was better than all the pharmaceuticals in the world.

"The Northern wilderness is terrible," Kent said in a letter to an esteemed art critic, Dr. Christian Brinton, written for publication. "There is discomfort, even misery, in being cold. The gloom of the long and lonely winter nights is appalling and yet do you know I love this misery and court it. Always I have fought and worked and played with a fierce energy and always as a man of flesh and blood and surging spirit. I have burned the candle at both ends and can only wonder that there has been left even a slender taper glow for art. And so this sojourn in the wilderness is in no sense an artist's junket in search of picturesque material for brush or pencil but the fight to freedom of a man who detests the petty quarrels and bitterness of the crowded world—the pilgrimage of a philosopher in quest of Happiness!"[24]

Much of the tone and tenor of *Wilderness* arose from the bonding of father and son. Like Huck Finn on the river, Rocky found freedom in many things: fox dens, hollow logs, starfish in the icy water. Together Kent and Rocky created "magic" kingdoms on the island, fantasizing about being marooned like the Swiss Family Robinson. Birds, they marveled together, were better swimmers than fliers along the windswept offshore islands. They drank hot chocolate, flipped buttermilk pancakes, read *Robinson Crusoe* aloud, memorized William Blake's poetry, sang Celtic ballads, explored headland coves, and sailed to remote blue islets. They collected driftwood for the evening campfire. Together they measured wind velocity with a new gadget picked up in Seward. They caught a little black-billed magpie (*Pica hudsonia*), caged it, and trained it to mimic words like a mynah. Out in the back corral, the Kents reluctantly tended goats when Olson went into Seward. (One afternoon an angry or scared goat got into the cabin, comically trapping Kent inside.) On a few evenings the full moon rose bold and blood-orange, magically illuminating every tree and rock. Rocky's indispensable textbook was J. P. Wood's *Natural History*. With the help of Audubon's *Birds of America*, the Kents were able to identify a red-throated loon, a couple of eider ducks, and a hooded merganser (*Lophodytes cucullatus*).

"The day has been glorious, mild, fair, with snow everywhere, even on the trees," Kent wrote in a journal entry. "The snow sticks to the mountain tops even to the steepest, barest peaks painting them all a spotless, dazzling white. It's a marvelous sight. Rockwell and I journeyed around the point today and saw the sun again. Tonight in the brilliant moonlight I snow shoed around the cove. There never was so beautiful a land as this! Now at midnight the moon is overhead. Our clearing seems as bright as day—and the shadows are so dark. From the little window the lamplight shines out through the fringe of icicles along the eaves, and they glisten like diamonds. And in the still air the smoke ascends straight up into the blue night sky." [25]

Fox Island had, briefly, been selected by the Biological Survey as an experimental fox-breeding station. But instead the land was leased to Seward's farmers—local businessmen like Hawkins. All over southern Alaska—particularly in the Aleutians—foxes were bred in captivity in the hope of producing fur pelts for the market. There were nonnatural foxes on 1 million acres of Alaska, on more than forty islands. The corral behind Kent's goat shed, in fact, had been built for fox breeding. Luckily, Roosevelt had created places like Saint Lazaria as fox-free zones, allowing bird species to survive. Nevertheless, a few feral foxes roamed freely on the island. For all of Kent's rhapsodizing about wild animals, most of the blue foxes on Fox island were being raised in captivity by Hawkins for money. Kent, in the end, didn't write anything substantial about fox propagation along Alaska's southern coast from Dixon Entrance all around to Attu (the most westward island of the Aleutian chain).[26]

Groups such as the National Audubon Society and U.S. Fish and Wildlife Service were concerned that the proliferation of foxes in Alaska would lead to the extinction of the Aleutian Canada goose (*Branta canadensis leucopareia*). Foxes, it turned out, particularly loved the cream-colored eggs of these geese. Starting in 1940, the U.S. Fish and Wildlife Service successfully worked to remove the unwelcome foxes from public lands. After a fifty-year effort the Aleutian Canada goose became one of the few species to return from the brink of extinction. Its recovery gave hope that other Alaskan species could rebound, with proper game management.

IV

W hen Kent's Wilderness *was published by* G.P. Putnam's Sons in 1920, it received splendid reviews. Kent had kicked out the doorjambs; he was a mystic who had flung himself into the Alaskan galaxy and returned with stories. The *New Statesman* went so far as to say that *Wilderness* was "the most remarkable book to come out of America since *Leaves of Grass*." According to the *Chicago Post*, Kent was a genius: "The artist who can put into the simplest drawings of a man and a little boy eating together at a rough table in a rough cabin all dear solidity of family and home life—that artist can make me bow my head before his sincerity." The perceptive Robert Benchley, in the *New York World*, thought Kent's sojourn on Fox Island was a magnificent artistic feat: Kent had brought back both wonderful prose and priceless illustrations, making him "the envy of all urbanites" in Greenwich Village.[27] Another astute critic—Martha Gruening—compared *Wilderness* to Paul Gauguin's memoir *Noa Noa*, which was about his first trip to Tahiti and was filled with descriptions of Polynesian mythology.[28]

Kent's *Wilderness* was a pioneering first-person narrative, promoting Alaska as an ecological retreat where city dwellers could find "OURSELVES— for the wilderness is nothing else." On every page of *Wilderness*, Kent paid homage to Thoreau and gave consumer-driven America the back of his hand. Like Thoreau, Kent kept a detailed list of all the provisions he brought to Fox Island and how much they had cost. Kent, a simplifier, was content with a sleeping bag, poncho, cooking pots, and paint kit. Whether he was studying otter tracks, marveling at the moonlight, or decorating a Christmas tree for his son, he made *Wilderness* a quirky hymn, a sudden burst of quiet celebration, for the offshore islands around the future Kenai Fjords National Park. In Alaska the drifter had found the true heart of the universe. "And now at last it is over," Kent wrote at the book's end. "Fox Island will soon become in our memories like a dream or vision, a remote experience too wonderful, for the full liberty we knew there and the deep peace, to be remembered or believed in as a real experience in life. It was for us life as it should be, serene and wholesome; love—but no hate, faith without disillusionment, the absolute for the toiling hands of man and for his soaring spirit." [29]

From Fox Island Kent moved to Vermont, where he completed his stunning series of Alaskan paintings—the windblown green sea, blue-golden hillsides, piercing gray mountains—along with intriguing character portraits of locals. Even as late as the 1960s Kent produced oil paintings of Alaska from the ink-and-brush drawings in his sketchbook. *Alaskan Sunrise* (1919), for example, is a minimalist scene of Resurrection Bay with a wide range of blinding blue and icy white hues; it conveys a barrenness that is humbling to contemplate. Somehow Kent made foreboding lifeless landscapes seem like uncrowded, untouched, unhurried holy land. Kent's painting *Alaska Winter* (1919) vividly shows the view from his cabin on Fox Island. Varied shades of blue-green capture the cold of Resurrection Bay and the mountains beyond. Long shadows and cool yellow light radiate from objects that warm the winter horizon and appear, mesmerizingly, from behind the stark, split trunks of trees. Doused in wintry light, the painting shows no humans, but only the brownish shadow of a lumberjack.

Encouraged by the success of *Wilderness*, Kent soon thereafter wrote two other books: *Voyaging* (1924) and *N by E* (1930); both did well. He also went on to illustrate a special three-volume limited edition of *Moby-Dick* for the Lakeside Press of Chicago. Kent's black-and-white pen, brush, and ink drawings of whales and Ahab were stupendous. Those first editions constituted a high-water mark in American book publishing. Kent's illustrations helped inspire the Great Depression generation to rediscover Herman Melville. Throughout the 1930s, in fact, Kent was as celebrated for his illustrations as Norman Rockwell of the *Saturday Evening Post*.[30] He was hired to illustrate classics such as the *Canterbury Tales, Candide, Beowulf*, and Boccaccio's *Decameron*.

The Norse side of Kent continued to ring forth. Believing that folk sagas were a window into cultures, he famously illustrated books about Paul Bunyan and Gisli of Iceland. Major magazines—such as Frank Crowninshield's *Vanity Fair*, Henry Raymond's *Harper's Monthly*, and Richard Watson Gilder's *Century*—commissioned his vivid black-and-white works. Even Kent's doodles were coveted in New York literary circles. One afternoon Kent was talking with Bennett Cerf, a founder of Random House. On the spot, he drew the colophon that Random House still uses. When Modern Library was created, Kent designed its logo, an elegant torchbearer. Eventually, no publisher felt adequate without a logo designed by Rockwell Kent. When Harold Guinzburg started Viking Press, for example, Kent produced its image, a ship.[31] "He was, indeed, so indefatigably busy at desk and drawing board," the Smithsonian Institution's *Archives of American Art Journal* noted, "that in the 1920s and early 30s his work was virtually inescapable."[32]

But it was *Wilderness*—the prose of a lonely seeker combined with bold illustrations—that has survived as a classic of travel literature. There was something noble about Kent at Fox Island, painting by day, drawing by oil lamp at night. By the 1960s, some readers considered *Wilderness* a second *Walden*. It is hard to describe the religiosity Kent had found in the wilderness at Resurrection Bay. The town of Seward honored him by painting a mural of his nautical map of Resurrection Bay—the frontispiece to *Wilderness*. Doug Capra, a ranger at Kenai Fjords National Park, hopes to someday rebuild Kent's cabin, which remains private property. Painters regularly make pilgrimages to the area to have their try at Bear Glacier. To Kent, the far north sky was "God's abode," with "truth and beauty emanating as the light from Heaven."[33]

For fifteen years after the publication of *Wilderness*, Kent, always full of pent-up passion, looked for an excuse to go back to Alaska. That opportunity finally presented itself in early 1935. The U.S. Treasury Department had commissioned him to paint two enamel murals for a post office in Washington, D.C. The idea was to demonstrate, in an impressive way, the far-flung services of U.S. airmail. Kent was to show Eskimos from Nome, Alaska, sending letters to a family in Puerto Rico, 5,350 miles away. So, suddenly, thanks to this commission, Kent found himself in Nome, in a frigid wind, looking for Arctic families to sketch under the graying sky. "Alaska in 1935 belonged, as much as a colonial country can, to 'the people who inhabited it': the miners and prospectors, the big merchants and little shopkeepers, the artisans and upper laborers; all white," Kent wrote. "It was no longer, as to a great extent was Greenland, the country of the aborigines. And although the Eskimo, to judge by what I saw of them in Nome and at my farthest north, Tin City, near Cape Prince of Wales, appeared to enjoy a greater material prosperity than the Greenlanders, their citizenship—politically, socially, and economically—was second or third class." [34]

Kent painted his mural, which he infused with the left-leaning political disposition of Diego Rivera. The explorer Vilhjalmur Stefansson, a consultant to Pan American Airlines, gave Kent some tips about the difficulties of aviation around Arctic Alaska, where long gravel spits and permafrost tundra were used as landing strips. The landscape was flat and mundane, and Nome did not even have a single attractive, tree-lined square. In Nome, Kent befriended George Ahgupuk, a talented Eskimo painter, who taught him about dogsledding. (Kent did Ahgupuk the great favor of arranging for him to have a gallery show in New York.) "I got every kind of information as to details and equipment and if, when I finished my picture, there is a single rivet in the dog harness out of place," Kent wrote to a friend, "it won't be my fault." [35]

Nervously, Kent unveiled his mural in September 1937 to a group of assembled journalists and bureaucrats. Everybody admired how amazingly he had captured Eskimo dogsleds and reindeer teams, and people bidding good-bye to their mail in the Arctic. All was well—until a few weeks later, when Kent was accused of having tried to foment revolution in Puerto Rico and Alaska, inciting indigenous peoples to break the chains of colonialism. Kent, citing the Bill of Rights in his own defense, said he was only encouraging people to be "equal and free" individualists. Only sheep could possibly believe in communism, colonialism, or corporations. Always his own man, Kent didn't like isms at all. Perhaps the poet Gary Snyder best captured the essence of Kent's mischievous, nonconformist mystique in his 1988 poem "Raven's Beak River at the End," written after a trip to Alaska:

> Raven-sitting high spot
> eyes on the snowpeaks,
> Nose of morning
> raindrops in the sunshine
> Skin of sunlight
> skin of chilly gravel
> Mind in the mountains, mind of tumbling water,
> mind running rivers,
> Mind of sifting
> flowers in the gravels
> At the end of the ice age
> we are the bears, we are the ravens,
> We are the salmon
> in the gravel
> At the end of an ice age
> Growing on the gravels
> at the end of a glacier
> Flying off alone
> flying off alone
> flying off alone
>
> Off alone[36]

Chapter Nine

. . . . . .

THE NEW WILDERNESS GENERATION

I

hile Rockwell Kent was living on Fox Island, Theodore
Roosevelt—who turned sixty on October 28, 1918—was dying.
A certain listlessness was evident. Physically spent, he often sat very still, his eyes glazed. Owing to deafness in his left ear, his
balance was off, and there were many other health issues. He had spent
some of the year at Roosevelt Hospital in New York City as a patient, receiving emergency surgery to remove abscesses in the leg and thigh. "I
feel as though I were a hundred years old," he wrote, "and had *never* been
young." [1] Adding to his general misery, his feet were so swollen from inflammatory rheumatism that he couldn't wear shoes. Gout, headaches,
and sinus congestion—he suffered from a host of discomforting afflic-
tions. One thing that cheered him up was receiving letters from readers who had enjoyed *Through the Brazilian Wilderness*. And he was pleased
that an utterance of his had been adopted as the motto of the twentieth-
century conservation movement: "The nation behaves well if it treats the
natural resources as assets which it must turn over to the next generation
increased, and not impaired, in value." [2]

Shortly after the armistice was announced on November 11, 1918,
with Germany surrendering unconditionally to the Allied forces, thus
ending World War I, Roosevelt again entered a hospital in New York; he
spent forty-four days there. At times he was incontinent. He had lost his
strength and felt like a broken-down engine that couldn't make it over the

next hill, an old gnarled oak about to come down.[3] His doctors wouldn't allow him to return to Sagamore Hill until Christmas Day. Writing for the *Kansas City Star* from his hospital bed, Roosevelt claimed that he was praying to God in his "infinite goodness and mercy" to give him a "speedy death."[4] But once he was back on Long Island for the holiday season, Roosevelt busied himself with reading William Beebe's *A Monograph of the Pheasants* and keeping up his lists of birds and wildflowers. "In it you say by inference that the grouse of the Old World and the grouse of the New World are in separate families," Roosevelt complained to Beebe in a letter, "although I believe that three of the genera and one of the species are identical."[5]

Harold Ickes made an appointment to see the Colonel, as he and others called Roosevelt, at his Manhattan office, only to be told that Roosevelt had been rushed to a hospital. Ickes found himself wondering whether to visit his hero's sickbed, perhaps offer a final good-bye, and cheer him up with Bull Moose stories, or to give the family privacy. He chose the latter. He came to regret the decision. On January 6, 1919, Roosevelt died in his sleep at Sagamore Hill of a lung embolism made worse by multiple arthritis. He also had serious heart problems.[6] Instead of mourning, Ickes, like many other Bull Moosers, reread Roosevelt's writings about how conservation had taught him to achieve peace in dying. "Nature is ruthless, and where her sway is uncontested, there is no peace, save the peace of death," Roosevelt had written, "and the fecund streams of life, especially of life on the lower levels, flows like an immense torrent out of non-existence for but the briefest moment before the enormous majority of the beings composing it are engulfed in the jaws of death, and again go out into the shadow."[7]

Funeral services were held on January 8 at Christ Church in Oyster Bay. The Army Air Corps dropped laurel wreaths over Sagamore Hill to start the day of national mourning.[8] Four hundred to 500 people attended the Episcopalian service to celebrate the ex-president's life. They buried him in Young's Cemetery, a village burial ground on a knoll situated between Sagamore Hill and downtown Oyster Bay. He was eulogized as the only American president who hadn't needed a crisis to be great. Today the grave is surrounded by the Oyster Bay National Wildlife Refuge: 3,204 acres of freshwater ponds, salt marshes, and subtidal habitats. "He was the most encouraging person in the world," said Edna Ferber, who would later write a novel about Alaska, *Ice Palace*. "The strongest character in the world has died. I have never known another person so vital, nor another man so dear."[9]

After Roosevelt's death, all the conservationist groups in the country, particularly those Roosevelt had been associated with at the uppermost level of New York society, offered ideas about how to honor him properly. William Temple Hornaday, for example, suggested placing a marble shaft, like the Washington Monument, on the highest point in Central Park.[10] Charles Sheldon wanted a second moose reserve (like the one on Fire Island) created in his honor in Alaska's Kenai Peninsula. Dr. C. Hart Merriam thought a new subspecies of bear should be named after the Colonel. President David Starr Jordan of Stanford University, a leading teacher of Darwin and a progenitor of Pinnacles National Monument, lobbied for a new national park in California, to be named after Roosevelt. The novelist Hamlin Garland, saying that a "mountain had slid from the horizon," wanted to name the Front Range of the Rockies after Roosevelt. "Death and Roosevelt do not seem possible partners," Garland wrote in his diary. "He was life abounding, restless life."[11]

A consensus soon developed that Roosevelt wouldn't have liked sad remembrances.[12] A few weeks after Roosevelt's death, the Grand Canyon was at last upgraded by Congress from a monument to a national park. The American Society of Mammalogists started publishing the *Journal of Mammalogy* as a quarterly aimed at researching and protecting mammals in the wild.[13] The Boone and Crockett Club urged that Sequoia National Park be renamed Theodore Roosevelt National Park to honor Roosevelt's conservationist ethos.[14] Secretary of the Interior Franklin Lane and Director of the National Park Service Stephen Mather both approved of this idea within a week of Roosevelt's death (but for bureaucratic and political reasons, the name was never changed). Eventually, in 1978, 70,447 acres of the Badlands near Medora, North Dakota, where the Colonel had spent time as a cowboy in the 1880s, would become Theodore Roosevelt National Park. As he would have wanted, buffalo and antelope were reintroduced into the park; his Maltese Cabin and Elkhorn Ranch sites were preserved as the "cradles of conservation."[15]

Gifford Pinchot decided that Roosevelt's death was an opportunity to inspire people to take action for wild Alaska. The Colonel, he believed, hadn't really died but like a big brown bear had lumbered into a deep winter sleep. Pinchot served on the Roosevelt Permanent Memorial National Committee, and he knew that the Roosevelt mystique would continue to influence a national audience for only so long. Pinchot wrote an aggressive article, "Overturning Roosevelt's Work," for the *Christian Science Monitor*, lambasting corporate Republicans who wanted to put Alaska's natural sites on the auction block. Concerned that Roosevelt's national forests in Alaska—the Tongass and Chugach—were going to be irreparably marred by private-sector entities searching for oil, gas, and phosphate, Pinchot reminded leaders that Roosevelt, in a message to the Fifty-Ninth Congress, had denounced the "looting" of public lands. Pinchot argued that the *real* memorial to Colonel Roosevelt would be for big business itself to renounce the molestation of Alaskan landscapes.[16]

Harold Ickes, a feisty, combative bureaucratic infighter, wanted to keep the Bull Moose conservation movement alive, and he succeeded. When Franklin D. Roosevelt was elected U.S. president in 1932, he selected Ickes as secretary of the interior. For the next eight years, Ickes always asked himself when reading documents: What would Theodore Roosevelt do? The answer was quite simple—promote the outdoors life, save parts of wild America, create wilderness areas, and properly manage forests and game for future generations to enjoy. Ickes had learned from the Colonel, who always promoted conservation, a central lesson: the U.S. government was the best steward of public lands—not the corporations or businesses that leased them for quick, short-term profits.

A roster of those who sought to thwart Roosevelt's conservation movement from 1901 to 1919 isn't worth a lot of ink. The Bristol Bay canneries Roosevelt had worried about succumbed to coastal erosion and fires.[17] There were also the Alaskan timber barons who tried to destroy the Tongass, politicians in Juneau who wanted to blast gaping holes in the Wrangell–Saint Elias, and reindeer and caribou breeders in Nome, ignorant of genetics. A group of U.S. senators from western states almost persuaded Congress to abolish the Chugach National Forest. Rich and powerful in their day, they've ended up in the trash can of U.S. history as exploiters of public lands. During his seven and a half years in the White House, Roosevelt outflanked the land skinners by withdrawing coal, minerals, oil, phosphate, forests, and waterpower sites from private ownership, and thereby saving wilderness from ruin for *the people*. Abusers of the land, when attacked by Roosevelt, curled up into a ball, afraid to be poked at under the glare of publicity. William Howard Taft learned the hard way what double-crossing Roosevelt with regard to Alaskan lands meant in raw political terms. Taft's allowing Alaskan coalfields to be exploited by the Morgan-Guggenheim syndicate, in essence, impelled Roosevelt to leave the Republicans to form the Bull Moose Party.[18] Taft is now remembered as a nearly bottom-rung president, lacking in executive skill.

"America has known over-concentrations of power before," David Brower, executive director of the Sierra Club, wrote in a foreword to *Wilderness: America's Living Heritage*. "Such men as Theodore Roosevelt, assuming a mandate summoning great courage, and deciding that he would rather wear out than rust out, came to grips with the graspers of power. He won that round. But graspers don't stay down, are not self-limiting, and are usually too insensitive to perceive the damage they do. The people have to speak."[19]

One old-school naturalist who truly grieved over Roosevelt's death was John Burroughs. Oom John, as Roosevelt had called him, purposefully avoided the funeral on Long Island; he felt unable to bear the spectacle of thousands of mourners lining up pro forma to stare at an ex-president's coffin. Burroughs waited for all the horse-drawn carriages and automobiles to leave Oyster Bay and then made his own journey from Poughkeepsie to Long Island with only an escort. Burroughs, now at least eighty, needed a walking stick to climb up the knoll to Roosevelt's grave. A drizzling rain cast a pall over the woods. A meditative Burroughs contemplated the burial site in silent reverie. Roosevelt had died; what more could be said? He had been a great man; with more humility, he might have equaled Lincoln. Now he was decomposing in the ground. For all his grandeur, Roosevelt never wanted a mausoleum. According to his instructions, he wanted to be buried among the living Long Island birds; his name was to be engraved on his simple headstone, along with his dates: "1858–1919."[20]

Burroughs knew that Roosevelt, whose appetite was insatiable, had tasted all of summer's bounty. Roosevelt was always talking about public service and the national spirit. He was a wilderness warrior. He was seldom neutral. Yet for all his ability to arouse people, Roosevelt was a calming force in the outdoors. Somehow he saw himself in a birch, a bear, or a bee. Yes, Burroughs was certain, Roosevelt had been *the* indispensable force in the fight for conservation in America from the Civil War to World War I. Unafraid to accept both God and Darwin, inspired by *On the Origin of Species*, Roosevelt had helped save birds, shores, rivers, lakes, mountains, mammals, fish, and forests. Certainly, Roosevelt knew the demonic side of nature, the brutal laws of the jungle, the crushing potential of instantaneous death by predator. But, more important, the fresh air and wonderful solitude of the outdoors would allow future American citizens to feel free.[21]

Walking silently away from Roosevelt's grave, carefully taking small steps to avoid slipping, Burroughs coughed. The trees were bare except for a stand of evergreens. Flicking his cane absentmindedly, Burroughs, with his wizened face and long, gray beard, seemed to be in a trance. Was this really the end of the road? Would the owlish eyes of Theodore Roosevelt no longer watch over the forests? Much as when Walt Whitman died, Burroughs realized that Roosevelt's spirit was *somewhere* . . . down the road, released from his grave, marching in a parade. And Burroughs knew that he himself was not long for this world. Feeling older than the Catskills, he did a lot of metaphysical thinking that winter. "More and more I think of the globe as a whole," he wrote, "though I can only do so by figuring it to myself as I see it upon the map, or as a larger moon. My mind's eye cannot follow the sweep of its curve and take in more than a small arc at a time. More and more I think of it as a huge organism pulsing with life, real and potential."[22]

Burroughs died in 1921, somewhere in Ohio, on a train traveling from California to New York. His last words were: "How far are we from home?"[23]

II

With the deaths of Theodore Roosevelt, *John Muir, and John Burroughs, the popular actors in the early environmental movement, the first thrust of the U.S. conservation crusade had come to an impasse. The stalwarts, however, forged forward with commitment and verve. Dr. E. W. Nelson, chief of the Biological Survey, for example, traveled to Alaska to establish an experimental laboratory in Unalakleet (at the head of Norton Sound just north of the Unalakleet River) to study parasites and diseases in reindeer. A veterinarian, a pathologist, and two grazing analysts were assigned to Unalakleet to investigate whether the reindeer browsing over huge spreads were killing native grasses. Two years later, the survey moved the domestic reindeer-caribou experimental station to Nome.[24] A major concern of Nelson's was the inherent genetic problems of native caribou breeding with imported reindeer from Norway and Russia.

Dr. C. Hart Merriam survived until 1942, collecting data on Alaskan bears as his lasting tribute to Roosevelt. With Muir gone, Merriam also focused his studies on California's Sierra Nevada, which extend 400 miles from Fredonyer Pass in the north (just west of Summerville) to Tehachapi Pass in the south (seventy miles northwest of Los Angeles). Like Nelson, Merriam continued his detailed taxonomic work on Alaskan species, with a scowl of distrust toward technology. The wildlife biologist Olaus Murie of Minnesota, always self-sufficient in the outdoors, became both Nelson's and Merriam's point man in Arctic Alaska, studying the great migratory caribou herds south of the Brooks Range.

When Charles Sheldon, the "father of Denali National Park," heard of Roosevelt's death, he felt discouraged. Without Roosevelt to rally the conservationists, many wildlife preservation initiatives in Alaska were bound to lose steam. Sheldon had hoped to bring Roosevelt with him to see the grizzlies of Denali; now that idea would never happen.

Sheldon turned more and more to his conservation-minded children to help him collect data for the Biological Survey and the Smithsonian Institution. He began seeing *everything* in terms of stewardship. His daughter Carolyn Sheldon, for example, published authoritative papers on Vermont jumping mice (genus *Zapus*).[25] His son William Sheldon started collecting new biological information on Dall and stone sheep, and made an expedition in China to conduct comprehensive research on giant pandas.[26] William went on to earn a PhD in biology from Cornell University. He later published a definitive work, *The Book of the American Woodcock*, with the University of Massachusetts Press, about the squat, short-legged shorebird whose range was the Atlantic coast and the Midwest in America.[27]

Anybody interested in wildlife and exploration in the 1920s eventually ended up spending time at Charles Sheldon's home in Washington, D.C. His library of works about the outdoors had more than 6,000 volumes; Roosevelt had called it the "choicest" in America. Yale University later acquired the rare books to form the core of a special collection. Regularly, Sheldon hosted dinners at his home for polar explorers such as Richard Byrd and Roald Amundsen. Peary had died in 1920, and Sheldon deemed it necessary to embrace the new generation of Arctic and Antarctic pioneers. But it wasn't all cold weather for Sheldon. Famously, he lived with the Sere Indians on Tiburon Island in the Gulf of Mexico, collecting artifacts and oral histories. He also became associated with the National Conference on Outdoor Recreation from 1924 to 1928, helping President Calvin Coolidge advance national wildlife policy in Alaska, and he joined forces with Professor William S. Cooper to stress the importance of a signed executive order to create Glacier Bay National Monument out of Muir's Inside Passage wonderland.[28]

After Roosevelt's death, Sheldon began corresponding intensely with George Bird Grinnell about fauna and flora. Both conservationists had been alive during the Civil War. And now, suddenly, it was 1920; automobiles had replaced horses, and new, younger, more technocratic types had entered the fields of wildlife biology and ecology. Taxidermy was fast becoming a lost art. But purposefulness, drive, and commitment never leave a person whose vocation or trade happens to be his lifelong passion. Together these outdoorsmen—Grinnell and Sheldon—remained determined to save Alaska's declining bear population, and to make sure that Admiralty Island would not be ruined. They had their work cut out for them. Secretary of the Interior Franklin Lane—appointed by President Woodrow Wilson—opposed the protection of bears. Sheldon also reported to Grinnell that the Alaskan legislature was lampooning members of eastern sportsmen's clubs like the Boone and Crockett Club as aristocratic New Yorkers who were out of touch with the hardscrabble north country. Alaskan newspapers derided bear protection as Hornadayism. "There are rumors that Hornaday is writing a pamphlet on the protection of the Alaska bear," a worried Sheldon wrote to Merriam on February 28, 1920. "If he does, this will finally prevent future possibility of ever agreeing with the Alaskans on the protection of it and will consider it on the basis of their dislike of him. I hope that these reports are not true."[29]

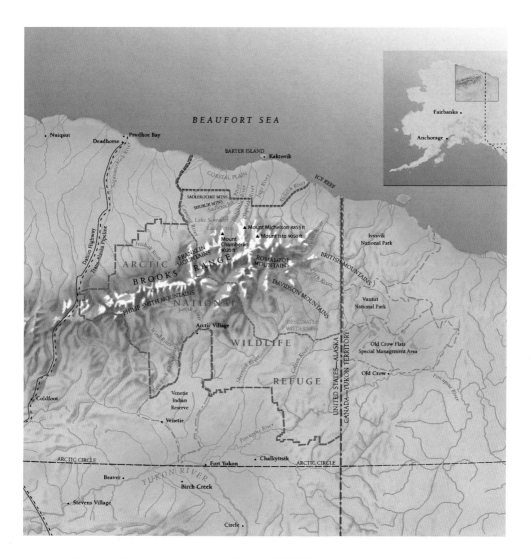

At 19.2 million acres, the Arctic National Wildlife Refuge supports the greatest variety of plant and animal life of any park or refuge in the circumpolar Arctic and is a crown jewel among Alaska's wilderness areas. On December 16, 1960, President Dwight D. Eisenhower established the original 8.9 million-acre Arctic National Wildlife Range. Twenty years later, President Jimmy Carter more than doubled its size, and established the Arctic National Wildlife Refuge. The coastal plain "1002 area" remains unprotected. Repeated attempts to open this area to oil drilling have been narrowly defeated.

Hulahula River valley, Arctic National Wildlife Refuge. © ART WOLFE.

Pacific loon nesting on the coastal plain. More than 180 species of birds migrate to the Arctic coast from six continents and all fifty states to feed on abundant insect life, and breed and raise young before returning to their native winter grounds.
© SUBHANKAR BANERJEE.

Because large herds of Porcupine caribou converge on the coastal plain to calve each spring, the region is known as the "American Serengeti." © FLORIAN SCHULZ.

The once-endangered musk ox is a relic of the ice age, and lives year-round on the Arctic Refuge coastal plain. Musk oxen give birth to their young between mid-April and mid-May, when the region is still fully covered in snow. The original Alaska musk oxen were exterminated in the late 1800s, but through conservation efforts beginning in the 1930s, the species is being gradually reestablished. © ART WOLFE.

A buff-breasted sandpiper engages in a courtship dance on the coastal plain, Jago River. These sandpipers migrate each year from Argentina to the Arctic Refuge to nest and rear their young. The species has a small world population estimated at only 15,000 birds, and has been identified as one of the top five species at greatest risk if there is oil development on the Arctic Refuge coastal plain. © SUBHANKAR BANERJEE.

Arctic wolf tracks appear overnight on the banks of the Canning River. © DAVE
SHREFFLER.

Above: Red fox hunting voles on the Arctic coastal plain during an early autumn snowstorm. Below: Willow ptarmigan in early fall on the north side of the Brooks Range along the edge of the mountains and the coastal plain of the Arctic Refuge.
© HUGH ROSE PHOTOGRAPHY.

Northern hawk owl hunting for voles in the boreal forest within the Arctic Refuge.

© HUGH ROSE PHOTOGRAPHY.

Polar bears are uniquely adapted to the harsh demands of the wild Arctic landscape, and are integral to the web of life that flourishes on the Arctic ice pack. © STEVEN KAZLOWSKI.

Subadults play-wrestle in the Arctic National Wildlife Refuge. Human consumption of fossil fuel has placed this fragile land and the polar bear in peril, and the continued survival of this mammal is uncertain. © STEVEN KAZLOWSKI.

Upper Jago River region, Alaska. © ART WOLFE.

Semipalmated plovers prefer rocky Arctic riverbeds for nesting sites. © ART WOLFE.

Caribou at Joe Creek, Arctic National Wildlife Refuge. © AMY GULICK.

Above: Caribou crossing the Kongakut River, Arctic National Wildlife Refuge.
© AMY GULICK.

Below: Oil development at Prudhoe Bay, west of the Arctic Refuge. © AMY GULICK.

Grizzly bear. All three species of North American bears (black, polar, and grizzly) range within the borders of the Arctic National Wildlife Refuge. It is a place of timeless ecological and evolutionary processes where one can experience solitude, self-reliance, and adventure. © FLORIAN SCHULZ.

Within the Biological Survey a feud developed. Sheldon had bitten his tongue instead of criticizing Hornaday's extreme animal rights rhetoric while the Colonel was alive. Although Roosevelt had prevented Hornaday from joining the Smithsonian Institution's safari in British East Africa (not wanting to deal with a loose cannon for months at a time), in 1910 he had firmly endorsed *Our Vanishing Wild Life* in the *Outlook*. Sheldon, an unrepentant hunter, thought Roosevelt had made a mistake linking himself with such an uncompromising maverick as Hornaday. Now, with Roosevelt gone, Sheldon tried to discredit Hornaday as being an irresponsible rabble-rouser with only a few good ideas about protecting seal rookeries. Sheldon worked hard as a lobbyist to build bridges. Hornaday, by contrast, was always accusatory, always at war, and he always used the sharpest language possible. Even though Hornaday had legions of enemies, he continued leading the wildlife protection crusade until his death in 1937.

Clearly, the deaths of Roosevelt, Muir, and Burroughs were a political setback for a conservation movement with Hornaday at the helm. While these three wilderness warriors were alive, there had been a sense that victory was certain, a radiant confidence that corporate despoilers would be contained. All Roosevelt had to do was shout *Those swine!* and the conservationists *felt* empowered, *felt* that history was on their side. Muir, through the Sierra Club, was influential and even feared: his every article or utterance seemed to be etched for the ages like the Ten Commandments. Burroughs, admired by *everybody*, was always able to get financial titans such as Thomas Edison, Andrew Carnegie, and Henry Ford to lobby Congress for bird protection laws—his clout (aided by his twinkling eyes of good faith) was strong, and his influence was compelling, even with profit-driven industrialists.

With these conservation leaders gone, the public debate over the value of wildlife in America degenerated. Presidential leadership for conservation during the 1920s, in fact, was anemic. The cause suddenly seemed out of joint with the antiregulatory spirit of the times. A popular belief in eastern business circles was that Alaska's Brooks Range and Arctic Circle were nothing but wastelands, frozen flats where only caribou and lemmings lived, valuable only if oil or gold could be extracted. Lacking any order except nature's own, the North Slope, according to the pro-development argument, could be divided, surveyed, regulated, mapped, and separated into homestead sections that anyone could own for a minimal fee. The U.S. Chamber of Commerce in Alaska promoted private ownership rather than forest reserves and wildlife reserves. Mount McKinley National Park, with its famous peaks, inviting to the eye, was accepted by Alaskan boomers because it would attract tourists to the railroad stop and curio shop of McKinley Station, a leg-stretch junction between Seward and Fairbanks with North America's tallest mountain looming in the near distance. But the rest of Alaska was available for the plundering of natural resources. In America during the booming 1920s, greed was king, and coal and oil were the prized sources of energy. Also, a new technology was being applied off the beach near Santa Barbara, California—offshore drilling. Oil speculators were starting to look for oil leaks all around Alaska's seas.[30]

Every decade in Alaska brought a new buzzword to promote industrialization and the conquest of the wilderness. During the Great War, the newest things in large-scale mining were hydraulic mining and dredging. Roosevelt had promoted both of these techniques to construct the Panama Canal. But now, in Alaska, wealthy absentee owners were buying up or leasing claims along rivers, shipping in heavy machines, and ripping into the land. The dredges were boatlike vessels that floated in artificially formed ponds. Using an array of steel buckets, they dragged gravel from the bottom of a pond, searching for gold. By the time of Roosevelt's death there were more than twenty-five dredges in the Seward Peninsula alone. A mill could process more than 12,000 tons of ore daily. By 1920 the Alaska Juneau Mining Complex along the Gastineau Channel was the biggest low-grade-lode gold mine on earth.[31]

And oil was starting to be discovered all over Alaska—good news for the territory. Between 1902 and 1933 twenty-seven new oil wells were dug. Eight of these failed to reach oil-bearing rock, and eleven were "no shows" (the term used at the time), but eighteen—all in the Katalla Slough claim—did produce oil. According to *Alaska Business Monthly*, the depth of the wells ranged from 366 feet to 1,810 feet. It was understood by conservationists that once Alaska became a desirable oil field, saving vast tracts of wilderness through congressional action or even by executive orders would be a far more difficult proposition.[32]

III

Aldo Leopold took the deaths of Roosevelt, Muir, and Burroughs just as hard as Merriam, Nelson, Hornaday, and Sheldon did. All three had, to one degree or another, been Leopold's inspirations. No longer would Leopold defer to anyone in his own area of expertise—he himself was the new front line. Quitting the Albuquerque Chamber of Commerce, he rejoined the U.S. Forest Service to help protect more than 20 million acres of the Southwest. Leopold's partially formed vision of roadless wilderness lands inside national forests started to take firmer shape. In 1922, he submitted a formal proposal to the chief of the U.S. Forest Service, William B. Greeley, to have the Gila National Forest of New Mexico administered as a wilderness area; it was approved on June 3, 1924. That same year Leopold, a father of four, moved to Madison, Wisconsin, and started working for the U.S. Forest Products Laboratory as an assistant (later associate) director. Daily, Leopold grew perturbed that in the 1920s, the idea that bigger was better held sway. What worried Leopold was the fortune seekers' insistence that having steam shovels create ditches to drain marshes dry or giant circular bandsaws to cut up sequoias was somehow a technological advancement for modern America. In a series of letters and articles, he described big companies as being blind to the ecological destruction they often wrought.[33]

Leopold felt that the market hunting in Alaska was reminiscent—morally—of what had happened to the Great Plains buffalo in the nineteenth century. Without proper game laws, Alaskan caribou and Dall sheep would vanish. His revulsion at such slaughter of wildlife deepened. Wildlife resources, he insisted, should be handed down to future generations undiminished. "It appears to be a fact that even in the remotest region of Alaska indiscriminate slaughter is spelling the doom of the game supply," he said, at around the time of Roosevelt's death. "No wilderness seems vast enough to protect wildlife, no countryside thickly populated enough to exclude it."[34] The U.S. Biological Survey urged Alaskan fur wardens to arrest and prosecute poachers, whose carnage amounted to criminality. "No people," Ernest Walker warned Alaskans in 1921, "should forget that it is their duty to pass into posterity all that can be saved of our wildlife, for future generations likewise have a claim to it." [35]

Following the lead of Roosevelt—who had created fifty-one federal bird reservations—Leopold started calling for new wildlife refuges to protect threatened species such as the ivory-billed woodpecker (*Campephilus principalis*). "It is known that the Ivory-bill requires as its habitat large stretches of virgin hardwood," Leopold wrote in an article in *American Forests*. "The present remnant lives in such a forest, owned and held by an industry as reserve stumpage. Cutting may begin, and the Ivory-bill may be done for at any moment. The Park Service has or can get funds to buy virgin forests, but it does not know of the Ivory-bill or its predicament. It is absorbed in the intricate problem of accommodating the public which is mobbing its parks. . . . Is it not time to establish particularly parks (or their equivalent) for particular 'natural wonders' like the Ivory-bill?" [36]*

While Leopold—like Sheldon—continued hunting, he had become an activist like Hornaday with regard to species protection. But, haunted after shooting a wolf in New Mexico and watching its eyes as it died, Leopold was repentant by the 1920s. Over his objection, roads had been constructed in the Gila National Forest to allow hunters easier access to deer. By killing off wolves to make the Gila "safe" for sportsmen looking for a few days of kicks in the controlled wild, Leopold had inadvertently robbed the Gila of its primeval wildness. Leopold, along with his wife, Estella Bergere, started hunting with a bow instead of using a rifle, as part of the concept of a "fair chase." And he worked overtime to save North American species from extinction. [37] Whether it was a refuge for the condor in California, antelope in Nebraska, grouse in Missouri, or spruce partridge in Minnesota, Leopold was for it. Dispelling the misperception of bears as predators to be eradicated, he promoted their abundance *everywhere*. "That there are grizzlies in Alaska," he wrote, "is no excuse for letting the species disappear from New Mexico." [38]

---

* In 1921, Leopold wrote an important article for the *Journal of Forestry*: "The Wilderness and Its Place in Forest Recreational Policy." He argued that every state needed to have at least one large wilderness area with no commercialism and no roads. Leopold was advocating "virgin stands," a forest policy that he believed offered human psychic renewal, in contrast to urbanization.

An ardent supporter of Leopoldian conservation in Alaska was Frank Dufresne. Nobody knew the Alaskan wilderness quite as intimately as Dufresne. *Brrrr* . . . was a regular condition in his life. The hyperactive, wiry Dufresne traveled 17,000 miles by dogsled to inspect herds of moose, caribou, seals, otters, deer, and walrus.[39] He lived for months at a time in solitude. He could predict the weather. And as early as April, before the bushes bloomed, he could tell whether it was going to be a good year for wild mountain cranberry, salmonberry, or rose hips. Dufresne first came to Alaska from New Hampshire to both hunt and protect big game. More naturalist than game warden, he ended up writing three influential books about his outdoors life: *Alaska's Animals and Fishes* (1946), *My Way Was North: An Alaskan Autobiography* (1966), and *No Room for Bears* (1965).[40]

What made Dufresne unique among agents of the U.S. Biological Survey were his elegant dispatches from the Arctic, coupled with his soldier's sense of duty. Influenced by Roosevelt and Sheldon, Dufresne wrote government reports with panache, as if he were submitting them to *The New Yorker*. They conveyed a sense of life cycles; of death from old age and disease; of January's hardships; of desolation. When he saw a raven or a magpie hovering overhead, he knew there was a fresh kill. "There comes a particular uncanny, deathly stillness in the air at seventy below zero," he wrote in his report of January 1924, to E. W. Nelson at the Biological Survey. "No wild thing seems to stir. . . . The heavy breathing of our dogs, the squealing of the sled runners and the crackling of our own breaths in the air sound loud and harsh and seem to be violating this brooding silence of the north woods. It seems we are the only things that dare move— But no! There in the riffling shallows of an open waterhole a tiny, grey bird dashed and flits about with all the grace of a flycatcher. . . . Our map tells us we are forty miles north of the Arctic Circle; our thermometer tells us it is seventy below zero, yet there is a frail little bird seemingly unsuited to cold weather having the very time of its life. It is, of course, the Water Ouzel, or Dipper. . . . It requires considerable steeling of one's conscience to blast that little life into eternity for the cause of science."[41]

Dufresne was collecting specimens for the Biological Survey by killing and tagging them. Because Dufresne was respected as a hunter—and everybody knew he was *the* Alaskan outdoorsman, amazingly adept with a gun or a coil of rope—many sourdough Alaskans listened to his pleas to squeal on poachers in the backwoods and to make citizen's arrests of game hogs. Regularly he reached out to fellow Alaskans about protecting both bears and salmon. Dufresne refused to travel with ultra-conservationists like those in the Sierra Club. Nevertheless, he recognized the essential role that such preservationist groups played in protecting wild Alaska. "In a way I believe we owe something to the ultra-conservationists," he wrote, "who, by the very unreasonableness of their demands, have rationalized the press of Alaska to assume the middle ground."[42]

Preservationists of the 1920s and 1930s in turn owed Dufresne a debt for holding the fort in Alaska, for methodically teaching citizens of the territory to recognize that their wildlife resources weren't limitless. By taking a good old boy's approach to being a warden, being part of the day-to-day Alaskan milieu, Dufresne helped conservation principles take firm root in outback towns and hamlets. At public forums, his firm persuasiveness—expressed on his face by something halfway between a grin and a scowl—was palpable. "Help us keep this kind of fishing," was his simple plea to civic groups. "Your own boy might want to come up here some day."[43]

Chapter Ten

o   o   o   o   o   o

WARREN G. HARDING: BACKLASH

I

Oil—*that was the new rush* in Alaska. Between 1910 and 1920, huge oil and gas reserves had been discovered at Elk Hills, California, and Teapot Dome, Wyoming. Appetites were whetted. Alaskan boomers believed that it was only a matter of time until oil was struck in their vast backyard, and that oil would make them as rich as John D. Rockefeller. Theodore Roosevelt's secretary of the interior, James Garfield, spoke for all ultra-conservationists when he described Rockefeller in his diary, now housed at the Library of Congress, as a cold-blooded reptile: "Never have I seen a more sinister, avaricious face—repulsive and deceitful. I disliked to shake his hand, but of course could not cause comment by not doing so. . . . I wonder if anyone—outside his family—really cares for him apart from his money."[1]

The election of Warren G. Harding of Ohio as the twenty-ninth president of the United States, in November 1920, deeply depressed Leopold, Sheldon, and Merriam. Harding, who had owned a newspaper in Ohio, believed that the pro-business Republican old guard had received a mandate vote—which was certainly true. He felt duty-bound to act on Rockefeller's principle that the only good oil field was a drilled one. No sooner had Harding been sworn in as president, on March 4, 1921, than he opened up public lands in Alaska for development. Taking aim at the Bull Moose conservationists, he issued Executive Order No. 3421, under which the U.S. Department of Agriculture was to abolish the designation of Fire Island in Alaska as a national moose refuge.[2]

But the conservationists' sense of muted desperation after Harding's election in 1920 didn't last long. Citizens in Wyoming, angered over corruption in government, demanded that Harding's secretary of the interior, Albert Fall, a known foe of the conservationist clique inside the U.S. Forest Service, be investigated for land fraud. Fall's shady dealings became known as the Teapot Dome scandal. The courts eventually decided that the Harding administration *had* illegally leased the U.S. Navy's petroleum reserve No. 3 in Wyoming (near a rock outcropping resembling a teapot) to Harry F. Sinclair of Standard Oil without competitive bidding. At that point Harding had been in office for barely a year. Teapot Dome was just another sleazy grab of public lands, like the Alaskan coal mines controversy of 1909 over which Pinchot and Ballinger feuded. The decent folks of Wyoming, however, wouldn't tolerate it. In 1921, Fall was indicted for conspiracy and accepting bribes. He was fined $100,000 and sentenced to a year in prison, earning the ignominy of being the first U.S. cabinet officer in history to serve a prison term for misdeeds in office. The oil fields were restored to the U.S. government by court order, and Teapot Dome remained the symbol of political corruption until Watergate in the 1970s.[3]

Although Teapot Dome captured the newspaper headlines, Alaskan public lands also suffered under Harding's pro-development administration. But plagued by various scandals, and looking for an escape from journalistic criticism in the spring of 1923, Harding scheduled a trip to Alaska. As the historian Thomas Fleming aptly put it in the *New York Times*, Harding wanted to "get [away] from the stench that was rising in Washington" over graft in his administration.[4] Harding, accompanied by three members of his cabinet—Herbert Hoover (State), Henry Wallace (Agriculture), and Hubert Work (Interior)—and others in the administration, went aboard the SS *Henderson*, steaming northward from Tacoma, Washington. Harding would be the first U.S. president to visit the Alaska territory. Excitement ran high in the territory because on February 27, by Executive Order No. 3797-A, Harding had withdrawn 23 million acres, extending from the Arctic Ocean to the Brooks Range, as Naval Petroleum

Reserve No. 4. Although the Naval Petroleum Reserve was too remote to be drilled, there were reports of oil seepage, and Harding was encouraging private companies to make claims there.[5]

The Harding party arrived in Seward on July 13. The Fairbanks *Daily News-Miner* called it "The Glory of the Coming." The first lady, Florence Harding, had also come on the tour, and the rumor mill was full of speculation that Harding was trying to mend a break caused by his adultery. Following World War I the Republican Party had started encouraging non-Native settlement in Alaska, and now Harding immediately started preaching the doctrine of prosperity to Alaskans. After a rally in Anchorage, he headed out to inspect the Chicaloon coalfields. Starting in 1914, Alaska's coalfields had been placed in public entry—a low bid would get an entrepreneur a lease for extraction. A gouging of Alaska was under way, particularly in the coal seams just north of Mount McKinley. Bored by the grand scenery, not even stopping to hear a bird trill when he visited the national park, Harding seemed indifferent to the blue skies and green woods of Alaska. No meetings with game or forest wardens were included in the itinerary. Even when a moose crossed the road, Harding yawned. Complaints from fishermen that the coal and timber industries were polluting the Gulf of Alaska fell on deaf ears. Harding never had any burning curiosity about the natural world. Talking to handpicked audiences, he implied that he wanted to hear the *kaboom* of dynamite across the last frontier. Someday Ketchikan would be bigger than Seattle and Fairbanks, a new Minneapolis at the top of the world. On July 15, Harding played the part of an engineer on a railroad run from Wasilla to Willow. And then he drove a golden spike with a maul at Nenana, a new railroad hub to symbolize the completion of the 470-mile line connecting Seward and Fairbanks.[6]

President Harding missed two hammer blows in driving the gold spike; but this event was actually a high accomplishment compared with his folly regarding his wardrobe. Listening to Admiral Hugh Rodman, a supposed climatologist, Harding told his entourage to wear heavy wool sweaters, parkas, galoshes, gloves—the whole array of winter clothing—even though it was mid-July. The temperature hit ninety-five degrees, and some members of Harding's party collapsed from heat prostration and dehydration. Knowing nothing about Alaska except the profitability of drilling, timbering, and mining had its downside. Meanwhile, reports reached Harding, as he crossed Prince William Sound in a naval ship, that many of his "Ohio gang"—cronies who used a green house on K Street in Washington, D.C., as their headquarters—were being indicted. This was unsettling to the president.[7]

Nevertheless, Harding pressed on with his Alaskan junket. What he hoped to convey in Alaska was his desire for mechanized progress in this last frontier. No longer were a pick and shovel needed to look for gold. Technology had turned the search into a corporate endeavor complete with large-scale machines and hydraulic mining techniques.[8] There was still wilderness to be conquered—lots of it. Harding, serving as a mouthpiece for big business, also announced his plan to take a ride on the Copper River and Northwestern Railway, owned by the Morgan-Guggenheim syndicate. This seemed a deliberate insult to Pinchot, who was now governor of Pennsylvania but who remained an ardent opponent of the syndicate and its development efforts in Alaska. As it turned out, the ride was canceled at the last minute. The first lady, who was five years older than the president, had a bad stomach and fallen extremely ill. A few years earlier she had lost a kidney, and this had caused her health to deteriorate in general.

Harding's Alaskan trip then took an awful turn. On the voyage back to San Francisco aboard the *Henderson*, he himself became gravely ill, possibly from shellfish poisoning. Severe stomach cramps overcame him. He felt clammy and dizzy. His usual ruddy complexion had gone sheet-white. Somehow the president managed to deliver a speech in Vancouver, British Columbia, before a crowd of 40,000 well-wishers. But once offstage he continued complaining of abdominal pains. He was not only sick but in a foul mood. When told that the ship had serious maintenance problems, and that water was flooding into a cargo compartment, Harding snapped, "I hope this boat sinks."[9]

Conspiracy-minded Alaskan boomers suspected, crazily, that some associate of Pinchot's, or a bitter fisherman, had deliberately poisoned the physically exhausted Harding. No evidence of such poisoning has ever been found; but a few days after becoming sick in Sitka, a spent Harding died in a San Francisco hotel suite on August 2. Reporters called it the "curse of Alaska." The official diagnosis was a heart attack, but physicians said that the possible bout of food poisoning might, hypothetically, have triggered the infarction. Progressives had long called for American seafood to be inspected by the Food and Drug Administration so that people wouldn't get sick from shellfish, but Harding and his associates had scoffed at the notion. A cross-country funeral procession took place, and then Harding's flag-draped coffin was placed in the U.S. Capitol's rotunda. A few days later Harding was buried in a mausoleum in Marion, Ohio. American conservationists were scathing in their assessment of Harding as a steward of the land. They believed that since the creation of the U.S. Department of the Interior in 1849, Harding had been its poorest custodian. But many Americans loved Harding.

II

Replacing Harding as president was Calvin Coolidge, perhaps most remembered for his famous line "The business of America is business."[10] In terms of personality, the taciturn Coolidge seemed the polar opposite of Roosevelt. But Coolidge, even with his belief in limited government, recognized that Roosevelt had been a force of nature. He praised Roosevelt's efforts to build the Panama Canal, the Great White Fleet, and reclamation dams throughout the West as great American work. As much as Coolidge abhorred Roosevelt's penchant for what he considered excessive federal spending and overtaxing, he felt that his predecessor's desire to protect America's wildlife and create national parks wasn't such a bad thing. Coolidge—despite his image of being as lifeless as a waxwork—was an avid fly fisherman; he spent much of the summers of 1926, 1927, and 1928 in waders, gleefully looking for trout.[11]

When, in 1924, Congress at last allowed Native Americans to become citizens, Coolidge had marked the event by wearing a feathered headdress; and he was glad when a Tlingit, William Paul Sr., became the first Native elected to the Alaskan territorial legislature. Coolidge thought natural resource management should be decentralized. He encouraged states to develop their own conservation plans. In a highly symbolic act, Coolidge worked to protect Alaska's moose population, perhaps demonstrating with this small gesture that he cared about the natural world. And on a summer fishing vacation in the Black Hills (Custer State Park),

President Coolidge suggested to the sculptor Gutzon Borglum that Roosevelt should be included on Mount Rushmore along with Washington, Lincoln, and Jefferson.[12]

Somewhat surprisingly, and to his everlasting credit, Coolidge *did* create (with Congress) five spectacular new national parks—Bryce Canyon (Utah), Great Smoky Mountains (Tennessee), Grand Teton (Wyoming), Shenandoah (Virginia), and Mammoth Cave (Kentucky). But he ultimately rubber-stamped Alaskan oil and gold development projects promoted by treasure seekers. Sportsmen's clubs, such as the Boone and Crockett Club, hoping to protect wilderness in Alaska, would be marginalized by his administration. Much like wetlands, swamps, and deserts, Arctic tundra was considered a wasteland by Coolidge—devoid of aesthetic value. Besides the scenic new national parks, Coolidge's most notable measure with regard, presumably, to conservation was changing the name of TR's Reclamation Service (which had become an independent agency in 1907) to the Bureau of Reclamation in 1923.[13]

While the eminent ecologist William Skinner Cooper was fighting to have Glacier Bay saved as a national monument, in homage to John Muir,[14] Charles Sheldon was still trying to shame Congress into being a good steward of Mount McKinley. In December 1920, Sheldon testified to a House Appropriations Committee that the national park desperately needed federal funding for more game wardens, more law enforcement, and tougher laws to prosecute poachers who hunted Dall sheep. His voice had an almost scolding quality, with no suggestion of humor. Once again Sheldon defended the idea of protecting brown bears in Alaska. For Sheldon, watching Harding and Coolidge try to undo so much preservationist work in Alaska was disheartening.

Sheldon died suddenly of a heart attack in Nova Scotia in 1928. He had been vacationing at his family's summer cabin when he collapsed. It was a hard loss for the movement to absorb. Sheldon was buried in Rutland, Vermont. Besides his crusade for Mount McKinley, he had been lobbying to protect North American antelope on the eve of his death. The Boone and Crockett Club, always ready to memorialize its leaders, joined with the National Audubon Society in purchasing 4,000 acres of Nevada's Great Basin and creating the Charles Sheldon Antelope Refuge in 1931. President Herbert Hoover, who admired Sheldon, had issued the executive order. Today the Nevada refuge has more than 575,000 acres administered by the U.S. Fish and Wildlife Service, protected land that constitutes one of the few intact sagebrush steppe ecosystems in America. The reserve was enhanced by the introduction of bighorn sheep in 1968.*

Dr. C. Hart Merriam, heartbroken at the loss of his dearest ally, made sure that Charles Scribner's Sons posthumously published Charles Sheldon's naturalist diaries as *The Wilderness of Denali* (1930). It became a classic of the "hook and bullet" genre. Denali's rugged mountains, glacial streams, boggy plateaus, rushing rivers, and green-blue glaciers had found their most enduring chronicler in Sheldon. From the plaintive whistle of the golden plover (which flew from Central America to nest in Denali) to a lordly moose (with sixty-seven-inch antler spears), Sheldon had captured the drama of an entire ecosystem for posterity to ponder. *The Wilderness of Denali* was a gift to the nation. Its publication inspired a new wave of preservationist sentiment for Alaska's mountain ranges, such as the Fairweather, the Saint Elias, and the Wrangell. Taken together, these Alaskan places formed a huge semicircle of more than 1,000 miles from the Sitka region to the end of the Alaska Peninsula. Inspired by *The Wilderness of Denali*, Franklin D. Roosevelt, sworn in as U.S. president in March 1933, immediately allocated federal funds to Mount McKinley National Park. From the grave, the "father of Denali National Park" had been heard.[15]

III

While 1928 was known in conservation circles as the year Sheldon died, the big national event was the election of Herbert Hoover as U.S. president that November. Hoover, from the outset, was a conundrum to conservationists. He was a Wall Street–big business Republican, and he believed strongly in deregulating business. His chamber-of-commerce attitude didn't bode well for conservation-

---

* Sheldon's voluminous personal papers are now housed at different locations: University of Alaska-Fairbanks, Smithsonian Institution, Dartmouth College, and the Boone and Crockett Club.

ists. But he was also an avid fly fisherman and an active leader in the Izaak Walton League. Hoover, in fact, was a true believer in "fish reservations." To Hoover—unlike Harding—the outdoors mattered a great deal. As secretary of commerce, for example, Hoover had considered the very existence of dirty and polluted water barbaric. Pushing forward tough antipollution laws with the zeal of Gifford Pinchot, he had called for the Bureau of Fisheries to end the "steady degeneration" in "commercial fisheries in the Northwest of Alaska." Speaking to the U.S. Fisheries Association in September 1924, Hoover said that he wanted to "cultivate a sense of national responsibility toward the fisheries and their maintenance . . . to make a vigorous attempt to restore the . . . littoral fisheries on the Atlantic Coast; to secure the prevention of pollution from sources other than ships both in coastal and inland waters; to undertake the reinforcement of stocks of game fish throughout the United States." [16]

Sadly, as U.S. president, Hoover refused to put his idea of "fish reserves" forward in a meaningful way. The Republican Party had become a hostage to corporate interests. Looking for a way to promote his big business agenda in Alaska, Hoover focused on reindeer farming, believing that the territory needed more slaughterhouses and packing plants to compete with the cattle stockyards of Chicago, Omaha, and Kansas City.

IV

If there was a symbol of Alaska's vanishing wildlife in the 1920s and 1930s, it was the bald eagle. When the Pilgrims first arrived on the curled toe of Cape Cod, 500,000 eagles soared in the American sky. But colonists blamed these raptors for disturbing livestock, and an open season commenced. "For my part," Benjamin Franklin had stated, "I wish the Bald Eagle had not been chosen as the representative of our country. He is a bird of bad moral character; he doesn't get his living honestly. . . . Besides, he is a rank coward." [17] This anti-eagle attitude spread to Alaska, where half of America's eagles lived. In 1917, the territorial legislature enacted a bounty on eagles in support of fishermen and fox farmers who claimed that the raptors were snatching their livelihood from streams, rivers, and lakes. "Our national symbol, sad to say," Peter Matthiessen wrote in *Wildlife in America*, "subsists largely upon carrion; its alleged depredations on the salmon of Alaska, like its other crimes, have been grossly exaggerated."[18]

From 1920 to 1940, the National Rifle Association (NRA) in Alaska was intensely promoting the shooting of eagles. Aldo Leopold, who was conducting game surveys of midwestern states as a private consultant, later took the matter up with the NRA's president, Karl T. Frederick. "We gun enthusiasts are constantly complaining of restrictive legislation on firearms," Leopold wrote. "Is it likely that the public is going to accord us any more respect and consideration than we earn by our actions and attitudes? . . . I would infinitely rather shoot the vases off my mantelpiece than the eagles out of my Alaska. I have a part ownership in both. That the Alaska Game Commission elects to put a bounty on the eagle, and not on the vase, has nothing to do with the sportsmanship of either action."[19]

In July 1933, Leopold accepted a new chair of game management in the Department of Agriculture and Economics at the University of Wisconsin. As the desolate news of eagle loss continued unabated, Leopold tried to stir public consciousness against bounty hunting of birds of prey in Alaska. Besides being glorious to look at, eagles were part of the web of life in Alaska. But those who considered eagles a nuisance continued slaughtering tens of thousands of these birds. To counter Benjamin Franklin's negative view of eagles, Leopold quoted Ezekiel 17, telling how one of the raptors broke off the top of a cedar "and planted it high on another mountain, and it brought forth boughs, and bare fruit, and was a godly tree." Although skeptical about Hebrew silviculture, Leopold used the Old Testament, when it was convenient, to give eagles a better image in the public mind.[20]

Coming to the rescue of the American bald eagles was the veteran women's suffragist and raptor conservationist Rosalie Edge. Long before Rachel Carson described the dangers of DDT in *Silent Spring*, her environmental manifesto of 1962, Edge, from her Hawk Mountain Sanctuary in eastern Pennsylvania—the first rehabilitation center for birds of prey—warned against the chemical companies, gun manufacturers, and logging conglomerates that were trying to exterminate these magnificent creatures. Edge was a fearless activist, a feisty, independent spirit who couldn't accept the idea of America without eagles, hawks, and owls.

A New York socialite reared with patrician values, Edge defended wildlife by writing stinging articles and giving lectures and speeches. When Roosevelt, Burroughs, and Muir died and the conservationist movement was losing its spirit, Edge came to the fore. In a fourteen-page profile, the *New Yorker* accurately described her as somehow resembling both Queen Mary and an excited pointer on the hunt (adding that her crusade to rehabilitate eagles was "widespread and monumental").[21]

What concerned Edge in general, however, was the degradation of nature by industry. Without her activism, it's doubtful that Congress would have created Sequoia–Kings Canyon National Park in California in 1940, or that developers would have been prevented from diverting Wyoming's Yellowstone Falls in Yellowstone National Park. Stoop-shouldered, with a face remarkably like Eleanor Roosevelt's, Edge refused to be complacent. She contended that huge companies and manufacturers were interested only in dollars. They weren't to be trusted when it came to protecting nature. In the West they had to be tightly regulated by the U.S. Department of the Interior, but it was often in cahoots with the companies to which it was leasing land. Teapot Dome, Edge believed, was merely the tip of the iceberg with regard to corruption occurring between the U.S. federal government and private corporations.

Edge also believed that conservation began at home, and she led a crusade to ban the shooting of hawks and eagles in the Kittatinny Ridge of Pennsylvania. Every year these birds migrated from Canada to roost along fish-rich streams in Schuylkill County. She used the word "sanctuary" to connote that protecting birds of prey had a religious or missionary element. Edge was by 1920 the new William Temple Hornaday, an indomitable protector of species. She frequently claimed that killing bald eagles was as sacrilegious as slashing Emanuel Leutze's painting *Washington Crossing the Delaware* would be. Every fall, Edge was disgusted by the annual *sparbenbarich*, a local term (derived from German) for a massacre of thousands of hawks. The unethical hunters, with dozens of dead hawks strewn about them, would proudly smile for photographers.

When Edge was in Paris, she received a provocative pamphlet written by Willard Van Name, W. Dewitt Miller, and Davis Quinn: "A Crisis in Conservation: Serious Danger of Extinction of Many North American Birds." According to these authors, eagles, hawks, and owls were being systematically wiped out in the United States. Edge was repelled: Wasn't the bald eagle our national emblem? Didn't owls help farmers by eating small rodents? Having fought for women's suffrage with Carrie Chapman Catt at her side, Edge knew something about grassroots activism and about winning battles. Dissatisfied with the Audubon Society, which was sitting on the sidelines, and critical of its active founder Gilbert Pearson, who seemed rather lackadaisical about protecting eagle, hawk, and owl populations and habitats, Edge founded the Emergency Conservation Committee (ECC). There were about 300 species of raptors in the world—including hawks, eagles, and falcons—and the EEC wanted them protected.[22]

Edge marshaled a number of leading scientists to defend wildlife. She also sued the Audubon Society for misrepresenting itself as protecting birds. She believed that the society had become compromised by trophy hunters, timber barons, the pesticide industry, and government bureaucrats on the take. A lawyer for the Audubon Society tried to humiliate Edge by calling her "a common scold." Years later, recalling how she had first learned of the insult, Edge scoffed, "Fancy how I trembled."[23] Edge's lawsuit sent a wave of fear through conservation societies, impelling them to support her action. Roger Baldwin of the ACLU helped her as the plaintiff; the ornithologist Frank M. Chapman documented her claims; and a court ruled in favor of the ECC.

If only in terms of grit, Edge became the most effective reformist champion of national parks and wildlife habitat preservation of her era. Biodiversity and ecology informed her public dissent. Reporters loved to interview Edge, whose candor had become legendary by the time Herbert Hoover was president. She was called the "Hawk of Mercy," and her Pennsylvania sanctuary for birds of prey attracted visitors from all over the world. She became a celebrity. Always beautifully dressed, with a silver dragonfly brooch on her lapel, Edge became the conservationist darling of progressives and inspired an outpouring of concern for the survival of birds of prey.

According to the biographer Dyana Z. Furmansky, the pamphlets that Rosalie Edge published for the ECC had a profound effect on both Franklin D. Roosevelt and Harold Ickes. Here was the old Bull Moose conservation spirit being dispensed by a deeply informed woman whose wit matched that of TR's first child, Alice Roosevelt Longworth. By representing her crusade as David versus Goliath, Edge easily won public sympathy. Smitten with her pluck, Ickes regularly summoned Edge to the Department of the Interior for friendly chats. A friendship developed between Ickes and Edge. "Their relationship became as uniquely symbiotic as the one she had developed with scientists and bureaucrats reluctant to advocate publicly on behalf of their unpopular views, and sometimes as secretive," Furmansky writes. "Edge understood that for conservation's new breed of national policy makers to stem the tide of nature's destruction, they needed help from the ECC. Its pamphlets built 'public support in advance of action,' so that leaders could point to how they were fulfilling the informed will of the people. The policy makers needed the fresh input from the policy shapers, and fresh input is what the ECC would give them repeatedly." [24]

But for most Americans of the 1920s, Alaska's declining bald eagle populations—Edge's widely disseminated ECC pamphlets notwithstanding—seemed very remote, of interest only to the conservation cult. Hawk Mountain was near New York City whereas Haines, Alaska, where eagles roosted by the thousands, might as well have been Greenland or Timbuktu. Once the Klondike gold rush had faded from memory, Alaska wasn't much in the news in the East, except for Warren Harding's fatal junket. The first motion picture ever filmed in Alaska, *The Cheechakos*, released in 1924 by the Alaska Motion Picture Corporation, bankrupted the company. To tourists, visiting Yellowstone or Yosemite seemed possible on a week's vacation. Going to Mount McKinley, by contrast, seemed to be a summer-long endeavor for which an outfitter was needed.

Yet, thanks to *National Geographic* magazine, there was a growing public fascination with Alaska's polar bears and snowy owls. The far north had its fans and produced some fads. Santa Claus had reindeer. Salmon was a favored dish in New York restaurants. The Isaly Dairy Company of Youngstown, Ohio, was marketing a square of vanilla ice cream dipped in chocolate and wrapped in icy-looking silver foil: the logo for these Klondike bars was a smiling polar bear, as cuddly as a teddy bear. Walt Disney, the great cartoonist, also had an eye on Alaska, gearing up to make the Academy Award–winning documentary *Winter Wilderness*, which starred polar bear cubs and wolf pups.

And Rockwell Kent, as irascible as ever, continued promoting the far north with his expressive Alaskan paintings. Always pushing nearer to the north pole, Kent eventually wrote three books about his adventures in Greenland: *N by E*, *Salamina*, and *Greenland Journal*. Perilous treks with dogsled teams in below-zero weather became his persistent theme. His strongest supporter was Marie Ahnighito Peary, daughter of the great explorer Robert Peary. As Barry Lopez noted in *Arctic Dreams*, Kent found Alaska and Greenland holy shrines at which sojourners discovered "Godlike qualities" in themselves.[25] To Kent, the gatekeepers of the Arctic paradise were bald eagles (the fact that Ben Franklin thought these prey birds had "questionable moral character" only increased Kent's admiration for them).[26] For the dust jacket of *Salamina*—the title was the name of his Eskimo housekeeper in Greenland—he drew an inspired portrait of a bald eagle defending the quiet world from industrialization and mechanized progress.[27]

Leopold, Edge, Kent, and other conservationists were continually infuriated from 1917 to 1953, because Alaska's territorial legislature established a bounty system for eagles. It was originally 50 cents an eagle and rose to $2 over the years. The Biological Survey joined forces with the defenders of wildlife and pleaded with the territorial government to rescind its open season on eagles. During those years well over 128,000 bald eagles were shot or poisoned. The crux of the problem was that Alaskan fishermen believed bald eagles were gorging on salmon, depleting rivers, streams, and bays. Every eagle nest found by a professional Alaskan fisherman was destroyed or ransacked. According to the territorial governor Ernest Gruening—a Harvard graduate and a former managing editor of the *New York Tribune*—the mass slaughter of eagles had "become more or less an established custom."[28] The National Rifle Association (NRA) called shooting bald eagles the "purest of all rifle sports."[29]

Giving scientific credence to the pleas of Leopold, Edge, and Kent was Olaus Murie of the Biological Survey, who had spent 1936 and 1937 writing detailed reports on the bald eagle populations of the Aleutian Islands. These Alaskan eagles courted in March and laid eggs in April. Hatching took place in June and fledglings emerged in late August. Most important, Murie (and Hosea Sarber) concluded from studying eagles' stomachs that salmon *weren't* essential to an eagle's daily diet. Alaskan fishermen were grossly exaggerating the situation, just as Florida's fishermen had exaggerated the situation with pelicans. Stepping up to defend the rights of bald eagles was Flying Strong Eagle of the American Indian Association. It was sacrilege, Flying Strong Eagle argued, to massacre the national emblem of America because of a dispute over fishing; eagles, he argued, were a sacred species. They were the embodiment of wilderness, freedom, and strength. "Although the evidence did not persuade Alaskan legislators," the historian Morgan Sherwood wrote, "over the years the eagle's case was heard sympathetically by a variety of outsiders." [30]

Triumph came on June 8, 1940, with the passage of the Bald Eagle Protection Act. Congress at last recognized that eagles were essential to Alaska's ecosystem. No longer could Lower Forty-Eighters pursue, shoot, poison, wound, kill, capture, trap, collect, molest, or disturb bald eagles for "any purpose." But there seemed to be no attitude of self-congratulation in ornithological circles regarding the legislation. Congress, although it prohibited the destruction of eagles in the states, exempted the Alaskan territory from this ban. [31]

Eventually, killers of bald eagles could be fined up to $10,000. But by the 1960s chemicals such as DDT were starting to wipe out the species in the Lower Forty-Eight. Alaska was not an important agricultural area and, consequently, much less DDT was released there, so its bald eagle population held firm. However, the continued reckless clear-cutting of Alaska's old-growth trees was troubling. Bald eagles built their nests up to eight feet across and seven feet deep on top of Sitka spruce and western hemlock. These trees continued to be timbered in the Tongass and Chugach at an unhealthy, even maniacal rate. This loss of habitat looked as though it might spell doom for the bald eagle. But committed activists started an effective campaign after World War II to rescue these magnificent birds from going the way of the dodo and the passenger pigeon.

Chapter Eleven

∘ ∘ ∘ ∘ ∘ ∘

BOB MARSHALL AND THE
GATES OF THE ARCTIC

I

W hen it came to translating conservationist ideas into preservationist action, spreading the idea of wilderness across the North
American continent, Robert Marshall had no peers. Born in
New York City the year Theodore Roosevelt became president—1901—
Marshall became the first university-trained forester to promote the
urgent need to save Alaska's Brooks Range and Arctic tundra from commercial despoliation. Marshall's father, Louis, was a high-priced constitutional lawyer, regularly dining with the Manhattan social set, but young
Bob became infatuated with "Knollwood," the family's summer camp at
Lower Saranac Lake in New York's Adirondacks. During his childhood,
his first heroes were Lewis and Clark, whose brave exploration into an
"unbroken wilderness" he wanted to imitate in Arctic Alaska.[1] Marshall's
boyhood hikes in the Adirondacks and his hero worship of James Feni-
more Cooper's buckskin-clad pathfinders were the genesis of what would
eventually become the Gates of the Arctic National Park and Preserve and
the Arctic National Wildlife Refuge. And in 1935, four years before his
death, Marshall cofounded with Aldo Leopold and six others The Wilderness Society, a nonprofit organization that has led the conservation movement in "battling uncompromisingly" for wilderness protection, helping to
save 56 million acres from commercial development in Alaska alone.[2]

Committed philanthropy came naturally to Bob Marshall, who was raised in Manhattan's upper-class world of comfort and ease. His father had routinely doled out five-digit checks to New York–based nonprofits defending minority rights, including the American Jewish Committee. Infuriated by anti-Semitism, Louis Marshall led ferocious civil rights campaigns. With the U.S. Constitution as his sword, Marshall regularly sued institutions that barred Jews, in particular, from membership. He believed American Jews had an ancestral obligation to end their silence and confront anti-Semitism head-on. His most famous showdown against the WASP establishment was fought over the Adirondacks' Lake Placid Club, founded by Melvil Dewey (originator of the Dewey decimal system and state librarian of New York). The club's wealthy patrons had waged a surprisingly fierce campaign to bar Jews from membership in their exclusive 9,600-acre resort. After a bitter stalemate and under extreme legal pressure from Marshall, Dewey was eventually forced to capitulate. Dewey, in the end, admitted that in the United States exclusion of Jews from private clubs should *always* be forbidden. "I have succeeded in getting Dewey's scalp," Marshall bragged to a friend. "The result is most gratifying."[3]

Impressed by Marshall's legal prowess, the *Jewish Tribune* soon described him as the fourth most influential Jew in the world, after Albert Einstein, Chaim Weizmann, and Israel Zangwill; Marshall was the only American among the top five.[4] Like most Jews during the progressive era, Louis Marshall saw Theodore Roosevelt as a stalwart champion of their cause. In 1906 Roosevelt had become the first U.S. president to appoint a Jew to a cabinet position: Oscar S. Straus, as secretary of labor and commerce. Unusually for a politician of his era, Roosevelt supported a Zionist state around Jerusalem.[5] Furthermore, when Roosevelt won the Nobel Peace Prize for mediating in the Russo-Japanese War, he donated part of his cash award to the National Jewish Welfare Board. For his part, Louis Marshall backed many of Roosevelt's conservation initiatives to protect state-owned forestlands in upstate New York. According to Marshall's well-constructed argument, the cutting of oak, elm, and spruce to get logs for sawmills should be balanced by the formation of permanent wilderness reserves. "Blister, rust, canker, and insects are infinitely less dangerous than *Homo sapiens*," Marshall declared, "who, whether he takes the form of a lumberman, or a tax title exploiter, a vandal, or a commercial hotelkeeper, is the real enemy of the forest."[6]

Young Bob Marshall—pleasant-mannered, funny, and enormously energetic—was a chip off the old block. A hyperactive, rugged athlete, Marshall wasn't so much a bookish prodigy as a full-bodied, ravenous enthusiast for learning. He had a bubbling intensity that suggested pent-up steam. Marshall's piercing eyes were certainly his most notable feature and helped distract attention from his protruding front teeth. As a second-grader (or thereabouts) he was already saying, parroting what adults had told him, that recklessly destroying a forest was akin to treason against humanity. He had an intense passion for the history of North American forests. He was able to quote the venerable John Burroughs verbatim. He declared that nature was a cure for the "strangling clutch of a mechanistic civilization."[7] He all but memorized Ralph Bonehill's *Pioneer Boys of the Great Northwest*.[8] And he even found Darwin's principles of natural selection easy to grasp.

Marshall's mother, Florence, died in 1916, leaving Bob in the hands of a succession of nannies and assorted help. He took emotional refuge in Knollwood, tramping around the deep woodlands with his brothers George and James, silently thinking in the mountain coolness. Like Meriwether Lewis at the Continental Divide, Marshall started naming places around Lower Saranac Lake: Found Knife Pass, Squashed Berry Valley, Hidden Heaven Rock.[9] Routinely he examined tamarack, white spruce, and red spruce, peeling off their coarse bark for closer scrutiny. The Adirondacks encompassed a variety of forest communities including conifer swamp, lowland conifer, hardwood conifer, northern hardwood, mountain conifer, and alpine—ideal for an aspiring forester. After spending twenty-five summers at Knollwood, Marshall would consider himself an amateur expert on the Adirondacks State Park ecosystem. Sparrows chirruping, basins hollowed for clear lakes, engulfing solitude, the distant sound of timbering, the swarming insects—all were relished by Marshall. Influenced by his father, he emphasized the concept of *reforestation*. The interconnected Adirondacks ecosystem had a mystical harmony that made perfect sense. The sheer physicality of the great forestland molded Marshall into manhood.

Congress had passed the National Park Service Act in 1916 to protect

America's natural wonders. Under the leadership of Stephen Mather, the National Park Service launched a public outreach effort to engage young people with the outdoors. Marshall heard the call. Between 1918 and 1934 he climbed forty-two peaks, all over 3,000 feet, some days hiking more than thirty miles. Adopting Verplanck Colvin, a post–Civil War surveyor of coniferous forests, as his new role model, he pledged his life to the "Forever Wild" movement in the Adirondacks. Alaska became far more than facts to be memorized for a geography class. He began wanting to hear bears growling and lynx screeching. He imagined the slants of evening light around Mount McKinley. Preparing for the ordeal, he started getting into tip-top physical shape. In September 1901 Theodore Roosevelt (who was then the vice president) had climbed to the top of Mount Marcy (5,343 feet). Marshall naturally followed in his idol's footsteps, camping atop the summit. "I love the woods and solitude," Marshall wrote in a school essay. "I like the various forms of scientific work a forester must do. I would hate to spend the greater part of my lifetime in a stuffy office or crowded assembly, or even in a populous city." [10]

Forgoing Ivy League schools, Marshall instead attended the New York State College of Forestry in Syracuse. (The environmental service school was founded, in large part, by his father.) Marshall aspired to the forestry skills of Pinchot, the all-seeing naturalist's eyes of Muir, and the stiff spine of Roosevelt. Although Marshall was of average height, he gave the impression of being shorter because his shoulders were stooped from too much reading. What made Marshall unusual among forestry scientists was that the spirit of Thoreau stayed with him as he worked at experimental stations throughout the Rocky Mountains. Dutifully Marshall kept notebooks of his outdoor hikes. He wasn't yet twenty when, during his free time, he developed a new trail system for the Adirondack Forest Preserve. Believing in the restorative qualities of the New York woods for city dwellers, he compiled a thirty-eight-page guidebook, *The High Peaks of the Adirondacks*, aimed at helping greenhorns enjoy boreal forests.

Upon graduating from Syracuse in 1924 with a BS in forestry (he was fourth in a class of fifty-eight), Marshall followed the Lewis and Clark Trail for a summer, up the Missouri River and across the Continental Divide to the Oregon coast. In letters home he extolled the beauty of Lolo Pass and The Dalles. He happily hiked around the Willamette Valley with bulging backpack and trusty compass. One look at the three great Pacific Northwest mountains—Hood, Baker, and Rainier—made him feel full of vitality.[11] Nature could, he understood anew, lord it over civilization. Everywhere Marshall rambled in the Cascades he tried to strike up conversations, saying, "Hello, I'm Bob Marshall" as if he were running for public office.[12] With low mists in the dimples between the hills behind him, he set up his tent on a vacant beach. Although he never mastered the art of canoeing, Marshall became an early advocate of hiking as a sport, routinely marching thirty or forty miles a day. Often he wore tennis shoes instead of heavy leather boots for these long rambles; the tennis shoes provided far better traction.

Footsore, Marshall returned to the East in the fall of 1924 and headed for Harvard Forest in Petersham, Massachusetts. In 1907, Harvard University—not wanting to be outshone by Yale University's world-class Forestry School—acquired a 2,100-acre tract of woodlands in north-central Massachusetts. All the new modern forestry techniques were being taught at this experimental station in Worcester County.[13] Approximately 1,000 types of trees grew in America, and Marshall was determined to study the seed-to-growth rate of them all. He could take classes in forest ecology, harvesting, and fire management, and he learned about a burgeoning field called forest recreation. He improved his skills with saw, wedge, and ax. Pine trees were examined as potential forest products for turpentine and resin. Thousands of Americans earned their living in the lumber industry. Timbering was big business. One common denominator of "big timber" and conservationists, Marshall learned at Harvard, was that both loathed the insatiably destructive bark beetle. But eradication of this common pest was about all the two feuding factions could agree on.

Marshall also learned about conservation land trusts, whose birthplace was in nearby Waverly, Massachusetts. In 1891 Charles Eliot, the son of Harvard University's president, led a successful effort to save two dozen ancient white oaks in Waverly along the sand hills of Beaver Creek. His logic was simple. Just as libraries acquired rare books and museums collected fine art, communities should treat groves of trees as a local heritage. Under Eliot's leadership, the United States' first regional land trust was established (today it is referred to as the Trustees of Reservations).

The Waverly oaks taught Marshall an important lesson: forestry preservation wasn't just about large-scale national reserves. It must also be
localized and community-based.[14]

Even though Marshall was studying at Harvard Forest, his life and attitude had been dramatically changed by the Cascade Range, the sweeping
Columbia River, the volcanic mountains, and south-central Washington's
evergreen stands. Full of exciting tales about the Pacific Northwest, Marshall started presenting himself as an honorary Sierra Club type to his
fellow students, who were enraptured by his stories of the utter solitude
to be found in western forests. Several pursuits were simultaneously important to Marshall. Besides forestry, he was enamored with phenology,
studying periodic biological phenomena such as flowering, breeding,
and migration, in relation to geology and climate. How the insect world
damaged forests was another enduring interest. His friends in New York
started calling him Mr. Silviculture or Little GP (after Gifford Pinchot).

Marshall earned his master of forestry degree at Harvard in the spring
of 1925. To celebrate, he headed to his beloved Adirondacks backcountry
to climb a few 4,000-foot peaks with stalwart friends. Shedding his suit
of black broadcloth for alpine wear, he eventually ascended all forty-six
mountains in the Adirondacks that were over 4,000 feet high. At the summits he studied unusual lichens and mosses. His high-altitude climbing
also led to the creation of a New York mountaineers' club, the Adirondack
Forty-Sixers. Whenever Marshall hiked, he took careful notes about the
health of thickets of spruce and balsam. The outdoors life became his
theology. Refreshed and invigorated from climbing so high that even
the treetops below were not distinguishable, Marshall reported for duty
at the U.S. Forest Service in Washington, D.C. He asked to be dispatched
to Fairbanks, Alaska, but found himself assigned to the Northern Rocky
Mountain Forest Experiment Station in Missoula, Montana.

Missoula was an attractive college town with coffeehouses, outfitters'
shops, taprooms, and a giant stone M emblazoned on the west face of Sentinel Mountain. The forestry club at the University of Montana was considered one of the best in America. And everybody in town, it seemed, had
harrowing stories to tell about March blizzards, sheep-thieving cougars,
or a rampaging grizzly named Old Ephraim. Marshall found himself feeling strangely at home among the blue-collar hunters and no-nonsense

ranchers. The aristocratic expression left his countenance, for in Montana he was a forest scientist, not merely a bright boy with a trust fund.

In Glacier National Park country, he learned that forestry was a hard taskmaster. Timbering had to be controlled along the Idaho-Montana alpine border, and forest fires were a menace to the thickly wooded region. Sparks from a fast-traveling train could cause more extensive fire damage than an arsonist. Ever since the "Big Burn" in 1910, fire lookouts had been constructed from Missoula to the Canadian border. In July 1925 Marshall was principally responsible for extinguishing more than 150 fires in Idaho's Kaniksu National Forest. Watching mountainsides of ponderosa pine and sweet cedar burn like a "nebulous planet" taught Marshall a lesson about the "unconquerable, awful power of Nature."[15] All the firefighting supplies available were hauled up the Priest River. But they weren't enough to save the western white pine forests of Mount Watson. More than 2,000 virgin acres were burned. Only after a "thousand man-days of labor" was the fire suppressed.[16]

From June 1925 to August 1928 Marshall ranged all over Montana and Idaho, in and out of every threadbare frontier outpost in the heart of the northern Rockies. Meticulously he applied his Harvard training to analyzing sensible methods for cutting and replanting. Wandering high into a jagged line of snow-covered peaks, he often would stay at a forlorn ranger shack or a lumber camp. The loggers gossiped about his long-distance hikes, which he took despite hailstorms and downpours; his adventures became local folklore. Because Marshall explored widely, he made a memorable impression. These were the days before *Outside* magazine; hiking forty miles a day just for fun was viewed as aberrant behavior. Many people thought Marshall must be engaged in a coming-of-age rite to prove his manhood. With his toothy smile, his young man's beard, and his pack full of all-seasons gear, he seemed outlandish. He'd grow a mustache for a week, then shave it off, then let it grow again. With ropes around his waist and a boyish grin, he was a most arresting presence.

Further baffling hard-bitten Missoulans was Marshall's pride in being an American Jew. Most Jews in Montana downplayed their identity; some changed their names. But no matter where Marshall was in the wild, he'd observe the Sabbath. In September 1925, on Yom Kippur, the Day of Atonement, Marshall hiked into the backcountry. The glory of autumn was on display; the aspen leaves looked like gold coins. He believed that nature was a synagogue, that fasting in a virgin pine forest along the Continental Divide allowed him to clear his mind of frivolous thoughts. Being mindful of God's wild creation was religious for Marshall. That observance of Yom Kippur, fittingly enough, made Marshall even more committed to devoting his life to forest preservation.

Throughout his months in alpine Montana Marshall corresponded with his family and friends about the Northern Rocky Mountain Forest Experiment Station. Nobody before or since has written about western Montana's green-gold somnolence with as much grace as Marshall, who described everything from jackrabbits to larch needles. And his personality as a naturalist was gaining tangible professional benefits. Between 1925 and 1928 he wrote several articles on forestry, lumberjack culture, and wilderness preservation. Using the Forest Service's national newsletter (the *Service Bulletin*) as his pulpit, Marshall sided with the case against roads that the young forester Aldo Leopold was making for Gila National Forest in New Mexico. Leopold and Marshall both wanted primitive areas of national forests to be declared *roadless*, free of all construction—vast tracts of public lands saved from the press of modern humanity.

Sometimes the death of one conservationist seemed to coincide with the birth of another. On September 21, 1928, Charles Sheldon died in Nova Scotia, Canada, leaving behind his voluminous unpublished Alaskan journals (known in the Boone and Crockett Club to be immensely rich in detail about the Denali wilderness). Sheldon had received modest public acclaim—he was called the father of Mount McKinley National Park. And he had a legion of friends, having served on the board of directors of the National Parks Association, the National Recreation Committee, and the National Geographic Society, and as chairman of the Commission on the Conservation of the Jackson Hole Elk. Indeed, his list of affiliations was so long that he probably lost track of them himself. Nobody, it seemed, had done more than Sheldon to save interior Alaska. The snowcapped mountains of the Alaska Range were his life story. There was a lyrical intensity to everything Sheldon did on behalf of Dall sheep and caribou. As when Muir died in 1914, Sheldon's death left a void in the conservation movement, particularly in Alaska—a void that Bob Marshall would soon fill.

Meanwhile, in the fall of 1928, Marshall enrolled in Johns Hopkins University's PhD program in plant physiology. His goal was nothing less than permanently altering the course of U.S. forest policy. Already, his landmark article in the *U.S. Forest Service Bulletin*—"Wilderness as a Minority Right"—was being acclaimed by nature enthusiasts of all stripes as a major intellectual breakthrough, a long-overdue articulation of the right to open space, miles upon miles in extent. Much of his time in Baltimore was spent in greenhouse laboratories conducting experiments with evergreen and conifer seeds. Companies such as Weyerhaeuser Lumber had destroyed great forestlands, and Marshall vowed to reverse this trend. Rising at five o'clock in the morning, Marshall would toil away until eleven at night, determined to learn the secrets of soil composition. Defining himself first and foremost as a conservationist, he believed scientists should engage in political advocacy: that is, more stringent federal regulation of "big timber." He published articles in the *Nation* ("Forest Devastation Must Stop") and in the *Journal of Forestry* ("A Proposed Remedy for Our Forestry Illness").[17] The League for Industrial Democracy issued a thirty-six-page pamphlet written by Marshall, promoting the undeniable virtues of public forests over private ones. *Minimal impact* became his creed.

Bored by his academic career but refusing to feel boxed in, the twenty-eight-year-old Marshall organized a scientific trip to the unmapped Brooks Range—named after Alfred Hulse Brooks, who served as chief geologist of the U.S. Geological Survey (USGS) from 1903 to 1924—to study the northern timberline. Independently wealthy, he was ready to leave academia and make a career as an iconoclast for wilderness preservation. He had been spending his downtime at Johns Hopkins poring over topographical maps of Alaska printed by the USGS. More than 100 million acres had not yet been mapped. The Arctic tundra held an immediate—and lifelong—fascination for Marshall. Wanting to work north of Fairbanks, Marshall set his sights on the central Brooks Range, which stretches west to east across northern Alaska and into Canada's Yukon Territory. Any stream north of this range flowed into the Arctic Ocean. The meandering Yukon north of Denali and south of the Brooks Range flowed into the Bering Sea (technically an extension of the North Pacific). Marshall, the forester, was interested in how black spruce flourished even in subzero weather. He hoped to plant spruce seeds, wanting to prove exactly where the North American tree line ended.

II

In *1929 Marshall made his first* trip to Alaska. Prior to this he had looked at a blank spot on a USGS map around the Koyukuk River and had wanted to fill in the vast white space. At long last, he got to see the northern lights. When he arrived in Fairbanks, wearing thick flannel shirts, insulated pants with suspenders, and a rainproof jacket, the bearded Marshall looked like an Amish model for an Abercrombie and Fitch catalog. Eager to get to work, he bought a ride on a plane to Wiseman, a mining hamlet located along the middle fork of the Koyukuk River in the central Brooks Range. What an adventure! Marshall's eyes popped in wonderment as he peered out the plane's window at the awesome scenery. The Brooks Range had no central feature like Denali or Saint Elias, but it was nevertheless breathtakingly magnificent. Its beauty, however, was subtle. The size and remoteness of the Brooks Range—all those peaks without names and never climbed—made it the wildest roadless area in North America. Marshall was surprised that there was so much sedimentary rock, and so little metamorphic rock. Clearly, these peaks had once been under a sea. If you climbed any peak you'd be almost certain to find marine fossils. It astounded Marshall to think that the Brooks Range was the northernmost part of the Rocky Mountains.

Only 127 people lived in Wiseman when Marshall arrived. Dusty, ramshackle, and without modern conveniences, the place was lost to the world, with not even a railway connection or a decent road to somewhere else. The centerpiece of the town was Pioneer Hall, where trappers, Eskimos, and prospectors congregated to sing old Canadian folk songs or dance the Hesitation Waltz (a two-step-count, forward-and-back waltz with a prolonged pause) late into the night. Isolation wasn't merely a part of living near the snow-blanketed Brooks Range; it was the all-encompassing reality. Nature was the reigning king in Wiseman, so remote from the daily rhythms of even Fairbanks that it might as well have been Tierra del Fuego. It was hard to make even a brief call out on a shortwave radio. But even though there was not a single railroad depot, not a single room wired for electricity, and not a single tin lizzie within 200 miles along a dogsled trail, and the closest accredited doctor was more than two hours away, with only a grudging concession made to modern medicine (pills), Wiseman was, in Marshall's estimate, "the happiest civilization on earth," a magical place where alcohol flowed freely to help people cope with the long dark winters.[18]

Wiseman had the appeal of a hillbilly moonshine hollow in Tennessee. The house joints were caulked with mud. The best one-room homes were made of hewn pine with rusty corrugated tin roofs. Everything looked haphazard and hurriedly constructed. West of town were the austere Endicott Mountains, an ideal wilderness. Hiking the range, Marshall felt elevated to previously unimagined spiritual and moral heights. In the wind-torn rawness, the desolate bleakness, he shed New York, Massachusetts, and Maryland to become, as far as possible, a primitive Alaskan. He hiked a lot. Building bonfires to stay warm, staring into the flames of resinous wood, Marshall thought of Baltimore as useless and mediocre compared with the north fork of the Koyukuk River.

Getting up at dawn, Marshall would camp around the Brooks Range with feelings of awe, disbelief, and tense excitement. He didn't even mind the weather locals called "freeze-up." Just as John Burroughs had the Catskills and John Muir the Sierra Nevada, Marshall now had the Brooks Range. With great preciseness he recorded every craggy ridge and glacier-carved valley in the notebook he always kept handy. Because so few outdoorsmen had actually lived in the Brooks Range, Marshall felt that he was discovering summits virgin to Euroamerican footprints. Being alone in the Brooks Range created a sense of total removal, a supernatural out-of-body experience, stripping his spirit from the body as in a Chagall painting. The Arctic terrain made him feel cosmically alone.[19]

Two imposing portals Marshall explored north of Wiseman—Frigid Crag and Boreal Mountain—soon became altars to him.[20] Watching clouds clear off the summits seemed holier than a prayer. Incalculable geological forces had clearly been working here. Previously unexplored by conservationists, this unforgiving part of the Brooks Range had no equal in America for rugged beauty. It bore silent witness to a great cataclysm that geologists still didn't quite understand. "His joy," Marshall's brother George recalled, "was complete when, standing on some peak, never before climbed, he beheld the magnificence of a wild timeless world extending the limit of sight filled with countless mountains and deep valleys previously unmapped, unnamed, and unknown." Standing in a blowing wind, Marshall declared these towering portals the "Gates of the Arctic." The name stuck. On December 1, 1978, Gates of the Arctic became a national monument. The four words—Gates of the Arctic—conveyed a sense of exploration, discovery, and freedom to conservationists all over the world. On December 1, 1980, 7.95 million acres of this majestic land- scape were officially designated Gates of the Arctic National Park (the second largest in the U.S. system). In Gates of the Arctic country, Mar- shall said, an outdoorsman could "get away from the rat race." [21]

Marshall's intense explorations in the Gates of the Arctic region is legendary in Alaska. He saw the Brooks Range as unique among all rec- reational assets owned by the U.S. government because it was pristine wilderness, with scarcely an industrial imprint anywhere. All around were peaks and glaciers over 9,000 feet in elevation. Like Charles Shel- don before him, he tested himself against rising rivers, the midnight sun, and hungry bears. Wearing caribou-skin parkas, boots, socks, and mittens, he looked like a Nunamiut hunter. Hiking for Marshall became an aerobic endeavor that set his heart and lungs pumping. Caribou liver became his staple food, a marvelous source of iron and protein. He eu- phorically marveled at the abundance of Arctic birds such as the semi- palmated plover (*Charadrius semipalmatus*) and spotted sandpiper (*Actitus macularius*). While Louis Marshall was in Switzerland leading an effort for Jews to resettle in Palestine, Marshall was discovering the Gates of the Arctic. And when word reached him that his seventy-three-year-old father had died in Zurich, he took the news stoically.

III

A few weeks after Louis Marshall's death came Black Tuesday: the New York Stock Exchange crashed. America was in a panic. The Great Depression had begun. For Marshall, who had been mov- ing toward socialism since his days at Syracuse University, the economic downturn was proof that capitalism was a flawed system. Companies such as Weyerhaeuser, Long Bell, and Pacific Lumber didn't give a damn about working families. The Great Depression—like World War I—strained the resources of the U.S. Forest Service in the West. Starting in the fall of 1929 the Hoover administration had the deteriorating economy as an excuse to dismantle the expansive federal reserves. The administration now argued that clear-cutting federal reserves brought jobs. The forest-products industry, predictably, agreed. Then Gifford Pinchot, now serving as governor of Pennsylvania, stepped in. Pinchot, who had considerable standing among foresters, called for an emergency meeting of America's best and brightest conservationists—including Marshall. They were to convene at Pinchot's home in Washington, D.C., in late January 1930. Pinchot was furious about a "wishy-washy" report issued by the Society of American Foresters pertaining to woodlands preservation.[22] To Pinchot, the report was disingenuous and was actually aimed at opening up public lands. Although he had retired from the U.S. Forest Service in 1910, Pinchot was not at all hesitant to oppose clear-cutting. Pulling an activist committee together, he tasked Bob Marshall and Ward Shepard with drafting "A Letter to Foresters." Marshall, a master of invective, gleefully lambasted foresters as infected by capitalistic greed. It was sacrilegious, Marshall wrote, for any *real* forester to condone the "spiritual decay" of America's forestlands. Later, Marshall would accurately describe this open letter as "the most radical action any forestry organization had ever taken."[23]

Pinchot's summit had an inspirational effect on Marshall. Clearly, Marshall was more socialist than Pinchot, but their political differences—romantic preservation versus utilitarian conservation—were inconsequential compared with the differences between them and the Hooverites they were fighting. Actually, Pinchot and Marshall were both visionaries, were both nature lovers, and were both undoubtedly among the most colorful figures in the history of U.S. forestry. "Governor Pinchot," Marshall declared, "is one of the most amazing men I have ever met. After 35 years of forestry battles, instead of being discouraged and cynical, he is entering this new fight with as much enthusiasm and interest as a boy of 20." Perhaps after the loss of his father, Marshall had found a surrogate in Pinchot. At the very least Marshall enjoyed sharing enemies with the legendary Pinchot. "He thinks Hoover, Hughes, and Mellon are all terrible," Marshall said, "believes in government ownership of natural resources, is strong for civil liberties and really is interested in everything a liberal should be." [24]

Energized by Pinchot, Marshall published a landmark conservationist article in the February 1930 issue of *Scientific Monthly*: "The Problem of the Wilderness." It was an immediate hit among conservationists. Marshall, who was just about to earn his PhD from Johns Hopkins, offered the notion that *wilderness* should be preserved for its aesthetic and spiritual values alone. He sounded much like Aldo Leopold. His message had elements of both prophecy and doomsaying. Echoing Thoreau and Muir, Marshall asserted that places like Arctic Alaska were far more valuable than a Rembrandt painting or Brahms symphony. (This was in line with the reasoning in Hornaday's *Our Vanishing Wild Life*.) "The Problem of the Wilderness" article fitted nicely with a new initiative by the Forest Service. Matter-of-factly identified as Regulation L-20, it was a new policy aimed at establishing "primitive areas" within existing national forests.

Upon earning his PhD in the spring of 1930—his dissertation was "An Experimental Study of the Water Relations of Seedling Conifers with Special Reference to Wilting"—Marshall set his sights on the Arctic Alaska watershed. It was one thing to extol the virtues of wilderness in *Scientific Monthly*. It was quite another to *demonstrate* those virtues by analyzing the positive effects pristine nature had on people living in a remote Alaskan village. The single-minded Marshall wasn't thinking about the Kenai Peninsula or the Alexander Archipelago. His mind was set on the land north of the Arctic Circle. Pulling together his interests in forestry and sociology, he decided to chronicle his firsthand experiences living among Eskimos and white settlers in Wiseman. He would escape the incessant noise of urban life and write a book titled *Arctic Village*. Marshall, a forester extraordinaire, was now poised to become the Margaret Mead of Arctic folk. By adopting the dual vocations of wilderness advocate and sociologist he would document how beneficial unspoiled wilderness was for nearby communities.

Like *Fortune* magazine's reporter James Agee living among poor Alabama tenant farmers (an experience he recounted in his urgent and timely 1939 masterpiece *Let Us Now Praise Famous Men*), Marshall hoped to dignify the people of Wiseman in his book *Arctic Village*. During the Great Depression, some Americans—even some aged southern tenant farmers—dreamed of moving to Alaska in order to survive the economic downturn. Why worry about grocery bills when, in Alaska, you could hunt moose and caribou? Drop a fishing line into any icy stream and reel in salmon, trout, and graylings. In industrial centers such as Cleveland and Pittsburgh, workers were earning 20 to 30 cents an hour; many such cities were also brutally cold in winter. Might as well move to Alaska, where game was plentiful. Although Alaska wasn't a hub of the New Deal, FDR would soon do a lot to help the territory prosper during hard times.

Marshall returned to Wiseman in 1930 and remained there for more than a year to gather firsthand observations for *Arctic Village*, making sure that even the most inarticulate resident wouldn't stay tongue-tied for long. His book depended on everyone's candor. Marshall was a careful listener. With their high cheekbones and pacific, far-seeing eyes, the Nunamiut Eskimos mesmerized Marshall. Speaking in near whispers, they told him how polar bears swam 200 miles for a fat seal, or why the ptarmigan was the hardiest bird alive (able to endure temperatures of minus fifty degrees Fahrenheit with apparent good cheer). Stories about dogsledding to the abandoned mining town of Coldfoot, once the largest community in Arctic Alaska, were favorites. Like all ghost towns, Coldfoot taught a lesson: that today's boom is tomorrow's bust. Marshall strongly believed that Alaska's natural resources should be developed *slowly*. His Eskimo friends saw nothing noble about pillaging everything valuable in one big gulp. Alaska's silver, gold, and copper were the result of aeons of natural processes. Modern man had no right to rape the land in a frenzy of greed and then leave open pits behind. Marshall had picked up a few suitable Eskimo colloquialisms and used them daily. But Inuit was difficult, and therefore Marshall's speech was never more than three or four broken sentences strung together. Before interviewing village Nunamiuts, Marshall would practice a few lines, eager to prove that he valued their distinctive culture.

Wiseman was a collection of hard-core people. Without roads, travelers had to arrive by either dogsled or a steamship from the village of Bettles. Marshall was, in a sense, the first professional visitor. Less than 5 percent of Alaska's population lived north of the Yukon River. It was *different* that far north. Residents all had a sense of being in touch with the base of life. The weather in Wiseman was coldest in January, but the darkest month was December (for thirty days there was no sunlight).

Luckily, Marshall had a lot of good books to keep him company during his twelve and a half months in Wiseman; his taste ranged from potboilers to law books to belles lettres. In his cabin, the complete works of Shakespeare, Plato, and Emerson shared space on the painted bookshelf. New nonfiction titles such as Joseph Wood Krutch's *Modern Temper*, George P. Ahern's *Deforested America*, and Gaston B. Means's *The Strange Death of President Harding* sat on a shelf behind the makeshift Franklin stove. "There is not a trace of the usual chaos of papers, books, magazines, gloves, snowshoe straps, and the like," Marshall wrote home, "but an immaculately clean oilcloth surface whereon I can spread the work of the moment without having first to shovel clear a simple space on which to set my papers." [25]

The artist Thomas Hart Benton was painting huge murals about American life in the 1930s, receiving commissions from, for example, the Missouri state capitol and the Chicago World's Fair. Marshall saw *Arctic Village* as his prose mural of one Alaskan frontier town, where everybody was his or her own master. Ancient peoples lived in perfect harmony with guys from Brooklyn and Philadelphia in search of freedom. Old sourdoughs from the Klondike gold rush of 1898 used to come over to Marshall's cabin to drink beer and play records like "Ol' Man River" and *Hungarian Rhapsody* on the turntable. Outside the wind might roar in seeming harmony with each crescendo. Moose stew and caribou steaks were favorite dishes. Homegrown turnips and potatoes were ladled out in heaping mounds from serving bowls. The midnight sun provided marvelous growing weather for certain vegetables, allowing farmers to produce giant cabbages weighing up to eighty or ninety pounds.

Political banter in Wiseman was usually aimed at the folly of Madison Avenue slicksters and the thieves at the House of Morgan. There were conversations about the Ku Klux Klan, the New York Yankees, agricultural prices, sexual escapades, and the mating habits of seals—nothing was off-limits. There was some decidedly populist bias against elites of any shape or form. There was an ingrained distrust of big government. "If them bastards would cut out some of their battleships and spend the money for aviation research," the gold seeker Vernon Walts complained, "we wouldn't have to finance people like the Guggenheims to give money to it." [26]

To honor his friends in Wiseman, Marshall started naming topographical wonders after them. Over time, U.S. Geological Survey maps accepted 164 place-names that he had conjured up in the Koyukuk region. In a fit of community patriotism Marshall named beautiful features Big Joe Creek, Ernie Creek, Harvey Mountain, Holmes Creek, Jack Creek, Kupuk Creek, Snowden Creek, and so on. Whatever trepidation his newfound friends had about this PhD from Baltimore asking questions about their sex lives and hunting habits vanished when they learned of the permanent high honor Marshall had accorded them in the Rand McNally atlas.

But Marshall's biotic journals from his expedition of July–August 1931 exploring the Alatna and John rivers are most treasured by outdoors types. Using a compass and old field guides, Marshall, accompanied by Ernie Johnson, carefully inventoried the mountain walls along the Arctic Divide. The serious-minded, scientific side of Marshall seemed to evaporate amid such magnificence. Pausing at Loon Lake—which he named because of the high concentration of Arctic loons—Marshall scribbled enthusiastic notes, which were published posthumously in *Alaska Wilderness* (1956). "Nothing I had seen, Yosemite or the Grand Canyon or Mount McKinley rising from Susitna, had given me such a sense of immensity as this virgin lake lying in a great cleft in the surface of the earth with mountain slopes and waterfalls tumbling from beyond the limits of visibility," he wrote. "We walked up the right shore among bare rocks intermingled with meadows of bright lichen, while large flocks of ducks bobbed peacefully and unmindful of us on the water of the lake, and four loons were singing that rich, wild music which they added to the beautiful melodies of earth. No sight or smell or feeling even remotely hinted of men or their creations. It seemed as if time had dropped away a million years and we were back in a primordial world." [27]

IV

Nobody has written about the beautiful desolation of the central Brooks Range with the love that Bob Marshall brought to *Arctic Village* (and later *Alaska Wilderness*). To Marshall, his time spent in the central Brooks Range, where clouds wrapped the serried peaks, was like witnessing all the snows of yesteryear in a single jaw-dropping glance. Even the best cameras couldn't capture the wavy glow of the northern lights. The sky could turn from cold gray to a huge shimmering curtain of flashes within an hour. The air was rent with silence. The frosty dew along the Koyukuk River had a distinctive purity. Every gorgeous vista in the Brooks Range seemed like a mirage. To bring the industrial order into such an ethereal Alaskan landscape would be a ruinous mistake. All of his forestry studies reached their apex here. Marshall wrote that the Arctic Circle was an experience that brought the "joy of physical exploration" into "mental continents." Just striking out across the flat tundra north of Wiseman in snowshoes, even in bitter subzero weather, exhilarated him. Every time he survived an avalanche or a washout, or was almost blown over by a cloud of snow, it made him understand how hardy the First Nation tribes were.

Marshall had found nirvana in the seeming *nothingness* around Wiseman. His sense of place, his affection for a specific locality, was focused on this serene Arctic region, where every quiet slope seemed to sing a hymn. Having adapted to the long Arctic winters, he felt privileged. The complete absence of machines gave Arctic life integrity. In a state of exaltation, Marshall declared Wiseman his enchanting community "200 miles beyond the edge of the 20th century." How tame the Adirondacks were by comparison! Many townsfolk in the Arctic Divide were poor but simply didn't know it. To Marshall the local elders had a dignity hard to find along the eastern seaboard. New Yorkers were self-centered by comparison. Broadening his source of names beyond Wiseman, Marshall started attaching Inuit terms to numerous sites he encountered in the Brooks Range: Yenituk ("white face") Creek, Pinnyanatuk ("absolute perfection of beauty"), and Karillyukpuk ("very rugged").[28] It was a world that was drawn in vibrant, sharp colors—a humbling world where the low tundra fauna burst with fresh growth, undeterred by durable permafrost. When he was back in New York, Marshall could close his eyes and imagine Wiseman set against a wide background of snow and smiles. The memory of his designations in the Gates of the Arctic filled him with joy.

Using the village of Wiseman as a sociological laboratory, Marshall put forward a theory that being surrounded by raw wilderness led to a marvelous "amount of freedom, tolerance, beauty, and contentment such as few human beings are ever fortunate enough to achieve." Where others saw desolation, Marshall saw Eden. Like many anthropological studies of the early 1930s, *Arctic Village* was influenced by Sigmund Freud's theory that it was unhealthy for humans to bottle up primal urges.[29] Every chapter presented the amazing frankness of the people of the upper Koyukuk. The farther north one went in Alaska, nature became greater and greater and man became less and less. The virgin Arctic wilderness, Marshall now argued, offered the opportunity for a sojourner "craving for adventure" to break into "unpenetrated ground, venturing beyond the boundary of normal aptitude, exerting oneself to the limit of capacity." His intimate sociological portrait of life in Wiseman—a hamlet on the edge of nowhere—was a pioneering work. As Roderick Frazier Nash pointed out in *Wilderness and the American Mind*, the words "nameless" and "trackless" and "unknown" were continually used to describe Alaskan landmasses north of the Arctic Divide.[30] Marshall, embracing each word, wasn't a member of America's wilderness cult. He personified it.

The fact that Marshall was writing *Arctic Village* didn't mean he had forgotten about upstate New York. On July 15, 1932, Marshall (along with his brother George) broke a world record, climbing fourteen Adirondack peaks in less than twenty hours. To Marshall, each peak was unique, with a personality of its own, sharp and green against the sky. There was something about Marshall's exhausting feat, however, that hinted at mania. In *The Adirondack Park: A Political History*, Frank Graham Jr. wrote that he found "something a little disturbing in all this bustling from one mountain peak to another. . . . Pull up a pumpkin and sit down for awhile, one wants to say to Marshall."[31]

Much as the novelist Thomas Wolfe wrote candidly about the citizens of Asheville, North Carolina, in *You Can't Go Home Again*, Marshall wrote—uncensored—throughout 1932 about adultery, casual gossip, and random quarrels in Wiseman. When Marshall's lawyer read an early draft of *Arctic Village*, the first word that came to his mind was *libel*. This clearly wasn't a travelogue. To avoid lawsuits, pseudonyms were quickly assigned to a few of the residents, who were also slightly disguised. And, feeling somewhat guilty, Marshall gave $3,609—half of his royalties—to the residents of Wiseman. By the time Marshall wrote a check for $18 to every adult in the village upon publication—even the dissolute idlers and wastrels—all was forgiven. The predictable upshot was that the dollar talked, even in the Brooks Range.

Marshall had started writing *Arctic Village* in earnest in a Baltimore apartment during the fall of 1931. Early drafts of chapters with titles like "Wilderness of the Koyukuk" and "The Wilderness at Home" were delivered as papers to the Society of American Foresters; these weren't vetted or bowdlerized by the U.S. Forest Service. Simultaneously, he continued to urge the U.S. Forest Service to understand that sport fishing, bird-watching, and hiking would, in the long run, bring in more money to local economies than manufacturing plywood or grazing Herefords. The U.S. government needed a long-term vision. With a first draft of *Arctic Village* completed, Marshall rented a room on C Street in Washington, D.C., and began writing a "recreation" report for the Forest Service. "One of the first things he did," his biographer James M. Glover wrote, "was compile a list of roadless areas remaining in the United States." [32] Vision was never in short supply for Bob Marshall.

When *Arctic Village* was published in 1932, it received positive reviews: all 399 pages were considered a testimony to the power of ahimsa, the concept of honoring all living entities. Rockwell Kent said, in the *New York Herald Tribune*, that the intimate portrait of Wiseman was "a classic of our native literature." In the *American Mercury* H. L. Mencken deemed the folks of the Koyukuk truly blessed because no theologians resided within 100 miles of the town center. Ruth Benedict of the *Nation* said that Marshall, in a low-key way, had written an "Arctic Middletown." Meanwhile, academics praised the new statistical information about the central Brooks Range that Marshall had interspersed throughout the text.

V

When the Democrat Franklin D. Roosevelt defeated Herbert Hoover in the 1932 presidential election, Marshall believed his ideas about wilderness and forestry might actually be taken seriously by the Department of the Interior. The election was a milestone in U.S. conservationist history. Wild Alaska had its best friend in the White House since 1909. FDR, in fact, was such a forestry buff that he called himself a "tree farmer." [33] A master talent scout, Roosevelt was open to all sorts of new conservationist ideas percolating up from the U.S. Forest Service. Philosophically, FDR wanted corporations regulated and natural resources protected. Wisely, he chose Gifford Pinchot—who in 1934 ran an unsuccessful campaign to be a U.S. senator from Pennsylvania—to become his forestry adviser. Having an acknowledged arbiter of issues regarding public versus private lands on the New Deal team boded well for wild Alaska. Immediately, Pinchot asked Marshall to write a memo on the state of national forestry policy.[34] Predictably, Marshall recommended a huge program to protect public lands in Alaska. In the controversial report Marshall stated flatly that private forestry had "failed the world over."[35]

Although Franklin D. Roosevelt was only a distant cousin of TR's, they shared a belief that conservation of natural resources was essential if America was to remain a great nation. Whereas TR's primary interest was wildlife, the young FDR defined himself as a forester. Born on January 30, 1882, in Hyde Park, New York, along the banks of the Hudson River, Franklin was enamored of all aspects of bucolic Dutchess County at a very young age. His 1,200 acres of green trees were his paradise. He learned how to nurture every square foot of his property. When the cornerstone was laid for his presidential library at Hyde Park in November 1939, FDR reflected on his abiding love for the Hudson River Valley.

"Half a century ago a small boy took especial delight in climbing an old tree, now unhappily gone, to pick and eat ripe seckel pears," he said. "That was about one hundred feet to the west of where I am standing now. And just to the north he used to lie flat between the strawberries—the best in the world. In the spring of the year, in hip rubber boots, he sailed his first toy boat in the surface water formed by the melting snow. In the summer with his dogs he dug into woodchuck holes in this same field, and some of you are standing on top of those holes at this minute. Indeed, the descendents of those same woodchucks still inhabit this field and I hope that, under the auspices of the National Archivist, they will continue to do so for all time."[36]

By the time FDR went to Harvard University in 1900 he presented himself as a tree farmer. In 1912 he started planting Norway spruce and Douglas fir all over Dutchess County as any good Bull Moose conservationist would do. Roosevelt, in fact, became chairman of the Forestry Committee of the New York state senate, personally planting 2,000 or 3,000 trees a year. As a hobby FDR would purchase land adjacent to his Hyde Park estate and play at being Gifford Pinchot. In 1929 he hired Nelson Brown, a professor at the New York State College of Forestry at Syracuse University, a program funded by Louis Marshall, to help transform Hyde Park into an arboretum. For his entire life, the deep glades of his hemlock woods were among his favorite places to contemplate political issues.

As governor of New York from January 1, 1929, to December 31, 1932, FDR put more than 10,000 unemployed men to work planting trees, managing forests, and stopping erosion. When Roosevelt won the presidential election in 1932, he asked Brown to create the Civilian Conservation Corps (CCC). Roosevelt's hobby was going to be an impetus for some elements of the New Deal. "F.D.R. saw the restoration of the land—the prevention of dust bowls and floods through soil conservation practices, the rotation of crops, the planting of trees," the historian John Sears wrote, "as intimately bound up with restoring the livelihoods of the people living on the land."[37]

When Roosevelt created the CCC on March 31, 1933, Bob Marshall celebrated. FDR had once told his aide Harry Hopkins that every boy should work for at least half a year in forestry; to Marshall, this was a very wise statement indeed. Within a few months 1,000 CCC camps were operational, offering employment to nearly 300,000 young men. A few weeks later FDR established the first CCC marine station in the southern Tongass National Forest; at long last TR's greatest accomplishment in Alaska received ranger boats and increased protection by wardens.[38]

Marshall—who never looked for financial opportunities beyond the strictures of a government salary—was thrilled by all the New Deal efforts made in Alaska toward parks, wildlife management, rangeland, and soil and water conservation, but he was distressed that the reclamation of *wilderness* didn't grab the president's attention. The Forest Service did continue to preserve "primitive areas" as stipulated in the L-20 regulations of 1929, but it wasn't ardent about enforcing the laws or actually prohibiting development.[39]

Harold L. Ickes, however, feisty and belligerent, was drawn to the wilderness movement. He was, politically, a Bull Moose conservationist, a throwback to the turn of the century when TR claimed that the vast open spaces were the great incubator of American democracy. "We ought to keep as much wilderness area in this country as we can," Ickes told a convention of CCC workers. "I am not in favor of building any more roads in the National Parks than we have to build. I am not in favor of doing anything along the line of so-called improvements that we do not have to do."[40]

President Roosevelt also established the Kenai National Moose Range in Alaska with Executive Order No. 8979 (just a few days after Pearl Harbor was attacked on December 7, 1941).[41] An editorial in *Seward Gateway* had called for a moose reserve on the Kenai Peninsula a decade earlier. And, in 1932, thirty-seven conservation-minded citizens from the village of Ninilchik petitioned Secretary of Agriculture Arthur M. Hyde to create a new refuge like the one on Fire Island. In addition to lobbying to create a Kenai National Moose Range, the Alaska Game Commission had issued a number of new hunting regulations throughout the territory. But boomers in mining towns like Hope and Sunrise objected strongly to the federal government's protection of moose. Moose was Alaska's regional meat, prepared marinated, used in casseroles or stews, and eaten in burger buns.

From 1932 to 1941 boomers fought against the Biological Survey, opposing an executive order to save Alaska's moose population. But when the Biological Survey was transferred from the Department of Agriculture to the Department of the Interior on July 1, 1939, the idea of a moose refuge gathered momentum. Under President Franklin Roosevelt's executive order the U.S. government started constructing military installations on Kodiak Island and at Dutch Harbor. As a trade-off, Ira N. Gabrielson, the director of the Fish and Wildlife Service, was able to persuade FDR to establish Kenai National Moose Range, encompassing 2 million acres. Furthermore, the U.S. Fish and Wildlife Service agreed to allow limited hunting and land leasing in the refuge—leading important "hook and bullet" nonprofits to support the moose sanctuary. With World War II dominating all aspects of American life, the Kenai National Moose Range seemed like a fine way to protect wildlife. But life isn't that simple. Richfield discovered oil on the Kenai Moose Range in the early 1950s and began demanding immediate exploitation of the field to obtain petroleum. A showdown over Alaskan moose was looming.[42]

Marshall continued to worry that the New Deal wasn't socialist. He was opposed, for example, to the federal government's building dams in wilderness areas of immense value as natural resources. But he cheered the

Department of the Interior for saving such treasured landscapes as the Sonora Desert of Arizona, Cape Hatteras in North Carolina, and Big Bend in Texas. Other national monuments—Zion, Death Valley, Joshua Tree, and Capitol Reef—were either expanded or upgraded to national park status later.[43] Besides the CCC, Marshall approved heartily of the Soil Conservation Service, whose aim was to stop erosion. Feeling that "big timber" was waning in the Pacific Northwest, Marshall tried to persuade his publisher to change the title of *Arctic Village* to *Those Bastards, the Lumbermen* (possibly he was joking). But his dedication never changed, from the first draft to the final proof. It read: "To the people of the Koyukuk who have made for themselves the happiest civilization of which I have knowledge."

Arctic Village greatly enhanced Marshall's career. Appointed by President Roosevelt as director of the Indian Forest Service, Marshall would travel around the country for about six months a year inspecting forests from Minnesota's Lake Superior to Arizona's Gila National Forest. Visiting reservations where pent-up frustration was increasing, he also helped Native American tribes reacquire forests stolen from them in the nineteenth century when treaties were broken. When presented with discrepancies in land title cases, he usually sided with the Indians. But he wasn't helping all the Native tribes. Marshall confronted the Navajo over their overgrazing of stock on the Arizona range. Wearing an old cotton workshirt, faded dungarees, and a straw hat, he didn't seem like a USDA Forest Service officer. Returning to his old hobby of collecting unique American place-names, the ever-studious Marshall learned fascinating words from various tribes. He marveled that the Chippewa in Minnesota, for example, had a particularly long word for cranberry pie: *muskegmeenanboskominnasiganeetibasijigunbadingwaybaquazyshegun*.[44]

Although *Arctic Village* sold only 3,000 copies, many people interested in Alaskan conservation read it, including the twenty-nine-year-old Mardy Murie. Olaus Murie, the husband of the buoyant, clear-minded Mardy, was considered the world's leading biological expert on Arctic caribou, and she had been assisting him to better understand the 11,000-year relationship between people and caribou in northern Alaska. A careful, college-educated note-taker, Mardy eventually wrote an elegant memoir, *Two in the Far North*, as a tribute to Arctic Alaska and the frontier spirit of its people. Marshall, while staying at a hotel in Chicago, received a warmhearted fan letter from her. "Of course I heard of you often and your fame still lingers in the Koyukuk as the most beautiful woman who ever came to that region," Marshall replied. "Jack Hood, Cone Frank, Pass Postlethwaite and Verne Watts each told me so, and they've seen them all come and go from Dirty Maude to Clara Carpenter. . . . Mr. Murie's fame also lingers in the Koyukuk."[45]

This exchange of letters was the beginning of a united front to preserve what would become the Gates of the Arctic National Park and the Arctic National Wildlife Refuge. As part of the New Deal conservationist program, Marshall soon held highly significant meetings with the Muries in Washington, D.C., and Moose, Wyoming. Following the death of Charles Sheldon, Olaus Murie—taking some time from the caribou—had also become the Biological Survey's leading expert on the Jackson Hole elk herd. Mardy Murie continued assisting her husband with his study of the caribou herds of Alaska and the Yukon. The three conservationists believed in one crucial point: that in the aftermath of Hetch Hetchy, wild Alaska was now *the* environmental battleground.

Pleased with the success of *Arctic Village*, Marshall began working on a follow-up, *The People's Forest*. Published in 1933, it was crammed with scientific data, and it charged that the reckless clear-cutting strategy of "big timber" was destroying American landscapes. The U.S. government's halfway policies were failing to arrest this destruction. No longer, Marshall argued, should Americans tolerate tracts of tree stumps in their communities. Private companies, he feared, were bent on ruining landscapes for quick profit and were unconcerned about the environmental devastation they casually left behind. From Marshall's perspective, it behooved the Roosevelt administration to acquire about 200 million additional acres of land. That would bring America's public forestlands to well over 500 million acres. According to Marshall it was time to "discard the unsocial view" that our woods belonged to lumbermen. "Every acre of woodland in the country," he insisted, "is rightly a part of the people's forest."[46]

As the historian James M. Glover noted, the publication of *The People's Forest* had the unfortunate effect of leaving Marshall somewhat marginalized as a utopian socialist.[47] There was a strain in the book that could be read as hostile to the whole capitalistic notion of land acquisition. Owners of large ranches in Wyoming and Texas considered Marshall as communistic as Leon Trotsky and as eccentric as Rockwell Kent. Bureaucrats didn't respect his clarion calls or his pejorative language. Franklin Reed, editor of the *Journal of Forestry*, said *The People's Forest* was "dangerous," an irresponsible plea for huge U.S. government reserves with virtually no concern about how such a preservationist policy would affect jobs in depressed areas.[48] Marshall then started a well-organized campaign to have Reed fired. In any case, by 1935 FDR started speaking in extremely eco-friendly terms. "It is an error to say we have 'conquered Nature,'" Roosevelt told Congress. "We must, rather, start to shape our lives in a more harmonious relationship with Nature."[49] In Vermont's Winoski River valley alone, more than 3,000 CCC workers were encamped for a "green" reforestation campaign.

Marshall proved to be a good bureaucrat, accounting for every penny spent; but his memorandums were often filled with belittling sarcasm and ridicule. Allan Harper, head of the Indian Bureau's Organization Division, warned Marshall that if he didn't want to be fired, he'd better consult a censor who would have the power to omit some particularly brutal phrases. What saved Marshall was the fact that Secretary of the Interior Harold Ickes admired his vision and boldness. Cantankerous, judgmental, and deeply devoted to conservation, Ickes was nicknamed Donald Duck by FDR because of his cartoonish temper tantrums. Horace Albright, head of the National Park Service, said that Ickes was "the meanest man who ever sat in a Cabinet office in Washington" but also "the best Secretary of Interior we ever had."[50] Ickes shared Marshall's foresight when it came to increasing primitive areas within America's national forests. "I think we ought to keep as much wilderness area in this country of ours as we can," Ickes told CCC workers. "I do not happen to favor the scarring of a wonderful mountainside just so we can have a skyline drive. It sounds poetical, but it may be an atrocity."[51]

In the summer of 1934 Marshall was appointed by Ickes to represent the Department of the Interior in creating an international wilderness sanctuary with Canada along the Minnesota-Ontario border. Marshall hoped for a roadless national park. Visitors would instead paddle canoes like the voyageurs of old to get around the "land of 10,000 lakes." Deeply disturbed because the National Park Service had been building turnpikes inside wonders like the Shenandoah Valley, Marshall worked over-time to develop a new roadless policy. From his perspective, the skyline drives built through the scenic center of Shenandoah National Park were a betrayal of the law of 1916 saying that both wildlife and scenery should remain "unimpaired." Plans for concrete thoroughfares through Great Smoky Mountains National Park also incensed Marshall. Roads would facilitate logging, mining, dam construction, and oil drilling. By focus-ing on roadlessness Marshall knew he could eventually win the battle to preserve wilderness.[52]

Ickes dispatched Marshall to Tennessee to recommend routes for a proposed new highway from Shenandoah to the Smokies. Cleverly, Mar-shall wrote an urgent missive asking Benton MacKaye, the father of the 2,000-mile Appalachian Trail, to meet with him in Knoxville. Mean-while, Mackaye had asked Harold C. Anderson, secretary of the Potomac Appalachian Trail, to come with him to Tennessee. Meeting at the An-drew Jackson Hotel in Knoxville, the three conservationists, recognizing that there was no lobbying group aimed at keeping public lands roadless, started an open-ended dialogue. They wanted to found a new nonprofit group like the Izaak Walton League or the Sierra Club. To these men the idea of a highway in the Smokies reeked of "bad planning." Marshall's vocal dissent, however, was construed by some members of the National Park Service as an "improper and ungracious attack."[53]

A few months later, in 1935, Marshall met in Washington, D.C., with Robert Sterling Yard (a creator of the National Park Service) and Mackaye to officially announce the founding of The Wilderness Society.

The nature photographer Ansel Adams once wrote that certain "noble areas" of the world should be left in as "close-to-primal condition" as pos-sible.[54] That was what Marshall wanted to happen in America. "All we de-sire to save from invasion," he declared, "is that extremely minor fraction of outdoor America which yet remains free from mechanical sights and sounds and smell."[55] Aldo Leopold, the most eminent wildlife biologist of the twentieth century, was brought in to become a cofounder of The Wil-derness Society.[56] "It will be no longer a case of a few individuals fighting," Marshall declared, "but a well organized and thoroughly earnest mass of wilderness lovers."[57]

The Wilderness Society was officially created at the Cosmos Club in Washington, D.C., on January 20, 1935. Professional, self-assured, and devoted to nature, these high priests of the wilderness planned to challenge huge corporations that were hungry for public lands to be opened up for lumbering, mining, and grazing. The Wilderness Society saw itself as focused on results. Saving roadless land areas was the binding motivation of this new nonprofit. The first paragraph of its four-page mission statement read as follows:

> Primitive America is vanishing with appalling rapidity. Scarcely a month passes in which some highway does not invade an area which since the beginning of time had known only natural modes of travel; or some last remaining virgin timber tract is not shattered by the construction of an irrigation project into an expanding and contracting mud flat; or some quiet glade hitherto disturbed only by birds and insects and wind in the trees, does not bark out the merits of "Crazy Water Crystals" and the mushiness of "Cocktails for Two." [58]

Under the enterprising leadership of Marshall and Yard, The Wilderness Society deliberately limited its membership. Approximately 500 dedicated fighters seemed about right. Compromisers weren't welcomed. "We want no straddlers," Marshall said. "For in the past they have surrendered too much good wilderness and primeval forest which should never have been lost." [59] The headquarters for The Wilderness Society was Yard's apartment at 1840 Mintwood Place in Washington, D.C. An advertisement-free magazine, the Living Wilderness, was issued; its main feature was an attempt to stop road construction in Idaho's Selway-Salmon river region and Washington's North Cascades and Olympic Mountains. By October 1935 Marshall was in southeast Utah fighting to maintain 1 million acres of roadless wilderness. A movement had begun.

Marshall was wise to cofound The Wilderness Society with seventy-four-year-old Robert "Bob" Sterling Yard. Born during the Civil War in Haverstraw, New York, Yard was an old-style gentleman, the kind of man who tipped his hat and never swore. He had graduated from Princeton University and become a leading journalist and editor in New York City. One of his closest friends had been Stephen Mather, a fine reporter who went on to become the founding director of the National Park Service.

Yard quit his career as a journalist to become the vital advocate of protecting wild and scenic America. Unlike Marshall, he had a calming personality that never grated on anyone.

A ferocious worker, Yard started a letter-writing campaign on behalf of The Wilderness Society that was stunningly impressive. Membership drives, public photographs, and lyceums were all part of Yard's programming agenda, based on the gospel of "wilderness salvation." He became the first editor of the *Living Wilderness*, perhaps the most important circular promoting Alaska's nature heritage in the Tongass and Chugach. "The spirit of the forest is American," he wrote in 1936. "It moves indomitably against all obstructions."[60]

With The Wilderness Society up and running, and Yard handling the daily logistics, Marshall advocated on behalf of Arctic Alaska. Capitalizing on his appointment as director of forestry for the Bureau of Indian Affairs from 1933 to 1937, and later as head of recreation management for the U.S. Forest Service, Marshall kept asking this question: Why not have Alaska's North Slope designated a wilderness area? In a report he wrote for the U.S. government in 1937, building on The Wilderness Society's mandate, Marshall called for "all of Alaska north of the Yukon River" (minus a small area around Nome) to be officially declared *wilderness*. There should be no roads or congestion, just wilderness with caribou herds roaming free and birdlife thriving as if industrialization had never happened. It would be a sublime place with brilliant patches of tundra and wildflowers. An Arctic refuge would be cathartic for city dwellers, a vast treeless landscape uncompromised by jackhammers, smog, or bulldozers. In the future, someone like Thoreau could wander on snowshoes to the northernmost Arctic, camping along the Hulahula or Kongakut river in June, and warding off mosquitoes by a dwindling campfire as he witnessed the surreal spectacle of the aurora borealis. Such tramping was a wonderful part of the American intellectual tradition.

When Marshall visited California's Sierra Nevada on a listening tour in 1937, he was appalled by what he saw: campgrounds filled with too many people and too much garbage. Many of the gorgeous places where John Muir had tramped were damaged by roads, commercialization, and pack stock. Brainstorming with the Sierra Club's president, Joel Hildebrand, Marshall wondered whether certain parts of California couldn't be preserved in "super-wilderness condition,"[61] particularly the area around Kings Canyon.

Saving Arctic landscapes as *wilderness* became Marshall's crusade in the late 1930s. The worth of Arctic Alaska, Marshall argued, was that "the emotional value of the frontier" could be preserved. In the Arctic, where rivers were made of ice, an explorer could have a mystical union with the creator. The Beaufort Sea coastal plain was still unknown to wildlife biologists. In the late evening Marshall, like another Clausewitz, plotted strategy for the wilderness. After flirting with various preservationist schemes for Alaska, he decided that even the unexciting fields of tufted cotton grass on the Brooks Range ("rock desert," a topography inhospitable to plants or birds) should be off-limits to development. Although not a bird-watcher, he described the golden plovers (*Pluvialis dominica*)— mottled black and white with a rich golden tinge on the back—that flew annually from Wiseman all the way to Patagonia. Having earned three academic degrees (including the PhD in forestry at Johns Hopkins), he hoped people might listen to his persuasive argument about leaving the Arctic wilderness alone. Still, Marshall had few illusions that launching a political movement to create an Arctic refuge would be easy, and he wasn't quixotic or overly romantic. Success, he knew, would come one bureaucratic step at a time. Independently wealthy since his father's death, Marshall underwrote a new map, approved by the U.S. government, of more than forty wild roadless areas, surveying those forlorn areas himself. Because of his relentless, focused energy, Marshall had faith that he was making an impact from the fringe of the Roosevelt administration.

Marshall went back to Alaska in 1938 to map and explore the upper Koyukuk region anew, in part to settle a bet regarding the source of the Clean River.[62] He carefully studied the calcium-rich soil of the tundra and also wanted to prove his theory about the effects of glaciation on the timberline. Spruce seeds he had planted eight or nine years earlier didn't sprout. The climate was too harsh. "My experiment," he wrote, "was a complete, dismal failure on both plots."[63] Stopping for lunch one afternoon at the side of a minor stream, Marshall marveled because nothing seemed to grow along its banks. Resorting to his habit of naming geographical landmarks off-the-cuff, he called it Barrenland Creek. He watched the aurora bolt like lightning across the sky, and this reenergized his campaign to save the Arctic refuge as wilderness. Somehow rivers like the Innoko (500 miles), Nowitna (250 miles), and Tanana (659 miles) had to escape the fate of becoming part of Harding's petroleum reserve. He would devote his considerable energy in Washington, D.C., to making the Arctic refuge happen. It was a life mission.

After visiting the Brooks Range again in 1939, Marshall consolidated all his ideas about the wilderness into an airtight proposal, which he brilliantly presented to a congressional committee. Convinced that saving wilderness was as American as Lewis and Clark, Marshall used terms like "pioneer conditions" and "the emotional values of the frontier." Boldly Marshall proposed all Alaskan lands north of the Yukon River be kept free of roads, pipelines, electrical wires, smokestack industry, and even farming. America was being given a rare second chance to establish something of permanent value: an American frontier. "Alaska is unique among all recreational areas belonging to the United States, because Alaska is yet largely a wilderness," Marshall told the congressmen. "In the name of a balanced use of American resources, let's keep northern Alaska largely a wilderness!"[64]

Owing to Marshall's testimony, *wilderness* was now the new concept in serious land conservation circles. Nobody during the New Deal era was doing more than Marshall to persuade the U.S. Forest Service and the Bureau of Indian Affairs to preserve wilderness in the public lands they managed. Then, on November 11, 1939, Marshall died of heart failure on an overnight train trip from Washington, D.C., to New York. To have such a bright star vanish at only age thirty-eight was devastating. The prospect that he would write more books like *Arctic Village* and *The People's Forests* had simply been assumed. Marshall, however, had known he had a serious heart problem. In preparation for sudden death he had made out a will bequeathing one-quarter of his $1.5 million estate to The Wilderness Society.

At Marshall's burial service in Brooklyn, scores of foresters from the departments of the Interior and Agriculture came to pay final homage to the great man. They pledged to continue Marshall's quest to protect Arctic Alaska. They agreed to devote their lives to protecting wild lands. A couple of lines that Marshall had written years earlier became the rallying cry for the burgeoning environmental movement. "As society becomes more and more mechanized," Marshall warned, "it will be more and more difficult for many people to stand the nervous strain, the high pressure, and the drabness of their lives. To escape these abominations, constantly growing numbers will seek the primitive for the finest features of life." [65]

The historical implications of Marshall's conservationist philosophy were monumental. Twenty-five years after his death, largely owing to his advocacy, The Wilderness Society helped pass the Wilderness Act of 1964. Such pristine locales as the Grand Tetons, Two Ocean Pass, and the Middle Fork of the Salmon River region of central Idaho were designated by Congress as *wilderness*. And, lo and behold, the Clear Water Country in Montana where Marshall had been a forester in the 1920s was likewise declared *roadless*. Also in 1964, more than 1 million acres in Montana officially became the Bob Marshall Wilderness. Only a few administrative cabins for trail crews and fire rangers were allowed. With Glacier National Park bordering it on the north, the Bob Marshall Wilderness remained a protection zone for grizzlies.* And Montana was just one example. More than 109 million acres of America are now designated *wilderness*. One Adirondack wonder was named Mount Marshall in the state system—probably the most fitting tribute of all to the proud "forty-sixer."

---

* Today the Great Bear Wilderness, the Bob Marshall Wilderness, and the Scapegoat Wilderness form the Bob Marshall Wilderness Complex, totaling more than 1.5 million acres.

Chapter Twelve

⚬ ⚬ ⚬ ⚬ ⚬ ⚬

THOSE AMAZING MURIES

I

Mostly it was Mardy Murie's ability to motivate people and hold them accountable by her steadfast decency of spirit that set her apart. To know Mardy was to love her: she was deeply humble, with eyes sharp but innocent, always elevating others to conscientious endeavor, never worried over whether she got her due credit. As a girl, Murie fell in love with Arctic Alaska's remoteness. She was intoxicated by the tearing wind. The wildlife and the desolation made her heart stand still. Though she received various honorary doctorates later in life for her pioneering work as a naturalist, she never grew smug or overbearing. Anybody who wrote to Mardy received the courtesy of a quick, handwritten reply. Affectionately known as the "mother of the American conservation movement," Mardy, who lived to be 101 years old, was a true activist, opening people's consciousness to the fragile beauty north of the Arctic Circle. In her old age, when her gray hair was braided into a bun and crows' feet framed her hazel eyes, three U.S. presidents—Lyndon Johnson, Jimmy Carter, and Bill Clinton—honored Mardy at White House ceremonies as nothing less than a national treasure, an embodiment of wild Alaska. Her kindness was intrinsic, but for all her gentleness of spirit, she smoldered like a fuse when oil and gas interests dared despoil her homeland in the far north. "Thanks in part to her work," Verlyn Klinkenborg wrote in the *New York Times*, "great swaths of land were set aside with a single presidential pen stroke."[1]

Margaret "Mardy" Thomas was born on August 18, 1902, in Seattle, and she would always maintain a strong identification with the Puget Sound area. She rented her first apartment near Pioneer Square, where lumberjacks skidded logs down Yesler Way into the bay, and would always remember the thunderous rumble. When Mardy was still a baby, however, her family moved to Juneau, Alaska: a community crowded on a slender strip of land between Douglas Sound and mountains that seemed to rise straight out of downtown. No roads connected the city to the world at large—visitors sailed or steamed into the harbor. Many of Mardy's earliest memories were of the gorgeous, thickly forested Juneau mountainsides, which catapulted up from the dark blue waters of the Gastineau Channel. Atop these sheer mountains was the famous Juneau ice field, an immense frozen ice mass from which dozens of bluish glaciers flowed, so that this spirited, prosperous seaport village was never drought-stricken. The Victorian mansions on Seventh Street attested to the fortunes made in Juneau during the gold boom.

Juneau was also the major fishing center of the Panhandle. Salmon and halibut were thick in the waters surrounding the city. Close by, bald eagles built stick nests in the spruce and circled overhead in their continual hunt for prey. All told, Juneau was a fine place to grow up in at the turn of the twentieth century, the foremost city of Alaska. "Juneau, on the mainland opposite the Douglas Island mills, is quite a village, well supplied with stores, churches, etc.," Muir wrote in *Travels in Alaska* in 1915. "A dance-house in which Indians are supposed to show native dances of all sorts is perhaps the best-patronized of all the places of amusement." However, Muir went on to note that the forests on Douglas Island were being "rapidly nibbled away" by "a large mill of 240 stamps." [2]

An eloquent photograph of Mardy at age four and a half, with a bright white bow in her curly brown hair, posed leaning on her right hand, shows the sparkle she would never lose. Just a month after this studio photo was taken, Mardy's parents divorced. Bruised by the savage quarrels and by her husband's betrayal, Mardy's mother, Minnie, left Alaska and went back to Seattle, taking Mardy with her. It seemed that Alaska would no longer be a factor in Mardy's life.

But then Minnie married a well-known, well-paid attorney named Louis B. Gillette, who in 1911 was assigned by a federal court to be assistant

U.S. attorney in the Fairbanks office. Congress had finally given Alaska its own civil and criminal codes, just as President Theodore Roosevelt had urged. A reform-minded conservationist, Gillette was responsible for bringing the rule of law to the last frontier, imposing federal standards regarding land claims, big-game poaching, and so on. Fairbanks was a rough-and-tumble outpost when Mardy's family arrived—they had left civilization behind in Seattle. The imposing Masonic Temple in Fairbanks couldn't disguise the essential character of the town. Every muddy lane reminded visitors that there was little indoor plumbing. Townspeople relied on well water, and mail arrived—if it arrived—by boat and dogsled. Thomas Edison's inventions had barely penetrated Fairbanks, although one three-story skyscraper had been wired for electricity. The local hero was Walter Harper, a Native sled driver, who climbed 20,320-foot Mount McKinley in 1913: he was the first to reach the summit.[3]

To get from Seattle to coastal Skagway, Alaska, and then to Fairbanks was a three-week ordeal involving five different modes of transportation. Because Alaska was practically roadless, river travel was the only reliable transport. More than 4,000 miles of Alaska's waterways were navigable by steamers. Mardy remembered a huge crowd gathered at the Seattle wharf to see her steamer, *Jefferson*, set off for Alaska. Taking a train from Skagway to Whitehorse, the family then boarded the *Sarah*, a sternwheeler, which, with its huge green-plush saloon, filled Mardy with glee. Mardy had expected to see stately mountains, but she was surprised by the wildflowers: whole gorges were filled with royal purple blooms. "For a nine-year-old girl, it was a time to watch the landscape unfold, to adjust to new ways of daily living, and to take her own measure of the frontier," her biographers Charles Craighead and Bonnie Kreps wrote in *Arctic Dance*. "Mardy's vivid childhood memories of the epic journey seemed to set the stage for her somewhat nomadic lifestyle; long before Alaska became a state, she would travel up and down the length of Alaska's southeast coast seven times and journey thousands of miles crisscrossing the Territory."[4]

Fairbanks—founded in 1901, when E. T. Barnette established a trading post at Tanacross on the upper Tanana River—was all about gold (although Native peoples had lived in the general area for thousands of years). The new town was named for Charles Warren Fairbanks—the U.S. senator from Indiana who successfully negotiated an Alaskan boundary dispute with Great Britain at the 1898 Quebec Conference. The ever-popular Fairbanks was elected vice president on the ticket with Theodore Roosevelt in 1904.[5] Judge James Wickersham, the most powerful politician in the territory, successfully promoted the new Chena River settlement he named "Fairbanks" after his own political mentor.[6] Somehow the settlement survived food shortages in 1903, a flood in 1905, and a fire that wiped out the downtown in 1906. (The fire burned fifteen blocks of buildings. Until the 1930s the entire town was built solely of timber logs with sawdust insulation.) The rumors of gold always came back. . . . Another strike . . . just one creek down . . . on a free claim . . . the mightiest vein of all. . . . Until they didn't.

Scraping out a living became a permanent condition in Fairbanks. During the winter, miners worked away relatively warm, compared with the surface temperature of forty to fifty below zero. Gold production rose dramatically, from $40,000 in 1903 to $9.6 million in 1909. In 1915 it became a permanent hub, when, after a lot of false starts, the construction of the 470-mile Alaska Railroad commenced. By 1920, however, with the gold rush fading, only 1,100 residents remained. Isolated from the Lower Forty-Eight, and from modernity, they resorted to gossip to keep themselves amused. Notably bad company could be found in the town's twenty-three saloons. The same could be said of its five clapboard churches. Booze ran freely—both in the red-light district and outside it. During the winters, the miners, pioneers, saloon keepers, preachers, and prostitutes all knew that the only way to endure until spring was to embrace the darkness. But come Independence Day, when Cushman Street was bedecked with the Stars and Stripes, Fairbanks seemed like a prosperous town in the Midwest, glad to be alive.

Fairbanks was, however, a close-knit community, brought together by the surrounding wilderness and by the unrelenting forces of nature. Snow was measured in feet, not inches. People shared yarns of encountering brown bears digging in their trash barrels and of caribou "turning blue" from the cold. Taxidermy allowed men to flaunt their hunting prowess. There was hardly a building in Fairbanks in 1912 that didn't have a moose or elk head on one of its walls. Bear rugs were mandatory in homes. Regularly in the spring, residents would go on hunting trips up the Chena River. There were no bag limits, and wild game was served in most restaurants. Nobody had yet divided the Arctic caribou into three distinct herds—the "Porcupine," "Central," and "Imperial"; sourdoughs saw all caribou as the same.

Everybody in Fairbanks—Native Alaskans and newcomers alike—also shared an intense reverence for the northern lights. The earliest descriptions involved lonely spirits and supernatural battles in the skies. Not until 1905, when a British physicist made the connection between the sun and the aurora borealis, was the phenomenon understood as something natural.[7] Some auroras gave off so much light that people could hunt by it. But mainly the aurora borealis was spoken about in hushed terms; only a few scientists analyzed the ionospheric gases being drawn in by the gravitational pull of the earth, causing a wavelike electrical discharge. In any case, as Mardy was growing up, on clear nights she looked skyward hoping to see the shimmering green aurora dominating the sky. An old piece of folk wisdom around Fairbanks was that the aurora borealis was as alive as a person; it whistled and cracked and seemingly came to scrutinize you up close.[8]

Mardy Thomas (she kept her real father's name) was by all accounts a little hoot, entertaining as well as being entertained when trappers regaled her with stories about the Arctic wind or timber wolves attacking frightened horses along the Yukon River. Much like Jack London, she began to romanticize Arctic wanderers: they might look shabby, but they had hearts of gold. Most of these prospectors were, in fact, drifters of ill repute, ne'er-do-well misanthropes unable to make an honest living in the Pacific Northwest or anywhere else. But their hobo tales were what mattered most to young Mardy. While other girls were enthralled by *Little Women*, Mardy was memorizing the poems of Robert W. Service and pondering survivalist tactics for blizzards. She craved adventure. At school she raised her hand so many times to ask questions that her teachers felt like muzzling her.

In Fairbanks circa 1912, self-reliance wasn't merely an idealistic principle in a dog-eared volume of Emerson's *Essays*. The residents relied exclusively on cordwood for heating; the forest belts were clear-cut. Logging, catching northern pike and whitefish through the ice, and building fires were necessary skills in a land where below-zero temperatures were routine. Sometimes even inside the best houses on First Avenue, the hearth wasn't warm enough to melt the outdoor snow from a guest's boots. Sometimes in Fairbanks it would be fifty or sixty degrees below zero for weeks at a time, causing the rubber tires on Model T's to shatter. Sleet blew sideways. Icicles were thicker than logs. There were no modern goose-down and nylon parkas; Alaskans swaddled themselves in wool and thick wolf fur. Both inside and outdoors, keeping warm was a full-time preoccupation. For children frostbite or hypothermia was nearly as common as a runny nose in the Lower Forty-Eight. Wood shacks often collapsed under the weight of snow. The cold forced everybody to eat more, because constant shivering burned a lot of calories. But Mardy never complained. Fairbanks had cast a spell on her. It was her home. The people needed large amounts of timber for buildings and water flumes—and for stern-wheeler riverboats until coal-fired boats arrived in 1925. Wood was harvested recklessly. As a result, the country for miles around Fairbanks was stripped of trees. Even the N.C. Company's power plant downtown was fueled by wood. Almost 19,000 cords of wood were burned annually for heat.

Citizens of Fairbanks adopted some customs of the Midwest—quick coffee, saving pennies, school spelling bees, bake sales, trick or treat at Halloween—but it all seemed staged. Individualism was the core value here—the kind of libertarianism that Ayn Rand would celebrate in *Atlas Shrugged* (only they were pro-God). Still, ironically, the isolation and the rigor of the climate fostered deep codependency here. Virtually all the children—including Mardy—had a Siberian husky as a pet. Without dogsleds nobody could traverse the snowbound country. The unity between man and dog belied the go-it-alone posturing. Sled dogs loved to be harnessed, and Mardy was accomplished at harnessing them. "When the trail was good at all, I'd stand on the handlebars; otherwise, I'd have to run," she recalled. "And those Alaska dogs were so eager to get into harness and go that you could hardly restrain them in the morning." [9]

Dogsledding was a part of Native life in Alaska long before the gold rushes of 1897 to 1898. Around Fairbanks when the Muries got married, parts of dogsleds were found in Athabascan archaeological digs on the outskirts of towns. During the gold rushes, however, outfitters shipped dogs by the thousands—German shepherds, Saint Bernards, samoyeds, and enormous mongrels—to work in Alaska. A mixed dog with no real pedigree was called a husky (and sold as a "thoroughbred mongrel").

These huskies weren't just for endurance mushing across rivers and gale-force blizzards. Copper miners used these hardy dogs as pack animals; they could easily pull five times their body weight. Others hitched them to wagons, buggies, and even boats. A common sight in Fairbanks while Mardy was growing up was dogsleds hauling firewood. The U.S. mail service gave yearly honors for the best dog mushing. The Nome Kennel Club organized the first Alaskan sled-dog races in 1908; it predated the Iditarod by sixty-five years.[10]

Perhaps because her stepfather was a stylish upper-crust lawyer, book learning came easily to Mardy. She was also blessed with social intelligence, and could make friends with nearly anyone—and especially with the restless seekers and backcountry idlers. Early on she decided that nomadic life was a virtue. Few people actually stayed in Fairbanks. Everybody, it seemed, was "striking out for the creeks" (a popular expression of prospectors and hunters). Along the mountain switchbacks were *promyshlenniki* (Russian traders looking for furs), stampeders from the Lower Forty-Eight intent on gold strikes, and Seattle businessmen seeking coal and copper.[11]

When Mardy turned fourteen, her father, Ashton Thomas, reentered her life. Like a "Wayfaring Stranger" in Carl Sandburg's *American Songbag*, one day he showed up at her door, wearing a brand-new suit, asking to be forgiven. At first Mardy didn't recognize him. Like some other Alaskans who had given up on gold, Thomas owned a salmon cannery; his was in Port Ashton, a handsome village of a few hundred people along Prince William Sound. Remarried, he wanted Mardy in his life. Seeing a chance for adventure, Mardy packed her suitcase and headed south to work in the cannery, with her mother's grudging permission.

It was 375 miles by dogsled or horse carriage from Fairbanks to Port Ashton. Open sleds made it an arduous journey; the wind would rip at the travelers. There were, at least, plenty of roadhouses along the route. Many inns sold vegetables and refreshments at stands. What was amazing to Mardy was the engineering involved in constructing a road through a deep wilderness. She couldn't believe men had been able to cut a five-foot trail in a cliff 1,000 feet above a raging stream. When the trail was misty with rain, plunging into the Copper River was a very distinct possibility. But Mardy relished every harrowing moment. Suddenly she understood that Alaska was far more than muddy little Fairbanks. Peak upon peak loomed over miles of wet, timber-rich mountains, purple immensities in the Pacific gloom that stretched all the way across the Aleutians to Japan. Mount McKinley's south summit had been climbed in 1913, and mountaineers from the Lower Forty-Eight were coming to the area, looking for the right pass or ravine to test their mettle.

That summer of 1918, with America at war in Europe, Mardy became a young adult. The hamlets along the Valdez Trail (now the Richardson Highway)—Salcha, Sullivan's Rapids, Big Delta—were sites of outdoors excitement. (This sled trail provided the only winter access to the Tanana Valley during the early decades of the twentieth century.) There was no end to the outdoors drama of the Valdez Trail, where a few hardy souls were even bicycling through the sixty-degree switchbacks and oxbows. Those on horseback spent every few minutes tightening the cinch for fear of falling down the mountainside. Mardy loved everything on the trail, from the trading stores' imitation totem poles to Mount McKinley's frozen grace. Spellbound, she vowed to climb McKinley someday. Many of Alaska's 3 million lakes were in the area where she traveled. Caribou herds dotted the swampy peat bogs and blue-green pasturelands. Animal tracks were studied on "bathroom" breaks in berry thickets. Kingfishers dived into waterways. Alaska wasn't just an icebox but also a green paradise teeming with wildlife. In Cordova, which was abuzz with the politics of coal, she boarded a Gulf of Alaska steamer and headed out into Prince William Sound to the offshore island of Port Ashton, officially part of Chugach National Forest.

Mardy's new family greeted her heartily. Exhilarated and feeling grown up, Mardy spent the next three months learning the Alaska fisheries business, and also learning to row and use a compass as she explored the sound's bays in a little boat with an outboard engine. The shoreline of the nearby Kenai Peninsula, where seine boats were working, was amazing; the mountains were vast and silent. "That first summer gave me a picture of that part of Alaska, a knowledge of camping stalls, and a respect for tide and storm," she recalled. "We went through all the islands and their enticing coves. We hiked to the upper reaches of many of the islands. We watched a fight between a large whale and a killer whale." [12]

After that summer, Mardy returned to Fairbanks full of Alaskan lore.

Suddenly, learning the territory's history and geography seemed important. Mardy wore boys' lumberjack shirts and wanted to understand the mentality of stampeders who drifted into Fairbanks from Dawson. She wanted to know why the Tanana Valley, of all places, was the "garden spot" of Alaska. She took notice of pine grosbeaks feasting on frozen buds and berries in the upland spruce forests and woodpeckers scouring for hibernating insects under dark trees. She wondered about the smell of ozone after a big storm, and about the propagation of moss. Wild Alaska was a unique mystery to her. "Curiosity," her stepfather said matter-of-factly, "that divine thing, curiosity. It will carry you when all else fails."[13] His words stayed with Mardy for the rest of her life.

Graduating from high school in 1919, the year Theodore Roosevelt died, Mardy enrolled at Reed College in Portland, Oregon. Down the 375-mile Valdez Trail she trekked again, arriving in Port Ashton a few weeks later. From there she caught a steamship to Seattle and then the train to Portland. Feeling carefree, Mardy explored the Cascades and the Columbia River and studied hard at Reed College. During the summer months she returned to Port Ashton to work as a cannery storekeeper, watching birds forage for fish whenever she had a free moment. Ashton Thomas's cannery business was doing extremely well. In his derby hat and three-piece suit, and with his pocket watch, he epitomized the successful Alaskan businessman. When Thomas decided to move to Boston for a year to develop better contacts in the seafood distribution industry, Mardy seized the opportunity to go with him, to experience the glamour of New York City and Boston. After two successful years at Reed, she transferred to Simmons College in Boston for her junior year.

II

Gathering her belongings in Fairbanks before heading east, Mardy was introduced to a handsome wildlife biologist, Olaus Murie. She was saucer-eyed at her first sight of him. Intense, steely, and bursting with talent, Olaus was in Fairbanks to be outfitted for an arduous trek into the Brooks Range by dogsled to study the habits of caribou in winter. On chaperoned dates Olaus told Mardy about his life as a wildlife biologist, camping under the spruce boughs and constellations. A pursuer of silence, he unfeignedly liked the privations of traveling where there were no roads but plenty of portages. Homelessness was his home. His precious dogs were all he usually had for companionship. Born in 1889—he was thirteen years older than Mardy—Olaus was blond and blue-eyed. Like so many great naturalists, he had been a bird lover since childhood. He was of Norwegian descent; his hometown was Moorhead, Minnesota; and his outdoors sanctuary was the Red River valley. "There were woods, birds, mammals," he recalled of his happy youth in Minnesota. "It was living close to the earth—you know what that does for you. Gee, it was wonderful."[14] Flushing out grouse from the prairie grasses was a favorite outing of his; he knew how to put meat on the dinner table. Olaus had attended Fargo College in North Dakota, but wanting to get out of the flatlands, he transferred to Pacific University in Forest Grove, Oregon. After graduating in 1912, he stayed in Oregon for two more years. He was employed as a field naturalist for William L. Finley (a state game warden and perhaps the best photographer in America affiliated with the Audubon Society).

Much like Gifford Pinchot, Aldo Leopold, and Bob Marshall, Olaus Murie took trees seriously and considered deforestation a curse. Determined to make his mark as a scientist in the Arctic, he headed to Labrador and Hudson's Bay on a paid assignment for the Carnegie Museum in 1914. Vilhjalmur Stefansson, the Canadian explorer and anthropologist, was starting to present Arctic habitats in a series of papers (in 1921 he would write *The Friendly Arctic*, a distillation of everything he had learned in below-zero temperatures, hoping to entice settlers to the north pole); but Murie was really the first serious biologist after Peary to adopt the Arctic as a laboratory. The Arctic, Murie believed, was very important to the new field of ecology. "Will we have the patience to understand what the northern part of the Earth has to offer?" Mardy Murie asked after traveling in the uncorrupted Arctic with Olaus. "Wherever we went in this country, there was something to see and wonder about. There were so many little things."[15]

During World War I, Murie served with the Army Air Corps balloon troops based in Fort Omaha, Nebraska; he was therefore something of an expert regarding the impact of wind on high-altitude vegetation. Murie believed that scientists needed empirical data about the varied wildlife in the Arctic biosphere. He was displeased that no teams of biological experts had been dispatched to either pole. Looking around the saloons of Fairbanks he saw sea otters and polar bears stuffed and mounted. For a moment, a hatred seemed to clog his blood. It was one thing, he believed, to kill a moose for a steak or stew. It was quite another to use the antlers as a hat rack in a tavern or bar. His feelings ran particularly strong when he considered the free-roaming caribou—called the Fortymile Caribou Herd—that lived southeast of Fairbanks.[16]

In 1920 Olaus got his big break. Hired by the U.S. Biological Survey, he was tasked with studying the migration routes of Alaskan caribou. Olaus's official title was assistant biologist and federal fur warden. He purchased a hooded oilskin poncho, thick wool socks, and the best snowshoes available from the mail-order catalogs. And romance was in the air. Before meeting Olaus Murie, Mardy Thomas had only a superficial appreciation of Alaska's great caribou herds. She knew that the Gwich'in ("people of the caribou") in the Brooks Range had prayed to the roving herds for 20,000 years. On dates with Olaus, Mardy now learned how caribou served this northernmost people's utilitarian needs. The reverence that the Gwich'in (or Kutchin) felt toward the caribou was like the Plains Indians' veneration of bison. Mardy had eaten caribou steak. She had worn caribou-skin boots. She had watched a hungry herd browsing on lichen in the tundra. She had heard caribou huff and hiss while being chased. When shot, caribou uttered a cry so anguished, so pleading, so terrified and mournful that Mardy winced with sympathy. In northern Alaska, caribou were as common as red squirrels. Mardy knew about caribou. But now she learned about their biological traits as if she were taking a college course. What Mardy liked most about caribou was that their *fatness* meant that at long last summer had arrived in frigid Alaska.

Olaus Murie soon taught Mardy more about the behavior of Alaskan-Yukon caribou. The U.S. Department of Agriculture had experimental stations at Sitka, Kodiak, and Rampart, aimed at trying to figure out how to grow vegetables in inhospitable terrain. However, no biologist was stationed in the Arctic. Murie volunteered for that duty. Charles Sheldon had an easy job in the Denali wilderness circa 1906, compared with Murie's work at subzero Arctic temperatures where woolly mammoths once lived. Olaus's brother was going to join him for his Arctic studies in 1922 but died of tuberculosis. Instead, his younger half brother, Adolph, joined him in Fairbanks to start a comprehensive study of the Alaskan-Yukon caribou. Working for the U.S. Bureau of Biological Survey, the brothers collected data on caribou in the noble tradition of Dr. C. Hart Merriam, head of the Biological Survey, a noble man who relished discovering new subspecies of North American mammals.

Olaus and Mardy fell in love. For Mardy, being the wife of a U.S. government caribou specialist from 1920 to 1945 meant that if their marriage was to work, a genuinely cooperative relationship would have to be formed. Olaus advised Mardy to take a series of business classes at Simmons College, on the theory that somebody would have to be the bookkeeper. While she was working to complete her degree at Simmons, Ashton Thomas suddenly died. Mardy was popular at Simmons, but there had always been some ridicule of the girl from Alaska who didn't curtsey. Now, lonely and lost, the nineteen-year-old Mardy was homesick for the dogsled trails and country waltzes of Fairbanks, for the lullaby of the wind and sleet that swept down from the Brooks Range. Her approach to God was based on communion with nature.

Returning home, Mardy started working as a clerk for the U.S. attorney. She lived with her mother, who was employed by the Bureau of Mines. On Sundays she sang "Rock of Ages" at church, almost on pitch. Olaus was in town getting dog teams ready for a run to the Koyukuk country. As an octogenarian, Mardy would reminisce about how, in this idyllic summer of 1922, she taught Olaus ballroom dancing and the standard hymns. But as winter began, he vanished like the sun, going off to inspect herds, rookeries, or dens.

Knowing his north country itinerary, Mardy would mail letters—filled with empty pleasantries—to the forlorn Yukon Territory towns where Olaus planned to stay overnight. Fort Yukon was essentially Murie's Biological Survey headquarters; it was about 110 miles south of Arctic Village. Mail was delivered by dogsled so infrequently that Mardy often got her beloved's letters in batches of four or five at once. Olaus had brought along an art kit (a souvenir from the infantry), and he drew wonderful ink illustrations of all the mice and birds he encountered for Mardy. "How I wish you were with me right now," he wrote in December 1922 from the Koyukuk Trail. "We are up on a summit, the night is silver clear, with twinkling stars and a pure crescent moon. I was out a moment ago to look at it and think of you at the same time." [17]

By the summer of 1923 it was clear that Olaus and Mardy were meant to be together. An overjoyed Mardy joined Olaus at Mount McKinley National Park, as his assistant on a caribou count for the Department of the Interior. Olaus had established a base camp on the upper Savage River, where at night they whittled sticks and told stories. Mardy thought of marriage as the art of two being one—they might as well get started in the Denali wilderness. But how to achieve marital harmony when the spouse's job is to disappear into the most remote reaches of North America on behalf of the Biological Survey? Before they could marry, both Mardy and Olaus decided it was essential to be better organized. Olaus would go to Washington, D.C., to officially submit his reports on the Yukon-Alaskan caribou. Mardy, who hadn't graduated from Simmons College, would enroll in the one-year-old School of Mines at Alaskan Agricultural College, soon to become the University of Alaska. In 1924 Mardy became the university's first female graduate.

Following her graduation in June 1924, Mardy prepared for an Arctic honeymoon. She would travel more than 800 miles down the Yukon River—which flows almost 2,000 miles from northern Canada to the Bering Sea—to the riverside hamlet of Anvik, where she would rendezvous with Olaus. They were to be married in a log chapel at three o'clock in the morning, under the midnight sun, on August 19. She had marked the all-important date with a star on her calendar. Mardy's trousseau was winter wear: fur parkas, wool mittens, and snowshoes. The couple's most essential equipment on the three-month trek into the Upper Koyukuk River terrain included a weatherproof tent and a portable Yukon stove. Accompanied by her mother and bridesmaids, Mardy left Fairbanks on the stern-wheeler *General J.W. Jacobs*. Their complicated rendezvous was successful, and Mardy and Olaus were married.

The Muries then began their honeymoon, dogsledding 550 miles into the central Brooks Range, far away from prying eyes. They went north up the Koyukuk River to the area from Allakaket to Bettles and beyond. Rain was frequent: a thin, chilly spitting that came with squalls of wind. Canada geese graced the sky. Clouds of mosquitoes orchestrated a faint hum, which marred the romanticism. Mardy created comfort in their outback camps. Up the A-frame canvas tent would go; at night it was closed tight except for a peephole for air. Morning was always the most magical time; just being alive was lusty. They breakfasted like cowboys on coffee and oatmeal. During the day Mardy chopped wood, smoked salmon, concocted caribou stew, and made a large sleeping bag bed for her and her new husband to share. Fish—pike, grayling, or lake trout—was often their favorite course at dinner. When the sled dogs got dirty, she brushed them. She had mastered the primitive arts of survival. Dutifully she kept a diary recording times, places, and temperatures. "I remember once saying to Olaus on our dogsled honeymoon, 'Whatever made you think I could do all this?'" Mardy recalled. "And he looked at me and said, 'Oh, I knew you could.'"[18]

Olaus was honeymooning, but he was also intensely studying the habits of North Slope wildlife from red-throated loons to moose browsing on buggy patches of tundra. For all his scientific expertise Murie had an old-school, almost primitive way of looking at wild things. Field naturalists of that time were encouraged to submit ink drawings with their official reports. Besides shouldering a rifle, Olaus carried with him an art kit that had a porcelain slide to mix the watercolor paint. A fine taxidermist, unhurried and precise, he also set small traps to catch and analyze subspecies of mice. While grizzlies eluded them, he carefully monitored Arctic birds such as tundra swans or ravens. Most important, he observed the great barrenland caribou (*Rangifer tarandus granti*) herds amid the mountains. All the way to the Beaufort Sea, the herds of caribou browsed on the tundra. Caribou in Alaska were distributed into about thirty herds; the Muries hoped to document a fair number of them.

Before the Muries, nobody had done proper reconnaissance on caribou for the Biological Survey in Arctic Alaska. By 1922 the domesticated reindeer industry in Alaska was booming, producing a bigger net profit annually than copper, gold, and silver mining combined. The hope in Alaska was that reindeer meat would become competitive with beef as a source of protein. The Biological Survey published articles such as "Reindeer in Alaska" (1922) and "Progress of Reindeer Grazing Investigation in Alaska" (1926). The Muries thought that this kind of analysis of reindeer farming was the job of the Bureau of Animal Industry; after all, reindeer were domesticated animals. The Muries saw reindeer as a threat to the indigenous caribou herd. Interbreeding caribou with reindeer was, to their minds, biologically unsound.[19] All of the Native Americans had their own reverent names for caribou; reindeer had been lumped together by early French voyageurs under the term *la foule*. Each Athabascan group had a loving name: *udzih* (Ahtha), *bidziyh* (Koyukon), *vadzaith* (Gwich'in), and *tutu* (Eskimos).[20]

On afternoons in the far north, while Olaus was out collecting, Mardy would wander around the banks of the Porcupine River, staring into the crystalline, sparkling water, amazed at the varied colors of smooth rocks. She'd comb through gravel bars, looking for animal skeletons to bring back to camp, daydreaming about writing an as-yet-untitled book about Alaskan Native tribes. Peregrine falcons (*Falco peregrinus*), a rather rare raptor, sometimes circled overhead. The river water was about sixty-two degrees Fahrenheit, too cold for a swim. Also, if she stripped naked, the mosquitoes would torment her. So Mardy contented herself with just hiking and pondering, thinking about ancient cave dwellers, caribou herds, and what bird might be magically flushed out of the low-lying bush, startled by her meandering. "Gravel bars are havens in the North Country," Murie wrote in *Two in the Far North*, "providing some refuge from the scourge of Alaska."[21]

Roughing it in the utterly wild Arctic had tribulations and frustrations. But both Olaus and Mardy considered the honeymoon a success, especially when they stared at the night stars together. North of the Arctic Circle you could dip your cup in a stream and drink the cleanest water on the planet. Bears, foxes, grouse, and other animals were fattening on the blueberries of summer. Because neither of the Muries had been overly concerned about physical comfort, everything that happened to them in the Koyukuk was fun and scientifically useful. There was enough wild land in the Arctic so that they never felt claustrophobic or hemmed in. The Muries had, they believed, been in harmony with the will of God. Now, returning to Fairbanks, they agreed that preserving the biological integrity of Arctic Alaska would be the duty of their lives. Spending time in the deep Arctic made them see every experience from birth to death with new eyes. When they built a log home in Moose, Wyoming, they carved over the upright piano the "Mardy and Olaus Murie Life Philosophy," developed during their honeymoon: "The wonder of the world, the beauty and the power, the shapes of things, their colors, lights and shades; these I saw. Look ye also while life lasts." [22]

The Muries were in high spirits after their Arctic honeymoon. As the temperature plummeted and the winter days turned short, they talked about the awesome Brooks Range to anybody who would listen. Back in Fairbanks, exhausted, they indulged in a week of well-earned sleep. They then headed to Washington, D.C., for the rest of the winter. Olaus wrote up the official report of the honeymoon expedition to be submitted to Dr. E. W. Nelson of the Biological Survey. The Department of Agriculture liked typed reports and Olaus, a dedicated bureaucrat as well as an intrepid biologist, handed in a batch of them. Mardy had gotten pregnant on the Arctic trek and was happy at the prospect of becoming a mother. That spring Olaus was dispatched to the Alaska Peninsula to conduct field research on brown bears. Unable to come with him because of the pregnancy, Mardy stayed with her mother in Twisp, Washington, where the Gillettes had relocated along the Methow River, unable to endure another bleak winter in Fairbanks. That summer the Muries brought a boy, Martin, into the world.

With Martin, the Muries returned to Washington, D.C., in the fall of 1925. Olaus had become celebrated as perhaps America's leading Alaska wildlife biologist—an honor that had previously gone to William Healey Dall, who died in 1927. Murie met regularly with fellow biologists at the Cosmos Club to compare notes about bears—and discuss the prospects for preservation in Alaska. Geologists were concluding that because of the shallowness of the Beaufort Sea, combined with ice-clogged harbors, drilling for oil in Alaska's North Slope wasn't economically feasible yet. Transporting petroleum from the Chukchi and Beaufort seas was deemed dangerous, unrealistic, and even foolhardy. Also, demand wasn't high. There was still plenty of oil left in the fields of Texas and Oklahoma.

That was the good news, from the Muries' perspective. On the other hand, the airplane was starting to have a profound effect on Alaskan life. Long dogsled runs could be replaced by two- or four-hour flights. In 1923 commercial service was established from Fairbanks. Territorial cities were now linked by daily flights. Two or three dozen smaller planes were also in operation, connecting far-flung mining camps. There were no radios or weather reports as support systems, however. A breakdown above the Arctic Divide was almost certain to be fatal. Pilots were literally on their own. When Christian missionaries took to the skies to spread the gospel, it was dangerous work. On October 12, 1930, two Catholic priests were killed in a plane crash near the village of Kotzebue.[23]

Steamers were also beginning to move logs from the uppermost timberline. And by the end of the decade the two-lane Steese Highway linked Fairbanks to Circle City on the Yukon River.[24]

While Alaska still had thick migratory caribou herds, market hunters were starting to drive up to Circle City to blast away at them. The country was too vast for a single game warden to patrol. Effective law enforcement under such circumstances was impossible. The rogues ran Alaska, dismissing federal authority every step of the way.[25]

Working at a feverish pace, Olaus submitted his encyclopedic study of caribou to the Biological Survey. There was a lot for the public to learn about their migratory patterns. Most Americans failed to appreciate the wonder of the caribou—or even to comprehend that reindeer were nothing more than domesticated caribou, or that caribou traveled longer distances than any other terrestrial mammal—up to 3,100 miles a year. When Olaus was assigned to observe the waterfowl along the Old Crow River in northeastern Alaska, Mardy insisted on coming along. She would carry the baby, Martin, like a papoose, slung on her back so as not to slow the expedition down. The Muries poled a scow up 250 miles of that muddy river, collecting the best field observations to date on the white-winged scoter (*Melanitta fusca*), pintail (*Anas acuta*), and American wigeon (*Anas americana*) as John James Audubon might have done. Olaus could only wonder what Audubon would have thought of Alaska's abundant birdlife.

Employees of the Biological Survey in Washington, D.C., considered it sheer lunacy to bring a baby along on such an arduous trek. Mardy, who managed to keep daily diaries of their adventure, retorted that the Inuit had been doing it for years. Mardy dutifully recorded everything from swarms of gnats to styles of moccasins. Her once soft skin became as hard as scar tissue. Whereas Mardy's honeymoon diaries had the feel of an accountant's ledger, her journals from the Old Crow had the feel of another Thoreau in the making. "The river was empty, the other shore just a thick green wall," she wrote. "At my back, behind the little tent, stretched the limitless tundra, mile upon mile, clear to the Arctic. Somehow that day I was very conscious of that infinite, quiet space. . . . We could see, far out over miles of green tundra, blue hills in the distance, on the Arctic Coast no doubt. This was the high point; we had reached the headwaters of the Old Crow. After we had lived with it in all its moods, been down in the depths with it for weeks, it was good to know that the river began in beauty and flowed through miles of clear gravel and airy open space."

What a combination! Mardy wrote prose poetry about Arctic Alaska. Olaus, sticking to empirical evidence, recorded all the facts about the fauna. Photographs were also taken and Olaus's wildlife drawings continued. Once they had conquered the Old Crow, anything was possible. They felt empowered. Much like John Muir, the Muries had a childlike passion for the wonders of nature. "I think we should go beyond proving the rights of animals to live in utilitarian terms," Olaus wrote. "Why don't we just admit we like having them around? Isn't that answer enough!" [26]

Another baby, a girl named Joanne, was born in 1927. The Muries decided to relocate outside Jackson Hole, Wyoming, with the Grand Tetons as their backyard. Olaus had been tasked by the Biological Survey to make a complete study of the famous 20,000-head elk herd, which, in his words, "had fallen on evil times." [27] Having filled their library shelves with first editions of all the important Alaskan books and reports, they devoured knowledge about every facet of Arctic Alaska.

Scientific expertise had its social advantages: the Muries' home soon became a virtual bed-and-breakfast for conservationists wanting to learn more about Alaska. Olaus continued his frequent business trips to Washington, D.C.; a third child (Donald) was born; and Bob Marshall's new "wilderness philosophy" became their guidebook. Their love story was famous in the conservationist movement during the dark days of Herbert Hoover.

And then Franklin D. Roosevelt became president. Suddenly, the White House cared about what *they* thought—imagine that. Throughout the New Deal years, 1933 to 1940, in fact, the Muries joined Bob Marshall as the world experts on Arctic Alaskan wildlife. Olaus—who was called the "father of modern elk management"—shuttled between Fairbanks, Jackson Hole, and Washington, D.C. He adhered to Charles Darwin's belief that "a man who dares waste one hour of time has not discovered the value of life."[28] Bringing the three children with them, the Muries also traveled in British Columbia and the volcanic Aleutians. Marshall came to stay with them in Wyoming to plot conservation strategy. Olaus's brother Adolph, a wildlife biologist himself, continued assisting in their pioneering studies of the great Arctic caribou herds and Jackson Hole elk. In 1940 Adolph published his landmark *Ecology of the Coyote in Yellowstone*, the first serious predator study in the history of the National Park Service.[29]

Being an ardent preservationist also had social drawbacks. Very few people want to discuss moose dewlap or black spruce seedlings over supper. To many Alaskans, the Muries were a bore. Moreover, the Muries' occupation didn't bring in dollars. So it was a memorable occasion when Olaus and Mardy visited Washington, D.C., in the early 1930s to have dinner with Bob Marshall, cofounder of The Wilderness Society. He was "full of enthusiasm and eagerness," as Mardy put it, to learn about grayling spawns, lagoon ice, gray whales (*Eschrichtius robustus*) delivering calves, and marshy cotton grass. Experts like an audience, and Marshall was a fine audience for Olaus and Mardy Murie. What Bob gave to the Muries, and what initiated a lasting friendship, was the momentum needed to win the fight to keep Arctic Alaska roadless.[30]

The Muries and Marshall weren't alone in their heartfelt concern for the fate of Alaskan wildlife. Aldo Leopold, for example, was sickened by the slaughter of bears. When the writer and photographer John M. Holzworth published an awe-inspiring book about the brown bears and bald eagles on Admiralty Island, Leopold entered the fray to save places rich in wildlife.[31] Leopold, in fact, urged that Admiralty Island should become a national bear reserve. Leopold also wanted the Katmai National Monument to be enlarged to encompass a feeding area for brown bears. (The highest density of brown bears ever recorded was at what is today Katmai National Park: 551 bears per 1,000 square kilometers.[32]) When the Alaska Game Commission, bowing under political pressure, came up with the policy that the only good bear is a dead bear, Leopold led a lobbying campaign on Capitol Hill, petitioning the special Senate Committee on Wildlife Resources to save Alaska's bears. "I personally lack firsthand knowledge of Alaskan conditions but I strongly lean to the belief that where commercial interests conflict with bear conservation, the former have been given undue priority," Leopold wrote to Senator Frederic

Walcott of Connecticut. "I favor the sanctuary and will strongly support any policy when your committee of others may evolve to not merely perpetuate the species, but to assure such perpetuation on the largest range in the largest possible numbers."[33]

All three Muries (along with Aldo Leopold and Marshall) became very active in The Wilderness Society when it was created in 1935. With a unified voice, they were determined to save Arctic Alaska for perpetuity by writing books and holding chautauquas. They had earned the right to preserve the unfathomable Arctic by loving it more than anybody else. When they spoke about Arctic sea ice being continually converted into fresh ice, they were believed because of their doctoral and master's degrees. When Marshall suddenly died, while only in his thirties, Olaus and Mardy stepped in to fill the void. What Mardy had learned best from Marshall was to attack conservation issues with a "pixie sense of humor." The fight to save natural resources in Arctic Alaska would be a long, hard struggle. But they were on the side of the angels. As Edward Abbey, author of *Desert Solitaire*, later wrote, "the idea of wilderness needs no defense, only more defenders."[34] Important landscapes throughout America that were unprotected—such as Arizona's Tumacacori Highlands, California's Eastern Sierra, Idaho's White Cloud Mountains, Oregon's Mount Hood, Pennsylvania's Allegheny Forest, and Washington's Wild Sky—had grassroots defenders, locals ready to stand in front of a timber truck or fight in courts for injunctions. In Arctic Alaska the Muries had replaced Bob Marshall as the frontline defense for the Brooks Range on both the local and the national level.

Unbeknownst to the Muries, there was a U.S. Supreme Court justice—William O. Douglas, a legal prodigy from Yakima, Washington, just turning forty—who was ready and eager to push the agenda of The Wilderness Society forward in very dramatic ways. Even when the Supreme Court was in session, he would often wear western-style shirts and pants. History would soon know Douglas as "nature's justice" for his relentless conservationist efforts to protect America's wild places. "I hiked, rode horseback, and took canoe trips through all parts of the United States and often related my experiences in public," Douglas wrote. "I became increasingly alarmed at the pollution of our rivers, at the darkening skies due to smog, at the silting of rivers due to overgrazing and reckless logging practices. I saw beaches despoiled by industry and Lake Erie turning into a cesspool. I saw highways destroying wilderness areas. I was shocked at the manner in which 'development' programs were ruining the wilderness recreational potential of the nation." [35]

Almost miraculously, the Muries cast off their grief over Marshall's death because Douglas was there to step into a new leadership role. A public intellectual, Douglas was always astutely political when it came to protecting wild America. Nobody before or after him championed the freedom to roam—the general public's right of access to wilderness—more enthusiastically than Douglas. "Commercial interests unrestrained by biologists, botanists, ornithologists, artists and others, who see the spiritual values in the outdoors," he wrote, "can in time convert every area of America into a money-making scheme." [36]

The inherent difficulty of mining or drilling in Arctic Alaska became powerfully clear to Americans on August 15, 1935, when the beloved humorist Will Rogers and his friend Wiley Post, a renowned aviator, crashed near Point Barrow. Rogers had named the little Lockheed Orion plane *Aurora Borealis* to honor the northern lights. After hunting and fishing near Fairbanks, they had decided to see the Arctic, which was just becoming popular in sportsmen's periodicals. Their aircraft crashed into the water, and their widely reported deaths were a warning to other enthusiasts that the North Slope airspace was as unpredictable as the roughest seas and that nature was still decidedly in charge. "When Will Rogers died with Wiley Post in 1935 in an airplane crash in Alaska, an important influence went out of American life," Douglas wrote in his memoir *Go East, Young Man.* "Apart from FDR, there have been no presidents in this century who could make America laugh. We need laughter for good health. I have left my saddle to the Will Rogers memorial in Oklahoma." [37]

Starting in 1937, Olaus Murie occupied a seat on The Wilderness Society's board. Like Douglas, he became known for his articulate dissent against proposals to build huge federal dams in Glacier National Park and Dinosaur National Monument, and against Rampart Dam on Alaska's Yukon River. Throughout the late 1930s the Muries also turned their attention to the Aleutian Islands, leading a reconnaissance mission (similar to the Harriman Expedition) to study sea otters and birds. "What a rich prize and privilege the assignment was," Victor B. Scheffer, who accompanied the Muries to the Aleutians, recalled. "Our chief mission was to make a 'wildlife inventory' of those treeless, nearly unpopulated islands that reach for 1,100 miles westward from the Alaska Peninsula." [38]

Chapter Thirteen

. . . . . .

WILL THE WOLF SURVIVE?

I

A damnable problem regarding Mount McKinley in the 1930s was that wolves were blamed for the decline of its Dall sheep population. Alaskans were at war with wolves, direct competitors for supper-time meat, and therefore hunted them relentlessly. Wolf dens were destroyed like rats' nests—best to kill the whole goddamn litter. Some grown wolves weighed as much as a 175 pounds and could run twenty-five miles an hour when chasing prey. They could eat a huge amount of meat—around 20 percent of their gross body weight. A wolf pack could devour a full-grown moose every two or three days. The "big bad wolf"—the term used in Walt Disney's 1933 animation *The Three Little Pigs*—was considered vermin, a wanton killer, best eradicated. Wolves were also villains in children's tales such as "Little Red Riding Hood" and Prokofiev's *Peter and the Wolf*. "They're all dirty killers," was a popular remark.[1] When a wolf appeared around a bend of the Yukon River or in the Gates of the Arctic, an Alaskan instinctually reached for a rifle.

In April 1939, the unassuming but revolutionary wildlife biologist Adolph Murie appeared on the scene in Alaska. He had recently published a landmark study, *Ecology of the Coyote in Yellowstone*, in which he argued that predator species such as coyotes and wolves were beneficial, not detrimental, helping to maintain healthy populations of other species. As one of the first biologists in the National Park Service, Murie moved to Mount McKinley to study the relationship between Dall sheep and gray wolves; in truth, he was working in the historic shadow of Charles Sheldon and was hoping to protect the threatened species.[2] Living in a hidden meadow and sometimes spending time in Sheldon's old lean-to cabin along the Toklat River, Murie wanted to change the "shoot at sight" mentality of most Alaskans with regard to wolves. Murie discovered that Alaska's wolves, for the most part, subsisted on old, injured, and diseased animals. Seldom would a pack or a loner raid a ranch; for one thing, there wasn't much livestock in Alaska.

When Adolph Murie was watching wolves in interior Alaska, sometimes in the biting cold, he must have been quite a sight: he often wore Indian snowshoes or knee-length leather boots, and he carried a week's worth of provisions in his backpack. He didn't have to hear a howl to intuit when a wolf was nearby. Like all good animal trackers and saddle tramps, he looked for obvious clues: paw prints, specks of blood, dung, broken branches, and decayed animal carcasses. He had developed a sixth sense for *Canis lupus*. Although Murie knew he was unlikely to be attacked by a wolf, he nevertheless carried a loaded automatic pistol in a holster as a precaution.

Murie also carefully analyzed many wolf-dog hybrids, of the kind Jack London described in *The Call of the Wild*. Alaskans had marvelous sled dogs that were a quarter-breed wolf; Murie could spot them easily because their muscular legs were longer than those of a typical sled dog. What worried Murie was that many wolves and mush dogs had body scabs—a sign of mange. He also worried about rabies and distemper; both viruses affect motor functioning. The decreasing number of wolves at Mount McKinley was also due to aggressive hunting and poisoning of them all around the perimeter of this national park. Because wolves in the Alaska Range were struggling to survive, Murie was overjoyed whenever he discovered a healthy den. "Wolves vary much in color, size, contour, and action," he wrote. "No doubt there is also much variation in temperament. Many are so distinctively colored or patterned that they can be identified from afar. I found the grey ones easier to identify since there is more individual variation in color pattern among them than in black wolves."[3]

Wolves were already considered predators of livestock throughout America, from the Rio Grande to the Beaufort Sea, and a sizable bounty was offered for wolf pelts, so Murie had his work cut out for him. To Alaskans, it seemed, killing wolves wasn't a sport but an imperative. Fur trappers, bounty hunters, and cattle barons all abided by the "culture code of the pioneer," which was to "kill what couldn't be dominated."[4] In the vernacular of the territory it was called "getting your meat."[5] The Denali wilderness, Murie would write, had a wildlife population almost as diverse as Yellowstone's; he was grieved by Alaskans' reckless attitude toward predators. It was Murie's mission to make sure that Mount McKinley maintained its original wolf population.[6]

The counterforce to Adolph Murie in Alaska was Frank Glaser (also known as the "Wolf Man"). Never before or since has Alaska produced such an efficient exterminator of wolves as Glaser. From 1915 to 1966 Glaser killed wolves. It didn't matter to Glaser whether a client was an Eskimo enclave wanting to save caribou from wolf packs or the U.S. Fish and Wildlife Service wanting to contain the spread of rabies; he would kill wolves for cash. His weapon of choice was his "coyote getter," a set device that fired cyanide if triggered by a coyote. But he had other tools of the trade as well. He could kill wolves with strychnine bait, a rifle, snares, and traps. "Frank was like an Indian at picking up a wolf sign that was all but invisible to me," Charles Gray, an Alaskan game warden, recalled. "He knew which clump of grass they urinated on and which little ridge they preferred to travel on. I was always in awe of his knowledge of wolves, for he was almost always right."[7] Glaser's motto, "The only good wolf is a dead wolf," summarized his hatred of predators.

Wilderness surveyors like Charles Sheldon also saw wolves as a menace to big game, "the chief enemy of the caribou," always "hovering about the feeding herd and following them around as they roamed, usually in a fairly well-defined circuit."[8] Hikers in the uninhabited Yukon Flats told of wolf packs shadowing them for days, seemingly looking for an opportunity to kill.[9] The Biological Survey itself turned against wolves throughout the 1920s and 1930s. Influenced by Vernon Bailey's booklet *Wolves in Relation to Stock, Game, and the National Forest Reserves* (1907), Merriam considered all predators lesser species than big game. Bailey, who married Merriam's sister Florence, wanted Alaskan wolves exterminated wherever they encroached on civilization.[10] "The fierce destructiveness of large wolves and of mountain lions," the 1924 Annual Report of the Biological Survey read, "both to domestic animals and game, is so great that it becomes a necessity to eliminate them from certain areas."[11]

Only a few weeks after Franklin D. Roosevelt moved into the White House in March 1933, Aldo Leopold published *Game Management* with Charles Scribner's Sons. In 481 pages, Leopold created the discipline of modern wildlife management. Pioneering in the burgeoning fields of systems ecology and genetics, he supported protecting wolves in ecosystems (albeit managed). A lifelong hunter, Leopold explained exactly why big game such as moose, Dall sheep, deer, antelope, and elk needed large habitats to survive properly. Filled with scientific charts and survey studies, interspersed with ideas espoused by Darwin and Malthus, *Game Management* introduced Leopold to the general public as the most distinguished conservationist of the New Deal years. He argued that predators such as wolves and coyotes were an essential component in any healthy ecosystem. To Leopold, the environment wasn't a marketplace commodity. It was a biotic community in which all living creatures belonged. "How shall we conserve wildlife," he asked, "without evicting ourselves?" [12]

Adolph Murie was thrilled to read about Leopold's philosophy of game management in his articles and surveys during the early 1930s. Murie had rarely, if ever, encountered such sound ecology expressed in such lean, elegant prose. Analytically Leopold's works mirrored his own thinking in *Ecology of the Coyote in Yellowstone*. Leopold's chapter "Predator Control," from Murie's perspective, was a weapon in the new effort to save wolves, cougars, coyotes, and bears from systematic extermination. What made Leopold such an important conservationist was his sense of judicial fairness, even though he was dubious about technological advancements that ate away at wild lands. Regularly Leopold, as if taking a poll, asked fellow wildlife biologists about wolf populations in various ecosystems.

"I do not find the coyote a bad fellow at all," Murie wrote to Leopold from Wyoming. "As far as the elk are concerned he is not nearly as big a factor as several other things. I will not go into detail here, but would point out that a considerable number of people enjoy the coyote in the hills, he is part of the environment, and his entire removal would make elk hunting less attractive to some people. I feel that if sportsmen and non-shooting conservationists could get together, progress would be so much more rapid. If sport could be placed on a higher plane, and some recent plans might work in that direction, nature lovers in general would be likely to help in game matters. We all have the same interests and must work together to accomplish anything." [13]

In 1897, Frederic Remington painted *Moonlight Wolf*, an eerie, frightening scene of a lone Great Plains wolf (*Canis lupus nubilus*) creeping around a corral in a blue winter snow. It is one of Remington's best works. In the dead of an Alaskan winter not much moved. But Remington's wolf doesn't hibernate—it hunts in the dark. What we don't see in this painting is the wolf being shot by the rancher or pulling down livestock. What happens is up to the viewer's imagination. Unfortunately, the Great Plains wolf that Remington painted had nearly gone extinct by the time Murie arrived at Mount McKinley in the 1920s. "Alaska is the last North American stronghold of the wolf," Barry Lopez wrote in *Of Wolves and Men*, "with Eskimos and Indians here, with field biologists working on wolf studies, with a suburban population in Fairbanks wary of wolves on winter nights, with environmentalists pushing for protection, there is a great mix of opinion. The astonishing thing is that, in large part, it is only opinion." Even biologists acknowledge, Lopez noted, that there are some things about wolves' behavior that you just have to guess at. [14]

II

Given the hatred for wolves in Alaska, protecting them was going to be a tall order. But Adolph Murie was up to the task. Much like his older brother Olaus Murie, Adolph (nicknamed Ade) had become well known in wildlife protection circles by the 1930s. He was raised along the Red River of the North, and his résumé revealed a man who couldn't sit still. After earning a BS degree in biology at Concordia College in Moorhead, Minnesota, Murie became a ranger at Glacier National Park. [15] His hope was to write a series of definitive scholarly papers on various North American mammals. In 1926 happenstance helped him pursue this goal. Professor Lee R. Dice, a pioneer in animal ecology, offered Murie a PhD fellowship at the University of Michigan–Ann Arbor. Murie decided to become an expert on the common deer mouse (*Peromyscus*), prey extremely important to understanding predators. There was one main advantage of starting this low on the food chain: nobody had done it before. Professor Dice—a mammalogist by training—was not only a pioneering American ecologist but also a geneticist.[16]

By the time Dice took Adolph Murie under his wing, he had made the University of Michigan a leading opponent of predator-control practices such as steel traps and meat laced with strychnine. The Bureau of Biological Survey had become the U.S. Fish and Wildlife Service, but eradicating predators—wolves, coyotes, and cougars—remained the policy of the federal government. As a U.S. government report (as mentioned above) declared in 1924, "The fierce destructiveness of large wolves and of mountain lions, both to domestic animals and game, is so great that it becomes a necessity to eliminate them from certain areas."[17] Such policies infuriated Dice. A bioprospector ahead of his time, Dice shamed the government's scientists for being more concerned about protecting livestock than wildlife.

Adolph Murie became a favorite student of Dice's. Not only did Murie complete his dissertation, in 1929—"The Ecological Relationship of Two Subspecies of Peromyscus in the Glacier Park Region"—but he was hired to revamp the University of Michigan's Museum of Zoology to reflect the ecological revolution.[18] Perhaps to demonstrate his adeptness at both extremes of the wildlife kingdom, he went from spying on field mice to assessing herds of the lordly moose. Slipping away from Ann Arbor during the summers of 1929 and 1930, Murie ventured north to Isle Royale (a thickly wooded island in Lake Superior teeming with unmolested wildlife). The moose population Murie encountered at Isle Royale was thriving. There were 300 moose on the island during the Great War—and by the time of the Great Depression the number had risen to 3,000.[19] Murie helped bring their population back.

During the late 1930s, Adolph Murie bounced around a lot outside Michigan. He collected 700 mammals in British Honduras (now Belize), emphasizing gophers and bats.[20] He spent time in Jackson Hole, Wyoming, with his brother Olaus and Olaus's wife, Mardy, watching moose herds browsing in the fields. He wrote the still widely influential book *Ecology of the Coyote* (1940). On a visit to Twisp, Washington, he fell in love with Louise Gillette (Mardy's stepsister); they married in Wyoming. Fox species—red, gray (*Urocyon cinereoargenteus*), and arctic—grabbed Murie's professional attention. Filling burlap bags with fresh fox scat, he analyzed its composition under a microscope back in Ann Arbor. In his 1936 study *Following Fox Trails*, Murie documented how red foxes would often kill shrews merely for fun, not to eat. This finding reinforced Murie's belief that backyard mesopredators—medium-size predators such as coyotes and skunks—were an essential part of any healthy ecosystem. Without them, garden pests such as shrews would become menaces. But he also promoted the aesthetic notion that foxes were charming creatures to watch up close. "The feeling of a woods is much improved by the presence of fox," he wrote. "It is good to know that the fox is present in a region for it adds a touch of wickedness to it, gives tone to a tame country."[21]

After nine years at the University of Michigan, and backed by the powerful sponsorship of Dice, Murie made a career change. Frustrated that animal ecology was being ignored by the U.S. government, he joined the new Wildlife Division of the National Park Service.[22] Murie was now in a position to help the greater western parks achieve something close to *natural conditions*. Murie's biological expertise could be applied to bring back species like wolves, cougars, coyotes, foxes, bobcats, lynx, minks, weasels, and otters. The days when the National Park Service had promoted picture-postcard tourism—when the outdoors experience was rigged in favor of the "Kodak moment"—were ending.[23] (During the early 1990s, in an article in *Wild Earth*, this discarded approach was famously called *ecoporn*.) Besides ranchers, Murie was at war with backcountry people who still hunted and trapped furbearing animals for pelts; this was a primary source of winter income.

Realizing that introducing wolves into a national park was going to be a long battle, Murie set his sights on Mount McKinley. It was unlike the national parks in the Lower Forty-Eight because the surrounding area had no organized stockmen's associations to protest. However, Alaska did have a bounty for predator species: $15 for every wolf and the same for every wolverine, not an insubstantial sum for a backcountry family. Finding ways to protect packs of gray wolves in the Denali wilderness from twelve-gauge shotguns and 30.06-caliber rifles would not be easy. During World War I surplus Springfield weapons had been sold to Alaskans, so the territory was well armed.

As a wolf ecologist, Murie had a sublime ability to watch wolves undetected by the packs in their dens.* He wrote notes about their sleeping habits, tail wagging, and long jaunts looking for prey. Because wolves have no predators besides humans and other wolves, Murie was able to creep within a few yards of their dens. When they suddenly became alert, however, their defense mechanisms were aroused, and their eyes did not miss much. Sometimes Murie would set up a movie camera to capture their behaviors, such as cubs catching mice and males sniffing each other in greeting. "The strongest impression remaining with me after watching the wolves on numerous occasions is their friendliness," Murie wrote. "The adults were friendly toward each other and amiable toward the pups, at least as late as October. This innate good feeling has been stronger marked in the three captive wolves which I have known." [24]

The Biological Survey and the Bureau of Fisheries were merged in 1939 to become the U.S. Fish and Wildlife Service of today. Around Juneau little snub-nosed government motorboats patrolled the Alexander Archipelago. On April 14 Adolph Murie dogsledded into the Denali wilderness and began a two-year stint studying the wolves of Mount McKinley. Using the log shack Sanctuary as his base camp—located twenty-two miles from the border of the national park—Murie started tracking wolf packs for preliminary insights. A cold spring wind whistled around him as he studied wolf stool for signs of Dall sheep hair. At high altitudes, his lips turned purple from the frigid temperatures. A new park road helped Murie survey a vast amount of Denali territory. Murie hiked nearly 2,000 miles that year, exposed to vicious spells of cold and heat, procuring data that he hoped would help the wolf survive in midcentury America. [25]

Using methods that he first developed at Yellowstone on behalf of coyotes, Murie carefully estimated the ages of Dall sheep supposedly killed by renegade wolves. He studied the tooth marks on the carcasses and analyzed the rams' horn rings. He also took climate change and varied diseases into account in his wolf studies. No one before Murie had undertaken such a serious biological study of North American wolves. The

---

* According to the raptor ecologist Joel E. (Jeep) Pagel of U.S. Fish and Wildlife, in Asia golden eagles are known to hunt wolves. In North America, however, golden eagles have never been seen to seize a wolf, although they do eat coyote pups.

popular author Ernest Thompson Seton had written a series of articles about wolves, but scientists dismissed these as fiction.* Egerton Young's *My Dogs in the Northland*—a memoir Jack London liberally mined for *The Call of the Wild*—dealt with domestic sled dogs and wolves. Stanley P. Young had cowritten a landmark work, *The Wolves of North America*, with color plates provided by Olaus Murie. In 1939 Stanley Young, who wanted wolves to survive only when "not in conflict with human welfare," was appointed senior biologist in the Department of the Interior's branch of Wildlife Research. But it was Murie who became the defender of wolves, submitting reports to the National Park Service urging it to end its wolf-control efforts.[26]

Through the 1930s Native Alaskans also started protesting against the slaughter of wolves, although they didn't speak with a unified voice. New Dealers sought to help Native Alaskan populations prosper during hard times. Congress allowed the Tlingit and Haida Indians, for example, to sue the U.S. government over tribal lands; this helped curtail market hunting in the territory. In 1935, Congress included all Native Alaskans in the Social Security Act. And, more helpfully in the long run, Secretary of the Interior Harold Ickes set aside lands for Native Alaskans as hunting and fishing sanctuaries. Wolves in these Native lands were safe from slaughter.[27]

III

World War II transformed Alaska, seemingly overnight, from a backwater territory to a major strategic asset that was well worth defending. Following the attack on Pearl Harbor the Japanese military actually seized two Aleutian islands—Attu and Kiska—as part of its North American campaign. The Roosevelt administration

---

* Olaus Murie, however, was a fan of Ernest Thompson Seton, who had been his literary hero during his boyhood. He once encountered Seton at an event in Washington, D.C., and said, "Oh, my, I know all your books. My friends and I grew up with them. We just lived *Two Little Savages*, along the Red River in Minnesota. We did everything you wrote about in there, and we built a tipi but we could never make the smoke go up right." Seton replied, "I never could either."

quickly established a military command in Alaska and moved defense forces to Adak, Dutch Harbor, and Cold Bay. Because the U.S. government censored news from Alaska in the 1940s, this campaign is called the "forgotten war." The United States engaged Japanese troops in the Aleutians from June 3, 1942, to August 15, 1943. The wind and fog were obstacles for both sides. Many B-24 pilots, in fact, described the Aleutian campaign as a three-sided battle waged between the United States, Japan, and the uncooperative weather.[28] Samuel Eliot Morison, in his magisterial volume *History of United States Naval Operations in World War II*, wrote that the Aleutians were the "Theater of Military Frustration."[29]

After World War II ended, the U.S. armed forces' presence in Alaska declined dramatically, from 152,000 troops in 1943 to 19,000 in 1946.[30] During the war, the U.S. government had spent more than $1 billion on Alaskan infrastructure projects like building the 1,523-mile Alaska-Canada Highway and modernizing Alaskan railroads. The overland road built from Delta Junction (southeast of Fairbanks) to Dawson Creek (in British Columbia) forever changed how Alaska's lands would be managed. Huge D-8 Caterpillar bulldozers with enormous cutting blades uprooted towering spruces. Gravel was hauled in dump trucks and other vehicles from alluvial riverbeds and hillsides. When it rained or snowed, four-wheel-drive convoys often got stuck in mud craters and impassable ravines.[31]

The U.S. Navy also dramatically improved Alaska's docks, wharves, and breakwaters. Pan American Airways had introduced a commercial link between Seattle and Juneau. Alaska was ready for business. As more and more civilians moved to the territory, the postwar movement to reinstate the timber industries swelled. Both of Theodore Roosevelt's great national forests—the Tongass and Chugach—were now under siege. During the war, also, the U.S. Bureau of Mines had started surveying Alaska's North Slope for oil. Reports of seepage between Cape Simpson and Point Barrow were becoming commonplace. Congress gave a $1 million grant to begin oil extraction work at the Naval Petroleum Reserve on November 4. A big question was whether oil drilling was feasible in the subzero Arctic conditions. By September 1945 Alaskans were saying that Point Barrow was located at one of the world's great oil fields. Secretary of the Interior Ickes wrote to an oil booster in Seattle, "Time alone will show whether there is oil . . . and, if so, what its quality and quantity may be." [32] But Ickes was merely stalling. The U.S. Navy was all over the North Slope, considering how best to drill Unimat Mountain along the Colville River. Seabees were busy with pipeline mitigation issues. By late 1945 Secretary of the Navy James Forrestal said that the U.S. government was ready to invest $150 million in Arctic Alaskan oil. Seabees drilled the first well that year.

IV

W hile the U.S. armed forces were defending Alaska from the Japanese during World War II and the U.S. Navy was promoting the drilling of oil wells, Adolph Murie worked on his book *The Wolves of Mount McKinley*. Every page was informed by his shrewd taxonomic analysis of the family Canidae. Focusing on wolves' home life, Murie created a sympathetic portrait. Wolves' fur was usually gray, or mostly gray, but could vary from white (in the tundra) to black. Their denning habits, pack frolics, cunning, and preference for sheep meat were all thoroughly analyzed by Murie. When *The Wolves of Mount McKinley* was published in 1944, Aldo Leopold deemed it the classic wildlife study of *Lupus*. Since his speech before the American Game Conference in 1935, Leopold had insisted that the U.S. government's "predator control" programs were wrongheaded. Echoing Leopold, periodicals such as *Audubon* and *Natural History* called Murie the world's foremost wolf ecologist.

Throughout the 1940s Leopold had been putting together *A Sand County Almanac*, a book about conservation that is equaled only by *Walden Pond* as a meditation on the need for the wild in our commerce-driven lives. Because of his strong scholarly bent and his "micro-knowledge" of silviculture, it is somewhat surprising that Leopold could write so philosophically and with such poetic grace. Leopold worked for thirteen years on the book and had written hundreds of articles in preparation for undertaking the task. Every line and every comma seems exactly right in this reflective memoir, which is also a work of natural history. Prefiguring the "deep ecology" movement of the 1960s, Leopold reasoned that land wasn't a commodity to be possessed. Instead humans should be caretakers of the Earth, and wild places should be saved. Famously, Leopold wrote the essay "Thinking Like a Mountain" in 1944 for inclusion in *A Sand County Almanac*. Filled with an anguished regret, Leopold told of a sad afternoon when, in Arizona's Apache National Forest, he killed a mother wolf and her pups. "In those days we had never heard of passing up a chance to kill a wolf," Leopold wrote. "In a second we were pumping lead into the pack, but with more excitement than accuracy: how to aim a steep downhill shot is always confusing. When our rifles were empty, the old wolf was down, and a pup was dragging a leg into impassable slide-rocks." [33]

What happened next became one of the most profound moments in the annals of the wildlife protection movement. Leopold had an epiphany in the Apache, a pang of conscience. All those bloody hunt stories, the machismo, and the random slaughter of species that came to signify the winning of the West seemed perverse as Leopold watched the pup drag its bleeding body toward cover, away from its mother lying dead in the dust. It was a scene of carnage. "We reached the old wolf in time to watch a fierce green fire dying in her eyes," Leopold wrote. "I realized then, and have known ever since, that there was something new to me in those eyes— something known only to her and to the mountain. I was young then, and full of trigger-itch; I thought that because fewer wolves meant more deer, that no wolves would mean hunters' paradise. But after seeing the great fire die, I sensed that neither the wolf nor the mountain agreed with such a view." [34]

Thanks to Murie and Leopold, national park superintendents started considering wolves an asset. Poisoning animals was now frowned on in public lands. After spending the summers of 1940 and 1941 studying Alaskan wolves in their dens, Murie refuted the popular perception of wolves as savage and morose.[35] Although far more elusive than bears or raccoons, wolves began to attract tourists, who would brave the cold, damp winds around Mount McKinley and train their binoculars on the horizon hoping to spot a pack. Murie had succeeded in changing the reputation of gray wolves from sheep-killers to wild dogs that maintained long familial ties with their pups. But he encountered negative reactions as well as accolades. When Murie returned to Mount McKinley in 1945 to continue writing articles about protecting wolves, the Alaska territorial legislature derided him. Outdoors groups such as the Tanana Valley Sportsmen's Association had preferred the conservationists of Charles Sheldon's era who saved Dall sheep; these sportsmen disliked the Minnesotan ecologist—Murie—who was bent on protecting wild wolves and who seemed to be trying to turn gray wolves into teddy bears. A grassroots countermovement, in favor of killing wolves, developed across Alaska; and Alaskans turned against the U.S. Department of the Interior as never before.

Murie was a nice Midwesterner, disdainful of overwrought conflict; he had no overweening desire to battle with Alaskans over wolf conservation. But reports that bush pilots were now shooting wolves from the air sickened him. Very discreetly, he returned to Mount McKinley to conduct more field research on wolves and to protect packs from being slaughtered. Murie was annoyed by the false dichotomy that forced a choice between Dall sheep and wolves. He thought it was childish that in the atomic age people still accepted the image of the wicked wolf presented in Aesop's fables and the Grimms' fairy tales. Determined to amass more scientific evidence, Murie would disappear into the trackless wild for weeks at a time, dutifully recording the *real* behavior of wolves, not the legends. Mount McKinley without wolves, Murie concluded, would be mere scenery.

Wonder Lake, near the base of Mount McKinley, became Murie's favorite place to watch the wilderness. In the spring of 1948, in preparation for a visit by the famous photographer Ansel Adams, Murie went to clean up a five-room log bungalow maintained by the National Park Service near Wonder Lake. Upon opening the front door, Murie found that grizzly bears had torn the place apart. Flour bins and pantry cupboards had been ravaged. The bears had also gotten into the basement and had ripped into boxes of army surplus Hershey bars. The bears had opened up cans of brown paint, tracking it throughout the bungalow. The basement windows were smashed. "The building was repaired," Murie wrote in *A Naturalist in Alaska*, "but the bear could not forget those chocolate bars."[36]

It took Murie an arduous day to make the bungalow bear-proof. Using a mop, he wiped away all traces of chocolate. A few days later, however, after a long day hiking the tundra observing wolves, an exhausted Murie went straight to bed. It was around midnight and still light outside. Murie, peering out of his bedroom window before drifting off, saw a grizzly running across the tundra headed right for his cabin. The bear, curiously spectral in the moonlight, circled the cabin, unable to find a way inside. Murie felt triumphant and went to sleep. "In a few minutes big chunks of wallboard were torn loose, and soon a hole was big enough to allow him to pass into the dining room beside the fireplace. He did not come the few steps down the hall to my bedroom, but sat down in front of the kitchen door. With his powerful paw he wriggled the doorknob, and soon I started hearing the rattle in my sleep. I awoke and heard the fumbling at the doorknob." [37]

Although 50 percent of a brown bear's diet consists of vegetation, bears were also known to bring down caribou in the soft snow. Around Mount McKinley, locals claimed that if you wanted to attract a bear, you should put chocolate on your porch. Brown bears were diurnal, but if they smelled even a whiff of chocolate, they could suddenly became nocturnal. [38] Now Murie, grabbing his rifle, prepared to shoot the intruder at Wonder Lake. But, perhaps sensing danger, the bear jumped out the dining room window and ran off. [39]

Murie and Leopold's ethos had made inroads in the Department of the Interior after World War II. A turning point for the protection of wolves in Alaska occurred in December 1945. A bill, H.R. 5004, was introduced in Congress stipulating that wolves *could* be protected around their historic range in Mount McKinley National Park, but only if their population was very strictly controlled. Conservationists—including Aldo Leopold—saw this bill as the first major step in protecting a predator. Because wolves dispersed over huge distances and easily colonized new habitats, a lot of federal land would have to remain unmolested in order for packs to survive. "The wolf has been demonized, defeated, and defended by humans," *National Geographic* declared. "It must now renegotiate its place in a changed habitat." [40]

From 1947 to 1950, Adolph Murie served with the National Park Service at Mount McKinley as resident biologist. [41] Ostensibly, his job was to control the wolf population and protect the Dall sheep, even in blizzards and crawling fog, but his real aim was to persuade the service to permanently ban shooting wolves within this park. America's largest national park would be a wolf haven. Olaus Murie (who had retired from U.S. Fish and Wildlife in 1944) and Mardy Murie (who was interested in writing a novel with a Siberian-Alaskan setting, to be titled *Island Between*) locked up their log home in the Tetons—which had become a mountain headquarters of The Wilderness Society—and temporarily moved to the Mount

McKinley area to help Adolph. They set up shop at a cabin in Igloo Canyon during the summer and lived at park headquarters in the winter, home-schooling their children. The Muries—all three of them—believed that *all* wildlife in Mount McKinley needed federal protection. Adolph would go on long patrols on snowshoes, always collecting biological data about the species' natural resilience and adaptability. He identified four principal vocal communications by wolves: howls, little whimpers, prolonged talklike mumbling growls, and a passionate talking bark. To quell local suspicion that he wasn't properly performing his duties—thinning out wolves to protect the Dall sheep—Murie selectively shot a few sick-looking older wolves. "Ade knocked off a couple of wolves," a former park ranger, Bill Nancarrow, recalled, "just so they wouldn't send Fish and Wildlife to start killing them." [42]

Victory was at last achieved in 1952 when Conrad Wirth, director of the National Park Service, unequivocally prohibited killing wolves at Mount McKinley. [43] The Muries went back to Wyoming, and young activists of The Wilderness Society, such as Howard Zahniser of Pennsylvania and Sigurd Olson of Minnesota, journeyed to the Murie ranch there to discuss wildlife protection strategy in the aftermath of the successful outcome at McKinley. [44] Hoping to continue the momentum, the park service now asked Adolph Murie to write a follow-up book, *The Cougar of Olympic National Park*. "If you could put out a publication on the Olympics featuring the cougar as well as you did the wolf," his brother Olaus wrote, "you will certainly have made a big mark in the conservation world." [45]

Adolph Murie wasn't enthusiastic, however. Some people, if lucky, discover an ecosystem that speaks to them spiritually—scattered woodlands in the Ohio River valley, for example, or a chasm like the Grand Canyon. Mount McKinley, to the Muries, as to Charles Sheldon before them, was a special place. Adolph decided to devote his life to protecting those 2 million acres* of interior Alaska, using the periodical *Living Wilderness* as his forum. When a biologist or an ecologist like Murie falls in love with

---

* In 1980, Denali National Park was expanded by 4 million acres. Today it encompasses a total of 6,075,107 acres. The original 2 million acres are commonly called the "old park" and are designated wilderness.

a treasured place, it usually occurs as a result of arduous fieldwork in all types of weather. This process—called "thrumming"—allows the outdoors enthusiast to feel the pulse of the ecosystem.[46]

A pedagogical change had occurred in U.S. natural history since the late 1920s. Collecting biological specimens in the field—i.e., shooting wildlife for mounts and studying skins—was now outdated. The new impulse emanating from places such as Woods Hole Laboratory in Massachusetts and the Scripps Institute in La Jolla, California, called for analyzing animals in their own distinctive habitat. Farley Mowat, the wildlife naturalist from Belleville, Ontario, who wrote nearly forty books, was already documenting Arctic Canada in realistic novels such as *People of the Deer* (1952). Rachel Carson's *The Sea Around Us*, published in 1951, celebrated starfish, coral reefs, squid, and dolphins. Famously, Carson, a biologist with U.S. Fish and Wildlife working at the Woods Hole Laboratory in Cape Cod, collected sea urchins along the coast and then released the creatures back into the ocean, without killing a single one. Her science writing at the time was instrumental in developing a public understanding of ecosystems. No longer did a trophy-lined wall (like TR's) signify a naturalist's prowess. Ecology was the new ethos: understanding the unity of an ecosystem, documenting the habitual condition of species.

And a new biological field of study—ethology, the science of animal behavior—was becoming extremely popular in the nature periodicals of the 1950s. New ethological approaches were necessary to deal with the ecological issues raised during the postwar era by the concept of "better living through chemistry." For instance, DDT was causing a decline of peregrine falcons, pelicans, ospreys (*Pandion haliaetus*), and eagles; habitats were lost when people moved to urban and suburban areas (there were fewer family farms and fewer people living close to the land); and there was a general loss of habitat. The scientific community was determined to learn how to better manage conflict between humans and wildlife. Instead of being killed as livestock predators, for example, coyotes were now being praised for their sonorous song.

Professor Lee R. Dice became president of the Ecological Society of America and also worked closely with the Ecologists Union (later the Nature Conservancy). Ever since Professor Cooper had successfully advocated for the creation of Glacier Bay National Monument, the Ecologists

Union having persuaded him to sponsor the idea, there was a real feeling of making a difference in Alaska. The early 1950s saw many great photographers of Alaska wildlife working in the temperate zone—in the Katmai, the Kenai, and Mount McKinley National Park. Was there a more beautiful sight in the world than Kachemak Bay from Homer Spit or the glacier lands from the John Muir Trail? Creatures such as wolves and grizzlies, moreover, were appealing to moviegoers of the 1950s, who were intrigued by Arctic lore. "People were accustomed to the idea that animals had a wide range of behavior and individual mannerisms," Thomas R. Dunlap wrote, in *Saving America's Wildlife*, of the postwar era. "They were used to the idea that people could establish links with animals."[47]

Meanwhile, however, a particularly objectionable form of hunting was being practiced in Alaska. Aerial hunting of wolves started in 1948 and became popular as a sport. Guns and planes were a wicked combination. Some hard-core Alaskan hunters would fly over the tundra and blast away at wolves with increasingly powerful automatic weapons. Two men in a plane might sometimes shoot ten to fifteen wolves in this way. The grim sport called aerial hunting attracted trophy hunters from all over the world, and the plane services catered to tourists. "Back in Kotzebue or Bettles or Fairbanks the story was embellished and hunters and pilots were congratulated for their bravery and daring," Barry Lopez wrote in *Of Wolves and Men*. "It is both ludicrous and tragic that the death of a wolf so cheaply killed confers such prestige."[48]

Throughout the 1950s and 1960s, Adolph Murie continued trying to ban the aerial hunting of wolves, but with only modest success. Not until 1959, when Alaska became a state, was there serious recognition that wolves had value. In 1963 they were upgraded to the classifications of fur bearer and big game. Bag limits and hunting seasons were established.[49] Yet state-sponsored aerial wolf gunning continued unabated throughout the 1960s. It was an ingrained behavior.

It wasn't until 1969, when NBC presented the prime-time documentary *Wolves and the Wolf Men*, that the public turned against this cruel practice. Eventually, the federal Airborne Hunting Act of 1971 forbade it. But numerous states-rights activists, including the future governor Sarah Palin, were unenthusiastic about the federal law, insisting that killing wolves was an all-American sport, a way of life in Alaska.

Chapter Fourteen

o   o   o   o   o   o

WILLIAM O. DOUGLAS
AND NEW DEAL CONSERVATION

I

Sitting at his desk in his U.S. Supreme Court office, William O. Douglas was swamped with legal work, including writing decisions on
such issues as why trees had standing and why wildlife deserved
legal rights to protected habitats. During his tenure as an associate justice of the Court—which began on April 17, 1939, and extended until November 12, 1975—the great civil libertarian would also become the most
historically significant pro-wilderness American political force since
Theodore Roosevelt. From the Great Depression to Watergate, Douglas
composed vivid prose sketches about the American valleys and mountain
ranges that had stolen his heart. The Olympics, Wallowas, and Brooks
Range consumed his imagination even when the Court was in session.
A glint in his eye indicated to his colleagues that he was thinking about
fly-fishing in the Middle Fork of the Salmon or on the Quillayute River.
Douglas, who had climbed in the high Himalayas, encouraged groups like
the Sierra Club and The Wilderness Society—he was an active member of
both nonprofit societies—to bring class-action suits against despoilers of
the American landscape. When Douglas received the John Muir Award
from the Sierra Club in June 1975, he noted that his "view" of "policy in
environmental matters" came from the "powerful influences" of Buddhism, Gifford Pinchot, Clarence Darrow, and John Muir. "I thought so

well of Muir and his works that in 1961 I wrote a book about him," Douglas boasted, "*Muir of the Mountains.*"[1]

In a series of books, articles, and letters, Douglas proudly argued that tramping around the unspoiled wilderness, as Muir had done, was part of a noble American tradition that dated back to the transcendentalists of Concord. What could be more American than rediscovering the natural world to offset urban angst? Wasn't it essential to leave some areas unmapped, so that wanderers could *get lost* in the wild? Shouldn't young Americans be encouraged to answer the "call to adventure" represented by white-water rivers, unbounded tundra, and dense forest reserves? Citizens needed retreats in the natural world from the degradation of city life. "The distant mountains make one want to go on and on and on," Douglas wrote after exploring the Brooks Range of Alaska in 1956, "over the next ridge and over the one beyond."[2]

Always an iron-willed individualist, Douglas was concerned that the *freedom* associated with exploring the wilderness, hitchhiking, backpacking, camping, and mountain climbing was being constricted by anti-vagrancy laws. (The novelist Kurt Vonnegut later supported this belief, saying that the Constitution protected our right to "fart around.") During the Great Depression, Douglas had been a hobo, traveling the rails from Yakima to Chicago, west to east, living out of a rucksack. Disappearing down the open road and shedding the shackles of the nine-to-five workday was—to Douglas's mind—an American right just as surely as free speech or equal education. Douglas worried that national parks like Yellowstone and Yosemite were being corporatized. Visitors in the mid-twentieth century encountered bumper-to-bumper traffic, gift shops, asphalt parking lots, uniformed rangers, and firework displays—and at Yosemite, the Hetch Hetchy valley had been destroyed by the construction of a reservoir. As Thoreau had complained in *Walden*, many stout-hearted Americans, seeking regeneration in wild places, were fleeing the "desperate city" only to arrive at the "desperate country."[3] What demon, Douglas asked, had possessed the National Park Service to turn natural wonders like Old Faithful into sites for gewgaw shops? What fools would hollow out a redwood tree in Mariposa Grove so automobiles could drive through it? "When roads supplant trails," Douglas wrote, "the precious unique values of God's wilderness disappear."[4]

Although he admired Pinchot, Douglas dissented, as he matured, from the whole concept of "multiple use" of natural resources. He saw Americans' mania for constructing roads in national parks and forests as "evidence of our decline as a people." Habitats for wildlife, he argued, should be left alone. All the national forests, as far as he was concerned, should be redesignated as wildernesses. Douglas, agitated, predicted that the world of 2200 would be choking on concrete, smog, industrial blight, and the withered wastelands left by clear-cut forests and oil spills. If Americans were wise, he believed, they would understand the importance of preserving roadless wilderness for its own sake: wilderness was more valuable than all the gold bars in Fort Knox. Without the possibility of escaping into the noiseless backcountry, the United States would become merely a tacky version of tourist-packed Europe. "There is no possible way to open roadless areas to cars and retain a wilderness," Douglas asserted. "This is one diabolic consequence of the 'multiple use' concept as applied. The Forest Service recognizes, of course, that the application of the 'multiple use' principle means that some areas must be devoted exclusively or predominantly to a single purpose. The difficulty is that, in the Pacific West, 'multiple use' in practical operation means that every canyon is usually put to as many uses as possible—lumber operations, roads, campsites, shelters, toilets, fireplaces, parking lots and so on."[5]

Repeatedly, throughout his life, Douglas rallied to the defense of pristine Pacific Northwest and Alaskan landscapes. During the 1930s it was the Olympics; in the 1940s, the Cascades; in the 1950s, the Brooks Range; and in the 1960s, the redwoods of California. As his biographer Bruce Allen Murphy noted in *Wild Bill*, Douglas helped launch the modern environmental movement in 1960 by dissenting to a denial of certiorari in a dispute over DDT being sprayed in Long Island.[6] Douglas, never idle, continually thought of legal ways to help save America from ruin. Later in Douglas's legal career, following the oil spill near Santa Barbara of January 28, 1969, he stoutly refused to let Union Oil get away with impunity for fouling the Southern California coastline from Goleta to Rincon, and all of the northern Channel Islands. Since his young adulthood, Douglas had fought to protect American wilderness and coastlines. Now, in 1969, more than 10,000 birds had died because of a faulty blowout preventer on Union Oil's platform A in Santa Barbara, and a furious Douglas wanted justice.

This oil spill impelled Douglas to put some of his long-held judicial beliefs into writing. He was, after all, the leading light of the wilderness movement. Douglas famously held, in a Supreme Court case, that trees, oceans, and rivers had legal standing. (Look up his dissenting opinion: *Sierra Club v. Morton*, 405 U.S. 727, 1972.) As a justice of the U.S. Supreme Court, Douglas had somehow found time to read an obscure essay by Christopher D. Stone in the *Southern California Law Review*: "Should Trees Have Standing?"[7] Stone, a former Supreme Court clerk, thought the article was a breakthrough argument on behalf of the environment. Douglas used Stone's argument to go after Walt Disney. In 1969, when Disney received approval to build a huge $35 million ski and swim resort at Mineral King Valley in Sierra Nevada courtesy of the U.S. Forest Service, Douglas dissented. What infuriated Douglas was that the state of California was going to build a twenty-mile asphalt road through the heart of Sequoia National Park to reach Disney's high-country resort.

Drawing on Aldo Leopold's ennobling notion of a land ethic, Douglas firmly believed that a sequoia tree, a barrier island, or a sand beach should be allowed to be a litigant. He wrote that "inanimate objects" about to be "despoiled, defaced, or invaded by roads and bulldozers and where injury is the subject of public outrage" could fight for their constitutional rights. Excoriating the U.S. Forest Service for being a patsy of the timber industry, Douglas maintained that before these "priceless bits of Americana (such as a valley, an alpine meadow, a river, or a lake) are forever lost or are so transformed as to be reduced to the eventual rubble of our urban environment, the voice of the existing beneficiaries of these environmental wonders should be heard." What mattered to Douglas was that flora and fauna had rights: "Perhaps they will not win. Perhaps the bulldozers of 'progress' will plow under all the aesthetic wonders of this beautiful land. That is not the present question. The sole question is, who has standing to be heard?"[8]

War against anything associated with Mickey Mouse had become a sport for Douglas. With typical brio, he called the "Disneyfication" of America a deleterious trend aimed at turning children into slaves of television. There was more magic in one's backyard woods or fields, Douglas believed, than in all the rides at Frontierland, part of the Disney theme park in Anaheim, California. The thought that Disney might build a $35 million resort in the Sierra Nevada, the heart of John Muir country, next to Sequoia National

Park, repulsed Douglas; he considered the very notion grotesque. And the fact that the resort was to be called Mineral King—in the land where redwoods ruled—added insult to injury. The Wilderness Society naturally concurred, deeming Douglas's opinion as "important judicial history." [9]

When the attorneys for the Sierra Club Legal Defense Fund adopted Stone's concept of environmental law—that if sequoias were going to be cut down, then they could indeed be plaintiffs—Douglas did the same. Both as a Supreme Court justice and as a public intellectual, Douglas fought to protect the Mineral King area from Disney bulldozers. His colleagues on the conservative Burger Court, however, saw this situation far differently. The other eight justices decided that the Sierra Club didn't have a genuine stake in the Mineral King resort and thus had no standing to sue. [10]

Douglas's stirring opinion in *Sierra Club v. Morton*, in fact, became a distillation of his lifelong convictions about preserving nature. By the twenty-first century it had been adopted as a manifesto by nonprofit groups including the National Audubon Society, Greenpeace, and the World Wildlife Fund. Robert F. Kennedy Jr., founder of Riverkeeper, recalled hiking, as a young boy, with Douglas along the C&O Canal in Washington, D.C., in the 1950s. "Bill was legalistically way out in front in his dissent," Kennedy said. "*Sierra Club v. Morton* has only grown in relevance. When the BP spill occurred, I immediately thought of that case." [11] Douglas's carefully crafted dissent is taught in classes in environmental law from Harvard to Berkeley.

> *The corporation sole—a creature of ecclesiastical law—is an acceptable adversary and large fortunes ride on its cases. . . . So it should be as respects valleys, alpine meadows, rivers, lakes, estuaries, beaches, ridges, groves of trees, swampland, or even air that feels the destructive pressures of modern technology and modern life. The river, for example, is the living symbol of all the life it sustains or nourishes—fish, aquatic insects, water ouzels, otter, fisher, deer, elk, bear, and all other animals, including man, who are dependent on it or who enjoy it for its sight, its sound, or its life. The river as plaintiff speaks for the ecological unit of life that is part of it. People who have a meaningful relation to that body of water—whether it be a fisherman, a canoeist, a zoologist, or a logger—must be able to speak for the values which the river represents and which are threatened with*

destruction. I do not know Mineral King. I have never seen it nor travelled it, though I have seen articles describing its proposed "development." The Sierra Club in its complaint alleges that "one of the principal purposes of the Sierra Club is to protect and conserve the national resources of the Sierra Nevada Mountains." The District Court held that this uncontested allegation made the Sierra Club "sufficiently aggrieved" to have "standing" to sue on behalf of Mineral King. Mineral King is doubtless like other wonders of the Sierra Nevada such as Tuolomne Meadows and the John Muir Trail. Those who hike it, fish it, hunt it, camp in it, frequent it, or visit it merely to sit in solitude and wonderment are legitimate spokesmen for it, whether they may be few or many. Those who have that intimate relation with the inanimate object about to be injured, polluted, or otherwise despoiled are its legitimate spokesmen.[12]

From the 1930s to the 1970s, any reckless clear-cutting in the American West got Douglas's dander up. He had seen the deep scars that this unsavory practice left on slopes: a mountaintop would be shaved bald and left with only debris; torrential runoffs of water then occurred, transforming a biosphere into a dead zone. Should the landscape surrounding Sequoia National Park be so cruelly scarred for the sake of a Disney park? The menace of hyperdevelopment was everywhere in the West. At a meeting of the U.S. Forest Service that Douglas once attended by happenstance in Wyoming, rangers were preparing to aerially spray chemicals to kill weeds growing on sagebrush land. "They roared with laughter when it was reported that a little old lady opposed the plan because the wild flowers would be destroyed," Douglas recalled with incredulity. "Yet was not her right to search out a painted cup of a tiger lily as inalienable as the right of stockmen to search out grass or of a lumberman to claim a tree? The aesthetic values of the wilderness are as much our inheritance as the veins of copper and coal in our hills and the forests in our mountains."[13]

II

Douglas was born in Maine, Minnesota, on October 16, 1898. His first name was William, but his mother insisted on calling him Orville, his middle name. When he was three years old his parents—

Julia Fisk Douglas and the Reverend William Douglas (a Presbyterian minister) moved the family to Estrella, California. They had heard that the California sunshine was good for the nerves and the elder Douglas had vicious stomach ulcers. However, Douglas's father died in 1904 from a botched ulcer operation. Julia moved her three children to Yakima, in the agricultural belt of south central Washington, to be near her sister. The Douglases moved into a tiny house a stone's throw from the Columbia Grade School. Unfortunately, Julia invested her small inheritance in a scheme to irrigate the Yakima valley; it failed; and crushing poverty fell upon the family. William, only seven years old, had to scrounge in the industrial yards of Yakima, collecting scrap iron in burlap apple bags to sell at a market. No menial task was beneath him. Seasonally, he picked fruit and threshed wheat. His biographers have claimed that his hard youth poisoned his trust in companies, rich people, and class privilege. But Douglas himself rejected this theory in his 1974 autobiography *Go East, Young Man*, saying that he never felt "underprivileged." In any case, though, at an early age he was an advocate for the underdog. (Douglas *did* admit that he sometimes felt wounded because God had placed him on the "wrong side of the railroad tracks.") [14]

Douglas's life was changed when he contracted polio as a child.* A doctor in Yakima predicted that he might be permanently paralyzed. All Douglas's mother could do was soak his legs in saltwater and get lower-body massages. When he returned to school, other children mocked him mercilessly; he was a puny misfit. So he started venturing outside Yakima, hiking the sagebrush trails and lava rock and backcountry, hoping to develop physical vigor. Ten miles soon increased to twenty. Every day Douglas could walk beyond the outskirts of town, high up into the Cascades, away from schoolyard taunts, learning the calls of birds, chatting with subsistence farmers and woodchoppers, singing old hymns like "Shall We Gather at the River?" He hiked through broad valleys and past anxious watchdogs. His shock of hair was fine and unruly. The more Douglas

---

* Some scholars believe that it is impossible to overcome polio. But the historian David Oshinsky, author of a Pulitzer Prize—winning work on polio, knows that this is indeed possible.

walked, the stronger his legs got. "The physical world loomed large in my mind," Douglas recalled. "I read what happened to cripples in the wilds. They were the weak strain that nature did not protect."[15]

Happiness engulfed Douglas whenever he was outdoors. Believing that fresh air was a curative, he started writing secret odes to the high lakes of the Wallowa Mountains, giving each a distinctive personality as if it were a new friend. When Douglas discovered Izaak Walton's *The Compleat Angler*, he became devoted to fly-fishing for trout. "And of all fly-fishing, the dry fly is supreme," Douglas said. "The dry fly floats lightly on the water, going with the current under overhanging willows or riding like a dainty sailor on the ruffled surface of a lake. It bounces saucily, armed for battle but looking as innocent as any winged insect that rises from underneath the surface or drops casually from a willow or sumac into a stream or pond." The sight of a trout rising never failed to make Douglas's heart stand still.[16]

Remembering his childhood fishing and the glory of sunshine, Douglas decided that his life, no matter what his employment was, would be centered on protecting America's fishing streams and forests. Conservation became his electric wire, which would produce the brightest sparks throughout his storied intellectual career. "Pinchot and Teddy Roosevelt were in my eyes romantic woodsmen," Douglas wrote in *Of Men and Mountains*, his 1950 autobiography, the first of several. "I did not then know about Pinchot's 'multiple use' philosophy, which, as construed, allowed timber companies, grazing interests, and even miners to destroy much of our forest heritage under the rationalization of 'balanced use.' I only knew that Pinchot was a driving force behind setting aside wilderness sanctuaries in an effort to save them from immediate destruction by reckless loggers. I was so thrilled by Pinchot's example that I perhaps would have made forestry my career had the choice been made in my high school days."[17]

Devoted to scholarship, Douglas received top grades at Whitman College in Walla Walla, Washington, a first-rate liberal arts institution where he was on a full scholarship. Now he started coming into his own, intellectually. While at college, he joined the Student Army Training Corps. But the clannishness of such outfits didn't really appeal to him. He adopted the stance of an iconoclast, a lone mountaineer, a skirt-chaser, an impatient doer eager to see the great wide world. As a hobo, he traveled from hopyard to forest camp to orchard to earn money during the Great Depression.[18] He was a young man willing to take risks—a fact historians should not ignore.

Upon graduating from Whitman College in 1920 with a BA in English and economics, Douglas became a high school teacher and debate coach. Thoreau, Emerson, and Muir became his inspirations. Impressed by their transcendentalist philosophy, he wanted to chase the sky and learn about every part of the wild Wallowa Mountains in northeastern Oregon. Pinchot stayed on his shoulder like a good angel, informing his views about the stewardship of forestlands. Douglas watched many of his friends in Yakima sinking into tedium, logging and mining for the minimum wage. Having licked polio, and having developed an iron will and newly strong legs, Douglas wanted much more out of life. Teaching English and Latin for two years at Yakima's high schools bored him. "Finally," he recalled, "I decided it was impossible to save enough money by teaching and I said to hell with it." [19]

Distrustful of the timbering promoted by Weyerhaeuser Lumber and worried about becoming an obsolete teacher in the Rattlesnake Hills range, Douglas found liberation from Washington's provincialism at Columbia University Law School. He was imbued with a Pacific Northwest belief in the power of mountains, stone, and rivers, and his train journey to New York City sounds like a drifter's ballad. In the summer of 1922, Douglas signed up to escort 2,000 sheep by rail from Wenatchee, Washington, to Minneapolis, Minnesota. In *Of Men and Mountains*, he told of sleeping in a dirty caboose, meeting rascals in boxcars, rattling along with Montana's fields and peaks flashing by outside the open train door. In Idaho he encountered a railroad strike. He feared the billy clubs of yard bulls, who were always trying to shake down the transients. Douglas had a vivid way of telling anecdotes about life along the train tracks and in the hobo jungles. "I needed a bath," he matter-of-factly wrote of the adventure, "and a shave and food; above all else I needed sleep. Even flophouses cost money. And the oatmeal, hot cakes, ham and eggs and coffee—which I wanted desperately—would cost fifty or seventy-five cents." [20]

Douglas unloaded the sheep from the railcars in the Minneapolis railroad yard, then bummed a train ride to Chicago, wanting to see Lake Michigan. He was appalled by the industrialization that had polluted the

Illinois air. Chicago wasn't the "City of the Big Shoulders" that the poet Carl Sandburg had described, but an urban cesspool: dilapidated buildings, noise, broken glass, and "dingy factories with chimneys pouring out a thick haze over the landscape." Loneliness engulfed Douglas in sooty Chicago, where the decibel level was too high, transforming him overnight into an environmentalist. Hungry, exhausted, homesick, bruised, frightened, and confused, he now placed a higher value on Yakima and Walla Walla than ever before. "Never had I missed a snowcapped peak as much," he recalled. "Never had I longed more to see a mountain meadow filled with heather and lupine and paintbrush." [21]

Eventually Douglas made his way to New York and enrolled at Columbia, working at odd jobs to pay the big-city bills. After his first year at Columbia, he was appointed to the staff of the law review. Nobody else attacked the law books with the same fervent hunger as Douglas. Harlan Fiske Stone, a dean of Columbia University who would later serve with Douglas on the Supreme Court, recognized that Douglas was a nonstop worker. Imbued with a libertarian spirit and deeply committed to the Bill of Rights, Douglas staked his reputation at Columbia on defending misfits, outcasts, drifters, migrants, the unemployed, the homeless, and tramps. Lonesome, forsaken people had a special place in Douglas's heart. What's more, his experiences in Chicago and New York led him to conclude that country folk needed legal protection from city slickers. Douglas was an anomaly at Columbia because he was already claiming that clean air and clean water were a constitutional right. What right did Chicago have to despoil Lake Michigan? What right did General Motors have to pollute the Detroit River?

"It seemed that man had built a place of desolation and had corrupted the earth in doing so," Douglas wrote of his arrival in New York City. "In corrupting the earth he had corrupted himself also, and built out of soot and dirt a malodorous place of foul air and grimy landscape in which to live and work and die. Here there were no green meadows wet with morning dew to examine for tracks of deer, no forest that a boy could explore to discover for himself the various species of wild flowers, shrubs, and trees; no shoulder of granite pushing against fleecy clouds and standing as a reminder to man of his puny character, of his inadequacies; no trace of the odor of pine or fir in the air." [22]

Douglas, determined to succeed and always in need of cash, spent three years as a tutor at Columbia, helping high school students prepare for the Ivy League. According to the historian James O'Fallon, editor of *Nature's Justice*, Douglas had "two criteria" for his ambitious pupils—that the "student be rich and stupid."[23] Regularly, when he was broke, he would borrow $10 or $20 dollars from friends; he never welched. Eventually Douglas, with only one year in law school remaining, had saved enough money—$1,000— to return to the Pacific Northwest. He needed the mountains and his mother needed him. Over the summer of 1923, he married Mildred Riddle in her hometown, La Grande, Oregon. For their honeymoon, the Douglases roughed it outdoors in the Wallowa range, catching trout, eating wild berries, horseback riding, and making love under the stars. And they went broke. "We blew," Douglas boasted, "my thousand bucks."[24]

One of Douglas's abiding traits was his recklessness with money. Even as Supreme Court justice he often had a gritty hand-to-mouth lifestyle. His cupboards were often bare. Never did Douglas trust the New York Stock Exchange: investment banking was, to his mind, legalized gambling. Washingtonians never knew whether Douglas could afford to buy a restaurant dinner in Chevy Chase or take a weekend trip to Virginia. Impending bankruptcy was a condition he actually embraced, if it meant freedom to think, hike, and have fun. Poverty never made him sullen. No business venture appealed to him except writing books. "Douglas preferred to invest in only one stock," his biographer James F. Simon wrote in *Independent Journey*: "William O. Douglas."[25]

Back at Columbia, Douglas was on fire. All his professors—Underhill Moore chief among them—were astounded at his intelligence. You could see it in his eyes. He could revise commercial law casebooks or explain the Pleistocene epoch with equal ease. Douglas worked hardest when taking on a big company, defending the people against a fat cat. Mischievously Moore unleashed Douglas against a Portland cement company that had supposedly cooked its books. But Douglas's belligerent attitude worried Dean Stone, who had just been confirmed to serve on the Supreme Court. The new justice selected Albert McCormack, a fine choice, to be his clerk, rather than the brilliant but wild Douglas. "The world was black," Douglas said of this snub. "I was unspeakably depressed that for all those years and all that work, I had so little to show for it. The one opportunity I wanted had passed me by." [26]

Douglas had a choice after graduating from Columbia: go back to Yakima to practice law or join a Wall Street firm. He did the latter. But Douglas was arrogant—and his voice was strained and defiant—when he was interviewing at New York firms. Famously, he was interviewed by John Foster Dulles, who would go on to become Dwight D. Eisenhower's secretary of state. Dulles, who tended to be pompous, was condescending. So Douglas turned the tables on Dulles: the interviewee started interviewing the eminent establishment lawyer. According to Douglas, to irritate Dulles even more, on his way out of the interview he tipped Dulles a quarter for helping him on with his coat. The job went to somebody else. But Douglas was hired by the prestigious firm Cravath DeGersdoff, Swaine, and Wood (later Cravath, Swaine, and Moore).

After only four months at Cravath, confused, like an athlete with a mild concussion, Douglas left New York and moved back to Yakima. "The only bird I ever saw was a pigeon," he complained of New York. "I longed for the call of the meadowlark, the noisy drilling of the pileated woodpecker, the drumming of the ruffed grouse." [27] He soon regretted the decision, however. Working his New York connections, he found a job teaching at Columbia. Douglas's legal career now soared. Yale University Law School wisely poached him. He became an expert on commercial litigation and bankruptcy. By the time Douglas was forty-one, he was an associate justice on the Supreme Court. From 1929 to 1934 he wrote five legal casebooks and almost twenty articles. What gave Douglas such authority was his wizard-like expertise on corporate reorganization and bankruptcy law. If a U.S. corporation got too big, Douglas always prepared to break it down to size. Working on Wall Street had made Douglas feel that some investment bankers were truly pathetic, preferring money to "love, compassion, hiking, or sunsets." [28]

With the election of Franklin D. Roosevelt in 1932, Douglas had an opening to positively affect the consciousness of his time. Main Street's anger at Wall Street had deepened since the stock market crash of 1929. At Yale University, where Douglas was the distinguished Sterling Professor of Law, he had already earned a reputation for his no-nonsense approach and for insisting that the federal government regulate big business to achieve transparency. When Congress passed the Federal Trade Commission Act of 1933, granting the Federal Trade Commission regulatory power over security sales, Douglas was tapped by the Roosevelt administration to head the Securities and Exchange Commission. He had few ties with the WASP establishment, but he formed an alliance with the Catholic tycoon Joseph P. Kennedy.[29] The entire Kennedy family liked the cut of Douglas's jib. At long last he had a sponsor. Other New Dealers also took a shine to Douglas; they included Abe Fortas, Tommy "the Cork" Corcoran, and Lyndon Johnson.

Insiders in Washington, D.C., were soon astounded by Douglas's love of the wilderness. Like a sudden storm, Douglas could take over a Georgetown cocktail party with his tales of the Pacific Northwest. In fact, the only Washingtonian whom Douglas truly revered was the aging Gifford Pinchot. The new secretary of the interior, Harold Ickes, and Pinchot were warring over policy for the national forests. Ickes was, in Douglas's words, a "bulldog battler" who was "hungry for bureaucratic power."[30] By 1939, Ickes had brought into the Department of the Interior the Bituminous Coal Commission, Bureau of Indian Affairs, Bureau of Fisheries, Bureau of Biological Survey, and Mount Rushmore Commission. Ickes now wanted to take control of the U.S. Forest Service from the Department of Agriculture. Pinchot objected, so Ickes went after him.

Years later, in his autobiography *Go East, Young Man*, Douglas attacked Ickes for reopening the feud between Ballinger and Pinchot of 1909–1910. It pained Douglas to think that Ickes had acted like a man motivated by envy and pettiness. "Ickes wrote that Ballinger had not been involved in a corrupt practice," Douglas fumed. "That was never the issue. The issue was whether private interest through subterfuge could defeat the public land policy. Bulldog Ickes would have been the first to attack any Ballinger of his day. In 1940 he was defending Ballinger only to attack Pinchot."[31]

III

T*he Alaskan wilderness movement thrived while* Franklin D. Roosevelt was in the White House from 1933 to 1945. When the president toured Washington's Olympics, in 1937, feasting on trout at the lodge and saying he never saw such grand trees in his life, he upgraded

the designation from national monument to national park. FDR understood more keenly than ever before Douglas's pleas for stricter wildlife
protections in Alaska and the Pacific Northwest. Conservation wasn't a
mere slogan during FDR's visionary presidency—it was a crucial part of
the New Deal. Under FDR's leadership the conservation movement was
appropriated from the Republican Party, and its tenets became central
to New Deal liberalism.[32] From the outset the Roosevelt administration's
natural resource team was impressive. How could anyone be better than
Harold Ickes as Secretary of the Interior or Jay Norwood "Ding" Darling
as director of the Bureau of Biological Survey? The 2.5 million workers at
the CCC planted more than 2 billion trees during its decade of existence.[33]
They also erected 3,470 fire towers and built 42,000 miles of fire roads.
Roosevelt also helped individual farmers reclaim eroded land. Working
with Roosevelt, Congress passed the Taylor Grazing Act of 1934 (shutting
down the public domain and putting grasslands under sound management); the Soil Conservation Act of 1935 (initially a nationwide program
of soil and moisture conservation); and the Act of July 22, 1937 (providing
administration of the National Grasslands).[34]

Another aspect of the New Deal was the WPA's sponsorship of painters
to capture wild America on canvas. Edwin Boyd Johnson, an Alaskan designer and muralist originally from Tennessee, was one of these painters.
He soon learned that painting wild Alaska was a daunting task. Mount
Kimball, the highest mountain in the eastern Alaska Range between Isa-
bel Pass and Mentasta, became for him what Mount McKinley had been to
the artist Sydney Laurence. Johnson's images of the bright orange-yellow
Mount Kimball closely resembled the work of Marsden Hartley. By having
the WPA pay Johnson to paint wild Alaska, the Roosevelt administration
ingeniously promoted the protection of places like Mount Kimball.[35] The
WPA also worked to establish a hotel at Mount McKinley National Park.
And grants were given to Skagway to help clean up the water system polluted by mining.[36]

Another important program by the Roosevelt administration in Alaska
was having Charles Flory, a forester, restore totem poles in the Inside
Passage. Flory had CCC workers begin an interpretive initiative on behalf
of Tlingit art near Juneau, Ketchikan, and Sitka. Negotiations were made
to have poles shipped to the restoration facility and then returned to the

appropriate Alaskan communities (sometimes as new features to attract tourism). Indian villages such as New Kasaan, Hydaburg, and Klawock participated. Roosevelt also allocated funds for a totem pole in Tongass National Forest.[37]

Roosevelt's concern for Alaskan wildlife—particularly marine species—was sincere. On April 18, 1939, the president had more than *doubled* the size of Glacier Bay National Monument, a tribute to John Muir. Professor William Skinner Cooper, one of the nation's most eminent ecologists, was teaching at the University of Minnesota when he heard this news. Marine areas teeming with Dungeness, king, and Tanner crabs were finally made off-limits to fishermen. Whole subtidal benthic communities, along with schools of Pacific halibut, rockfish, lingcod, Pacific cod, sablefish, and pollock now had protected Alaskan nurseries (although a limited amount of fishing was allowed until the 1970s).[38] Muir's glaciers may have been receding, but federal protection was intensifying.

For Cooper, the doubling of Glacier Bay National Monument meant that the complexes of plant life thriving around the terminal of receding glaciers could be properly analyzed by biologists. Because Glacier Bay had more than 220 bird species—half of all American birds—the National Audubon Society considered Executive Proclamation 2330 Roosevelt's grandest conservation effort yet. For the Sierra Club, it was the fulfillment of John Muir's vision. The Alaskan communities of Haines and Gustavus now prospered as gateways to glaciers and wildlife. (People in Haines started boasting that their town—the Chilkat Indian community Muir wrote about in *Travels to Alaska*—was founded by the great naturalist.) All of Glacier Bay's geographic provinces would remain protected, owing to Muir's early advocacy and Cooper's dogged lobbying.[39] (But there was no guarantee that the glaciers wouldn't melt.)

During the 1930s, while pushing for the Lake Clark region to become a national park or wilderness reserve, Frederick Vreeland, through the Camp Fire Club of America (CFCA), promoted the idea of allowing Native Alaskans exclusive reindeer breeding rights. Even since the missionary Sheldon Jackson imported a herd from Siberia to Amaknak Island, domesticated reindeer had been raised in Alaska to pull sleds and serve as a high-protein food source. By the 1930s they were a big business for Alaska (there were an estimated 640,000 reindeer in the territory). Vreeland hoped that if Alaskan Natives ate reindeer, as ranchers ate cattle in the Lower Forty-Eight, then the big game wouldn't be shot out. On September 1, 1937, Congress, with the approval of the CFCA, passed the Reindeer Act. Not only were Natives given exclusive reindeer breeding rights, but in the future they would earn concession rights. The interbreeding of caribou (wild) and reindeer (domestic) sometimes caused disease, but Vreeland had succeeded in protecting the Lake Clark caribou from overhunting.[40]

One New Deal conservation program that significantly affected Alaska was the Duck Stamp Act (its official title was the Migratory Bird Hunting Stamp Act of 1934). Ding Darling was a Republican, but his commitment to the biological conservation movement was not inhibited by his party affiliation. A Bull Moose at heart, Darling was brought into FDR's administration to serve on the President's Committee for Wild Life Restoration (along with Thomas Beck and Aldo Leopold). By 1935, Darling, a cartoonist who had won two Pulitzer Prizes, took over as head of the Biological Survey. Although he served for only eighteen months in this post, Darling was deemed the best friend that Alaskan ducks ever had.

With the Great Depression persisting, and with no signs of recovery on the horizon, Darling had to find creative ways to promote the protection of migratory birds. Putting aside his usual satirical wit, he designed an elegant blue-and-white duck stamp.[41] Anybody age sixteen or older who wanted to legally hunt a duck was required to purchase a stamp. The stamps raised a lot of money, and just in the nick of time. In 1934, migratory waterfowl had reached a low of about 27 million. Alaska was a huge part of this problem. Market hunters were devastating Alaska's largest migrant birds. Throughout the territory the prevalent attitude was "If it moves, shoot it." Two-thirds of all trumpeter swans—the largest waterfowl in the world—nested in Alaska. In all the Lower Forty-Eight, only the Yellowstone ecosystem was a stronghold for swans. For hundreds of years trumpeter swans had been slaughtered for their feathers, which made the best quill pens in the world. The British royal crown, for instance, signed every document with a trumpeter swan pen.

To the surprise of President Roosevelt, the duck stamps designed by Darling were a hit with Congress and the private sector. Capitalizing on his celebrity as a cartoonist, Darling raised millions to help protect migratory birds. The term "duck stamp," however, was misleading: Darling's program also printed stamps of geese and swans. Although Darling designed the first stamps, an annual art contest was soon instituted. Every year new winners were chosen.

When scholars write histories of the U.S. Fish and Wildlife Service, the duck stamp program is usually considered ingenious, and a high-water mark. The stamps became coveted collectors' items. During Darling's tenure revenues from the duck stamp were $635,001 in 1934 and $448,204 in 1935. In 1953, long after Darling had retired from government, he reflected on why the duck stamp program had worked. "Of course you understand that I am not nearly so much interested in the preservation of migratory waterfowl as I am in the management of water resources and the crucial effects of such management upon human sustenance," he told *Reader's Digest*. "Wild ducks and geese and teeter-assed shore birds are only the delicate indicators for the prognosis for human existence, just as sure as God made little green apples." [42]

Despite the fact that the nation was at war between 1941 and 1945, Roosevelt did his best to protect the flyways and nesting areas of Darling's beloved American birdlife. Glacier Bay was just one of a number of examples. Around the time of the attack on Pearl Harbor, when America was focused on military preparedness, he received a blueprint for a major new U.S. Army artillery range to be constructed in Idaho. A lifelong bird-watcher, Roosevelt rejected it. He sided with the bird-watchers over the army. "Please tell Major General Adams or whoever is in charge of this business that Henry Lake, Idaho, must immediately be struck from the Army planning list for any purposes," he wrote to Secretary of War Henry L. Stimson. "The verdict is for the trumpeter swan and against the Army. The Army must find a different nesting place." [43]

Groups like the National Audubon Society, Sierra Club, CFCA, and Izaak Walton League had an ally in Franklin Roosevelt. No longer was saving wild places considered fringe philanthropy. Also, John D. Rockefeller Jr. became the greatest conservationist capitalist of all time (only Ted Turner, the founder of CNN, comes close). Regularly, Rockefeller donated multimillion-dollar checks to help create Acadia National Park in Maine and the Grand Teton National Park in Wyoming. He considered this a Christian, gentlemanly thing to do. His family had taken something (oil) from mother Earth, and therefore he wanted to give something back to her.[44] Working closely with Horace Albright of the National Park Service, Rockefeller would pay for cleaning up environmental eyesores and industrial blight. He wanted America's special wilderness places to be roadless. "I never had any doubt about the existence of a divine being," Rockefeller said. "To see a tree coming out in the spring was enough to impress me that the fact of God existed."[45]

With impressive political acumen, FDR brought together Bob Marshall (a democratic socialist), Harold Ickes (a Bull Moose), Ira Gabrielson (a bird enthusiast), and John D. Rockefeller Jr. (a capitalist) to protect America despite the ordeals of the Great Depression and World War II. When Louis Brandeis retired from the Supreme Court in March 1939, Roosevelt appointed Douglas—the fierce environmentalist and opponent of Wall Street—to fill the seat. With Joseph Kennedy cheering him on, Douglas became the youngest justice in American history. When Franklin D. Roosevelt died in Warm Springs, Georgia, on April 12, 1945, Douglas was profoundly grieved. He believed that Roosevelt had struck the right notes of progressivism with the New Deal programs.

Unfortunately, Roosevelt's adroit conservationism was not continued by his successor, Harry S. Truman. Truman was indifferent to forestry and to protecting predators. Regarding Alaska, Truman time and again sided with miners—not with conservationists; he liked working people, not endangered species. Within a year after becoming president, Truman criticized Secretary of the Interior Harold L. Ickes for being too radical a conservationist. Ickes had claimed that a California oilman, Edwin Pauley, who was then treasurer of the Democratic National Committee, had tried to bribe him with $300,000 to allow offshore drilling near Santa Barbara in 1944. The payoff was to be a campaign contribution for Truman. Ickes wrote defiantly in his diary, "I don't intend to smear my record with oil at this stage of the game even to help win the reelection of the President."[46]

Ickes's resignation on February 13, 1946—in protest against Truman's appointment of Pauley as undersecretary of the navy—was a severe setback to the wilderness movement. The announcement took place in the auditorium and was at the Department of the Interior at the time the largest press conference in U.S. history. Ickes was loved and trusted by reporters; Truman was not. "I don't care to stay in an administration," Ickes wrote in his diary, "where I am expected to commit perjury for the sake of the party."

President Truman had first offered the post of secretary of the interior to William O. Douglas. From a conservationist's perspective, Douglas would have been an outstanding choice. Undoubtedly, he would have promoted wilderness in Alaska; he was firmly opposed to the U.S. government's poisoning of wolves; and he was averse to allowing domesticated animals to graze on public lands around Mount McKinley. For Douglas, in fact, Alaska was America's "last opportunity" to "preserve vast wilderness areas intact." [47]

By the time Truman had become president in April 1945, Douglas was a significant political presence. He had the tight-lipped look of a naval officer; some people said he resembled the pugnacious James Forrestal, or James Cagney. He was physically fit and had appealing wrinkles around his eyes. Douglas was so progressive-minded, his critics said, that he would have liked to be martyred in the Haymarket Riot. "I worked among the very, very poor, the migrant laborers, the Chicanos and the I.W.W.'s who I saw being shot at by the police," Douglas said. "I saw cruelty and hardness, and my impulse was to be a force in other developments in the law." [48]

Not known for either understatement or reserve in his personal life, Douglas was a force to be reckoned with in Washington, D.C., during the 1950s. Waking up at the crack of dawn, Douglas, a prodigious worker, would leave his home—at 4852 Hutchins Place, in the Palisades neighborhood—for a walk along the C&O Canal. After feeding his border collie, Sandy, he would be off to Capitol Hill. Lawyers arguing cases at the Supreme Court dreaded his piercing blue eyes, which were as keen as those of a condor. Unlike most Supreme Court justices, Douglas kept his opinions short and readable by a layperson. He was proud of his northwestern upbringing, and socialites in Georgetown knew that he might very well wear hiking boots to a black-tie dinner. Given both his personal austerity and his judicial stature, it was quite a coup when The Wilderness Society recruited him to join the movement for roadless, primitive lands. Douglas, in fact, became a filter through which U.S. senators and congressmen first learned about the new idea of "leave it alone" conservation.

During the 1950s in Washington, D.C., a popular comment along the C&O Canal was "There goes Justice Douglas." An article in the *Living Wilderness* called him "the most famous living American walker." [49] Wearing blue jeans and a work shirt, Douglas would walk along the canal daily, rain or shine, averaging ten to twenty miles a day, in protest against a motor park-way, which had been promoted by the *Washington Post* and the *Times Herald*. People would sometimes actually blink their eyes in disbelief: that was the nation's most famous jurist over there, with a walking stick. The threat to the towpath had become for Douglas the symbol of what was wrong with American life, and the canal was being used for sewage. He challenged the editors of the *Post* and the *Times Herald* to come and see wild nature there with him, to simply say no to motorized traffic. When the Potomac River was filled by spring rains, and young trees were blooming along its banks and birdlife was all around, Douglas believed that hikers could be transported back to the 1850s when horses and mules towed barges. "The river must be cleaned up and made pure again," Douglas wrote in the *Living Wilderness*. "Then campsites, fireplaces, pure drinking water, and sanitary facilities can be provided under the auspices of the National Park Service. That will be the best use of the Canal and the Potomac—far better than needless water storage of high-priced electric power." [50]

Douglas understood from his earlier long hikes in the Cascades that the richest Americans were those who had learned to let the nation's most treasured landscapes alone. Douglas believed that hard work was good for the soul but that no person should become a machine. Nonconformity, now and then, was a sign of a healthy mind. Loafing in nature made the senses keen. Why lead a life of quiet desperation when you could reel in salmon from Puget Sound or see an owl in The Dalles? Good behavior, to Douglas, was overrated. Exhilaration and voluntary poverty were far preferable to the gilded cage of a life of dull comfort. While he perhaps went a bit far with some of his judicial opinions regarding conservation, Douglas wasn't very different from a lot of Pacific northwesterners or the Depression-era boys who had a penchant for the outdoors. Perhaps because his father had been a minister, Douglas was quick to see all of life's blessings. As he aged his skin became weathered. There was nothing mystical about Douglas's outdoors world; unlike the Comanche he did not pray to buffalo, and unlike the Buddhists he did not meditate on moun-tains. He was simply the most brilliant person Yakima ever produced, and he lived to walk thousands of miles. Like Thoreau in *Walden*, he believed the "swiftest traveler" was one who "goes afoot." [51]

In 1946 that other great hiker and forest lover, Gifford Pinchot, died at age eighty-one at Grey Towers, his home in Pennsylvania. If Douglas had his way, Pinchot's face (along with John Muir's) would have been carved on Mount Rushmore, but others in Washington, D.C., had long considered Pinchot an irrelevant relic. At the funeral, Douglas reassured Cornelia Pinchot, the widow, that he would continue fighting for America's forestlands. She uttered the truest line ever about her husband: "Conservation to Gifford Pinchot was never a vague, fuzzy aspiration; it was concrete, exact, dynamic." [52]

Douglas, who felt he could be most useful to the burgeoning environmental movement in the Supreme Court, declined Truman's offer to make him secretary of the interior. He wrote his outdoors memoir *Of Men and Mountains* in 1950—a must-read for those in the up-and-coming field of environmental law. Working his back channels, he pushed for the National Park Service to take over vast areas of the Washington coast. All wildlife legislation of the era would cross his desk. Meanwhile Julius A. Krug—a Democrat from Madison, Wisconsin—became secretary of the interior.[53] Krug quietly went about slowing the rapid pace at which the department had operated under Harold Ickes. Krug's philosophy was based on the fact that *people* voted in elections—not wolves, cougars, or foxes. During the Truman administration, in fact, not a single national park was authorized.[54] Nor was there any expansion of the area of existing national monuments in Alaska. Truman didn't give a damn about nature. Douglas was the torchbearer for the Rooseveltian cause throughout the big debates of the 1950s over the Alaskan wilderness. Always ready for a fight, Douglas hoped that Americans would create an environmental protection agency to bust pollutors.[55]

Like many conservationists in the Pacific Northwest, Douglas viewed Alaska as an extension of Washington state. The Tongass and Chugach were sacred national places, steeped in Teddy Roosevelt's and Gifford Pinchot's lore, which weren't going to be destroyed for the benefit of the extraction industries. Douglas was never going to let them be ruined—any man who could overcome polio could surely square off against pollut-ers. The fact that Douglas had refused the post of secretary of the interior didn't mean that he had relinquished his Muirian duty to protect America's natural heritage. Never would he let Alaska become Chicago.[56]

Chapter Fifteen

∘ ∘ ∘ ∘ ∘ ∘

ANSEL ADAMS, WONDER LAKE,
AND THE LADY BUSH PILOTS

I

Visitors to Alaska arrived by plane in record numbers in the early
years of the cold war, some of them understandably apprehensive
about flying over the seemingly endless procession of Alaskan
mountain ridges. Lower Forty-Eighters felt minuscule at an altitude of
10,000 feet, peering through their little windows at clouds larger than
lakes. Madcap turbulence often caused the planes to rattle and rumble
like storm-tossed ships on a vertiginous sea. Then there was the memory
of Will Rogers, who had been killed in a plane crash in Alaska. Although
the photographer Ansel Adams didn't care for aviation—having lost a few
close friends to crashes—he wasn't afflicted by acrophobia. Adams knew
that flying was the only way to hopscotch around Alaska's immense area
and to be enlightened and awed by its extremes. Because Alaska's paved
road system in the late 1940s was confined to populated places, air travel
was the only feasible mode of transportation. Adams wrote that while flying was an "unnatural environment for man," it was, in truth, the only
"practical way" to "visit many of the areas I wanted to photograph." [1]

In 1942, Adams had traveled in the Pacific Northwest, photograph-
ing the rocky alpine slopes and glacier-capped summits of the Olympic
Mountains towering upward from greater Seattle against the Pacific sky.
This majestic panorama, fresh with the smell of rain, inspired some of

Adams's best photography. He shot ocean waves smashing into cliffsides and Piper's bellflowers growing in the crevices of rock outcroppings. However, while Adams recognized the Hoh and Quinault rain forests of the Olympics as botanical wonders, he craved glaciers and taller peaks. His intuition told him that Glacier Bay and Mount McKinley were the places to be. He also craved the light of the far northern skies. The Olympics were too low—foothills, compared with the Alaska Range. None of the major peaks in the Olympics were higher than 8,000 feet. "Imaginatively inclined," Adams recalled in *An Autobiography*, "I felt Alaska might be close to the wilderness perfection I continuously sought."[2]

Sometimes dreams come true. Alaska exceeded all of Adams's expectations. His excursions in 1947 and 1949 left him with cherished memories and enduring photographs (even though the weather had fluctuated between bad and awful). Building on the artistic photos Edward Curtis had taken of Alaskan landscapes during the Harriman Expedition of 1899, Adams used airplanes, helicopters, snowmobiles, jeeps, boats, canoes, and hiking boots as a means to a keeper shot; he was able to capture places like Mount McKinley and Glacier Bay in dramatic light.

Born in 1902 to upper-class parents in San Francisco, Adams became committed to photographing wild America after hiking in Yosemite National Park as a fourteen-year-old. Adams was flabbergasted to learn that tectonic plates had once pushed up piles of rocks that were now called the Sierra Nevada. "The splendor of Yosemite burst upon us and it *was* glorious," Adams recalled of his trip of June 1916. "One wonder after another descended upon us. . . . There was light everywhere. . . . A new era began for me." Adams's father soon thereafter bought his son a Brownie camera. Young Ansel was off and running, constantly searching for the right natural scene. "I believe photography has both a challenge and an obligation," he wrote of his own philosophy, "to help us see more clearly and more deeply, and to reveal to others the grandeur and potentials of the one and only world which we inhabit."[3]

Much like John Muir, his hero, Adams started wandering in the Sierra Nevada looking for picture-perfect vistas. Anxious to help save the Yosemite wilderness, he joined the Sierra Club. Occasionally he wrote articles for the *Sierra Club Bulletin*. His art introduced Yosemite to the general public, increasing consciousness about the old-growth redwoods of

Mariposa Grove and the priceless vistas from Glacier Point. Yosemite, it seems, had aroused all his subtle creative strains. In 1934 Adams, determined to protect Yosemite for perpetuity, joined the Sierra Club's board of directors; he remained active there until 1971. Following the lead of Alfred Stieglitz, who believed photography should be as high an art as painting, Adams adopted a variety of new lenses, determined to reveal Yosemite profoundly. Mountain landscapes, captured by the wide-angle lens, enraptured him.[4] *Monolith, the Face of Half Dome*, taken in 1927, was his first *visualization*—that is, he visualized the photo before it was shot, determining its essence in a quasi-scientific yet romantic way.[5] "My photographs have now reached a stage when they are worthy of the world's critical examination," Adams declared in 1927. "I have suddenly come upon a new style which I believe will place my work equal to anything of its kind."[6]

Starting in the early 1930s, Adams rejected the notion that his photographs were "pictorial"—a dreaded word used in Henry Luce's magazines *Time* and *Life*. Instead, Adams, with other West Coast photographers including Edward Weston, Imogen Cunningham, and Willard Van Dyke, formed Group f/64, championing so-called "straight" realist photographs. The group's name was derived from the smallest lens aperture on large-format cameras, which gives the greatest depth of field with maximum definition from foreground to background. They preferred pioneer western photographers like William Henry Jackson to New York's avantgarde.[7]

The way Adams photographed the West—his spiritual command of the landscape—allowed Americans to better appreciate their wilderness heritage. Adams's photograph of McDonald Lake in Glacier National Park, for example, helped increase the number of family visits to northwestern Montana. Starting with his first book, *Sierra Nevada: The John Muir Trail*, Adams regularly published his black-and-white landscapes of Yosemite in various popular formats including wall calendars. With a black beard and a broad-rimmed floppy hat, and dressed like the young Muir, Adams worked at his trade wherever high-country light met rock. Secretary of the Interior Harold Ickes read Adams's *Sierra Nevada* and marveled at the exquisite photography, amazed that the young Californian had so elegantly captured the mountainous Kings River Canyon region, where giant sequoias were found along with ponderosa pine, incense, cedar, and white fir. Awestruck by the book's nobility, feeling as if he were on a raft going down the Kings, Kaweah, and Kern rivers, Ickes brought *Sierra Nevada* to the White House to show to his boss. President Franklin D. Roosevelt wouldn't give it back. The New Dealers now considered Adams a favorite artist.[8]

Ickes wanted to make his mark at the Department of the Interior by creating a new kind of national park in the era of dust bowls, soil erosion, and wildlife depletion. Building on Bob Marshall's ideas about wilderness and relishing Adams's photos, he envisioned a vast John Muir–Kings Canyon Wilderness Park. When he went to Capitol Hill to take up the matter, he soon discovered that nothing had changed much since Hetch Hetchy Valley was dammed in the 1920s. Developers in California still wanted concrete water reservoirs, open grazing, timber clear-cuts, and ski resorts. Ickes showed Adams's book *Sierra Nevada* to congressmen and insisted that a roadless park was the "new way," but he faced strenuous opposition from the Republican Party. The lengthy process of compromise that followed included a great to-do over the park's name. Ickes was eventually forced to drop the name John Muir (California's businessmen still considered Muir a rabble-rouser), and Republicans didn't want the word *wilderness* on any piece of legislation.

Instead of emphasizing the fact that the Kings Canyon region was home to Sierra black bears, Ickes stressed the 200 Native American archaeological sites. On March 4, 1940, President Roosevelt signed legislation creating Kings Canyon National Park.[9] "Because it was a roadless park, and because of his disability, Roosevelt would never be able to see Kings Canyon in person," the historians Dayton Duncan and Ken Burns write in *The National Parks*. "Instead, he contented himself with following John Muir's trail through the photographs of Ansel Adams."[10]

Ickes now hired Adams to work for the Department of the Interior. Ickes paid him $22.22 a day to go all over the United States, visiting dozens of national parks and monuments, shooting images to bring back to Washington, D.C., for public display. Ickes hoped Adams's photographs would be analogous to the WPA guidebooks. The bombing of Pearl Harbor by the Japanese and the United States' entry into World War II lent urgency to Adams's project: his photos showed America's treasured landscapes—

landscapes surely worth fighting for. The forty-four-year-old Adams called his nature photography "emotional presentations" for the troops. Adams also taught soldiers at Fort Ord, California, photography and escorted troops around Yosemite Valley.

By 1945 Ansel Adams was almost a household name in America, the nation's most respected photographer. Nobody could match his achievements. Adams helped found the journal *Aperture*, showcasing up-and-coming photographers and promoting the newest camera techniques and equipment to the general public. Yosemite would remain Adams's special place, but a visit to Glacier National Park in 1941 (paid for by Ickes) set off a craving for Alaskan landscapes. Adams started looking for a way to experience the far north, and an opportunity arose in 1942 when the John Guggenheim Memorial Foundation offered him a fellowship grant to explore national parks with his celebrated lens. Seldom has a grant been so wisely allocated. Adams was convinced that in Alaska his ideal of the wilderness would evolve to new heights.

Adams didn't travel light. Wherever he went, he took his eight-by-ten camera, lenses, filter sets, Graflex cameras, and three specially designed pods—so many accessories, in fact, that to list them all would fill a page.[11] Adams, a consummate professional, was determined to capture the essence of every U.S. national park and of many national monuments. He looped through the Southwest to take photographs at Joshua Tree, Organ Pipe, and Saguaro. He had a new 1946 Pontiac station wagon, and he put a lot of tires and miles on it as he drove down to Big Bend National Park, where the Rio Grande flows like a lazy serpent.[12] A few of those first-round photos of national parks—taken in Arizona, Utah, Nevada, New Mexico, and Texas—appeared in *Fortune*, accompanied by an article by Bernard De Voto. De Voto, through his essays in *Harper's Weekly*, published during the Truman-Eisenhower years, did a good job of filling the void left by Marshall's death and Ickes's retirement. He became perhaps the best publicist for protecting America's public lands against powerful stockmen. Secretary of the Interior Julius A. Krug—Ickes's successor—in a rare burst of inspiration, appointed De Voto to the National Park Advisory Board. What struck De Voto about Adams's pictures, he later revealed in *The Western Paradox*, was the absence of living creatures. "For myself, I had a particular admiration for photographs of Ansel Adams but it struck you with force that the Adams landscape was sterile, a human figure in it would have been discordant to the point of sacrilege," De Voto wrote. "Say as much as you please about the landscape of time beginning, or of the world before time, the more accurate remark was that it was the landscape before life, without life, the landscape of death." [13]

Adams hoped his photographs would encourage Americans to visit their national parks. Conservation, he believed, would succeed only if everyday folks had memorable experiences in nature. Indeed, Adams's work did encourage an entire generation to look at wild America with fresh, neo-romantic eyes. Statuesque saguaro cacti, half-frozen lakes, roaring waterfalls, storm-filled skies, towering redwoods, slate outcrop-pings, wintertime orchards, lone peaks, nameless rocks, and black suns were all part of Adams's own interpretation of America the beautiful. "What I call the Natural Scene—just nature—is a symbol of many things to me, a never-ending potential," Adams wrote to his friend Ted Spencer in February 1947. "I have associated the quality of health (not merely in the physiological or psychological sense) with the quality and moods of sun and earth and vital, normal people. . . . The face of most art reminds me of a human face, bewildered, wide eyed, with a skin of pallor and pimples. The relatively few authentic creators of our time possess a resonance with eternity. I think this resonance is something to fight for—and it takes tre-mendous energy and sacrifice." [14]

It was this belief that the "national scene" had infinite possibilities for a photographer that Adams brought with him to Alaska just a few months after writing to Spencer. Like Rockwell Kent's son Rocky, Adams's fourteen-year-old son, Michael, accompanied him to Alaska's national parks and monuments during the summer of 1947. They would spend six weeks together in Alaska. They drove up U.S. Highway 101 from San Francisco to Seattle, parked in a garage, and boarded the steamer SS *Washington* to Juneau. They traveled along the Inland Passage, stuffing themselves on the buffet food, just as Muir had done decades before. An immense bombardment of thunder and bolt lightning left them enthralled, as if it were a fireworks display. "I was deeply affected by my first glimpse of the northern coasts and mountains," Adams recalled in *An Autobiography*. "The rain did not depress me; it was clean and invigorating, and the oc-casional glimpses of far-off summits gave promise of marvels to come." [15]

To facilitate Adams's travels, Ickes had asked Governor Ernest Gruening of Alaska to open the territory for the famous photographer. Everything in Alaska, Gruening told Adams, would be put at his disposal. Gruening had been editor of the *Nation* during the Harding years, lashing out regularly against the administration, and was pleased, twenty-five years later, to be Alaska's territorial governor, able to defend stupendous southeastern Alaskan landscapes from reckless development. But Adams was irate because Gruening had also vigorously advocated for construction of the Rampart Dam across the Yukon River, which if completed would have been an environmental tragedy. Having been an official with the Department of the Interior in the 1930s, Gruening knew all the special sites of Alaska, and laid them all out for the Adamses to enjoy. For Adams it was a golden opportunity to see Alaska's far-flung wonders with professional forest rangers and biologists as guides. The Department of the Interior, eager to promote Glacier Bay and Mount McKinley, thought that Adams would be an ideal publicist. So upon Adams's arrival in Juneau, Governor Gruening (who was also promoting statehood) fêted him. One gorgeous black-and-white photo of Mount Saint Elias or Admiralty Island, it was understood, could do more to increase tourism than a warehouse full of brochures.

Gruening put an amphibious two-engine Grumman Goose at Adams's disposal. The pilot, a wildlife officer, having suffered wind, rain, and dizzyingly high altitudes for decades, called the plane the Flying Coffin. After a shaky takeoff, Ansel and Michael's nerves stabilized. They soon enjoyed flying low over Alaska's coastal waters, landing in bays where their pilot inspected commercial fishing boats, ensuring that the crews hadn't exceeded their catch limits. A lover of gadgetry, Adams enjoyed studying the plane's instrument panels; the cockpits of planes used in the coastal areas of Alaska were quite different from those operating in the interior and the Arctic. From this bird's-eye view, Adams took a series of distant color shots of "Mount Saint Elias floating in the clouds." These "personal" photos remained, as late as 2010, in the personal collection of Michael Adams, unseen by the general public.

In his correspondence Adams was boyishly enthusiastic about flying over the Brady Icefield, Mount Fairweather, and Icy Strait. All high peaks were mysterious to Adams. Often, Adams wore a Brooks Brothers sports jacket, a white shirt, and a plain tie; he didn't like people who turned native. He was balding, and his broad forehead was perhaps his most recognizable feature. Adams's well-trimmed beard suggested a tweedy college professor. Alaskans soon learned that the always alert Adams was a master at interpreting the moods of their landscape; he would interrupt conversations to point out when a cloud drooped or the sun turned fierce. The Earth had been created long ago—in the flash of a starburst, he thought—and his calling was to turn the creator's magic flashes into framable high art. The whole labyrinth of human consciousness could be found, Adams believed, in a single blade of grass or a fallen rock. "The quality of place, the reaction to immediate contact with earth and glowing things that have a frugal relationship with mountains and sky," he wrote, "is essential to the integrity of our existence on this planet."[16]

In Juneau and Fairbanks, there would be many stories of Adams flying around Alaska in the Grumman Goose and landing on water. The six- to eight-passenger plane regulary landed on lakes and bays so that Adams could compose quick pictures. He ordered his pilots to swoop low toward the ground to gain a better angle on sunsets and wildfires. One afternoon, Adams's plane nearly crashed when the right-hand landing gear malfunctioned. But as Adams had predicted, the death-defying maneuver helped him find the perfect pink-and-purple rose light, a light that infused the blue-green-gray-white landscape with grace, making for memorable photos. "We crisscrossed the Coast Range many times, exploring deep valleys, lakes, passes, and peaks," Adams wrote in *An Autobiography*. "The shadows lengthened and the golden light on the snowy mountains intensified."[17]

A great photographer like Adams will spend weeks, even months, in search of the perfect picture. For that reason, Mount McKinley was a formidable challenge. It seemed arrogant in its immensity, stubbornly denying photographers access to its inner secrets. Perpetually snow-capped, the 20,000-foot, wind-bitten peak simply defied the power of Adams's thirty-five millimeter Contax lens. Even if a photographer had perfect conditions of light, shadow, and wind, it was difficult to capture such bulk, even with a wide-angle lens. Patience was necessary if the goal was to create *the* definitive photo of America's tallest peak. Many photographers, unaccustomed to the thin air at high altitudes, suffered dizzy spells near this peak. Adams, however, took to Mount McKinley, climbing it and plotting his strategy. Since McKinley was three times higher than Yosemite's Half Dome, he realized that the task would be three times as difficult.[18]

Tourists coming to Mount McKinley National Park during the summer of 1947 were often unsure how to approach the steep summit. Somehow simply driving along the park's road was unfulfilling. Thus a booming business began; companies offered hourlong air trips around the peak. The tourists were thrilled to experience McKinley from above; and motion picture crews, it was widely thought, could capture the mountain in this way far better than still photographers. But Adams did not want his defining photograph of McKinley to be an aerial shot. He wanted to capture the spiritual essence of the entire Denali wilderness from the ground. If he could reveal the sublime beauty of Yosemite's waterfalls and the Point Lomas seascapes on his plates, he should be up to the challenge of McKinley.

Like Charles Sheldon, Adams believed the park's proper name was Denali. But whatever the place was called, everything about it proved difficult. On the train ride to McKinley Station, for example, a steady rain made the rails slippery, and the conductor almost collided with a full-grown moose. "The rain finally stopped," Adams wrote later, "the rails dried, and the brakes worked. We passed several busy repair crews; the melting permafrost frequently causes the rails to sag, creating a continuous maintenance problem."[19]

Eventually, Adams and his son reached the diner at McKinley Station for a meal that tasted of cardboard. At night, mosquitoes filled the air. The Adamses were tired even before their adventure had begun. After a good night's sleep, the two headed ninety miles into Mount McKinley National Park in a flatbed Ford truck with camera equipment piled in back. They had been given the key to the ranger station at Wonder Lake—where they discovered that bears had recently broken into the storage bin and wreaked havoc. The bruins had eaten huge boxes of U.S. Army K-ration chocolate. "It was a kick to me as a kid to see the muzzle imprint of a bear on the window glass," Michael Adams recalled. "They had made quite a mess." To get a feel for Wonder Lake the Adamses hiked a steep switchback trail where the wind bore down on them with a vengeance. Oddly, the remote landscape reminded them both of Death Valley. It looked like the lifeless landscape Bernard De Voto had noted in *The Western Paradox*, although here they had insects to contend with. As a connoisseur of light, Adams was keenly aware of changes in the weather and of wind velocity in particular. Now, swatting bugs at one o'clock in the morning and dealing with the strange reality of the midnight sun, he felt the pressure of accomplishment. Adams was above all else a professional.

They found an ideal panoramic view of McKinley from just above Wonder Lake. Adams set up his tripod, but it was hard to determine the best angle for the shot. The right light would last at that angle for only two or three minutes. For a photographer seeking the perfect frame, sometimes in rain and fog, it was an ordeal. "The scale of this great mountain," Adams admitted, "is hard to believe."[20]

In his letters, Adams complains of the insistent rain. Visibility was awful. Fat mosquitoes swarmed around them in huge clouds. The bugs even insinuated themselves between the film and lens. When Adams developed his photos of McKinley, many had been ruined by mosquitoes, which showed up, looking like cartoon airplanes, within the frame. Adams was "disgusted" with himself for not being able to get the perfect picture shot. But in truth, he was being too hard on himself. One of his black-and-whites—*Mount McKinley and Wonder Lake*—would be considered a modern masterpiece, easily the equal of his own *Monolith, The Face of Half Dome*, and *Clearing Winter Storm*. Taken with a telephoto lens (twenty-three-inch focal length on an eight-by-ten format camera), *Mount McKinley and Wonder Lake* shows the mountain and a few swiftly moving clouds, giving a shifting, otherworldly effect.[21] The grayness of the scene, the slight blurring, the melting snow seem to have been heaven-sent. Adams had hit his mark.

Michael Adams never forgot the moment when his father shot *Mount McKinley and Wonder Lake*. They had set up a tripod on a hillside, had doused themselves with citronella to ward off the bugs, and were waiting for the right moment, sheltered by a few stunted spruce trees. A near-silence peculiar to Alaska permeated the air—a vibrating, void-like hum. Mount McKinley conveys rock-hard permanence, silent and untouched— it is in fact a place where no one has ever dwelled in the winter months. Clouds shift rapidly around the summit; looking up at McKinley for too long can actually induce motion sickness. Michael also remembered the glare of the moon, and the colors swirling at dawn and dusk. From their ridge, they patiently waited for *it*—the flashing moment when, as the novelist Jack Kerouac declared, everything becomes valiantly understood. All around them was rolling tundra; and on Wonder Lake the ripples reflected and distorted light. It was hard to tell whether the light was falling or rising. "We both knew the moment," Michael Adams recalled. "It was really something special. We had been to a lot of national parks, seen a lot of sights, but this was beyond amazing."[22]

Judging art can be a matter of personal taste. Nobody has a monopoly on opinion. But it is safe to say that in Alaska Adams produced one of the greatest modern landscape photographs. In *Mount McKinley and Wonder Lake*, shot at 1:30 A.M. in July 1947, with an ethereal light on the lake, Adams managed to make the summit, the tallest mountain in North America, seem an auxiliary to the Denali wilderness that surrounds it. The contrast in the photo between peak and lake is sharp. The tonal effect is that mountain and sky are both subservient to the lake.

In the 1930s, Adams had perfected his "zone system," a pragmatic method of achieving high vision by "controlling exposure, development, and printing, incisively translating detail scale, texture, and tone into the final image."[23] This process became his preoccupation. Put in layman's terms, Adams had professionalized the art of capturing the changing nature of light and how it sweeps over a landscape. "The zone system is designed to eliminate guesswork," Robert Hirsch explained in *Seizing the Light: A History of Photography*, "and give photographers repeatable control over their materials so that the outcome can be predicted (that is, previsualized)."[24]

Years after taking *Mount McKinley and Wonder Lake*, Adams explained his process in *Examples: The Making of 40 Photographs*. For aspiring nature photographers, for whom natural light is *everything*, the book remains a paragon of the art form. In it, Adams revealed how he debated whether to use a red 25-A filter but ended up going with a deep yellow 15, which served to suppress foreground shadows. In total, Adams shot three fine eight-by-ten images of Denali. Half an hour later, at 2:00 A.M., clouds had enveloped the peak and the light no longer radiated so expressively off Wonder Lake.[25] Night at last fell over the summit. There would be other impressive compositions by Adams in the coming years—*Moon and Half Dome* (1960) and *Rock and Surf* (1951) are often cited—but none ever matched the haunting presence of his 1947 Alaskan masterpiece.[26]

Clearly, Adams represented an ideal blend of empathy with the outdoors, artistic visualization, mathematical calculation, intense patience, wizardry with a camera, and proficiency in the darkroom. He was a master of nonanimal nature. In Alaska, having hauled his equipment up a steep incline with only his son to help him, he was determined to succeed. Undeterred by the intermittent downpours, he captured the frozen splendor of McKinley at an instant in summer. He had waited for the miraculous moment, with all the elements aligned just right, and clicked. It was a matter of mathematics and heart. Somehow he had captured both the "spectacular" and "quiet still life" of Mount McKinley.[27]

II

From Mount McKinley National Park, Adams and his son headed to Fairbanks and had a plentiful meal. Then they boarded an airplane headed to Juneau and went on to capture the natural essence of Glacier Bay National Monument. Unlike McKinley, where any photographer knew what to aim at, Glacier Bay didn't have a centerpiece. As a warm-up exercise Adams took minimalist still-life shots: a blade of grass, a veiny leaf, smooth rock faces—the elements of nature at Glacier Bay. Working in black and white, Adams was more interested in geometric shapes than in the wildflowers amid the ice such as yellow paintbrush, blue nootka lupine, or red dwarf firewood. Adams's image *Trailside*, for example, a botanical composition of ferns taken outside Juneau in a rain forest, was a work of modern art in its utter simplicity and lack of ornamentation. In its own way, it prefigured the abstract expressionist paintings of Mark Rothko. From Adams's perspective, kelp beds, besides being a crucial habitat for sea otters, became a work of minimalism to equal examples by Donald Judd or Carl Andre.

When the weather held up, Adams aimed for the Gustavus Forelands, the monument's largest glacial outwash plain, located near the entrance of Glacier Bay. He had little success and felt frustrated. When the light wasn't right, he read books and chatted with the fishermen who worked in Cross Sound. Agents of the National Park Service gladly ferried Adams about the park, enabling him to get close to the brittle surfaces of Margerie Glacier and Johns Hopkins Glacier. But Adams felt that his creative output from Glacier Bay was thin. There was no green fern light to create magic. Somewhat embarrassingly, all Adams had to show from the outings into the whipping fog was a handful of gray negatives. "The weather was so bad that Ansel got very few pictures," Michael recalled. "It was sort of an abortive trip."[28]

A professional nature photographer in the 1940s was, by definition, also a professional traveler. Every day Adams was being hustled off in planes or motorboats in pursuit of *it*. A nonstop roamer, he enjoyed this aspect of his vocation. While Carmel was his home, and tides were his timepiece, his spirit was footloose. Creatively, he was never at ease. Constantly worried about the light, in need of a strong assistant to help with the heavy lifting, and with a meteorologist's understanding of shifting winds and tides, Adams, cameras in hand and dangling around his neck, was a distinctive figure in postwar America. His visual intelligence was probably comparable to that of an eagle or a hawk. But all his comings and goings led to occasional accidents. He had been lucky not to collide with a moose at Mount McKinley, and he had a serious mishap at Glacier Bay. One afternoon while unloading gear from a seaplane, Adams dropped the suitcase holding his shot film into a few feet of cold water. Upon opening the case he found that water had indeed seeped in and damaged his work. He felt ill. All he could do was wait to get to Seattle and send the damaged film to Pirkle Jones in San Francisco, a wizard whose forte was repairing damaged film. "I was naturally quite worried about them," Adams recalled, "but thanks to Pirkle's care only a few were irreparably damaged; my prized Mount McKinley negatives were perfect."[29]

Adams wasn't through with Glacier Bay National Monument. Determined to get a better series of photographs of the mountains, forests, glaciers, and seascapes in the famous Inside Passage, and with the Guggenheim Foundation continuing to pay his expenses, Adams returned to Juneau in the early spring of 1949. He wrapped himself up in a U.S. Army surplus parka to stay dry, but weariness and boredom consumed him. He found himself cursing the bad weather. Determined to shake the rainy-day blues, he visited Muir Inlet on the eastern arm of upper Glacier Bay, where Muir had indeed camped in 1879. Adams took note that the calving glacier had receded seventeen miles since then. Adams himself now practically glowed at seeing the glacier glisten in the effervescent mist, a kaleidoscope of light reflecting off the bluish-green ice. Chunks of ice collapsed into the frigid waters. To Adams, life seemed to thrive in the waters around Glacier Bay. A feeling of creative exuberance swept over him. Catchmen's basins were filled with mussels and crabs and starfish. "This harsh land," Adams wrote, "is blessed by the beautiful northern light and the constant, cleansing rain."

This time Adams was working with the experts of the U.S. Geological Survey who were studying the territory's more than 100,000 glaciers. For days Adams traveled with them in planes and helicopters, learning everything possible about glacial systems, from why ice flowed down the valley to the process of firnification. It was impossible for an intelligent man like Adams to inspect a glacier's terminus and not be overwhelmed by its titanic force. A number of the seventeen tidewater glaciers Adams visited were calving, dropping huge hunks of ice into the waterways with thunderous splashes. Adams found it mind-boggling that the Stikine Icefield blanketed more than 2,900 square miles along the Coastal Mountains that defined the U.S.-Canadian border. "In Alaska," Adams wrote, "I felt the full force of vast space and wilderness."[30]

The question facing Adams at Glacier Bay was exactly what constituted the essence of the national park. How could he get one perfect shot, a representative glimpse into such a spread-out, diverse ecosystem with thousands of varied natural features? Instead of aiming his camera at Muir Glacier, Adams took his best photographs by shooting a chunk of ice jutting out of a bay like a colossal piece of crystal. He titled the composition *Grounded Iceberg*; it's included in the oversize hardback edition of *An Autobiography*. This is not one of Adams's great landscape photographs, but it aptly captures the sensation of a water world, of the isolation, frozenness, and summer thaws that are characteristic of Glacier Bay National Monument.

That summer, Adams once again fell in love with Alaska's steep mountains and intricate waterways. A connoisseur of rain, because it often scrubbed the sky, he now complained it "RAINS AND RAINS AND RAINS AHHHHHH PLOP!"[31] Overall, however, his letters to friends in the Lower

Forty-Eight reveal a boisterous enthusiasm for Glacier Bay that almost equals his passion for Yosemite. For example, here is his letter of June 25, 1949, to his friends Beaumont and Nancy Newhall:

Dear B & N,

WHAT A FLIGHT TODAY! Was in Grumman Amphibian which was dropping loads of supplies to advance base of Juneau Ice Field Expedition. . . .

We crossed and re-crossed 600 square miles of glaciers and ice fields, and encircled the most incredible crags and spires I ever imagined. Bearclaw Peak rises sheer 5000 feet above the ice. We flew around it about 1000 feet distant!

Pictures will help to describe it! The rear door was open to permit dumping loads by parachute. I am full of fresh air, spray on the take-off, noise, but simply unbelievable scenery.

I am afraid Alaska is the Place for me! I am NUTS about it.

Best to you and all our friends,

Ansel[32]

When Adams's retrospective opened at San Francisco's Museum of Modern Art in June 1949, his Alaskan photographs generated considerable excitement. (He himself was in Alaska and missed the opening.) *Mount McKinley and Wonder Lake* was a standout. Everybody, it seemed, agreed that the shots of McKinley had a rare originality: minimalism meeting romanticism in the forlorn Alaskan Range. Like Muir before him, Adams used his photographs to encourage tourists to visit Alaska with their own cameras in hand. He wanted everyone to experience the national parks. In Alaska many ridges remained unclimbed. A new consultant for the Polaroid Corporation, Adams urged amateurs, the core of the conservation movement, to try to capture Mount McKinley and Glacier Bay in their own photographs. The rewards of Alaska, he would tell students at the Ansel Adams Yosemite Workshop (an intense, short photography program held annually in California's premier national park beginning in 1955), were life-changing. As the new oracle for the Sierra Club and a true disciple of Muir, Adams knew that only *seeing* Alaska would lead to *saving* the last frontier. Echoing Horace Greeley's "Go west, young man," Adams said to the postwar generation, "Go to Alaska, folks, and bring a camera." [33]

III

Two female pilots—*Virginia "Ginny" Hill and Celia Hunter*—followed Adams's advice. Because they became lifelong friends during World War II, when they served in the Women Air Force Service Pilots (WASP) corps, Hill and Hunter are almost always written about together in histories of Alaskan conservationism. Both were born and raised in Washington state; they had conservationist values instilled in them when they were girls; they opened Camp Denali together to promote what is now called ecotourism; and in the late 1950s they fought dramatically to save Arctic Alaska as a U.S. National Wildlife Refuge. "Do we really want," Ginny would ask, "to make Alaska over in the image of Los Angeles?" [34]

The WASPs represented those can-do outfits that later led the journalist Tom Brokaw to call the World War II generation the "greatest." After Pearl Harbor there had been a serious shortage of pilots for small planes. General Hap Arnold, chief of the army air forces, decided to recruit women pilots. The idea was to train women to do all the domestic aviation—transporting cargo from warehouses to bases, for example—while the men engaged in combat missions in the European and Pacific theaters. Both Hunter and Hill entered the program. "We became known as flyer girls," Hill recalled. "We towed targets for live air-to-air gunnery, testing aircraft . . . whatever we were asked to do." [35]

Luckily for historians, Hill kept a marvelous scrapbook of her experiences in WASP. It was filled with newspaper clippings, postcards from Texas and California, and photos of the women pilots. One document confirms that she got her pilot's license on March 31, 1943; earned $1,800 annually; and was affiliated with the 319 AAFFTD. There is a *Life* cover story about women in the sky, and there are lots of letters home. "Something new in army discipline—a girl in our platoon was reprimanded by the C.O. for knitting while she marched," Hill wrote on February 19, 1943. "She had a ball of yarn stuffed in the leg pocket of her 'zoot suit' and was blithely knitting on, purling too, while she marched to and from mess. We

Above: John Muir, circa 1902. The great naturalist's trips up the Inside Passage of Alaska in 1879 and 1880 inspired popular interest in glaciers. LIBRARY OF CONGRESS. Below: "Tombstone to Extinct Species" (1913) is an illustration by William T. Hornaday from his revolutionary book Our Vanishing Wild Life. Hornaday helped launch the modern endangered species movement. LIBRARY OF CONGRESS.

Clockwise, from above left: Former president Theodore Roosevelt examining gopher tortoises in Gulf Florida. As the Bull Moose Party's candidate for president in 1912, he vigorously campaigned to protect American wildlife. HARVARD UNIVERSITY. *Theodore Roosevelt's snowy owl (Bubo scandiacus), shot on Long Island during the 1870s, is part of the permanent collection at the American Museum of Natural History.* AMERICAN MUSEUM OF NATURAL HISTORY. *Gifford Pinchot, director of the U.S. Forest Service from 1905 to 1910, helped protect the Tongass and Chugach national forests in Alaska.* COURTESY OF THE LIBRARY OF CONGRESS.

Above: Colonel A. J. "Sandy" Macnab (right) and Frederick K. Vreeland (left) aboard the SS Admiral Watson en route to Anchorage in July 1921. Together they explored the Lake Clark region of southwestern Alaska. COURTESY OF LAKE CLARK NATIONAL PARK AND PRESERVE. Below: Charles Sheldon in front of his cabin at Toklat River, just across from the mouth of what is today Sheldon River (named in his honor). The upper Toklat River is located in today's Mount McKinley National Park. KARSTENS LIBRARY.

*Margaret "Mardy" Murie, often called the "grandmother of the conservation movement,"
with her husband, Olaus Murie, the "father of modern elk management," upon their
return from their Arctic honeymoon in January 1925. The Muries led the grassroots effort to
protect Arctic Alaska in the 1950s.* U.S. FISH AND WILDLIFE SERVICE.

Above: Robert Marshall, cofounder of The Wilderness Society, with Native people from the Brooks Range of Alaska. His memoir Arctic Village *(1933) brought national attention to the Alaskan frontier. He is considered the founder of Gates of the Arctic National Park. Below: Lois Crisler's memoir* Arctic Wild *(1956) became* White Wilderness *(1958), a popular documentary by Walt Disney Productions. Crisler led the campaign to protect Alaskan wolves from aerial hunting, bait-trap poisoning, and government extermination.*
UNIVERSITY OF WASHINGTON.

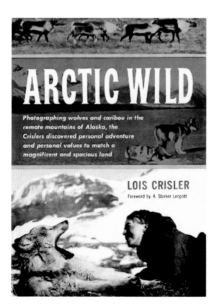

Above: The photographer Ansel Adams on a steamer ride up the Inside Passage of Alaska in 1947. His ethereal photographs of such sites as Mount McKinley, the Tongass, and Glacier Bay attracted ecologically minded tourists to wild Alaska throughout the cold war era. MICHAEL ADAMS PRIVATE COLLECTION. *Below: Ansel Adams's flawless* Mount McKinley and Wonder Lake *(July 1947). This photograph has come to define the surreal beauty of the Denali wilderness.* ANSEL ADAMS COLLECTION.

Clockwise, from above left: The artist, illustrator, and author Rockwell Kent, circa 1920. His outdoors manifesto Wilderness (1920) is considered the Alaskan equivalent of Henry David Thoreau's Walden. THE ROCKWELL KENT ESTATE. Aldo Leopold, beloved author of A Sand County Almanac (1949), dedicated his life to protecting American forests, prairies, and open spaces. As cofounder of The Wilderness Society, Leopold promoted roadless land tracts. THE WILDERNESS SOCIETY. William O. Douglas, a Supreme Court justice from 1939 to 1975, was a fierce advocate for saving the Brooks Range of Alaska and the Arctic National Wildlife Refuge. His memoir My Wilderness, published in 1960, promoted the value of preserving the Alaskan tundra. YAKIMA VALLEY MUSEUM. Virginia "Ginny" Hill Wood, bush pilot, at age eighty-five, holding up a picture of herself as a young woman in the WASPs. Wood was a cofounder of the Alaska Conservation Society and helped save the Arctic National Wildlife Refuge in 1960. VIRGINIA HILL WOOD PRIVATE COLLECTION.

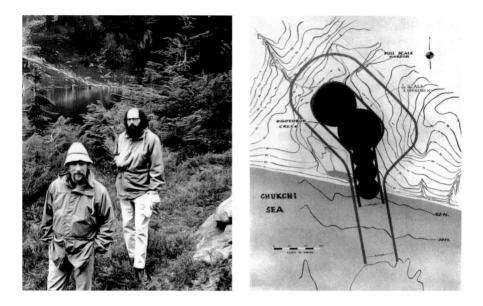

Clockwise, from above left: Gary Snyder (left) and Allen Ginsberg (right) hiking together in the Sierras. As part of the beat generation, they promoted ecology in the 1950s and beyond. Snyder eventually wrote a cycle of poems about Alaska. COLUMBIA UNIVERSITY. *The Atomic Energy Commission's Project Chariot wanted to create an oil port in the Chukchi Sea by detonating five thermonuclear devices. Environmentalists such as William O. Douglas and Virginia Wood successfully derailed the plan.* THE ALASKA CONSERVATION SOCIETY. *President Dwight D. Eisenhower (left) and Secretary of the Interior Fred Seaton (right) created the Arctic NWR on December 6, 1960. Seaton is among the most underrated secretaries of the interior in U.S. history.* DWIGHT D. EISENHOWER PRESIDENTIAL LIBRARY.

The Tongass National Forest was established in 1907 by President Theodore Roosevelt. At close to 17 million acres—about the size of the state of West Virginia—the Tongass is the largest forest in the U.S. Forest Service system. Approximately 70,000 people live in several dozen communities throughout the region, with commercial fishing and tourism driving the economy. After five decades of industrial-scale clear-cut logging in the Tongass since World War II, the U.S. Forest Service is transitioning out of harvesting old-growth forest for timber. U.S. FOREST SERVICE. MAP DESIGN: ANI RUCKI.

Coastal temperate rain forests are rare, covering just one-thousandth of the Earth's land surface.
© AMY GULICK.

The Tongass National Forest contains nearly one-third of the world's remaining old-growth coastal temperate rain forest, and the largest reserves of old-growth forest left in the United States. © AMY GULICK.

Above: One of the world's densest populations of black bears (Ursus americanus) *lives on Kuiu Island in the Tongass, with three to five bears per square mile.* © AMY GULICK. *Below: About 40 percent of the Tongass National Forest is not forested, and consists of glacier ice fields, alpine tundra, wetlands, and water.* © AMY GULICK.

Every spring, humpback whales migrate to southeast Alaska, where food is abundant. They come primarily from their calving and breeding waters in Hawaii, some 2,800 miles away. © AMY GULICK.

Above: With more than 4,500 spawning streams, the Tongass National Forest supports all five species of Pacific salmon—chinook, coho, sockeye, chum, and pink. © AMY GULICK. *Below: More than fifty species feed on salmon, including bald eagles, bears, wolves, sea lions, orcas, ravens, and people. The abundance of salmon in the Tongass supports some of the world's highest densities of bald eagles, brown bears, and black bears.* © AMY GULICK.

Owing to the bounty of both the forest and the sea, the Native peoples of the Tongass region developed one of the most complex indigenous hunting-and-gathering societies in North America. Today, 10,000 Tlingit, Haida, and Tsimshian live in southeast Alaska.
© AMY GULICK.

More than 5,000 islands of the Alexander Archipelago make up much of the Tongass National Forest, emphasizing the interconnectedness of the marine and terrestrial ecosystems. © AMY GULICK.

are treated and trained just like the Air Corps Cadets but once in a while signs of the feminine gender pop up." [36]

Hill was a cutup, always spoofing the WASPs, doodling for fun, and writing racy (for those days) doggerel. Ginny couldn't stand to be bored. She liked to joke that she and Celia were "Daring Young Girls" on the "Flying Trapeze." But the scrapbooks also revealed Hill to be an excellent organizer. Every scrap of paper she saved was pasted in her fat maroon book and perfectly aligned. And she was considered one hell of a pilot. She was a master of the fundamentals of aviation, and cockpit procedure was second nature to her: fasten seat belts . . . unlock controls . . . check gas . . . Hill would usually fly out of Seattle to Portland, Yakima, and Spokane. The Northwest was her official beat. Walt Disney had published a WASP songbook for which he drew the cover cartoon himself: it was a wide-eyed little girl with aviator goggles. Hill knew all of Disney's tunes by heart, singing her way up and down the Pacific Coast. "Usually there was nothing down below," she said, "but mountains, forests, or water." [37]

Both Hill and Hunter were annoyed by a weird law that wouldn't allow women pilots out of Seattle to ferry military planes any farther north than Great Falls, Montana. "We ferried them from factories clear across the U.S.," Hunter recalled, "but 'sorry, gals, turn them over to the men here' and they got to fly them on the Northwest Staging Route through Edmonton, Fort Nelson, Watson Lake, and Whitehorse to Fairbanks."

The male pilots, rubbing in the sexist rule, used to tease Hill and Hunter by saying that Alaska was for *real* pilots, that the fog and sleet were not for the fainthearted female. These taunts stuck in the women's craw. After the war, Hunter and Hill concocted a scheme to borrow two planes and fly to Fairbanks. They were like mountain climbers wanting to reach the top of Mount McKinley. Alaska . . . all that space below . . . the "great land" from the bird's-eye view of the cockpit. Even though Hill's plane was not really airworthy, they named the aircraft *Lil' Igloo* and took off for the wild blue yonder. It took them twenty-seven days to fly from Puget Sound to Fairbanks. They landed on January 1, 1947, in a blizzard. The *Fairbanks Daily Mirror* recorded minus fifty degrees Fahrenheit; what a way to start the new year!

Hunter and Hill celebrated their successful flight and were greeted in Fairbanks with good cheer. The only problem was that they were snowed in for weeks. "We were two babes in a man's world," Hill recalled. "We were bored. We saw a posted sign that read 'Skiing: Women Wanted.' Well . . . I grew up in the snow and figured *why not*." [38]

On a ski mountain, Hill met her future husband, Morton "Woody" Wood. A U.S. Army veteran of the famous Tenth Mountain Division, Wood had seen combat in the Battle of the Bulge. After the war Wood, an expert mountaineer, took classes at the University of Alaska. A forestry major at the University of California and later the University of Alaska in Fairbanks, he would eventually become a park manager at Mount McKinley National Park. "He asked me on a date to a downtown diner and dance," she recalled. "All the guys around Fairbanks were rough. He was a gentleman. I was hooked. We got married and made Alaska our lives." [39] From the beginning of Ginny's marriage, her life still included her best friend, Celia Hunter.

Ginny Hill and Celia Hunter fell head over heels in love with Alaska. Just as Ansel Adams wanted to share his photos of Mount McKinley with the world, Ginny recalled wanting to have all her good friends fly with her over the 20,000-foot peak. For a while Celia worked as a flight attendant on the first trips by Alaska Airlines to Kotzebue and Nome. Meanwhile, the newlywed Woods bought a used Cessna 170, believing that nature tourism would soon become a big business in Alaska. Woody worked for the U.S. Department of the Interior for a while, but the pay wasn't good. He also earned his pilot's license, with Ginny acting as instructor. Together they started taking people to Fairbanks on aerial tours of Alaska. "Ansel Adams had opened things up with his photography of Alaska," Ginny recalled. "Everybody we took just couldn't believe Denali from the air. There wasn't anything like it in North America." [40]

Influenced by Adams, American families started planning to spend summers in national parks like Mount McKinley and Glacier Bay. The Woods joined forces with Celia Hunter and opened Camp Denali in 1952, building their own rustic cabins not far from Wonder Lake and shipping in equipment from Fairbanks. Camp Denali was like a rustic Adirondacks village in the heart of frontier Alaska. "The connection with the land was important," Morton Wood recalled. "It was important to us and important to our guests." [41]

Camp Denali became a hit with tourists. Once Denali Highway opened in 1957, linking Richardson Highway to McKinley Park, a new wave of tourists came by automobile to see America's tallest peak. The McKinley Park Station Hotel, which had opened in 1939, was more service-oriented, with picture-perfect window views by a communal fireplace. What the Woods and Hunter achieved at Camp Denali was an old-style log camp (right down to the cabin doors, with wood and leather pulls). It was a rustic retreat where Ansel Adams's *Mount McKinley at Wonder Lake* could be seen *for real*. The combination of Adams and the WASPs opened up interior Alaska to tourists as never before; the money was in nature photographs, not the extraction industries.

Nobody before or since Adams has ever taken such luminous photographs of America's treasured landscapes. His 1947 composition *Moon and Mount McKinley* has adorned numerous calendars and greeting cards. There is no such thing as a "dated" photograph of Alaska by Adams—his images are all flawless and eternal. It's as if Adams had made himself part of the vast Denali wilderness. If you stayed in a cabin at Camp Denali long enough, you became part of the experience of the place. A new postwar generation was seeking to get away from the suburban doldrums and to discover America's national parks. "You must be able to touch the living rock, drink the pure water, scan the great vistas, sleep under the stars, and awaken to the cool dawn wind," Adams wrote. "Such experiences are the heritage of all people."[42]

o   o   o   o   o   o

PRIBILOF SEALS, WALT DISNEY,
AND THE ARCTIC WOLVES OF LOIS CRISLER

I

W*alt Disney, a veteran of World War I*, wanted to help the United States strike back against the Japanese following the bombing of Pearl Harbor. Patriotically, he put his film company at the disposal of the U.S. War Department. Working with the director Frank Capra, he made films for the Army Signal Corps's series "Why We Fight," which explained America's rationale for going to war.[1] When a patrol-torpedo boat squadron asked for a cartoon insignia, Disney gladly obliged without remuneration and quickly produced an image of a mosquito carrying a torpedo on its back. That morale-boosting, comical mosquito became very popular in Alaska, where the actual insect was a menace.

Other outfits in the armed forces soon wanted their own insignia, and Disney's studio was inundated with requests. Stationed on Kodiak Island in December 1941, for example, was a forlorn naval base with only seventeen minutes' worth of ammunition. The base was run by the Western Defense Command in San Francisco under the leadership of General Simon Buckner, tasked with protecting the Aleutians from a Japanese attack. When, in June 1942, the Japanese landed 8,600 troops on Kiska and Attu islands, Buckner's mission to thwart them became a national security priority. This was the first occupation of U.S. soil by a foreign country since the War of 1812. Over the coming months Allied aircraft dropped more than 7.5 million pounds of bombs on these two Alaskan islands, forcing the Japanese to retreat westward. Alaska was becoming an important theater of war, though that is now widely forgotten. The Western Defense Command, however, didn't have a logo for its Alaska Defense Command (ADC) as 1943 began.

Disney—a lifelong lover of the northern fur seals that congregated on the Pribilofs—now entered the picture. During World War II, the humans on the Pribilofs had been evacuated, but, to Disney's consternation, the seal harvesting continued unabated.[2] Disney drew a cute, frisky-looking seal, balancing the letters ADC on its nose, for the soldiers and sailors to enjoy. In the background was a bright orange-yellow midnight sun. As of late 1943, the patch became the symbol for ADC (although it was never officially approved).[3]

Throughout the 1950s, Walt Disney pioneered in making nature films about Alaska. What most interested the general public was distinguished naturalists who hand-reared the wild animals they were observing. The thrill for audiences, particularly children, was watching a fierce animal like a wolf or bear become a family friend. Disney had struck gold with Jiminy Cricket and Mickey Mouse in cartoon format and would also make money with the True-Life Adventure documentary series, offering wildlife up-close. Serious conservationists of the 1950s weren't particularly fond of Disney's domestication of wildlife as a way of attracting converts to ecology. It smacked of "nature faking." But wolves were being exterminated in Alaska (sometimes by airborne hunters), so the Muries decided to collaborate with the husband-and-wife team of Herb and Lois Crisler, who they knew were defenders of wildlife.[4] And before the Crislers, Alfred and Elma Milotte—also a married couple—had moved to the Bering Sea to document the rituals of Alaskan fur seals on the Pribilof Islands for a pioneering Disney documentary.

Ever since the young Walt Disney left Kansas City for Hollywood in the 1920s, his urge to make movies about Alaska was intense. Once Disney reinvented animation as an art form with *Snow White, Pinocchio, Fantasia*, and *Dumbo*, he produced the True-Life Adventure nature documentary series, which brought in many young recruits to the modern environmental movement. Disney's contribution to conservation was that he helped sensitize the general public to the beauty of fragile ecosystems (such as deserts, swamps, and tundra) and of animals (such as bears, cougars, and seals). A Disney comic book series, published in collaboration with Dell, featured such "charismatic" (that is, appealing) animals as flamingos and seals. Disney, in fact, had dispatched Alfred and Elma Milotte in 1940 to document Alaska as the "last frontier wilderness." When the Milottes returned to Hollywood with more than 100,000 feet of film, proud of having captured everything from climbers on Mount McKinley to lumberjacks in the Tongass, Disney balked. "Too many mines," he complained. "Too many roads. More animals. More Eskimos."[5]

Feeling that they had wasted the better part of a year in Alaska, and hoping to salvage the project, the Milottes wrote to Disney about possibly doing a film on saving the Pribilof Island seals. Disney seized on the offer. Theodore Roosevelt's old friend David Starr Jordan, former president of Stanford University, had been a coauthor of *The Fur Seals and Fur-Seal Islands of the North Pacific Ocean*, about a legal battle of the late nineteenth century aimed at stopping the Russians and Japanese from slaughtering seals.[6] (Jordan, a Darwinian scholar with a PhD from Indiana University, had been the commissioner in charge of Fur-Seal Investigations in the 1890s.) Disney hoped the Milottes could make a documentary that showed what wonderful, playful animals the Pribilof seals were. Disney was a staunch defender of seals—which had been his favorite animals ever since he watched them frolic at the Kansas City Zoo. The director of the U.S. Fish and Wildlife Service, Ira N. Gabrielson, who replaced Ding Darling at the Department of the Interior, had published *Wildlife Refuges* with the Macmillan Company in 1943, to great acclaim. The book was his counterpart to the duck stamp. Gabrielson wrote vividly of Alaska's great seal herds: "birth and death, breeding, living, fighting," he said; "the drama is continuous."[7]

Disney loved drama. Off the Milottes went to the Bering Sea, motion picture equipment in tow, to live with fur seals on the principal Pribilof islands of Saint George and Saint Paul. The fog on the Pribilofs had caused other filmmakers to abandon working there. Seal hunting was supposed to have been banned on the islands (except that Aleuts and Indians were allowed to kill a few seals for subsistence). However, the Fouke Fur Company of Greenville, South Carolina, had contracted with the federal government to process seal pelts. The Milottes also encountered organized poachers. To Disney, the seal lover, the Pribilofs were the "Galápagos of the North," a nirvana for naturalists. Sea urchins in tide pools of the Bering Sea interested Disney, not Aleuts removing blubber and meat from a seal pelt during the canning process. The black mound of Sea Lion Rock, about 100 yards away from Reef Point, the southernmost tip of Saint Paul, was magical to him, not Aleuts in Saint Paul boiling *oosik*, the penis bone of the walrus. Many of the 3 million seabirds on the Pribilofs, during migration season, came all the way from Asia to live with the seals and sea lions. When the Milottes sent back a box of film footage for Disney to look at, they received a two-word telegram back in reply: "More seals." [8]

The prolific wildlife of the Pribilofs sometimes made the islands hard to protect. Sea otters, for example, were hunted to near-extinction during the nineteenth century in this far-flung part of the Bering Sea. While the lives of the approximately 150 Aleuts who lived on the Pribilofs were interesting, Disney thought their activities were too bloody for children to watch: a lot of finback whales were being sliced and diced. To Disney, the northern fur seal had the physical features and playful demeanor that kids loved. Pups were shiny black when born but soon turned an appealing silver gray. Pupping season! That is what Disney wanted to capture on film. During World War II, the Pribilofs had been evacuated by the Aleuts, who were worried about a Japanese invasion. But when they came back in 1944 they resumed slaughtering seals. With the approval of the U.S. government, more than 117,000 seals were slaughtered annually in the years after World War II. [9] Disney rejected the notion that these smart, highly inquisitive mammals should be treated like fish. His documentary on the Pribilof seals would help stop the carnage.

The Milottes learned the hard way that Disney meant what he said about promoting seals. Stubbornly, they continued to include Alaskan fishermen in their film narrative. They tried dealing with the struggles the Aleuts had undergone as serfs under Russian rule in the nineteenth century. Disney had been in Ireland for a while and didn't oversee the Pribilof project carefully. When he finally looked at a rough cut, he flew off the handle. He wanted more seals! Goddamnit! He wanted film of the seals' drama that Gabrielson had written about in *Wildlife Refuges*. Taking direct control of the project, Disney had the Aleuts omitted completely from the film. Accompanied by his eleven-year-old daughter, Sharon,

Disney flew to the Pribilofs himself in August 1947. Seals were the stars, not humans. Seeing these Pribilof seals would be a significant memory for millions of kids. Roy Disney, his brother, recalled that Walt wanted to see the seal herds "firsthand" and get a better "idea of Alaska," so as to make the documentary himself.[10]

What Disney's distributor, RKO, didn't understand about *Seal Island*, as the documentary was called, was that the creator of Donald Duck and Goofy was an ardent conservationist. Disney was on a mission to help save Alaskan wildlife. However, RKO, standing up to him, simply refused to distribute *Seal Island*, insisting that without the human drama of Eskimos, hunters, and loggers struggling to survive in Alaska, the movie was sure to be a flop. Incredulous RKO executives asked, "Who wants to watch seals playing house on a bare rock?" Refusing to abandon *Seal Island*, the determined Disney ended up renting a theater in Pasadena on his own dollar. With a powerful musical score by Jim Algar replacing dialogue, Disney showed his film for one week. This qualified *Seal Island* for an Academy Award in the category "documentary short subject." A couple of months later, to the amazement of RKO, it won an Oscar.[11]

For Disney, the Academy Award was a triumph. He told his brother Roy to take the statue over to RKO and whack the executives "over the head with it." The award also persuaded RKO to distribute the film, and to concede that frolicking Alaskan seals could indeed be a box office hit. Disney was now eager to showcase Alaskan wolves and polar bears as animals worth saving. A secret to Disney's success with nature documentaries was the music that accompanied *Seal Island* and other films such as *The Living Desert* (1953) and *The Vanishing Prairie* (1954). Disney, it seemed, was also willing to fabricate the habits of seals, bears, and wolves for entertainment value. "He had found a way," his biographer Neal Gabler wrote, "to combine entertainment with education."[12] Some critics didn't mind this. Certainly children loved seeing seals anthropomorphized, goofing around with each other like kids at a playground. But others, including the esteemed film critic Richard Schickel, detected fraud. "The tone of a Disney nature film is nearly always patronizing," Schickel wrote in *The Disney Version*. "It is nearly always summoning us to see how very nicely the humble creatures do, considering that they lack our sophistication and know-how."[13]

In any case, *Seal Island*—released in 1948, when Ansel Adams was photographing Alaska's national parks—was a transformative moment in the
conservation movement. Using slow-motion and time-lapse film techniques, Disney allowed moviegoers to see life from an animal's eyes. As a
rule, there were no humans in the True-Life Adventure films. Disney was
on a mission to protect wildlife. In 1942 his animated film *Bambi: A Life in
the Woods*—based on a book of 1923 by Felix Salten (pen name of the Hungarian journalist Siegmund Salzmann)—had done more than all of John
Muir's books combined to turn American popular culture against deer
hunting. All the woodland animals in Disney's *Bambi*—Owl, Thumper (a
pink-nosed rabbit), Flower (a skunk), and so on—found a place in children's hearts as cozy friends. The death of Bambi's mother was described
by the film critic Pauline Kael as one of the most emotionally devastating
scenes in film history.[14] Woodland ecology got a boost from *Bambi*, just as
Alaskan wildlife conservation did from *Seal Island*.[15]

Never has a film done more to promote wildlife protection than Disney's *Bambi*. Whittaker Chambers had been responsible for translating
Salten's original book from German into English. The seasons of nature,
from showers in April to leaves falling in November to ponds freezing in
January, are dealt with magically in the animated feature film. The horror of deer being chased by hunters provided *Bambi* with harrowing moments. *Bambi* was nominated for three Academy Awards—"Best Sound,"
"Best Song," and "Original Musical Score." Deep human emotions are
touched when Bambi's mother dies. Affected by the intense music, moviegoers became angry at the snarling dog pack that created havoc in the
once idyllic wild animal kingdom. But Raymond J. Brown, editor of *Outdoor Life*, sent a curt telegram to Walt Disney, furious that law-abiding
hunters were being portrayed as "vicious destroyers of game and natural
resources."[16]

During the coming decades, hunters would object to the "Bambi complex," "Bambi factor," and "Bambi syndrome." But advocates of protecting
wildlife, such as the Crislers and the Milottes, approved of the cartoon
creatures. Disney had Americanized Bambi as a whitetail for an American audience; in Salten's novel the characters were roe deer. Deer ecology was an important philosophical premise in the film. To present forest
life realistically, Disney had dispatched his artists to Maine's Baxter State

Park for six months. They lived among the animals in order to properly sketch head movements and sleeping patterns.[17] The columnist George Reiger of *Field and Stream* said that, overnight, Disney had turned hunting into a grim endeavor: "once Bambi is raised in status from mere deer to Jesus Whitetail Superstar, man's hunting of deer becomes a crime comparable to the prosecution of Christ."[18]

II

Disney also became associated with *Lois* and Herb Crisler. They were going to help the "save the wolf" movement, which Adolph Murie had long promoted at Mount McKinley, much as the Milottes had helped with the Pribilof Island seals. The Crislers had lived for months at a time in the Olympic Mountains. Herb once spent thirty days in the Olympics without food or a gun; he was *hard-core*. Together they worked to save the Roosevelt elk, the larger "kings" of the Olympics. By vocation, Herb, a native of Georgia, was a motion picture photographer, and Lois was an English instructor at the University of Washington. (Her master of arts thesis had been on "Santayana's Definition of Beauty."[19]) After the Crislers got married, the Olympics became their living room. Both were excellent skiers. The Crislers' homestead was Hume's Ranch, a ranger station on the Elwha River. They were hired by the U.S. government to build isolated fire lookouts and hunting shelters. During the winter of 1942–1943, the Crislers served as Aircraft Warning Service lookouts on Hurricane Ridge in the Olympics. Wherever the Crislers went in the Olympics, they took photographs of the gorgeous backcountry. In the winter—following even the worst snowstorms—the Crislers would go skiing. Eventually they traveled around the country showing home movies and slides of their adventures in Washington state's high country. To supplement their income, Lois wrote articles about wildlife for the *Port Angeles Evening News*, the paper of the Olympics region in Washington. She also worked on a memoir, "Gift from the Wilderness" (still unpublished).

The Crislers' big break came in 1949, when Walt Disney decided to purchase film footage of Roosevelt elk playing in the Olympics for his company's nationally televised show. The public loved the segment, filmed entirely by Herb Crisler and titled "The Olympic Elk." Building on that success, Walt Disney Productions hired the couple to film big-horn sheep at Colorado's Tarryall Peak and grizzly bears in Alaska. Herb's thirty-five-millimeter motion picture of Colorado rams charging each other in gladiatorial combat was stunning. In pursuit of photographic quarry in Alaska, the Crislers traveled by dogsled, whale boat, canoe, and bush plane, and on snowshoes.

"Your film, which you presented at the Annual Meeting of the Wilderness Society Council on the Olympic Peninsula, was an inspiration to us all," Olaus Murie, director of The Wilderness Society, wrote from Moose, Wyoming. "The photography was of course excellent, and you have a happy choice of subject matter. The artistry and warmth of feeling which permeates the whole film presented the wilderness quality of Olympic National Park in a manner that can hardly be excelled. You are doing a great service for the American people in showing this film and I only wish it were possible for all the millions of Americans to have the privilege of seeing and experiencing its inspiration." [20]

The Crislers were on their way to becoming stars. Herb would spend time with Walt Disney, regaling him with folk songs and poems and pioneer tales like a backwoods bard. No longer did Herb and Lois have to sleep in flophouses. In Detroit they stayed at the Statler, and in New York at the Waldorf-Astoria. At one studio meeting in Hollywood, Disney told the Crislers about a coyote den in his backyard; he found coyotes charming. Lois recognized that in Disney, an Eisenhower Republican, the wildlife protection movement had a stalwart ally. "We could see that to a lot of people in the United States, the wilderness that we take for granted up in the Olympics," Lois wrote, "was becoming one of the choicest things they could contact." [21]

Wildlife documentaries were coming into vogue, and the Crislers were leading the filmmakers. Unlike a Hollywood set production, the great outdoors offered a wildlife photographer plenty of elbow room. But lugging a ninety-pound camera up hills and switchbacks was physically draining work. Capitalizing on their growing fame, Lois started writing scripts; these had an overdrawn, dramatic narrative and seemed more like Dashiell Hammett than like John Burroughs. Lois could make a bluebird feeding its chick an insect into an event of grand importance. "The fawn got itself somehow down the steep bank and onto the lake," she wrote in a typical passage; the style is exhilarating yet clipped. "It crossed the flat whiteness, now hurrying, now seeming very tired and standing still. But it drove itself on."[22]

When the opportunity presented itself, Lois wrote serious articles—one was about a rare marmot (*Marmota olympus*) in the Olympics—for *Natural History*.[23]

With regard to wolves, Herb and Lois Crisler were walking, talking zoological encyclopedias. Lois was fascinated that in 20,000 B.C. southern Europeans were drawing wolves on cave walls. Besides the mandatory books of wolf biology in her library, she had underlined references to wolves in such literary classics as the epic of Gilgamesh and the *Iliad*. According to Lois, Jesus Christ used wolves in parables to emphasize moral principles; Pliny the Elder gave a pseudoscientific account of wolves in *Natural History*; *Beowulf*—the oldest important narrative poem in English—had a wolf as the heroic protagonist who kills the monster Grendel; and Shakespeare also mentioned wolves in his plays with noticeable regularity. Lois Crisler liked to point out to people that wolves used to be beloved animals, part of the wild kingdom, not beasts to be exterminated. It infuriated her that the average American mistakenly thought a wolf's howl was menacing. "Like a community song, a howl is a happy occasion," she explained to the general public. "Wolves love to howl. When it is started, they instantly seek contact with one another, troop together, fur-to-fur. Some wolves . . . will run from any distance, panting and bright-eyed, to join in, uttering, as they near, fervent little wow, jaws wide, hardly able to wait to sing."[24]

For eighteen months, the Crislers holed up in a rustic plywood cabin called the "Crackerbox" along a forlorn rivulet in the Brooks Range, sometimes in driving snow, encircled by a curtain of mountains, befriending one of Alaska's thirty-two caribou herds. Lois wrote a memoir about the experience among the Central Arctic caribou herd on the North Slope along the Killik River (a tributary of the Colville River that is the eastern boundary of the National Petroleum Reserve). The memoir, *Arctic Wild*, was published in 1956 and was dedicated to the "Wolves of the Arctic Tundra and to Those People Who Will Act to Preserve Life and Habitat for Them." Justice William O. Douglas deemed *Arctic Wild* "one of the most exciting wilderness books in the English language."[25] Ansel Adams,

Mardy Murie, Margaret Mead, and many other prominent conservationists considered *Arctic Wild* a historic breakthrough in the wildlife protection movement. The Crislers had formed a bond with a wolf family. Every morning wolf pups would jump on the Crislers, lick their faces, and howl for an hour.[26] Lois's chiseled, elongated face and her braided hair tied in a bun were shown in photos throughout the autobiography, and the book made her a celebrity. "What *do* I want?" she asked herself. The answer: "To be where the people that walk on four legs are."

Crisler's *Arctic Wild* certainly wasn't the first serious book about Alaska's wolves—it had been preceded by Adolph Murie's *The Wolves of Mount McKinley* and Stanley P. Young and Edward A. Goldman's *The Wolves of North America* (both published in 1944). But Crisler, by using the first-person narrative style, brought the family life of wolves to a general readership in a touching, loving, and respectful way. Eight years later, the Canadian biologist Farley Mowat would publish *Never Cry Wolf*, to great acclaim; it clearly superseded *Arctic Wild* as literature. But during those crucial years of 1956 to 1960, when the fight to save the Arctic was particularly intense, it was Lois Crisler who most troubled the anticonservationists in Alaska.

Nothing, in fact, infuriated her opponents more than the fact that Lois Crisler was teaching wolves to cuddle, nurse, and howl. To the average Alaskan, wolves were useful only for their pelts. "Lois's accounts of the wolves' howling and using their incisors to finely lift her eyelids while she slept truly portrays the remarkable abilities of wolves," the wolf ecologist David Mech later recalled. "No doubt such descriptions helped recruit a large number of people into the ranks of wolf admirers."[27]

The biologist Rachel Carson of the U.S. Fish and Wildlife Service was elated with Crisler's *Arctic Wild*. A warmhearted correspondence ensued between the two women throughout the late 1950s. Carson's articles of the 1930s and 1940s about marine ecosystems—which had appeared in the *Baltimore Sun*, the *New Yorker*, *Field and Stream*, and *Yale Review*—encouraged Crisler to write about wolves. When Carson's *The Sea Around Us* was published by Oxford University Press in 1951, Crisler sat mesmerized, reading it over and over again. Whenever Lois felt lonely or depressed in the Olympics, Alaska, or Colorado between 1955 and 1963, she wrote to Carson. "We live in a [Silver Spring, Maryland] house that is too large for us, especially since my mother's death, and it would be a joy to entertain you," Carson wrote to Crisler. "We can promise you the song of mockingbirds and cardinals, and by mid-March we might even manage the beginnings of our frog chorus."[28]

Although wolves were the stars of *Arctic Wild*, the 50,000-head Central Arctic caribou herd (so named in the 1970s) came in a close second. In Alaska every caribou herd on the North Slope claimed its own calving area, which was a fair distance from other calving areas. Because the caribou had large concave hooves, which made wide imprints in the tundra soil, they were relatively easy for a Disney camera crew to track. Newborn calves weighed only thirteen pounds. With their pretty suede-soft gray coats, these caribou were as appealing as Bambi. The crush-crush of Arctic caribou on frozen tundra, clumsy calves clinging to their mothers' protective sides, captivated Crisler, who wrote that the mass migrations "beat like a pulse through our time."

At first, the Arctic seemed to the Crislers barren of wildlife—an almost empty land. There were no throngs of caribou or packs of wolves. The Dall sheep came down to the rivers only during the winter months. Although the Crislers were well-known wildlife photographers, regularly giving slide shows on college campuses and at corporate retreats, they had assumed that the Brooks Range was like the Rockies, only colder. But once the Crislers sat still, didn't look so hard, and actually lessened their expectations, a kingdom of wildlife appeared before them. Little voles were burrowing in the sedges. Asian bluetooths fluttered along the rivulets. Ptarmigans flushed *put-p-p-p* from the willows, turning from white to brown as the seasons dictated. Grizzlies patrolled streams, waddling away only when they picked up the scent of man. Perky eider duck mothers were followed by a single-file parade of youngsters. "There was a miraculous fact about this deadly white wilderness: it was alive!" Lois Crisler wrote. "Animals lived here and found food."[29]

Ostensibly, the Crislers were going to follow the caribou's migratory trail north of the Arctic Circle throughout the deep summer of 1955, as Charles Sheldon had tracked Dall sheep; but Disney had another idea. Why not adopt wolf cubs and raise them? As entertainment, tracking caribou in the golden Arctic light—despite the cute newborns—was boring. Raising wolves, by contrast, had immediate box office appeal. So, with money from Walt Disney Productions, two cubs were purchased from an

Eskimo—a male, Trigger; and a female, Lady. (The names conjured up both Roy Rogers's horse and Disney's cartoon feature film *Lady and the Tramp*.) By day, Lois would observe wolverines—capable of bringing down prey five times their size—wading along sinuous creeks and gorging on caribou meat. At night, with willow bushes crackling away in the cabin fireplace, Lois would cuddle with the adorable wolf cubs. In her journal, Lois described being a mother to the wolf pack. She claimed that wolves, an extremely sociable wild species, "smiled" and "talked" and "read my eyes!" The concept was anthropomorphic, the film was filled with embarrassing hyperbole, and the raising of wolves was morally questionable. Nevertheless, the Crislers succeeded in their quest to make wolves more beloved the world over. "Wolves are not a menace to the wilds but orgies of wolf hate are," Crisler wrote in *Arctic Wild*. "Wolves themselves are a balance wheel of nature."[30]

The historian Vera Norwood has written insightfully about Lois Crisler in *Made from This Earth: American Women and Nature*. While admiring the Crislers for their advocacy for wolves, Norwood nevertheless raised smart questions about the ethics of the Disney film. Was this proper holistic ecology? To Norwood these habituated wolves were no better off than those behind bars at a zoo. Scenes of the Crislers releasing the wolves back into the wild only to have them scratch at the cabin door, seeking hearth and home, seemed cruel. One follow-up episode was unambiguously wrong. When Herb Crisler realized that two pet wolves weren't generating enough entertainment value, Disney's cameramen raided a den and swiped five more pups for Lady and Trigger to raise. The Crislers justified this raid by saying that bounty hunters would soon have slaughtered the pups.[31] For real biologists, the Crislers were hard to take. But *Arctic Wild*, the memoir by Lois Crisler of their experiences in the Brooks Range, did make people think about the north country and about wolves. William O. Douglas (grumpy about the Disney film), the *New York Times*, and Rachel Carson all praised it as an educational work ideal for young people—and their approval alone was worth a lifetime of accolades for Lois Crisler. Disney ended up marketing the documentary as the feature film *White Wilderness* and also produced educational shorts from the footage, such as *Large Animals in the Arctic* and *The Lemmings and Arctic Bird Life*.

After filming *White Wilderness* for Disney in 1956, the Crislers took their four wolf pups home with them to Tarryall Peak in Colorado. They had no other choice. Because these wolves had been domesticated, they had never learned to hunt. Releasing them into the wild would have meant certain death. Killing them wasn't an option. So the Crislers got government permits to keep them in Colorado as pets. Harper and Brothers advanced Crisler money to write a follow-up memoir—*Captive Wild*—about raising and breeding wolves at Crag Cabin, their ranch.[32]

Never known for holding his tongue, William O. Douglas lambasted Disney for the irresponsible nature-faking stunts in *White Wilderness*, though he was careful not to hold Lois and Herb Crisler responsible for the staged material: "In my time Walt Disney did more than anyone to distort and depreciate our wildlife," Douglas wrote. "He had a wolverine fight a bear to death. Animals, other than men, do not follow that course. They have conflicts but soon withdraw. Disney got the wolverine to fight the bear by starving both animals for weeks in a Los Angeles zoo. The battle actually took place in a movie set in the city."

In his memoir *Go East, Young Man* Douglas intensified his criticism: "Disney showed rams of the mountain-sheep family charging each other, their foreheads clashing to the tune of the 'Anvil Chorus,'" he scoffed. "They charge, of course, but in between charges they rest, walk around, paw the earth, and the like. They do not follow the pattern of a Hollywood dancing troupe."[33]

While the Crislers were raising wolves in Colorado, Frank Glaser, the wolf hunter, was still using his "coyote getter" from Seward to Nome. On most mornings he loaded cyanide into his "coyote getter" (which was set off by animals attracted to a bait station) and headed out into the wild. He was also given access to a plane, making it easier for him to slaughter wolves far and wide. But a change was occurring in Alaska. Glaser, once considered an Alaskan hero, was starting to be viewed by the general public as a menace. Glaser's idea of success was discovering a wolf den and slaughtering the pups; this practice was now frowned upon by an increasing number of Americans. Still, Eisenhower's secretary of the interior, Douglas McKay, presented Glaser with a Meritorious Service Award for controlling predatory animals. Glaser moved around Idaho, California, and Oregon for a while in the 1950s. But there weren't enough wolves to slaughter in other states, so he moved back to Anchorage. One night Glaser heard wolf howls in the distance. He seethed with rage. "They don't belong in town," he fumed. "They'll kill dogs, and a lot of kids are running around too. I'm worried about them. Those wolves have to be killed." Upset by the fear that wolves were going to lay siege to Anchorage, Glaser telephoned Dr. Louis Mayer for psychological help. "Mayer made a housecall," Jim Rearden recalled in *Alaska's Wolf Man*, "talked with Frank, gave him a sedative."[34]

III

B y a happy coincidence, *1956 brought* another milestone publication that to many Arctic conservationists transcended Lois Crisler's writings. The brother of The Wilderness Society's cofounder Bob Marshall brought out, with the University of California Press, a posthumous work by Marshall, *Alaska Wilderness: Exploring the Central Brooks Range*. Whereas Bob Marshall's *Arctic Village* had dealt with the citizens of Wiseman, *Alaska Wilderness* offered meditations about the Upper Koyukuk drainage system to the Gates of the Arctic wilderness. Every page offered wisdom and enlightenment. Suddenly Marshall's voice was alive again, nearly two decades after his death in 1939. He described battling Squaw Rapids below the mouth of the Glacier River and recounted "a furious blizzard" swooping down upon him and freezing his party's "cheeks and necks." It all made for riveting outdoors reading. *Alaska Wilderness* included scientific data and drainage maps, and it had a revivifying effect on Adolph, Olaus, and Mardy Murie. There were also twenty-two photos taken by Marshall in the Arctic. *Alaska Wilderness* was a welcome reminder of what was at stake in saving the Arctic Range from road construction and industrialization.

Hoping to arouse the Arctic preservation movement, Justice Douglas jumped at the opportunity to review *Alaska Wilderness* in The Wilderness Society's periodical *Living Wilderness*. "This is America's last frontier, as yet untouched by man," Douglas wrote. "Bob Marshall saw them by plane, by foot, by dogsled. His account is an enduring one. It tells why this great area should be preserved in perpetuity as a wilderness area."[35]

Alaska Wilderness particularly advocated roadless areas, and Douglas absolutely agreed. "This is a book for every man and woman who loves the wilderness," Douglas said. "While it will bring back some echoes of one's own experiences, it will remind even the expert that he yet has much to learn about the wilderness on our frontier. And it will help marshal public opinion to preserve the Brooks Range as a Wilderness, keeping it forever free of roads, lodges, and filling stations."[36]

Disney's movie *Winter Wilderness*, based on the Crislers' experiences in the Arctic, wouldn't come out until 1958. But before even a single frame was seen, conservationists knew it would put the Frank Glasers out of business. *Bambi* and *Seal Island* had already convinced conservationists that Walt Disney was the best publicist the wildlife protection movement had had since Theodore Roosevelt. Having Justice William O. Douglas as an advocate for the wilderness, ready to protect Arctic Alaska, was also good, with Robert Marshall gone. Help for wild Alaska also came from the pioneer Arctic archaeologist J. Louis Giddings, whose forte was the prehistory of northwestern Alaska. For the first time First Nation tribal history was being treated seriously: Giddings's research made the notion of populations crossing the Bering Land Bridge respectable.[37]

Throughout the 1950s Disney was a die-hard supporter of both President Eisenhower and the wildlife protection movement. While America was going through the processes of suburbanization, bureaucratization, and the emergence of what William Whyte called the "organization man," Disney's Alaskan adventures were a journey back to the frontier. Eisenhower, for his part, considered himself a "Disney man," and with good reason—Disney solicited campaign contributions and held fund-raisers for the Republican Party. According to his biographer Neal Gabler, the conservative Disney also put bumper stickers on the car he used on his Hollywood lot, endorsing Richard M. Nixon for president in 1960 over John F. Kennedy.[38] It might very well be that Disney's steadfast support of Arctic preservation and the Pribilofs influenced President Eisenhower's Alaskan land policies. If the extremely popular Walt Disney thought that families might someday want to see polar bears and seal herds in Alaska, then who was Eisenhower to question his intuition?

. . . . . .

Chapter Seventeen

THE ARCTIC RANGE AND ALDO LEOPOLD

I

The Wilderness Society's cofounder Aldo Leopold set the tone for saving Arctic Alaska. When Leopold died in 1948 while fighting a wildfire, *A Sand County Almanac*, his poetic meditation on protecting and renewing land, was not yet published; the typed manuscript remained on his desktop at his home in central Wisconsin. Luckily for the conservation movement, his son Luna, recognizing the importance of this work, had it published by Oxford University Press the following year. Sales were minimal, but conservationists immediately grasped that Leopold had written a tour de force. Rooting through his father's file cabinets, Luna organized another volume of Aldo Leopold essays and journal entries as *Round River*. It was published in 1953. For conservationists during Eisenhower's two-term presidency, these two texts were gems to be cherished. Leopold's words were quoted throughout that decade to protest against the construction of unnecessary dams in the Pacific Basin region. Regarding Alaska, Leopold's call to keep places "wild and free" was a rallying cry for the small band of determined conservationists.

Pragmatically recognizing that every farm woodland by necessity yielded lumber and fuel, Leopold urged his countrymen to recognize that what was on top of the land was more valuable than what was underneath the soil. "The wind that makes music in November corn is in a hurry," Leopold wrote in *A Sand County Almanac*. "The stalks hum, the loose husks whisk skyward in half-playful swirls, and the wind hurries on. In the marsh, long windy waves surge across the glassy sloughs, beat against the far willows. A tree tries to argue, bare limbs waving, but there is no detaining the wind. On the sandbar there is only wind, and the river sliding seaward. Every wisp of grass is drawing circles on the sand. I wander over the bar to a driftwood log, where I sit and listen to the universal roar, and to the tinkle of wavelets on the shore. The river is lifeless: not a duck, heron, marsh hawk, or gull but has sought refuge from the wind."[1]

For Mardy Murie, reading *A Sand County Almanac* was a profound experience. Nowhere was the wind Leopold rhapsodized about purer or more forceful than in her own beloved Arctic Alaska. Like the northern goshawks, common redpolls and gulls, she felt invigorated by torrential gusts. The Arctic wind in springtime was her life force, her muse, her harmonic revelation of the cosmos. Sobering, enlivening, and somehow bitingly wise about the ancient universal secrets, wind velocity was the power source of the ages. And to Mardy the drafts in the Brooks Range were particularly intoxicating as they swept down chillingly from the North Pole, always making her spirit feel whole again. Although the Arctic Range was difficult to get to in the 1950s (transportation consisted mainly of small planes landing on gravel bars), it offered a monumental experience. A hiker by predisposition, Mardy knew that rivers like the Kongakut, the Canning, and the Hulahula would someday be popular with river runners.

In 1946, Mardy had spent time with the studious Aldo Leopold during a meeting of The Wilderness Society held at her home in Wyoming. When Leopold spoke, conservationists paid rapt attention, and Mardy knew he was the most far-seeing conservationist present—smoking cigarettes, wearing a white dress shirt with a pale necktie, squinting behind his rimless glasses while talking, calmly swapping information with Olaus about the biotic world. There was something noble about his low-key style. Leopold's nerves were always steady; verbally, his passion was muted; a steely integrity emanated from his clear blue eyes. To have left behind, in dying, such an elegant meditation as *A Sand County Almanac* was an act so lovely that it seemed preordained.

Reading Leopold's epitaph to the extinct passenger pigeon, Mardy thought of the fate of Arctic Alaska's birds such as the snowy owl and the willow ptarmigan. To Leopold the passenger pigeon "was the lightning that played between two opposing potentials of intolerable intensity: the fat of the land and oxygen and air." When Martha—the last passenger pigeon—died in captivity at the Cincinnati Zoo in 1913, the Audubon Society mourned. "Yearly the feathered tempest roared up, down, and across the continent, sucking up the laden fruits of forest and prairie, burning them in a traveling blast of life," Leopold wrote. "Like any other chain reaction, the pigeon could survive no diminution of his own furious intensity. When the pigeoners subtracted from his numbers, and the pioneers chopped gaps in the continuity of his fuel, his flame guttered out with hardly a sputter or even a wisp of smoke." [2]

If the passenger pigeon, once 1 billion strong, could go extinct, what of Arctic Alaska's polar bears, caribou, and willow ptarmigan? What of the shorebirds that bred along the coastal plain of the Beaufort Sea: the American golden plover, semipalmated plover (*Charadrius semipalmatus*), lesser yellowlegs (*Tringa flavipes*), wandering tattler (*Tringa incana*), spotted sandpiper (*Actitus macularius*), whimbrel (*Numenius phaeopus*), surfbird, least sandpiper (*Calidris minutilla*), Baird's sandpiper (*Calidris bairdii*), Wilson's snipe (*Gallinago delicata*), and red-necked phalarope (*Phalaropus lobatus*)? It wasn't enough for Mardy and Olaus Murie merely to count caribou for the U.S. Fish and Wildlife Service in Arctic Alaska. They would have to fight to save the Arctic Range along the Beaufort Sea, as Bob Marshall had done with the Gates of the Arctic and as Leopold had done in the Gila wilderness. They needed to lobby the Department of the Interior not to build roads in the Arctic, because changes in drainage patterns adversely affected habitats. [3]

What Leopold most admired about Mardy Murie was her confidence that someday U.S. citizens would stand up and say no to the obsession of the "harassed world" with industrialization. While other conservationists grew discouraged by toxic smokestacks and coal-burning power plants, Murie continued to simply marvel at the unmarked Arctic, where the aurora borealis beamed forth hope. [4] Her touchstone place was the 200-mile Sheenjek River, which flowed south to the Porcupine River from the highest peaks of the eastern Brooks Range, joining the Porcupine just northeast of Fort Yukon, Alaska. Anybody rafting down the smooth Sheenjek—which had only a few Class II rapids—had a good chance of seeing some of the 123,000-strong Porcupine caribou herd, because this herd often partly wintered in the Sheenjek valley. (The largest caribou herd was the 500,000-strong Western Arctic group, which ranged the National Petroleum Reserve.) Cradled by the Davidson Mountains, the Sheenjek was also the water's edge for Dall sheep, grizzlies, moose, and beavers.

Among Olaus and Mardy Murie's close friends, only Starker Leopold (Aldo Leopold's son, a professor of zoology at the University of California), Lowell Sumner, and George Collins knew the Sheenjek River well. The Muries realized that Mardy herself would have to spread the word about it, as John Muir and Ansel Adams had done for Glacier Bay. Certainly, Alaska needed farms, paved highways, modern industries, and mineral development—but the wilderness that made the territory unique should also be protected. The Muries hoped that saving a vast portion of the Arctic along the Canadian border could be promoted by national conservation groups (the Sierra Club, The Wilderness Society, Audubon Society, Izaak Walton League, etc.) and by a new local nonprofit, the Alaska Conservation Society. The Muries felt that the Glacier Bay area was being overrun with tourists on cruise ships (which Mardy called "floating nursing homes"). The Arctic needed to be preserved for the Gwich'in people and for true "Leopoldian" outdoors types. "Thoughtful people both in and out of Alaska were concerned, for the Age of the Bulldozer had arrived," Murie wrote in *Two in the Far North*. "Scientists like Starker Leopold, Lowell Sumner, F. Fraser Darling, and George Collins, who had recently traveled in Arctic Alaska, began writing and talking to Olaus."[5]

II

C ollins—*taking advantage of the momentum created* by the publication of Darling and Leopold's *Wildlife in Alaska*—thought about how best to protect the Arctic from despoliation. His foot-numbing explorations in the region (frostbite was in fact a constant risk) weren't holidays on the tundra, but despite the hardships he amassed reams of biological data for the Department of the Interior. Ideally, Collins concluded, the Gates of the Arctic area would become a national park. But the Arctic Range along the Canadian border—particularly the scenic Sheenjek River—should be designated a roadless wilderness where not even tourism would be promoted. There would be no gateway villages to the Arctic Refuge—nothing like Gatlinburg or Jackson Hole. The Wilderness Society's concept of "roadlessness" would be established in Arctic Alaska. Otherwise, tracked vehicles would wreak havoc on the tundra.[6] "While we were out in camp with Leopold and Darling, we had many discussions about this park idea," Collins wrote to the Arctic archaeologist Louis Giddings. "Every one of us came to the same conclusion—that a national or international park is the only solution. No other form of land use is a sufficient guarantee of security in our opinion."[7]

What the Arctic conservationists were proposing was a national park (or wilderness area) four times greater than Yellowstone. Flying over the Yukon Territory, both Collins and Sumner began thinking of a vast international park. As coauthors of a "Progress Report," Collins and Sumner wrote that the Arctic Refuge had to remain free of "artificial disturbance," and sportsmen's activities there would need to be strictly controlled. Theodore Roosevelt had saved the Grand Canyon by means of the Antiquities Act of 1906 for "scientific reasons." Similarly, the early cold war generation in Alaska, inspired by *A Sand County Almanac*, wanted a baseline virgin ecosystem to compare and contrast with other Arctic Circle lands damaged by the industrial order. The Sierra Club's president, Benton MacKaye, wrote in *Scientific Monthly* that an "Arctic Park" would be a "reservoir of stored experiences in the ways of life before man."[8]

If The Wilderness Society liked using the word *wilderness*, the National Park Service was invested in the notion of *primeval lands*. It was, at face value, a semantic issue. *Wilderness* seemed more commonplace than *primeval*, which harked back to efforts at the turn of the twentieth century to protect ruins such as Mesa Verde in Colorado and Chaco Canyon in New Mexico. The Oregon Caves had been saved in 1909 as a national monument, with local boosters calling themselves "cavemen." The Izaak Walton League (the premier anglers club) and the Federation of Western Outdoor Clubs both supported an "Arctic Wilderness" because it was based on honoring "primeval values." Numerous Darwinian scientists and Arctic Eskimo leaders in the early 1950s used the term *primeval* to explain the evolution of *Homo sapiens*. "We hurt because we see the land being destroyed," Trimble Gilbert, an Arctic chief, lamented. "We believe in the wild earth because it's the religion we're born with. After 10,000 years our land is still clean and pure. We believe we have something to teach the world about living a simpler life, about sharing, about protecting the land." [9]

In November 1952 Collins and Sumner offered the Department of the Interior a twenty-three-page paper, titled "A Proposed Arctic Wilderness International Park" and illustrated with handsome photographs. A version of this report appeared in the *Sierra Club Bulletin* as "Northeast Alaska: The Last Great Wilderness." "Unless an adequate portion of it can be preserved in its primitive state," the report claimed, "the Arctic wilderness will soon disappear." [10] Because no single country owned the north pole, it made sense to the authors to form a collaborative agreement with Canada. After all, polar bears, caribou, and wolves didn't recognize artificial borders. Ottawa hadn't yet dealt in earnest with Washington, D.C., about the "one habitat" concept; but President Eisenhower was a hero in Canada for having staged the Allied invasion of Normandy on D-day during World War II. While he was not a conservationist himself, Eisenhower had a vision of working closely with Canada on building the Saint Lawrence Seaway, which would link the Great Lakes to the Atlantic Ocean; and with regard to conservationist proposals, he was known to be cautious and slow, but *not* automatically opposed. Perhaps the Arctic International Park could be sold to Eisenhower as a bilateral initiative between two members of NATO?

Only half a year later, Collins and Sumner abruptly changed their minds about the international park. The Eisenhower administration was going to be pro-development in Alaska. The Department of the Interior might approve a Gates of the Arctic National Park—with extensive recreational facilities for visitors to the central Brooks Range—but it was unquestionably opposed to Bob Marshall's concept of wilderness simply for the sake of wilderness. Aging New Dealers were being retired from Interior, and it became clear that Eisenhower was more friendly toward Humble Oil than Franklin D. Roosevelt had been—and that this attitude would affect Alaska's wilderness. Advocates of wilderness at the Department of the Interior had been pampered by Harold Ickes. Now, greed and shortsightedness, two threats to conservationism, had returned to the forefront of the American public lands system, where they had been in the 1920s. The cold war was on, and the CCC had been dismantled. Minerals were in and mallards were out; and the president of General Motors, "Engine" Charlie Wilson, proclaimed that what is "good for General Motors is good for the country." [11]

But President Eisenhower—who tremendously respected the legacy of Theodore Roosevelt—wasn't an unreasonable man. That would prove vital for the conservation movement. There were murmurs at Interior that Eisenhower wanted to keep the cold war out of the Arctic and Antarctic, that he was considering international treaties to protect the poles. Also, Disney's film *White Wilderness*, about the Crislers' wolf pups, was being edited for release in 1958. Disney had also optioned Ernest Thompson Seton's book *Lobo, the King of Currampaw* to be made into a pro-wolf documentary filmed in New Mexico. [12]

III

I n the early 1950s, following the publication of *A Sand County Almanac* and *Round River*, the U.S. government, for the first time, earnestly pondered how to save Arctic Alaska. However, with the Korean War being fought and Senator Joseph McCarthy of Wisconsin looking for communists under every bed, Alaska was a low-priority issue. But to conservationists the time seemed near, if it hadn't exactly arrived, when millions of acres in the Arctic should receive permanent protected status. While the Crislers were making their film with Disney and Ansel Adams was measuring the light around Mount McKinley, many well-to-do conservation societies had a newfound interest in Arctic preservation. Photographs of the Brooks Range—impressive summits, monotone shoulders, and empty white spaces—appeared in the glossy pages of *National Geographic*. Readers could almost hear the booming wind. Robert Marshall's *Alaskan Wilderness* became a cult work within the conservation community; Supreme Court Justice William O. Douglas handed out copies to office visitors as if the book were his business card.

Capped by the gaunt summits of the Brooks Range, the inviolate Arctic offered timeless permanence in a postwar era characterized by transience and consumerism. Sir Frank Fraser Darling, a Scotsman whom the Sierra Club called the "Einstein of ecology," joined with the New York Zoological Society (which Theodore Roosevelt had helped found) to advocate protecting the Alaska-Yukon Arctic as a counterpart to Africa's Serengeti, centered on the Porcupine caribou herd. Darling worked with Starker Leopold to publish the landmark *Wildlife in Alaska: An Ecological Reconnaissance.* Comprehensive in approach, this book explored the interconnectedness of caribou herds, wolf dens, snowy owls, brown bears, and the entire North Slope. Darling and Leopold believed that the U.S. Department of the Interior had a "national responsibility" to save this primeval animal range, marine sanctuary, and nourishing landscape. Each American generation since TR had its own rendezvous with the wilderness, and Arctic Alaska was suddenly the landscape of the moment. Because Alaska was still a territory, without influential U.S. senators to represent it, the Interior Department could be directed to parcel out vast wilderness reserves relatively easily. The big question was which agency would be the best steward of Arctic Alaska.

Collins, head of the Alaskan Recreation Survey, had traveled far and wide across the territory in the mid-1950s, being flown around the North Slope and island-hopping in the Aleutians. He was a walking field guide to Alaska, able to predict ice and virga. By plane, he surveyed 147 Alaskan sites, from Bristol Bay to Clark Mountain to the Beaufort Sea, for potential protection by the National Park Service. Sometimes in flight Collins encountered the mysterious fata morgana (a mirage caused by layering of intensely cold or cool air against the water, sea ice, or land). His comprehensive 1955 report—*A Recreation Program for Alaska*—was aimed at widening tourists' opportunities for bird-watching, hiking, cross-country skiing, river rafting, and mountain climbing. According to the historian Roger Kaye in *Last Great Wilderness*, Collins—a career officer in the National Park Service from 1927 to 1960—envisioned a "fuller range of wildland values" through "transcendental and romantic concepts and new perspectives" promoted by conservationists such as Bob Marshall, Aldo Leopold, and the Muries.[13] Collins believed the postwar rush to over-timber, over-mine, and over-drill in Alaska had to be thwarted. Already the wildlife biologist Lowell Sumner was warning the Department of the Interior that spraying DDT would kill Alaskan lakes and forests as well as the insect hordes it was aimed at. Nature was under attack; there was an increased risk of species extinction and overexploitation of natural resources in Alaska.

Collins was born in Saint Paul, Minnesota, in 1903. Many of America's most effective environmentalists came from the upper Midwest. Aldo Leopold and William Temple Hornaday were from Iowa. So, too, was Congressman John F. Lacey, who from 1892 to 1906 did more than any other U.S. politician except Theodore Roosevelt to protect wildlife by means of federal legislation. Besides the Muries, Sigurd Olson (a staunch wilderness advocate and biologist), Gaylord Nelson (a Democratic senator and founder of Earth Day), and Joseph Hickey (a prominent wildlife ecologist) all came of age in Wisconsin. If you grew up in Wisconsin, you could explore Leopold's shack in Sand County and Muir's childhood home, Fountain Lake Farm, as historical landmarks of conservation. In photos of Collins as a young man growing up in Wisconsin, he has the look of Gene Autry, but with bushier sideburns. Usually Collins kept the top button of his checkered shirts fastened, as if he might want to attach a bolo tie at a moment's notice. Collins was a master of surveying the public domain and offering plans for preservation. "George had a hilarious sense of humor," Ginny Wood recalled. "And whatever he wrote about Alaskan lands was absolutely true, solid geography. He wasn't a lot of hoey." [14]

Encouraged by Horace Albright, head of the National Park Service, Collins had a conservationist résumé in the Lower Forty-Eight that helped make him highly effective in wild Alaska: serving as superintendent of Lassen Volcanic National Park in California; working as a ranger at the Grand Canyon from 1930 to 1935; running a CCC camp at Lake Mead (which had been created by Boulder Dam in 1936); establishing a district office for the National Park Service in Santa Fe; protecting the Channel Islands off the coast of Oxnard, California; and overseeing the survey that saved Point Reyes National Seashore. But Collins isn't praised in college courses in environmental history, for a single reason: he supported the construction of Glen Canyon Dam.

Glen Canyon Dam was indeed folly. In 1956, the Upper Colorado River Storage Bill was introduced in Congress. For $756 million, a huge dam would be built near Page, Arizona. To environmentalists, damming the wild Colorado River was sacrilegious. The construction area—along the Arizona-Utah border—constituted some of the world's most gorgeous canyon scenery. Governor J. Bracken Lee of Utah, however, declared that the

Glen Canyon Dam was "just the beginning of a long range program that will build up the West." [15] Eventually the bill was passed, and construction began on one of the largest reclamation projects in American history. President Dwight D. Eisenhower announced the official construction of Glen Canyon Dam—which formed Lake Powell—on October 15, 1956, by pushing a remote control at the White House, triggering an immense explosion in the Southwest. Huge hunks of Glen Canyon's west wall tumbled down thunderously. [16]

The Sierra Club, which had stopped Echo Park Dam, was silent about Glen Canyon Dam, evidently influenced by people like Collins and cognizant that Arizona would get 6 percent and Utah 13 percent of the electricity generated by the blocked Colorado River. "Glen Canyon died in 1963, and I was partly responsible for its needless death," the club's executive director, David Brower, lamented in his autobiography, *For Earth's Sake.* "Neither you nor I, nor anyone else, knew it well enough to insist that at all costs it should endure. When we began to find out, it was too late." [17] But the feisty novelist Edward Abbey had known that Glen Canyon was the Colorado River's "living heart." For decades he protested against the dam—and against the men who promoted it, like Collins. Abbey's novel of 1975, *The Monkey Wrench Gang*, begins with protesters dropping a huge black plastic banner showing a lightning-like crack down the dam as if the concrete were ruptured and crumbling. And the local Navajo predicted that the sandstone holding the dam in place couldn't last more than fifty years; nature would someday liberate the Colorado River. [18]

Still, Collins did a lot of good in Alaska. By drafting recreational plans for the territory he proved that there were ecologically responsible ways for tourism to be a boom industry in Alaska. His ordering of a biological survey of Katmai National Monument—known primarily for its volcanoes—led to new knowledge that the area was among the best brown bear refuges in the world. And his recognition that today's Arctic NWR was, in fact, one of the greatest wildlife corridors in North America earned him a place on the Alaska Conservationist Hall of Fame honor roll. [19] "That is the finest place of its kind I have ever seen," Collins said of the Arctic Range. "It is a complete ecosystem, needs nothing men can take to it except complete protection from his own transgression." [20]

Collins—who was famously photographed with his two Saint Bernard dogs at his side when he was in Alaska, and bundled up in fur-lined parkas for long Arctic treks—was enamored of the central Brooks Range. Looking eastward toward the Yukon Territory border, Collins pronounced his determination to create an "Arctic International Wildlife Range"—a pure wilderness zone not subjected to sabotage for the sake of oil, gas, or coal, but intended for "the everlasting benefit and enjoyment of man."[21] Collins obtained money from the National Park Service to prepare a survey on the potential boundaries of an Arctic park; his plea for restraint pertaining to oil development in the Arctic was being taken seriously in Washington, D.C.

Empowered by Marshall's influential book *Arctic Village* (and Frank Dufresne's *Alaska's Animals and Fishes*, published in 1946), Collins was starting to think in the same long-range ecological terms as The Wilderness Society. Sumner was in full agreement with Collins, stating that Alaska's Arctic Range needed to be protected "unhindered and forever," like Mount McKinley or Glacier Bay. A lover of the great outdoors, Collins was becoming part of what the historian Roderick Nash called a "national intellectual revolution" to save the Alaskan wilderness at all costs.[22] "We saw the fallout of having a Park, or whatever you want to call the area, divided by an international boundary when you had so many migratory species, both marine and terrestrial, that used both sides of the line," Collins explained. "We didn't know what to call it. We used such terms as 'conservation area.' Generally it was a park to us, always and still is. . . . The scenery was enthralling. It was simply stupendous, beyond description, absolutely magnificent."[23]

The fact that The Wilderness Society was making progress in protecting Alaskan landscapes, however, didn't mean that the U.S. Fish and Wildlife Service was filled with leaders like Mardy, Olaus, and Adolph Murie. Unecologically-minded Alaskans still saw wolves as vermin and seal fur as desirable clothing; many government agents agreed with them. The wild salmon in the Copper River were running too thin for comfort; the Wrangell and Saint Elias mountains were in need of federal protection. Magical places like the Matanuska valley, of which the village of Palmer was the hub, were hell-bent on allowing surface coal mines.

North of the Brooks Range, there were signs within the U.S. Fish and Wildlife Service that Alaska's territorial game wardens thought

Bob Marshall had exaggerated the allure of the Arctic. An example was Clarence Rhode, the half-knowing, half-uncaring director of the U.S. Fish and Wildlife Service in Alaska. Rhode mistakenly invited Sumner on a friendly trip to survey the Arctic. Sumner saw it as a fine opportunity to count caribou on the springtime tundra, but he soon found himself shocked and disgusted. Members of the service's delegation shot at wolves from airplanes whenever they were lucky enough to spot four or six trotting across the permafrost. Because the Arctic was flat and sparsely wooded, shooting the wolves was relatively easy. And these biologists were killing simply for sport, and later, in camp, bragging about their kills. Sumner developed a deep enmity toward Rhode: Where was the fair chase ethos? How could men of science be so ignorant?

Sumner returned to Fairbanks and thereafter cast a cold, skeptical eye on the directives of the U.S. Fish and Wildlife Service. His own view of the Arctic, he now realized, was more in line with that of the Inupiat Eskimos and Athabascan Indians than with that of the Truman administration. Clarence Rhode's employees, he now knew, had outdated ideas about controlling predators. And Rhode himself, only marginally interested in wolf ecology, was especially proud that the stockmen's associations, market hunters, and oil, coal, and ore developers of the Alaska territory considered him an ally in subordinating nature. That was a hard-won honor for a federal employee in Alaska. Sumner began a campaign against Rhode and in favor of creating a huge Arctic Range reserve—something that would far exceed Mount McKinley National Park in protected acreage. As a start, Sumner collaborated with Olaus Murie, the director of The Wilderness Society, about saving Arctic Alaska, saying he felt strongly that it was "one of the most spacious and beautiful wilderness areas in North America."[24] Throughout the early 1950s Sumner, who did not flinch from being a maverick, went after Rhode relentlessly. His journal is peppered with sharp, condescending remarks about Rhode's ignorance of the biological sciences. Sumner was convinced that Rhode wanted wolves exterminated to placate the politicians in Juneau. "My impression is that F & W's policies are those of game farming of all wildlife," Sumner wrote. "It seems to me that at the hands of our Government the Arctic is a very perishable place."[25]

So Sumner made his dissent and made it forcibly. And if he wasn't changing bureaucrats' minds, he was certainly galvanizing conservationists: he was admired by many wardens for courageously slapping Washington, D.C., awake. But this was clearly a rearguard action. Rhode boasted that in 1951 his service killed 287 Alaska wolves, and he promised that the number would rise. Furthermore, a future governor of Alaska, Jay Hammond, boasted that he had shot 300 wolves from his plane in a single month.[26] An aggressive new effort to poison wolves was under way in the Brooks Range. Rhode had approved dropping strychnine-laced bait in the Arctic, and he saw no reason why cyanide charges—mines—shouldn't be buried in springs near wolf birthing areas in the Brooks Range. Native Americans complained, but to little avail, that strychnine "bombs," tossed from planes, were also devastating wildflowers, caribou, and so on. "The wolf is universally hated in Alaska," Larry Meyers explained in the magazine *Alaskan Sportsman*. "It is hated with an intensity which seems to be handed down from our primordial ancestors—an instinctive hatred tinged with fear."[27]

IV

Although they weren't consulted about it in any meaningful way by the U.S. government, the Gwich'in Nation of Northeast Alaska and Northwest Canada wanted the coastal plain along the Beaufort Range permanently protected. They called the area *Iizhik Gwats'an Gwandaii Goodlit* ("The Sacred Place Where Life Begins"). Boldly the Gwich'in Nation started standing up to oil companies, protesting against strychnine, and opposing the mining of the Arctic Range. The coastal plain they knew was the birthplace of the Porcupine caribou herd (where 40,000 to 50,000 calves were born annually). Journeying across the range, maps in hand, Collins sought the best borders for his envisioned international park. Quietly he observed with field glasses a huge herd. The Gwich'in villages were located along the migratory route, and to the Gwich'in people the caribou represented life itself. They drew on the herds for clothing, tools, medicines, and food. These 8,000 Native people started demanding equal rights for Arctic residents in the 1950s.

What to do about the Gwich'in? That concerned both Collins and Sumner. There was a saying that if "Gwich'in retained a part of the caribou heart, then the caribou would, in turn, retain a part of the Gwich'in heart." [28] In other words, the people and the caribou had a symbiotic relationship: the fate of the Porcupine herd would determine whether the people's distinctive culture survived. Creating a national or international park didn't make sense to Collins. Glacier Bay National Monument had struggled with how to handle issues of hunting and fishing in a preservationist site. Collins knew he had to honor traditional Gwich'in subsistence living in whatever designation was chosen for the Arctic Range. "We had a tradition of hunting and prospecting," Collins explained. "We had international interests to consider. . . . It was felt in the service and in the department, I think, that national park status wasn't quite the thing for this one." [29]

Environmental activists seldom have enough political power or money to make changes—but they often know how to write. And there is no question in reading the reports of Collins and Sumner about the Arctic—unofficial documents not cleared through the Department of the Interior—that the campaign for Arctic preservation was promoted in mid-1951. Collins and Sumner would seize every advantage, work both sides of the aisle, and be essentially shameless in pursuing the goal of saving the northernmost third of Alaska. All this effort, however, could take them only so far. In the end, the American people would have to demand that Arctic Alaska be saved. A coalition of the Sierra Club, the Audubon Society, the National Park Association, and The Wilderness Society (among other nonprofits) would have to work for the Arctic Range. Operating in their favor was the fact that Alaska was still a territory. Around Anchorage, however, the movement for statehood was gaining momentum. Both Collins and Sumner now believed that conservationists could start lobbying Capitol Hill with a quid pro quo in mind: statehood for Alaska *only* if a sizable part of the Arctic became a nature reserve where the new wilderness philosophy would be honored.

Toward the end of his life even Theodore Roosevelt—the great hunter himself—wrote four or five essays on the advantages of wildlife photography over rifles. Ansel Adams wandered around Denali in 1948 taking amazing photos of Mount McKinley. Very few photographers, however, trekked up to the Arctic, because special equipment was needed in such cold country. On the North Slope the sun never set from May 10 to August 2. And from November 18 to January 23 the sun never rose. For visual artists, this meant that the sun didn't get high over the horizon; so they got low-angle light with distinct shadows. Add to the situation nameless valleys, stark mountains, and needle-sharp rocks, and very few people volunteered for Arctic duty. Only a few hardy photographers, such as Richard Harrington and Bates Littlehales, have made art from the Arctic. But Walt Disney Productions had discovered Lois Crisler—the author of *Arctic Wild*, for whom the "wolf's call" was so powerful that "nothing else would do but to look deeply into its eyes on its home ground"—and people were starting to think about Alaska. As Starker Leopold noted, Robert Marshall emphasized the topography of Brooks Range whereas Crisler focused on "a great living whole, with its proper animals going about their business." [30]

The Crislers were smart to focus on Arctic wildlife. For unlike redwoods or oaks, waterlogged muskeg depressions, filled with mats of decayed vegetation and moss, hadn't yet found defenders. While botanists might marvel about large areas of Arctic ground displaying arrays of geometric shapes called ice wedge polygons, it wasn't the sort of ecosystem that garden clubs held raffles to help protect. While a few photographers snapped close-ups of birdlife along the Beaufort Sea, aerial shots of the Arctic showed that endless cycles of freezing and thawing had caused the ground to crack in patterns similar to dried mud. Clearly, in the "big cold" decomposition had outraced accumulation. While caribou roamed the valleys and arctic grayling overwintered in deep pools, it was the *stillness* that was the real natural attraction. North of the Yukon River was like Washington and Oregon combined, without many human footprints. And there were a lot of thermals in the ever-changing sky.

The bond that kept all the Arctic Alaskan activists together was Olaus Murie—and he was very sick. In 1954 he was diagnosed with miliary tuberculosis (the disease his brother Martin had died of in 1922). Olaus headed to National Jewish Hospital in Denver—the best hospital in America for respiratory illnesses. For fifteen months he underwent experimental antibiotic treatment, determined to breathe without tubes. Never financially well off, constantly living hand-to-mouth, Mardy found employment as the secretary of the Denver office of the Izaak Walton League. [31]

All the pharmaceuticals in the world didn't offer the curative power of fresh air. Once the Muries returned to Moose, Wyoming, they reconnected with friends in The Wilderness Society for a conference at Rainy Lake, Minnesota. They became preoccupied with protecting Arctic Alaska. Coughing constantly, clearing his throat of phlegm, Olaus believed that he had one great act left in him and that, with death knocking on his door, cautious activism no longer made sense. He also started looking for young recruits. The Muries now were going to help Herb and Lois Crisler get their "white wilderness" preservationist message to college students. Furthermore, the Muries would help organize expeditions to Arctic Alaska with employees of the Department of the Interior. Olaus believed that if U.S. politicians actually spent a week in Arctic Alaska in late summer, when the blueberries were ripe and the fireweed was blooming, camped along a gravel bar or in a field of wildflowers, they would never dream of opening up the Brooks Range or coastal plain along the Beaufort Sea for development by the extraction industries. The Muries' ideals about the wilderness were now being translated into direct action as never before. And the Muries had the spirit of Aldo Leopold to bolster them.

Chapter Eighteen

∘　∘　∘　∘　∘　∘

THE SHEENJEK EXPEDITION OF 1956

I

Throughout the late 1950s, the Muries were lobbying intensely on behalf of the Arctic Refuge. When they approached Henry Fairfield Osborn, president of the New York Zoological Society and Conservation Fund, about helping them organize an expedition to the Sheenjek River in 1956, he funded it at once. Ever since Theodore Roosevelt had helped found the New York Zoological Society in 1895 it had probably worked harder and more thoughtfully than any other organization on protecting North American big game. In Arctic Alaska the great caribou herds were threatened, so Osborn was more than ready to finance the expedition. If time allowed, Osborn wanted to come along and explore the limestone peaks and narrow side valleys along the Sheenjek. In addition to the New York Zoological Society and Conservation Fund, the Arctic expedition was sponsored in collaboration with The Wilderness Society and the University of Alaska–Fairbanks.[1]

The Muries were hoping that 8.9 million acres in the northeastern corner of Alaska would be declared the Arctic National Wildlife Range. (The name was changed to Arctic National Wildlife Refuge in 1980. Within the U.S. Department of the Interior it was known as the Arctic NWR.) To The Wilderness Society, this huge range represented the only "undisturbed portion of the Arctic" that was "biologically self-sufficient." When talking to Osborn and others, Olaus would rattle off all the mammals—grizzly, black, and polar bears; caribou; Dall sheep; moose; wolverines;

and other fur-bearing creatures—that lived on the plain of the Beaufort Sea. With the Naval Petroleum Reserve occupying 23 million acres along the Arctic Ocean—to be developed as an oil field owned by the U.S. government—it seemed only fair for the Eisenhower administration to establish the Arctic Refuge. There, scientists could study an "undisturbed natural arctic environment" and outdoorsmen could hunt and fish.[2]

What Theodore Roosevelt had done for the Great Plains bison in South Dakota, Oklahoma, and Montana Murie was hoping to do with the caribou of the Brooks Range. Starting in 1920, he would work with the Biological Survey to make this happen. There were the Porcupine herd, whose calving ground was the coastal plain of what would become the Arctic NWR; the Western Arctic herd, a 500,000-head herd in what would become the National Petroleum Reserve (in an area known as the Utukok uplands), grazing atop 2 trillion tons of coal (9 percent of the world's supply); and the Central Arctic herd of 30,000 to 60,000, which roamed between the Colville and Canning rivers. Murie, it seemed, had a vision of the Great Caribou Commons remaining intact along the Brooks Range so that future generations could experience its primordial grandeur.

The Muries had chosen well in making the Sheenjek River their symbol of Arctic Alaska. There were hundreds of valleys just as beautiful, but the Sheenjek had Last Lake—a good place for pontoon planes to land—and was among the last great wilderness areas in America. Because of the perpetual summer sun, a twelve-hour hike was possible, through some of the most impressive big country anywhere. Olaus told Osborn that their trip would be a "sample adventure," a weeklong hike to see snowcapped mountains, blue lakes, and white spruces. Clucking ptarmigan, hungry bears, and gray wolves would be moving conspicuously through the landscape. Mardy believed that any decent person who spent a week on the Sheenjek during the summer months would be compelled to ask Congress to create a national park or ask President Eisenhower to sign an executive order offering permanent protection. "I sit here on this soft mossy slope above camp, writing. The writing has been very erratic because of those who live here," Mardy wrote in her Sheenjek River diary on June 3, 1956. "I have watched a band of fifty caribou feeding back and forth on a flat a quarter mile away; ptarmigan soaring and cluck-clucking and giving their ratchety call, all about tree sparrows so close and unafraid; cliff swallows hurrying by; Wilson snipe and yellowlegs calling, grey-cheeked thrushes singing."[3]

The Sheenjek expedition consisted of Mardy and Olaus Murie, Dr. Brina Kessel (an ornithologist at the University of Alaska), George Schaller (a graduate student at the University of Wisconsin–Madison), and H. Robert Krear (a postgraduate student at the University of Colorado). All the members agreed that banning mining or drilling in the Arctic Range was of the "utmost importance."[4] According to Schaller, the Sheenjek River was symbolic of everything The Wilderness Society stood for: good science, exploration, and conservation. "I've traveled in many parts of the world," he said, "in the most remote wilderness, and I don't think people in the United States realize what treasure they have, because there is very little remote wilderness left in the world."[5]

The weather was unpredictable along the Sheenjek River during the short summer. When the Muries led the expedition in June, one day the temperature was twenty-nine degrees Fahrenheit. Two weeks later the thermometer rose above eighty degrees. Along the glacially formed pothole lakes in the valley floor, every hour could bring a contrast. Olaus had brought a motion picture camera, which he aimed at caribou; it would be helpful for The Wilderness Society's presentations at college campuses. Much of the scenery in the Sheenjek valley was reminiscent of A. B. Guthrie Jr.'s novel *The Big Sky*. But for long stretches the Murie party hiked over soggy muskeg as if doing penance for being biophilic; nothing was easy in the Arctic. Mardy decided that Sheenjek should mean "Land of Contrasts" (rather than "Dog Salmon," its actual translation). On some days the Muries trapped mice to study and made borings in spruce trees to measure growth rates. One morning a grizzly visited the camp, and the mosquitoes followed. But overall, the "sample adventure" was working out idyllically.

Getting politicians in Washington, D.C.—or anybody—to care about Arctic Alaska in 1956 wasn't easy. But the Murie expedition had a stroke of luck when William O. Douglas confirmed that he would join the expedition on June 29, along with his wife, Mercedes Hester Davidson. (Their addition made the expedition a party of seven.) Olaus had hiked along the C&O Canal—the 180-mile waterway trail from Washington, D.C., to Cumberland, Maryland—with Douglas, amazed by Douglas's knowledge of birds, his astounding stamina, and his conservationist convictions. Douglas had fought to save the old towpath canal as a refurbished National Historic Park instead of allowing a concrete highway or a dam at River Bend just above Goat Falls, which would have flooded a section of the trail. Residents of Washington, D.C., have been grateful for his advocacy of the C&O Canal ever since. Douglas, an expert on land policy issues, continually thought of ways to protect the shrinking American wilderness from industrial ruin. As Douglas prepared for the trip to the Brooks Range, he was mulling over how best to draft a Wilderness Bill of Rights. "To Douglas," the legal scholar William H. Rodgers Jr. explained, "those who canoe or hike or backpack or ride horses or climb mountains deserve protection no less than that extended to religious minorities."[6]

Olaus knew that Douglas, who had hiked in the Cascades and the Olympics, disdained being pampered on the trail. The primitive conditions on the expedition—no pavement, no roads of any kind—would appeal to his desire to escape from the nation's congested capital during the humid summer months. The unanswered question was whether the justice's wife (his third) would be able to tolerate the backcountry conditions. Friends of Douglas had a theory that if a wife couldn't handle his arduous campouts in the Pacific Northwest, then he'd dump her.[7] "Trim, petite, blond, every hair in place, chic gray flannel suit, nylon hose, brown calf loafers," Mardy wrote, describing Mrs. Douglas. "But I needn't have worried! The first thing she said to me was 'I've got my blue jeans and rubber pacs just like you said, as soon as I can get into our duffel.'"[8]

For too long, William O. Douglas's judicial brilliance, intense manner, poetic demeanor, outdoors heartiness, uncluttered mind, environmental prescience, and landmark legal decisions have been neglected by historians. Because Douglas had a rather unconventional personal life, including numerous wives and numerous affairs with Supreme Court interns, gossip has often prevailed. But Douglas represented much that was good, true, and durable in America. Never did he fritter a day away with nothing accomplished. Hikes, to Douglas, were a productive time for *thinking*. During the cold war, nobody else fought to protect the Bill of Rights with the same ardor as Douglas. During his thirty-six years on the Supreme Court, Douglas—misleadingly pigeonholed as a New Deal liberal—was the truest western libertarian of his era. Time and again he was the best friend working people had on the Supreme Court. Douglas always defended the unemployed, the homeless, the freakish, and the contrarian against the abuses of both big corporations and big government. Ben Franklin or Thomas Edison surely would have understood his feisty unorthodoxy. Nobody would have been a better guide on the Lewis and Clark expedition than Douglas. The U.S. Army's lawyer Joseph Welch eventually embarrassed Joe McCarthy in 1954 by asking whether McCarthy had "no shame" in pursuing supposed communists; but Douglas had attacked McCarthy from the outset, accusing him of trampling on both procedural rights and the First Amendment. "The great danger of this period is not inflation, nor the national debt nor automatic warfare," Douglas wrote in the *New York Times Magazine.* "The great, critical danger is that we will so limit or narrow the range of permissible discussion and permissible thought that we will become victims of the orthodox school." [9]

Douglas had appropriately titled this article "The Black Silence of Fear." The narrow thinking of the Republican right annoyed him to no end. Luckily for America, by the early 1950s Douglas's shoot-from-the-hip voice had become unrestrained. While Douglas held no brief for Marxist-Leninist philosophy, he understood how essential it was for the Supreme Court to defend freedom of thought at all costs. Douglas predicted an Orwellian nightmare if American teachers, for example, were silenced and forced to adhere to official dogma. Yet Douglas, for all his virtues, made a series of bad choices regarding whether Julius and Ethel Rosenberg should be executed—as they were on June 17, 1953. He refused to fight for their lives: in the end, he had no tolerance for spies.

On the other hand, Douglas got the disaster in Vietnam right from start to finish. His 1953 book *North from Malaya* warned the Eisenhower administration not to get bogged down in Southeast Asia along with the French at Dien Bien Phu. *North from Malaya* was Douglas's third book on his "traveling social conscience" (as his biographer James F. Simon put it). Douglas was prophetic about the limits of U.S. intervention in the third world. He would have made a terrific secretary of state. All of his "magic carpet" trips took place while he was on the Supreme Court. Friends used to joke that there must be five William O. Douglas look-alikes because he seemed to be everywhere at once. Journalists and book reviewers often praised Douglas for being the most literary Supreme Court justice since

Oliver Wendell Holmes Jr. "The eye-to-ear witness reporting," the chief White House correspondent for CBS, Eric Sevareid, wrote of Douglas in the *Saturday Review*, "is magnificent."

Douglas brought along to the Arctic all his acuity, and his global perspective. While the Muries didn't know much about the Rosenbergs or Vietnam in 1956, they were keenly aware that Douglas might hold the key to persuading President Eisenhower to sign an executive order creating the Arctic National Wildlife Refuge. When Douglas asked, "You want to go for a walk?" power brokers quickly grabbed their hats. Only Woody Guthrie was a more celebrated tramper than Douglas in 1956. Bringing his tackle box with him, using mostly a light rod and dry fly, Douglas had fished Silver Creek in Idaho and the Rio Grande in Texas and everywhere in between. "I would rather hook a one-pound rainbow with a dry fly on a 3½-ounce rod," Douglas wrote, "than a four-pounder with bait or hardware."[10]

Douglas was a crusader for protecting treasured landscapes. Using the *New York Times* and the *Washington Post* as his forum, Douglas argued wholeheartedly that conservationists had to battle to save forests, lakes, canyons, and rivers from industrialization. For a CEO, dealing with Douglas on environmental protection laws had all the appeal of shaving with a blowtorch. Scolding, steely-eyed, and intolerant toward polluters, Douglas was always willing to be a lone vote on the Supreme Court when a case involved protecting America's natural heritage.[11] For a long time he dreamed about exploring the tussock tundra, which swept across Arctic Alaska and which reminded him of the Scottish moors. "I had seen this tundra on an earlier trip stretching from the north side of the Brooks Range to the Arctic Ocean," Douglas wrote. "That tundra, though differing in botanical detail from the tundra of the Sheenjek, has the same general appearance. It is in the main a dwarf-shrub heath marked by tussocks, and it runs for miles and miles."[12]

Prior to the Sheenjek Expedition, Douglas had been in fairly regular touch with Olaus Murie about finding new energy sources for America before all the rivers were dammed and the glaciers melted. This was another one of his hobbyhorses. No matter how long he lived in Washington, D.C., he remained a western individualist more comfortable in Goose Prairie, Washington, at the Double K Ranch than in the "marble palace" (as he called the Supreme Court). "We pay farmers *not* to produce certain crops," Douglas asked Murie. "Why *not* pay the Army Engineers *not* to build dams?"[13]

Olaus concurred with this idea, because he believed that hydroelectric power would become obsolete in the coming decades. As Douglas had made clear in *My Wilderness*, he wanted America to shake off its addiction to fossil fuels. "We are, indeed, on the edge of new breakthroughs that will open up sources of power that will make it unnecessary, and indeed foolhardy, to build more dams across our rivers to *produce power*. Hydrogen fusion, with an energy potential that is astronomical, has not yet been mastered. But it certainly will be. Solar energy, though not yet available by commercial standards, is in the offing. Nuclear fission already exists and promises energy supplies."[14]

Seldom has America produced a man more unnervingly prescient than Douglas. While the politicians of the cold war era were counting nuclear stockpiles and the agriculturalists were spraying crops with DDT, Douglas was envisioning a future in which U.S. citizens would find themselves estranged from the land, sadly living in what Michael Frome called "a shell of artificial, mechanical insulation." The great tragedy of postmodern America, Douglas believed, was that our children had lost contact with the environment. "We allow engineers and scientists to convert nature into dollars and into goodies," he said. "A river is a thing to be exploited, not treasured. A lake is better as a repository of sewage than a fishery or canoeway. We are replacing a natural environment with a synthetic one."[15]

Few American politicians look out for the long-term public welfare anymore—Douglas did. In the herd of sheep in Washington, D.C., Douglas was an iconoclastic visionary who never had a dull thought. The gossips of Georgetown tried to attack his character, mocking him for his divorces, scoffing at his promotion of Arctic Alaska, belittling him for including a long riff about the rattlesnakes of eastern Washington in his memoir. Conventional wisdom was tough on Douglas. But in the end he was one of the great men of the twentieth century, a champion of individual rights and of freedom of speech in a world dominated by corporate thinking. Fearless in his appraisals and always aware of the big picture, he asked the key questions about the arrogance of the industrial-military complex, angry that technocrats, in defiance of God, thought they could conquer nature with concrete monstrosities. Douglas believed that being outdoors in clean air reduced eye irritation, helped the respiratory system, and kept the blood pressure down. Even plants in offices, he said, reduced human stress.

"We have no conservation ethic," Douglas wrote in dismay of the U.S. government's refusal to rein in corporate abuse of landscapes and waterways. "Individuals in the bureaucracy understand it; but few bureaus practice it. America is dedicated to the dollar sign and the pressure of the Establishment on any of these bureaus is overwhelming. We get our oxygen for breathing from the green plants. Who is the guardian of the rate of combustion versus the rate of photosynthesis? Certainly no one in Washington, D.C." [16]

Some other Supreme Court justices have seemed to become parched, dull husks, but Douglas was always alive to the wind, sky, and grass. Donning a Stetson hat and western-style coat, insisting on going without a necktie, Douglas looked like a frontier character. "Bill was a genius and a visionary," Charles Reich, a law clerk to Justice Hugo Black, said. "He had the ability to take you to the top of the mountain and show you the entire vista of future issues, but then you would come down from the mountain, and lose sight of what you had seen. He never did." [17] Some critics tried to impeach Douglas because he wrote a controversial piece for the journal *Evergreen* (which published the work of rebels like Jack Kerouac and Terry Southern) or gave too many public speeches for compensation. But no matter how hard his opponents tried, they never did remove Douglas from the bench. Senator William Langer of North Dakota, late in life, came up to Douglas and wrapped an arm around him. "Douglas, they have thrown several buckets of shit over you," Langer said. "But by God, none of it stuck. And I am proud." [18]

II

Outdoors excursions, especially in the expansive North, are usually jolly when the weather cooperates and people share an interest in the ecosystem. The Sheenjek Expedition of 1956 was one of those trips on which people consider even cones of dried mud and cotton grass worth discussing. Hiking across the tundra was like walking on a sponge—it was hard to get into a rhythm because of ground squirrel holes or clumps of lichen. For once, in the roadless Arctic Range, afforestation was discussed instead of deforestation. Everybody was measuring everyone else's depth of spirit—not the accoutrements of success. Justice Douglas had no higher rank than tin plate cleaner after supper. Regularly Douglas deferred to Schaller on talus slopes; to Krear on the grizzly's hunting habits; to the Muries on caribou calving; and to Kessel on ring-billed gulls. There was never a pecking order when Douglas was in the wilderness. Also, to Douglas complaints were a tedious nuisance for everyone and undermined the serenity essential to endurance while camping. Decades of hiking had taught Douglas a basic lesson about the outdoors: be humble and do your proper chores. "I heard horrible stories of the mosquitoes of Alaska and went prepared with head nets," Douglas recalled. "But I never used them. There are mosquitoes—many of them. Even after a frost—one of which we experienced—new crops of mosquitoes are born. They swarm up out of the marshland and tundra. They are not too bothersome when the wind blows." [19]

Early on the expedition Mardy Murie, wanting to be gracious, said, "Justice Douglas, will you have some soup?" Furrow-browed, he glowered at Mardy, as if insulted, and said coldly, "Bill." A little while later Mardy innocently said, in her cheeriest voice, "Justice Douglas, can I make you a cup of cocoa?" Clearly perturbed that she hadn't gotten the message the first time, he gave her his blue gaze treatment and a single syllable: "Bill." Some evenings Douglas would pour a little bourbon into his hot chocolate to help him stay warm.

Meals on the Sheenjek Expedition weren't fancy, but the party ate like kings: caribou steaks, cheese rice, and corned beef, with blueberries, Fig Newtons, Jell-O, and angel food cake for dessert. Douglas was particularly interested in hiking to wherever ice presented itself. With field glasses he also scoured the Arctic landscape looking for the great bull caribou, which Bob Marshall had described. Up close—down on his hands and knees—Douglas examined lily plants, buffalo bush berries, and poppies. With field glasses he watched a fox eating blueberries. Douglas found bog cranberries—a tiny creeping plant with thin stems that threaded its way over sphagnum moss and was ideal for making jam. The fields shimmered in the fresh Arctic air. "What impressed me most," Murie recalled in *Two in the Far North*, "was the far-ranging interest of this man of the law. What a divine thing curiosity is!" [20]

The Muries had timed the expedition perfectly until about the second or third week in June. The rivers in the Brooks Range were snow-fed for part of the year, but then, about the time of the Douglases' arrival, the waterways of summer would be fed either by springs or by rain runoff. The largest river in the Brooks Range—the Colville—was far to the northwest of the Last Lake camp. The Sheenjek was a south-side river that drained south into the mighty Yukon River. It was lined with black spruce, birch, and alder brush (as thick as bamboo). When Douglas caught grayling along the Sheenjek, he'd cook them at night with alder wood, perfect for smoking fish. "These grayling, which run up to three pounds or more, are not prospering," he wrote. "Their small heads and broad-beamed bodies make them seem a bit awkward compared to our streamlined rainbows. But whatever they lack in grace they make up for in food. Their flesh is white and their thick steaks cook up into a sweeter and more delicious dish than any trout I have sampled." [21]

Douglas understood that there was a thread that began with Theodore Roosevelt and ran to Charles Sheldon and the Muries in Alaska. Saving the Brooks Range and the coastal plain of the Beaufort Sea aroused a kind of tribal passion in serious outdoors enthusiasts. They believed that this part of Alaska was the biological heart of North America. Although George L. Collins liked to use the term *recreation*, the word was inadequate to describe the hardiness and intensity of the Sheenjek expedition. All day long, well into the evening, the members kept busy identifying birds and wildflowers. Each party member believed deeply that Arctic Alaska belonged to the wildlife. Philosophically, the members were all aligned with the Gwich'in elders. As the Muries and the others set up base camps and collected bones and antlers among the caribou calves, the Arctic made them feel like little cogs in the huge machine of the modern world. The humbling effect of feeling small helped to develop character. Forget the judge's black robe: Douglas was nothing more than a grain of sand or a falling leaf.

There is no transcript of the conversations that took place between Justice Douglas and the Muries when they camped together in the Arctic

Range. But since everybody in the Sheenjek River party considered himself or herself a New Deal liberal, any banter about President Eisenhower couldn't have been complimentary. After all, Eisenhower had meant it when he said on the campaign trail in 1952 that he planned to restore the Republican Party's land policy in the West to help business. As president he had cleaned house, removing New Deal conservationists from the Department of the Interior. Without much concern about pension plans, he retired longtime employees of the National Park Service early. Friends of "big oil" and "big timber" were brought into the Forest Service. The attitude at both Interior and Agriculture favored leasing public lands. But new U.S. senators—like Hubert Humphrey of Minnesota—stepped into the picture, promising to give new lands protected status. Congressmen were defending wild places against an administration bent on helping the extraction industries in the West. Crunching across the tundra, putting on rubber boots to cross creeks, Douglas embodied the ethos of *A Sand County Almanac*. Getting an Arctic tan—neck-up, elbows-down—Douglas would talk, while hiking, about "man's responsibility to the earth." [22] At least, the Federation of Western Outdoor Clubs—influenced by Bob Marshall's *Alaska Wilderness*—urged Congress to create a "National Wilderness Preservation System." [23]

Justice Douglas and the Muries were particularly disturbed that Douglas McKay, a Chevrolet dealer from Oregon, had been confirmed as secretary of the interior. He was called "Giveaway McKay." In Alaska alone he had opened up the Tongass, the Chugach, and even TR's federal bird reservations to oil and gas leasing. The Arctic, to McKay, was worthless except as an oil field. McKay had learned to be genial from selling Chevys to customers; but his undersecretary, Ralph Tudor, was ruthless and enamored of Joe McCarthy—a narrow-minded conservative who wanted to purge the Department of the Interior of "wilderness screwballs" and "rabid New Dealers." When Justice Douglas and the Muries, along with numerous conservation groups, vociferously disapproved of desecrating Dinosaur National Monument by building a dam at the confluence of the Green and Yampa rivers, McKay retorted that wilderness "punks," communist types, cared more about Colorado's rivers than they did about hardworking people. David Brower, executive director of the Sierra Club, testified before Congress against McKay, showing photos of what had happened to Hetch Hetchy. "If we heed the lesson learned from the trag-edy of the misplaced dam in Hetch Hetchy," Brower argued, "we can pre-vent a far more disastrous struggle in Dinosaur National Monument."[24]

Certainly political conversation was in the air that summer of 1956, but none of the seven on the Sheenjek expedition gave many details. The jour-ney had an unexpectedly spiritual feel. In Athabascan-Inuit cosmology, animal species like the bear and the caribou were once humans. To cut down a white spruce or to shoot a trumpeter swan for no *essential* reason was considered a crime against the creator.[25] Perhaps Olaus Murie—who considered exploration the most profound intellectual activity known to man—summed up the Sheenjek River experience best when he simply wrote, "Here we found nature's freedom."[26] The short summer intensified the awareness that warmth in the Arctic was only a brief respite from the cold, that light was always followed by a deep, long darkness. This mood, Murie knew, dominated the land and everything living in it.

All the members of the expedition did publish articles in various pe-riodicals, including *Alaska Sportsman* and *National Parks*. Mardy Murie would use her Sheenjek diaries quite extensively in her memoir, *Two in the Far North*, published in 1962. Olaus had taken fine photographs of the Sheenjek Valley over the summer and was prepared to give public slide presentations throughout the Lower Forty-Eight. Olaus had collected cutting-edge biological information about Arctic Alaska to share with the U.S. Fish and Wildlife Service.[27] The Muries, in fact, had so much fun that they considered the expedition their second Arctic honeymoon.

Living Wilderness published a detailed account of the Sheenjek expe-dition of 1956 under the heading "Alaska with O. J. Murie." Murie began by praising Dr. Brina Kessel of the University of Alaska for document-ing eighty-five birds, but it was the spirit of William O. Douglas that per-vaded this account. Clearly, having a man of Douglas's eminence on the Sheenjek River was extremely encouraging. "I was impressed with the sincere motivation of this author of books such as *Of Men and Mountains* and *Almanac of Liberty*," Murie wrote. "And I feel fortunate in having on our Supreme Court a man of his honest outlook, and one who so loves the mountains and virile outdoor living."[28]

Clearly, Douglas had been enraptured by the snowcapped Brooks Range and the virgin Sheenjek River. His upbeat report on the Arctic Range as a wilderness area had a dramatic effect on the entire conservationist community. "This is—and must forever remain—a roadless, primitive area," Douglas said, back in Washington, D.C., about what became the Arctic National Wildlife Refuge, "where all food chains are unbroken, where the ancient ecological balance provided by nature is maintained."[29] George L. Collins expressed the prevailing opinion in conservation circles when he observed that Douglas's participation in the Sheenjek expedition was crucial, because that "goofy bird" from the Supreme Court had a name that was "sterling" and "magic" in the corridors of power in Washington, D.C.[30]

Douglas had left the Sheenjek Valley convinced that it should be preserved as a primitive park. It was an Arctic Eden where whales blew, grizzlies stalked, and caribou roamed freely. If President Harding could make a National Petroleum Reserve for the navy in 1923, Douglas didn't see why President Eisenhower couldn't declare an Arctic Range by 1960. Back in Georgetown, Douglas, who always wanted to keep the public estate out of corporate hands, started writing *My Wilderness: The Pacific West*. Its opening chapter was about the Sheenjek expedition with those amazing Muries. With Aldo Leopold and Bob Marshall gone, Douglas, a man of keen political instinct, knew he had to step up his own advocacy. Presidents dating back to Benjamin Harrison and Grover Cleveland had favored creating new forest and wildlife reserves on their way out of office; it gave them a few final good deeds for the historians to tally. Collins, Douglas, Sumner, and the Muries were all calling for an Arctic Wildlife Range, as were Alaskans such as Virginia Wood and Celia Hunter.

"The Arctic has strange stillness that no other wilderness knows," Douglas wrote of his experience on the Sheenjek expedition. "It has loneliness too—a feeling of isolation and remoteness born of vast spaces, the rolling tundra, and the barren domes of limestone mountains. This is a loneliness that is joyous and exhilarating. All the noises of civilization have been left behind; now the music of the wilderness can be heard. The Arctic shows beauty in this bareness and in the shadows cast by clouds over empty land. The beauty is in part the glory of seeing moose, caribou, and wolves living in a natural habitat, untouched by civilization. It is the thrill of seeing birds come thousands of miles to nest and raise their young. The beauty is also in slopes painted cerise by a low-bush rhododendron, in strange mosses and lichens that grow everywhere, and (to one who gets on his hands and knees) in the glories of delicate saxifrage, arctic poppies, and fairy forget-me-nots. The Arctic has a call that is compelling. The distant mountains make one want to go on and on over the next ridge and over the one beyond. The call is that of a wilderness known only to a few. It is a call to adventure. This is not a place to possess like the plateaus of Wyoming or the valleys of Arizona; it is one to behold with wonderment. It is a domain for any restless soul who yearns to discover the startling beauties of creation in a place of quiet and solitude where life exists without molestation by man." [31]

∘ ∘ ∘ ∘ ∘ ∘

Chapter Nineteen

DHARMA WILDERNESS

I

T*he silence is so intense that* you can hear your own blood roar in your ears but louder than that by far is the mysterious roar which I always identify with the roaring of the diamond of wisdom," Jack Kerouac wrote in *The Dharma Bums.* "The mysterious roar of silence itself, which is a great Shhhh reminding you of something you've seemed to have forgotten in the stress of your days since birth." Kerouac had never made it to Alaska on any of his cross-country treks in North America. But his 1958 novel, *The Dharma Bums*, based on his hikes in northern California and the Pacific Northwest with the laid-back poet Gary Snyder (Japhy Ryder in the novel) brought the wilderness movement to a whole new audience. Insisting that poets needed to learn the biological names of trees, plants, and animals, Snyder became a major voice for making ecology interdisciplinary.[1] Not since Muir had America produced a visionary so innovative in defense of wild nature as Snyder. "Is it all lost?" Snyder asked about nature in the atomic age. "Was it ever real? A world where men and women, trees, grasses, animals, the wind—were at ease with each other's songs?"[2]

Snyder was born in San Francisco in 1930, but his family moved to Lake City, a suburb of Seattle, when he was two years old. To survive during the Great Depression his parents had turned to subsistence farming: milking cows, mowing hay, collecting eggs, picking apples, and chopping cedar. Eventually Snyder's family moved to Portland. Shortly thereafter his parents divorced. As a teenager Gary was hired by a newspaper, the *Oregonian*, as a jack-of-all-trades. Like so many Depression-era children on rural farmsteads, he learned to survive economically on very little. He never shrank from a hard day's work. When Snyder was fifteen, in the summer of 1945, he climbed the volcano Mount Saint Helens. The next year he climbed Mount Hood. By the time Snyder turned twenty-two he had climbed Mount Hood many times. Extreme mountaineering was Snyder's favorite sport. He loved to climb. To Snyder, reaching a summit was an expression of ultimate freedom. His two summers as a fire lookout, on Crater Mountain (1952) and Sourdough Mountain (1953)—which together, in 1968, became North Cascades National Park—helped him contribute a new wilderness ethos and an ecological aesthetic to the cultural phenomenon known as the Beat Generation.

Snyder had a scholarly bent and an intense interest in Native American history, and he managed to win a scholarship to prestigious Reed College in Portland. Earning A's in English literature, an all-around excellent student, he spent his free time at the nonprofit Mazamas clubhouse on the top floor of Portland's Power and Light Building. The Mazamas sponsored alpine hikes and climbs all over the Pacific Northwest from Mount Baker in Washington to Mount Shasta in California. Snyder, using the club's library, studied the history of mountaineers in the Cascades, learning useful information from their firsthand accounts.[3] Since 1894 the club had been a leader in conservation in the Pacific Northwest, fighting to save Crater Lake and the North Cascades from over-timbering. Snyder also joined the satirical Regressive Party (whose slogan was "Back to the Neolithic").[4] The only *real* politics that Snyder and his friends engaged in was trying to get William O. Douglas to run for president in the Democratic primary in Oregon.[5] "Marshall, Yard, Douglas, and those guys were my animating force," Snyder recalled. "I joined The Wilderness Society at seventeen. And I received *The Living Wilderness*, which automatically came with membership. I was already mountain climbing with the Mazamas Club of Portland. Bob Marshall was a socialist, with very liberal ideas, and everything he had written about roadless areas made absolute sense to me. It still does."[6]

What Snyder admired about The Wilderness Society was that it worked closely with Native Alaskans and other allies to ensure that local voices were heard in the public debate over public lands. Snyder, even in his teens, wanted to prevent "big timber" from taking over the entire Alaskan territory and Pacific Northwest. Reading about the early explorations of the Rocky Mountains during the 1850s, Snyder came to admire rough-hewn mountain men such as Jim Bridger; they were intrepid, and they knew how to "read" nature as the Cayuse or Paiute did. But Snyder saw the "second wave"—the stockmen, timbermen, mine operators, and sheep ranchers—as pillagers and despoilers. They bought and sold nature's wonderful patrimony.

While Snyder was growing up, between 1947 and 1951, The Wilderness Society and the Mazamas Club were leading a campaign to designate the Cascade Mountains (from Mount Saint Helens in southern Oregon to the Skagit Mountains in north Washington and up to British Columbia) as Ice Peaks National Park. But many Washingtonians saw Ice Peaks as a land grab by Harold Ickes. Bob Marshall, along with Ferdinand Silcox, director of the U.S. Forest Service, insisted that this park would protect the Cascades from desecration. Marshall, working for the U.S. Forest Service, was able to save parts of the northern Cascades in the early 1930s: Glacier Peak Recreation Area (230,000 acres) and North Cascades Primitive Area (800,000 acres). But politicians in Oregon and Washington couldn't or wouldn't take on the lumber giant Weyerhaeuser. By the time Snyder climbed Mount Saint Helens and Mount Hood—which can be described as sentinel towers of the Pacific Northwest—a postwar housing boom was under way, timber was in high demand on the market, and the concept of Ice Peaks National Park was shelved.[7]

In 1951 Snyder earned his BA in literature and anthropology from Reed College; he then continued drifting around the Pacific Northwest in blue jeans and a zip-up rain jacket, working as a camp counselor, carpenter, and logger. Sometimes he would look for red-winged blackbirds (*Agelaius phoeniceus*) and owls in the Columbia Slough, using Roger Tory Peterson's *A Field Guide to Western Birds*.[8] His nomadic yearnings were inspired by Woody Guthrie's life and music. At heart Snyder was an itinerant poet with a deep love for mountain trails and for the Industrial Workers of the World, or IWW—the Wobblies—and their lore. Much like Bob Marshall, he was equally comfortable with bookish academicians and with working-class people whose creed was self-sufficiency. Open-minded, uncorrupted by conformity or by the consumer culture, Snyder labored as a timber scaler at the Warm Spring Indian Reservation in central Oregon and—determined to be a poet like Robinson Jeffers and William Carlos Williams—started developing a new, sparse style of poetry: no word was wasted. He pledged to treat the planet with respect, as the North American Indians did, and he had an intuitive understanding, reinforced by his long treks into the North Cascades, that earth was a holy, living being, a single entity. But his poetry was also informed by biology, forestry, socialism, Buddhism, Paul Bunyan, and Native American customs. Snyder treated animals with particular kindness and gentleness, like such earlier, pioneering advocates of animal rights as John Quincy Adams, John Burroughs, and Henry Bergh. One of his close friends at Reed College was Martin Murie, whose parents were Mardy and Olaus Murie.[9]

In June 1952, the twenty-two-year-old Snyder started working for the U.S. Forest Service at Marblemount, Washington, in the northern Cascades, where there was evidence of ancient volcanic upheaval in all directions.[10] This was the Skagit district of Mount Baker National Forest (sometimes called America's Alps). The North Cascades had about 300 glaciers; only Alaska had more. Having experienced many YMCA summer camps at Mount Saint Helens and many trails in Columbia National Forest (renamed Gifford Pinchot National Forest in 1949), Snyder was erroneously convinced that to be a fire lookout was rather easy work, far simpler than scaling timber: that he would get to live in splendid isolation in the Cascades, would call headquarters on a Motorola PT 300 radio if he saw smoke rising from a distant burn, and meanwhile would read books for pay. But Snyder's dream soon came up against reality. "Boy," a forest ranger warned him when he showed up for duty, "you have no idea what you've gotten yourself into."[11]

At one time, many youngsters wanted to be Daniel Boone or Kit Carson—outdoorsmen who could track a whitetail deer (*Odocoileus virginianus*) and survive in a blizzard. Snyder's boyhood idols were John Muir and Ernest Thompson Seton. The Sierra Club had done a marvelous job of presenting Muir as a lovable long-bearded prophet of the wild kingdom. "Muir inspired me, as a lad, on the practical level of boldly going out and staying longer in the woods with less gear, and having the nerve to do solo trips," Snyder recalled. "So I did (for example) some lengthy trips in the

summer of 1948 in the mountains north of Mt. St. Helens in the Washington Cascades, including some third-class rock scrambles."[12]

Snyder had been assigned to the Granite Creek guard station, high up
near snowcapped Crater Mountain, for the summer of 1952. In the winter, its rocks looked like blocks of ice and there was scant vegetation. But
in the summer, this part of the North Cascades was invigorated with life.
To get to the little ranger shack at Granite Creek, Snyder had to hike fifteen miles from the roadhead, into the primeval forest. The job called for
an outdoorsman, able to clear trails through thickets, chop wood, and
haul in hay from settlements lower down the mountain. Forest rangers
throughout the Cascadian interior laughed at the skinny kid from Reed
College, who was still trying to grow his first beard but who had actually volunteered for the desolate fire tower in the North Cascades. Over
the summer, Snyder learned that miners and loggers were marvelous
characters but poor stewards of the ancient forests. All over Washington, Weyerhaeuser—one of the largest pulp and paper companies in the
world—was engaged in speed-logging, which was profitable for communities that relied on timber. (However, the area where Snyder was—the
upper Skagit—was too steep for Weyerhaeuser to menace.) Worse, the 655-
foot Ross Dam—constructed between 1937 and 1949—was ruining the environment of the North Cascades in order to generate electric power for
greater Seattle. Snyder later wrote in *A Place in Space* about his worries
over "mineral exploitation" and his wish that miners and loggers could
learn to "make deeper connections to the earth."[13]

Snyder's home in the North Cascades was a cedar-log miner's cabin
dating from about 1920 but since remodeled. It belonged to the prospector Frank Beebe, who had looked for gold nuggets along Ruby Creek and
then, desperate for a paycheck, had lit out to work with the fishing fleets
of Alaska. After a few years bouncing around among salmon canneries on the Kenai Peninsula, Beebe gave up on Alaska and drifted back to
the Cascades. He remodeled his little cabin on Granite Creek and tried
breeding ermines and marten, with no luck. The U.S. Forest Service soon
hired him as a lookout. Upon retiring in the 1950s, Beebe moved to Bell-
ingham, Washington, a lumber town that was also a gateway to magnificent beauty.[14] His cabin at Granite Creek was now fitted out to become
the "guard station." Snyder was enthralled with the primitive conditions

and was proud to call this shack home. Every night, after an arduous day's work, Snyder would stay up late reading Po Chü-i—a poet of the Tang dynasty—by oil lamp. Nearby Canyon Creek was his companion, offering cool water as a salve. And Bob Marshall's teachings about the wilderness stayed with Snyder that summer. The bulk of his job was to make sure the backcountry hiking trails were free of debris. The young poet was fast developing his own outdoors philosophy. A journal entry from 1953 expresses his antagonism toward the "chop-chop-chop" concept of managing timberlands: "Forests equals crop / Scenery equals recreation / Public equals money. . . . The shopkeeper's view of nature."[15]

Although Snyder didn't write much poetry in the North Cascades, he kept a subtle, intelligent literary journal, parts of which were later published as *Earth House Hold*.[16] To his great surprise, one of his first poems, "A Berry Feast" (1952), had become a favorite among his core friends in San Francisco. In it, he had written about coyotes' mischief and a "neat pile" of bear scat found on "the fragrant trail."[17] This long comical poem, first published in 1957 by the *Evergreen Review*, was celebrated for its ethnopoetic merging of traditions—Native American; Asian; *The Living Wilderness*—and helped develop the ecological dimension of the beat generation during the 1950s.[18] Poetry, Snyder would tell the *Anchorage Daily News* in the 1970s, was another way—like science—of seeing the natural world as it truly is. "Buddhism is one of the few religious and philosophical systems on a world scale that asserts the ethical value of the nonhuman," he said in that interview. "What Buddhism contributes to environmental politics is a profound spirit of compassion. In the Buddhist's view, everything in the world has value, has authenticity. Ultimately, this goes beyond humans and animals and is an attitude of regard toward rocks, plants, clouds. Do you objectify and commodify the world when you look at it? Or do you see it as worthy, as beautiful, as full of its own intrinsic value?"[19]

There was nothing flaky about Gary Snyder. Even while tramping around America he found time to take graduate courses at Indiana University. He always considered nature a cure for the depression induced by society. To Snyder solitude was to be relished, not merely endured. And people should always be treated with generosity, kindness, and *namaste*, a bow of respect. Ever since childhood, Snyder had valued secret hiding places in the deep woods, as if he were a hobbit.[20] Inspired by Japanese mentors, he taught himself Zen meditation and, as noted above, read Chinese poetry of the Tang dynasty. Perhaps remembering the Portland YMCA credo "I am third" (God is followed by loved ones and then by oneself), Snyder perfected the art of humility.[21]

Snyder had also become infatuated with D. T. Suzuki's *Essays in Zen Buddhism*. By day he protected the North Cascades; by night he read Suzuki. A central message of Suzuki's was, "In Zen there must be *satori*; there must be a general mental upheaval which destroys the old accumulations of intellectuality and lays down a foundation for a new faith."[22] Snyder and his best friend, the poet Philip Whalen, liked to exchange Zen traditional sutras, koans, and sermons by Buddhist teachers. So here was Snyder, working for the Forest Service in Washington state and dreaming about the silent world of places like Alaska and finding inspiration in the Heart Sutra, the Diamond Sutra, and the Lankavatara and Surangama sutras. Before Snyder, few poets had considered the western wild places from a Buddhist perspective. Primitive, roadless areas, Snyder now believed, emptied the mind. A desolate peak, to him, became a prayer mat. Sitting cross-legged, he repeated the old mantra, *Om mani padme hum*, and drank green tea from a handle-less cup.[23]

It was during this first summer in the North Cascades that Snyder read the Platform Sutra of Huineng. The contemplative Huineng would have made a good member of The Wilderness Society. Born into a minority clan in southern China, he became a Buddhist wanderer, avoiding envious monks, sleeping in caves, and cunningly eluding pursuers. His philosophical reflections became known as *prajna* wisdom. Snyder took from him the notion that wilderness wasn't a commodity and that universities weren't the real places of learning. The North Cascades were Snyder's college; and enlightenment could be found in a spruce branch, a smooth rock, or a butterfly. Huineng taught awareness of all living things. And the combination of reading Huineng's meditations and being alone in the North Cascades freed Snyder from money-driven America.

After a few weeks in the cabin at Granite Creek Snyder packed his rucksack with provisions—including brown rice and soy sauce—and headed up Crater Mountain. His work as a lookout was about to commence. He would be watching 3 million acres of measureless mountains. His nickname, given to him by a district manager of the Forest Service, was "the

Chinaman." Snyder wore the appellation as a badge of honor. He carefully studied old-growth conifers, huge stands of Douglas fir, and ponderosa pine stretching high up into the sky. Alaska had taller peaks above the timberline, but in the North Cascades a climber saw 7,000-foot summits jutting up like teeth. Stepping as sure-footedly as a Dall sheep through snow piled up against boulders, Snyder became one with the mountain; the immense void of the North Cascades engulfed him. "Aldo Leopold uses the phrase *I think like a mountain*,'" Snyder recalled. "I didn't hear that until later, but mountain watching is like mountain being or mountain sitting. How do you watch a mountain? Nothing's going to happen in any time frame that you can consider—except the light changes on it. And so that was my mountain watching." [24]

Snyder felt that at the top of any desolate peak, the trail died and a dreamscape began. Summits were the end of the earthly road. Nature was supreme. On top of Crater Mountain, in the clear air, with only his two-way radio to connect him to the world, Snyder served admirably as fire lookout, but he was also filled with thoughts. His visions from Crater Mountain soon became an impetus for the beat generation, a spiritual reawakening based on a nonconformist attitude toward the military-industrial complex of the 1950s. Around this time, David Brower of the Sierra Club and Howard "Zahnie" Zahniser of The Wilderness Society toured the North Cascades, with the photographer Philip Hyde. Together they conceived of the "American Alps" campaign to save the Washington range as a new national park. [25]

Zahniser had become a legend in The Wilderness Society. While others went on hikes and picnics in the Sierra, he stayed deskbound. An advocate of the Adirondacks "forever wild" movement, enacted through legislation in 1895, Zahniser committed himself to protecting wilderness for "the eternity of the future." [26] A bureaucratic infighter, one of the sharpest lobbyists in Washington, D.C., Zahniser ceaselessly championed creating wilderness areas on public lands. Starting in 1935, he wrote a column for *Nature*. In 1945, he was asked to be executive director of The Wilderness Society; it was a post he kept until his death on May 5, 1964. Four months later, President Lyndon Johnson signed the Wilderness Act of 1964—a milestone in land protection—originally drafted by Zahniser; 9.1 million acres were saved as "untrammeled by man" zones.

Snyder called his lookout shelter—a prefabricated structure built by the CCC at 8,128 feet—"Crater Shan," Chinese for high point. Emptying himself of ego and pretension, he basked in its utter commonness. Snyder recognized anew, in the North Cascades, that money-consciousness, the reigning motivator in postwar America, was counterproductive. Withdrawing national forests from preservation, he feared, would lower water tables and accelerate the process of erosion. "Who can leap the world's ties," Han Shan had asked in a poem Snyder later translated. "And sit with me among the white clouds?" [27]

Snyder relished his Zen hermitage. He kept his ax sharp. Chinese calligraphy and meditation were part of his daily regimen. Insatiably he read the texts of Mahayana Buddhism. Some mornings his little shelter was awash in fog. On a clear day, however, he could almost see the Hope Range of British Columbia in the far distance. As an old Zen saying went, everything was "blue heaped on blue." In the center of his cabin was an Osborne fire finder, a rotating dish map with a peep sight; it could see over far ridges in all directions. Snyder hung Tibetan prayer flags on his walls. After having climbed Mount Hood numerous times, he had developed a pantheist attitude toward mountains as living entities; Aldo Leopold would have approved. Snyder was disdainful of the "hostile, jock Occidental mind-set" prevalent in Europe and the United States, the idea that mountain climbing was an act of *conquering*. "I want to create wilderness," Snyder was fond of telling friends, "out of empire." [28]

II

Deeply attuned to his surroundings, Snyder learned, that summer in the North Cascades, how strange being alone in the wild can be. Unlike Robinson Jeffers, the great nature poet of the California coast who enjoyed interacting with seabirds and raptors more than with people, Snyder, perhaps because he was reading Buddhist texts on Crater Mountain, craved people when he came down from his lonely post. The essayist and novelist Edward Abbey, in *Abbey's Road*, wrote of his own experiences as a paid fire lookout in the Southwest: "Men go mad," he said, "in this line of work." Abbey imagined a married couple getting assigned by the U.S. Forest Service to fire-watch together in the North Cascades:

"Any couple who survives three or four months with no human company but each other are destined for a long permanent relationship," he wrote. "They deserve each other." [29]

Committed to forestry, Snyder signed up to be a lookout again in June 1953; this time Sourdough Mountain was his assignment. Joining Snyder that summer in the North Cascades was another graduate of Reed College, Philip Whalen, whom Kerouac described in *The Dharma Bums* (under the name Warren Coughlin) as "a big fat bespectacled booboo . . . a hundred and eighty pounds of poet meat." [30] After serving in the U.S. Army after World War II, Whalen visited the Vedanta Society in Portland, his hometown, and became interested in eastern religions. Whalen had brought with him to Sank Mountain Ezra Pound's *Cantos* and William Blake's *Poems*, and he bragged of "absorbing" vitamins out of these volumes in the North Cascades. Also, Snyder had introduced him to D. T. Suzuki's books on Zen. Snyder and Whalen—who talked by radio from their respective peaks—were paid a handsome $700 a season for being lookouts. At Sourdough, as at Crater, Snyder had an Osborne fire finder in the middle of the all-purpose room. "Sourdough Mountain is very sweet," Snyder recalled. "It's a beautiful alpine environment." [31]

Snyder brought with him to Sourdough Mountain in 1953 a rucksack full of his own dharma literature that included Daito Kokushi's *Admonition*, William Faulkner's *Sartorius*, and Margaret Mead's *Coming of Age in Samoa*. Like Rockwell Kent on Fox Island, Snyder kept a detailed chart of William Blake's cosmology in his cabin. In Snyder's journal of 1953 is a passage from Blake's *The Marriage of Heaven and Hell*: "If the doors of perception were cleansed everything would appear to man as it is, infinite. For man has closed himself up, till he sees all things through narrow chinks of his cavern." [32] According to the biographer John Suiter, author of *Poets on the Peaks*, Snyder wrote next to this passage a simple, "Ah." [33]

The question Snyder and Whalen were asking that summer of 1953 in the North Cascades was whether modern societies were capable of living in harmony with nature. Did Americans have the ability to say no to the extraction industries? Would man destroy the planet Earth and move on to a different solar system? L. Ron Hubbard and the Scientologists thought so. World War II had brought new mechanized terrors—culminating in the atomic bomb. Many lovers of Earth wondered whether the apocalypse was at hand. Whalen, who became a Zen monk in 1973, believed that wilderness sanctuaries, where *quiet* ruled, were essential to rejuvenate an America that Henry Miller had derided as an "air-conditioned nightmare." Whalen wrote poems with the sparse energy of Basho's in the early stages of *zazen* (Zen Buddhist meditation). During his time in the North Cascades, Whalen wrote poems that would later be collected as *Canoeing Up Carbarga Creek: Buddhist Poems 1955–1986*, most of them concerning nonattachment as the mind drifts through the cosmic world.[34]

The modernist poet Robinson Jeffers cast a constructive spell over the thinking of both Snyder and Whalen. A Pennsylvanian by birth, Jeffers had gotten married in 1913 and constructed the granite Tor House and Hawk Tower in Carmel, California, overlooking the Pacific Ocean. At the core of Jeffers's long verse narratives, some resembling Greek tragedies, was his philosophical belief in *inhumanism* (the idea that humans were egoists: self-centered and unable to grasp the "astonishing beauty" of the natural world). Jeffers wanted poets to shift the emphasis of their verse from "man to notman," and urged the "rejection of human solipsism and recognition of the transhuman magnificence." Jeffers's poetry— particularly lines such as "long live freedom and damn the ideologies" (from "The Stars Go over the Lonely Ocean") and "I'd sooner, except the penalties, kill a man than a hawk" ("Hurt Hawks")—pointed toward a new distrust of political authority and from an embrace of religious instinct that included respecting wildlife.

Amid fears of radiation and of McCarthyism, reading Henry David Thoreau's *Walden* on Sourdough Mountain must have been reassuring to Snyder. Thoreau held the key to the wilderness: *solitude*. He knew the feeling of "total removal" found at the top of the world because he had explored Mount Katahdin (in Maine) and Mount Greylock (in Massachusetts). As he wrote in *Walden*, the most interesting dwellings in America were the "humble log huts" and "cottages of the poor." Snyder, who felt himself part of the Buddhist cosmos, was happy living in exactly this type of primitive structure. The new environmental consciousness that Snyder hoped would sweep America during the 1950s seemed to come from a single line of Thoreau's: "A man is rich in proportion to the number of things he can afford to let alone."[35] Snyder, like The Wilderness Society, wanted to see the North Cascades left completely untouched by commercial development. Ironically, San Francisco became the urban center where this Thoreauvian philosophy found a suitable home. All around this area were natural mysteries: seal rocks, redwoods, multicolored pebbles shimmering like jewels on the ocean beaches. There was a certain pioneer "island mentality" in San Francisco—a sense that this city, bounded on every side by wilderness or the Pacific Ocean, was the end of the road.

Snyder hoped that the wilderness cause, supported also by the Mazamas Club, would take hold in both high art and pop culture on the West Coast. Instead of an elite movement—in which members of the U.S. Forest Service and the U.S. Fish and Wildlife Service wrote memos about primitive roadless areas, and people like the Muries occasionally had their work published in *Scientific Monthly* or the *Sierra Club Bulletin*—Snyder envisioned a revolution of youth consciousness would reject industrialization. Fear of nuclear annihilation and toxic pollution was the root of his thinking: some kinds of technology were not to be trusted.* Howard Zahniser wrote in *The Living Wilderness* that wild places like the Cascades and Arctic Alaska weren't a "disparagement of our civilization." Rather, they were "admiration of it to the point of perpetuating it." [36] Echoing Leopold, proponents of roadless wilderness like Zahniser spoke about a nature *aesthetic* instead of using the outworn terms preferred by the National Park Service: "scenic" and "wonder." [37] When Mardy Murie complained that Americans had an insatiable need for "comforts and refinement and things and gadgets," she was saying much the same thing. "Where is the voice to say," Murie asked, "look, where are we going?" [38]

When Snyder came down from Sourdough Mountain in the fall of 1953, full of pleasant thoughts, he moved to Berkeley. Hungering for further enlightenment, he enrolled in courses on Japanese and Chinese culture at the University of California–Berkeley. The Bay Area was swirling with creative energy. The dean of West Coast poets, Kenneth Rexroth, had recently published *The Dragon and the Unicorn* to great acclaim. Snyder thought it a great book.

After two years of intensive study Snyder needed a break. Wanting

---

* Snyder did like computers. He even wrote a poem for his Macintosh, designed by Apple Inc.

to connect with the spirit of John Muir, Snyder worked on a trail crew in Yosemite National Park in June–August 1955, writing his fine poem "Riprap" (first published in 1959 as the title poem of *Riprap*). Snyder later explained that *riprap* meant a "cobble of stone" that was "laid on steep slick rock to make a trail." He had learned it from master trail builders in the Sierra. To construct these stone trails took the skill of a mason and the precision of a surgeon. Snyder was paid $1.73 an hour working around Pate Valley and Pleasant Valley. Always frugal with money, he planned to spend a couple of months in San Francisco and then take a steamer to Japan to study with Zen Buddhist masters. And he started thinking a lot about Alaska: "My sense of the West Coast," Snyder said, "is that it runs from somewhere about the Big Sur River—the southernmost river that salmon run in—from there north to the Strait of Georgia and beyond, to Glacier Bay in southern Alaska. It is one territory in my mind. People all relate to each other across it; we share a lot of the same concerns and text and a lot of the same trees and birds." [39]

III

In the fall of 1955, Gary Snyder and Allen Ginsberg became fast friends in San Francisco. They were something of an odd pairing. Certainly, Ginsberg had a more urban disposition, writing poems about his Jewish roots, such as *Kaddish* in 1961. But Ginsberg was a fierce critic of Moloch. Rejecting the notion of America as a monoculture, Ginsberg chastised industry, whose "factories dream and croak in a fog" and whose "smokestacks and antennae crown the cities!" [40] Since the early 1920s lead components had been mixed into petroleum as antiknock agents, regardless of the toxic effects on humans. Ginsberg was aghast. Until his death in 1997, Ginsberg enjoyed hiking with Snyder in California and the Pacific Northwest. One afternoon, in the fall of 1965, they were exploring around Washington's Glacier Peak Wilderness Area, walking in rhythm with the chant "Hari Om Namo Shiva." Snyder had a Vandyke beard and a crew cut and wore a mountaineer's cap. Ginsberg had long curly hair flowing down from his balding head. A little group of fishermen looked at them incredulously. Ginsberg walked up to them. "Hello," he said, extending his right hand. "We are forest beatniks." [41]

Ginsberg and Snyder's friendship began during the fall of 1955. Ginsberg was taken with Snyder's calm, scholarly way. "He's a head, peyotist, laconist," Ginsberg wrote to a friend, "but warmhearted, nice-looking, with a little beard, thin, blond."[42] The poet Kenneth Rexroth, a polymath who had a regular arts-culture show on KPFA-FM, had booked them together for a reading at the Six Gallery in San Francisco. This art cooperative was run by young painters from the San Francisco Art Institute, who threw a poetry party that launched the beat movement on the West Coast. (From 9 a.m. to 5 p.m. the gallery was an auto repair shop.) Ginsberg had arrived in the Bay Area bearing a letter of introduction from the poet William Carlos Williams, making the acquaintance of Kenneth Rexroth, and bragging about his French-Canadian friend Jack Kerouac from Lowell, Massachusetts, whose novel *The Town and the City* (1953) had marked him out as the new Thomas Wolfe. For the reading at the Six Gallery, Rexroth was asked to be the master of ceremonies, as a gesture of respect for his many years of mentoring poets in San Francisco. Four Bay Area poets were asked to read: Gary Snyder, Michael McClure, Philip Whalen, and Philip Lamantia. Snyder was excited to share the stage with Rexroth, whose poems, including "Another Spring" and "Toward an Organic Philosophy," expressed his own wilderness ethos.

On October 7, 1955, the night of the famous reading at the Six Gallery, more than 150 people showed up, in a festive mood. Wine bottles were passed around. With the exception of Lamantia, who read poems by a deceased friend, the participants focused on the theme of humans reconnecting with nature. Philip Whalen contributed the comical "Plus Ça Change," which kindly mocked Americans' fear of touching each other, a reaction attributed to "alienation conditioning." Kerouac, who was working on his novel *On the Road*—about his cross-country trips in the late 1940s and early 1950s, often with his delinquent friend Neal Cassady, sat Buddha-like on the concrete floor of the Six Gallery, hooting and hollering, slugging down wine, as the "forest poets" read their compositions.

If the United States faced a spiritual crisis in 1955, McClure believed, it was because many Americans insisted that animals didn't have souls. To most Alaskans, for example, harpooning a whale, shooting a wolf on ranch property, and slaughtering polar bears for fun were economic propositions. McClure, whose poetry combined biology with mysticism,

challenged the reckless treatment of wildlife in his long poem "Point Lo-
bos: Animism." Biologists and physicists admired his poems. Drawing on
the scientific writings of Ernst Haeckel, who argued that all living entities were sacred, McClure hoped to teach Americans to treat ecosystems
with reverent respect. Native Alaskans, for example, thought themselves
equal to the polar bear, perhaps even inferior, but never better. "What
I was interested in was the intersection of science and poetry," McClure
recalls. "There was too much distance between them, when in reality they
have a lot in common."[43]

The breakthrough poem at the Six Gallery was McClure's "For the
Death of 100 Whales." McClure said that slaughtering whales was im-
moral. In April 1954, *Time* magazine had published an article about how
the U.S. troops stationed at a NATO airbase in Iceland had gone on a ram-
page, slaughtering whales en masse with machine guns. They killed 100
whales, causing a wave of blood to ooze across the choppy waters. Making
artistic use of this troubling story, McClure claimed that the cold-blooded
killers were the troops, not the innocent whales, and protested against
the carnage. The poem chastised the "mowers and reapers of sea kine";
the closing verse was:

OH GUN! OH BOW!
There are no churches in the waves,
No holiness,
No passages or crossings
From the beasts' wet shore.[44]

IV

When Allen Ginsberg, *bespectacled and brazen,* took the stage at
the Six Gallery, the bohemians in attendance whooped like
warriors. His underground reputation for poetic drama had
preceded him. While Ginsberg wasn't a nature poet, his long signature
poem "Howl"—exploding with shamanistic prophecy[45]—was a bardic
condemnation of modern city life, a fiery indictment of society's destructive forces. In *A Sand County Almanac*, Leopold had written that when a
wolf howled, it was "an outburst of wild defiant sorrow, and of contempt

for all the adversities of the world."[46] This was the insurgent Ginsberg at the Six Gallery, chanting with conviction, "I saw the best minds of my generation destroyed by madness, starving hysterical naked/dragging themselves through the negro streets at dawn looking for an angry fix."[47] With this apocalyptic poem, a new American consciousness—a paradigm shift—was happening.

Ginsberg's reading of "Howl" was the highlight at Six Gallery. His sizzling words would ricochet from San Francisco to Singapore and beyond for the next decade. Some critics believe the beat generation was born that evening, with Ginsberg boldly putting the modern condition on trial. But Kerouac didn't see it that way. Long before Ginsberg chanted "Moloch," other poets—such as William Blake (in "London") and T. S. Eliot (in "The Wasteland")—had expressed the same ideas. The real breakthrough, Kerouac's keen poetic ear told him, came from the last reader: Gary Snyder.

Rocking back and forth, mesmerized by every line, Kerouac thought Snyder's "A Berry Feast" (later published in *The Back Country*) an important statement of human love toward animals. McClure's "For the Death of 100 Whales" seemed fueled by anger, which never solved much, whereas Snyder exuded a love of bears and coyotes. When Kerouac wrote about the event at Six Gallery in his 1958 novel *The Dharma Bums*, he described Snyder (the character Japhy Ryder) as a "great new hero of American culture." Kerouac intuited that Snyder represented an avant-garde new way—actually a revivification of an ancient way—of looking at nature holistically. "And he had tender lines, lyrical lines, like the ones about bears eating berries, showing his love of animals and great mystery lines about oxen on the Mongolian road showing his knowledge of Oriental literature," Kerouac wrote of Snyder. "And his anarchistic ideas about how Americans don't know how to live, with lines about commuters being trapped in living rooms that come from poor trees felled by chainsaws (showing here, also, his background as a logger up north)."[48]

Snyder shared with Ginsberg the belief that atomic bombs would destroy the world—that this genie had to be put back into the bottle. The most controversial line in Ginsberg's *Howl and Other Poems* came from "America": "America go fuck yourself with your atom bomb." It was unclear whether the obscenity laws of the time allowed such language to be put in print. But the American Civil Liberties Union (ACLU) agreed to defend City Lights Books, which had published *Howl and Other Poems* (with an introduction by William Carlos Williams). It was the U.S. Supreme Court justice William O. Douglas, always for freedom of speech, who insisted that books like *Howl* had to be protected by the First Amendment against would-be censors. "None of us wanted to go back to the gray, chill, militarists' silence, to the intellectual void—to the land without poetry—to the spiritual drabness," McClure wrote in *Scratching the Beat Surface*. "We wanted to make it new and we wanted to invent it and the process of it. We wanted voice and we wanted vision." [49] At its core, Ginsberg's "America" was a burlesque of the nuclear arms race between the United States and the Soviet Union.

On November 1, 1956, when "America" was published in *Howl*, Ginsberg didn't know that the U.S. Atomic Energy Commission (AEC) was establishing the "Plowshare Program" to "investigate and develop peaceful uses for nuclear explosives." An Inupiat from Point Hope Village, Alaska, would watch anxiously from a bluff as two men in a boat started unloading supplies on a spit of land jutting out into the Chukchi Sea. Before long, other Inupiat would gather around the boats asking, "Who are you?" The answer baffled them: the visitors were "surveyors" of the AEC.[50]

The AEC had chosen a site at Ogotoruk Creek, about thirty miles southeast of the Inupiat Eskimo village of Point Hope, as a nuclear test ground. Rumors swirled through Point Hope about the planned detonation. Would the residents get radiation sickness? What was the timetable? Would the people be paid reparations?

The truth was that the AEC did plan to detonate an atomic device, 100 times more powerful than the bomb used at Hiroshima, in Arctic Alaska. Ground zero was Ogotoruk Creek. The scheme—which later became infamous—was called Project Chariot. Edward Teller, father of the hydrogen bomb, was overseeing the project. As the director of the Radiation Laboratory at the University of California–Berkeley, Teller publicly announced the program on June 9, 1958. The AEC would detonate a 2.4-megaton atomic device on the northwestern coast of Arctic Alaska. According to Teller, there were two reasons for the explosion: to stay competitive with the Soviet Union, and to create a deep-water hole, which could thereafter be used as the Arctic harbor for the shipment of coal and oil extracted on the North Slope.[51]

Following Teller's stunning announcement, Lewis Strauss, the feisty chairman of the AEC, asked for 1,600 square miles of land and water in Arctic Alaska to be withdrawn from the public domain. Teller himself came to Alaska to promote another supposed reason for Project Chariot: jobs. Alaska could become an oil producer like Texas or Saudi Arabia. New federal funds would come pouring into Alaska. Traveling around Alaska to win support from various chambers of commerce, Teller promised that "the blast will not be performed until it can be economically justified."[52] Doctor George Rogers, an Alaskan economist, recalled having breakfast with Teller in Juneau that summer. "He gave me the pitch again [for Project Chariot]," Rogers recalled. "Then I said, 'Well, the Native people, they depend on the sea mammals and the caribou.' He said, 'Well, they're going to have to change their way of life.' I said, 'What are they going to do?' 'Well,' he said, 'when we have the harbor we can create coal mines in the Arctic, and they can become coal miners.' "[53]

But many Alaskans asked smart questions of Teller as he went around the territory. Undaunted by his well-earned fame as a nuclear scientist, they wanted answers: Wouldn't it take decades for such a port to be operational? How would the money generated trickle into working people's bank accounts? Meanwhile, the national conservation groups seized on Project Chariot as the worst idea ever conceived by mankind. Albert Einstein called it lunacy. Alliances of concerned citizens were organized to save Arctic Alaska from becoming a nuclear testing ground. "I was running the Camp Denali lodge when I learned about Project Chariot," Virginia Wood recalled. "This was a turning point for me. I knew we'd have to organize against the Project. That was beautiful country up there, the homeland to the Native Alaskans! I voted for Eisenhower. . . . I think. But I knew this one was wrong. That whole Arctic area needed to be left alone."[54]

The AEC was surprised by the backlash against Project Chariot in Alaska. Because the territory was preparing for statehood in 1959, the assumption was that only the Inupiat would complain—and they didn't matter in Washington, D.C. Recognizing that a potential economic boom wasn't a compelling argument, the AEC shifted gears. John A. McCone, now chairman of the AEC, testified in Congress before the Joint Committee on Atomic Energy that they were seeking an alternative to the Alaskan harbor because they couldn't find a corporate partner.

The AEC now went back to the drawing board. What was needed, they determined, was a Project Chariot Environmental Studies Program. Being out of tune with the ecology movement, the AEC had underestimated the impact Lois Crisler and Walt Disney had made on the American psyche with regard to Arctic Alaska. The environmentalists had depicted Project Chariot as bombing polar bears, caribou, seals, and whales—species the American people cared deeply about. The AEC had gotten ahead of itself. When Teller went around Alaska, he repeatedly claimed that the fish around Point Hope wouldn't be affected, that nuclear testing wouldn't be harmful to humans, that there would be no seismic shock, and that the people of Japan had already recovered from radiation sickness—none of which was true. Teller, for all his talents, may not have been entirely sane.

A group of scientists at the University of Alaska, led by William Pruitt, stepped up to dispute the AEC's scenarios. Never resorting to emotionalism, giving only the biological facts, Pruitt correctly noted that the food chain in the Arctic was hypersensitive and fragile. Caribou became his Exhibit A. Recent nuclear fallout in the Pacific had already affected the tundra; North Slope caribou suet in the late 1950s had a level of strontium seven times higher than the cattle in Texas or Oklahoma. Because caribou grazed on lichen and other rootless plants, the amount of nuclear dust they ingested was extremely high. They ate radioactive lichen "straight up," before it was integrated with other earth compounds. The same scenario applied to many of Alaska's migratory birds.[55]

Once Professor Pruitt had presented these counterarguments in a public forum, the Inupiat angrily entered the debate. Caribou meat was the staple of their lives—material, cultural, and spiritual. On the North Slope, the Gwich'in people had a creation story, passed down for 1,000 years, that the caribou had absorbed a chunk of human heart and the Gwich'in, reciprocally, held a piece of the caribou heart in their own bodies. In this way, each would always know what the other one was doing. Their relationship went beyond symbiosis; they were one. Upon felling a caribou, Gwich'in hunters offered a prayer of appreciation to their brother species, immediately biting into the heart at the "kill spot" to show honor and gratitude. That was the burden and joy of Gwich'in history. Would Gwich'in hunters get radiation sickness, after Project Chariot, from eating caribou heart? If the caribou died off, would the Gwich'in also die? Furthermore, because the caribou were so far-ranging, the impact of the project would be broader. Caribou migrated more than 500 miles around Alaska each spring, and not only the Gwich'in depended on them for sustenance. All the North Slope tribes who relied on caribou as a food source would become ill.

With emotions running so strong, the Eisenhower administration ordered the AEC to tone down the rhetoric. While Project Chariot wasn't canceled, it was "deferred." Still, rumors circulated in the beat and Native underground in the late 1950s that the U.S. military had injected Eskimos with radioactive iodine-131 as part of a research program to learn whether soldiers "could be better conditioned to fight in cold conditions."[56] Evidence for this claim is rather scant. But in any case many Native Americans in Alaska were feeling empowered to fight for the ecological integrity of their region.

There is no paper trail to clarify what President Eisenhower thought of Project Chariot; he may have pulled the plug on it himself. Douglas L. Vandegraft of the U.S. Fish and Wildlife Service believed that Eisenhower had a quasi-purist view of the Arctic and Alaska; in fact, he wasn't keen on seeing either the north pole or the south pole developed for economic purposes. What interested Eisenhower was atomic energy for peaceful purposes. Project Chariot, however, was too dangerous—and absurd.[57]

The 1950s were a time when faith in science—and the urge to explore new frontiers, using new technological developments—was soaring. The United States had sent a Jupiter-C rocket into space for the first time in 1956; and in 1957 the Soviets launched the satellite Sputnik. Despite the cold war, a remarkable event occurred in December 1959. President Dwight D. Eisenhower led the way to set Antarctica aside as a scientific preserve. All militarization of Antarctica was banned. This agreement—promoted by the United States—was considered the first major arms control treaty of the cold war. Forty-seven countries concurred in making Antarctica a sanctuary. Perhaps Eisenhower wanted to do the same with the Arctic?

V

Ginsberg's poem "America"—epitomizing the spirit of the First Amendment and the impulse to "speak truth to power"—was clearly applicable in Alaska. But what Kerouac loved most about the reading at the Six Gallery was how Snyder made the coyote—the Native American trickster figure—into a protagonist. With suburban developers chasing the coyote out of its homelands in the West, Snyder placed *Canis latrans* on a hillside, wiser than humans, scoffing at the idiocy of clear-cutting, bulldozing, and despoiling the natural world: "The Chainsaw falls for boards of pines/Suburban bedrooms, block on block/Will waver with this grain and knot,/The maddening shapes will start and fade/Each morning when commuters wake—/Joined boards hung of frame/A box to catch a biped in." As Snyder's Coyote watches a "Fat-snout Caterpillar, tread toppling forward/Leaf on leaf, roots in gold volcanic dirt . . ." all he can say is "Fuck You!"[58]

But Snyder doesn't end "A Berry Feast" with the long-suffering coyote losing out to what the poet Lawrence Ferlinghetti, publisher of City Lights Books, called "the omnivorous corporate monoculture."[59] Instead, the deity Coyote, after grievances accumulate, watches the world being restored, as a new generation adopts The Wilderness Society's ethos of leaving nature alone: "From cool springs under cedar/On his haunches, white grin long tongue panting, he watches: Dead city in dry summer, Where berries grow."[60] The Coyote and the poets themselves were messengers, perhaps fools, certainly brilliant trickster figures filled with creative power; their ideas about ecology were revolutionary. "The idea of saving wilderness for wilderness's sake came from West Coast consciousness," the poet and co-founder of the Fugs (a rock band) Ed Sanders recalled. "Ginsberg was the first one to use universe in poems. But it was Gary Snyder who taught us to think in terms of river systems, not boundary lines."[61]

Besides Thoreau, Blake, and Zen, the West Coast beats also developed an affinity for old Rockwell Kent. Considering himself a conservative, an ascetic, and a political socialist, Kent became a target for Senator Joseph R. McCarthy. When forced to testify before a Senate investigations subcommittee, Kent took the Fifth Amendment, refusing to state whether or not he was a communist. Once the most popular illustrator in America, Kent now found himself blacklisted, and his *Wilderness* was removed from libraries as subversive literature. New York galleries and museums in the late 1950s refused to show his Alaskan work. Defiantly, Kent donated his paintings, illustrations, and manuscripts to the Soviet Union. Today many of his Alaskan paintings and illustrations are on permanent display at

the State Hermitage Museum of Saint Petersburg.[62] Only one painting—
his portrait of Virginia Hawkins—remained in Seward, Alaska.

Ferlinghetti—publisher of *Howl and Other Poems*—also had a fierce
ecological consciousness in the 1950s. However, he was concerned more
about Malthusian theory than about the wilderness per se; he considered overpopulation "the root of all the other ecological problems." Why
were rain forests in the Tongass being destroyed? To make more houses
for people. Why might Point Hope be bombed? Because an oil port was
needed to fuel people's vehicles. Why was air pollution becoming a health
hazard in Los Angeles? Because more automobiles were needed. "No
matter what subject you brought up in the 1950s," Ferlinghetti recalled,
one "can trace it back to overpopulation. This is the basis of all ecological problems." Ferlinghetti, through City Lights Books, provided an open
forum to any ecologically minded poet seeking to promote environmental awareness. His getaway home in Big Sur became a haven for talented
artists who wanted to contemplate sea, forests, and air. Working closely
with McClure—who developed a friendship with the British molecular
biologist Francis Crick, one of the codiscoverers of the helical structure
of DNA in 1953—Ferlinghetti published what some scholars consider the
first true ecological periodical in America: *Journal for the Protection of All
Beings.* "What Alaska had going for it," Ferlinghetti believed, "was that
unlike California, it hadn't been overrun with people. Nature still had a
fighting chance." [63]

Crick was also a Malthusian. But what attracted him to McClure was the
almost molecular swirl of vivid words and surreal images in McClure's
poems about nature. McClure also seemed almost intuitively able to understand key concepts about human consciousness, and he and Crick
shared an interest in peyote. "The worlds in which I myself live," Crick
said, "the private world of personal reactions, the biological world (animals and plants and even bacteria chase each other through the poems),
the world of the atom and molecule, the stars and the galaxies, are all
there; and in between, above and below, stands man, the howling mammal, contrived out of *meat* by chance and necessity. If I were a poet I would
write like Michael McClure – if only I had his talent." [64]

Loving people so much, always needing human company, Snyder shied
away from Malthusian fretting and from poetry inspired by DNA. As a

warmhearted Buddhist, he didn't feel like telling people not to breed. In 1956 Snyder moved to Japan to study on a scholarship at the First Zen Institute of America. Often, he lived in an ashram. The monastic life suited Snyder fine—for short spells. But his wanderlust soon compelled him to get a job on the oil tanker *Sappa Creek*, traveling to Ceylon, Guam, and Istanbul. In the western Pacific in 1958 Snyder, aboard the tanker, wrote the four-verse poem "Oil." He was full of fear and dread about the planet's future, when "hooked nations" would need "long injections of pure oil." [65] America, he believed, was a society of petroleum junkies. Maybe—who knew?—Snyder later mused while visiting Alaska, the internal combustion engine would become obsolete. As Snyder wrote in his poem "Energy Is Eternal Delight":

We need no fossil fuel
Get power within
Grow strong on less. [66]

Chapter Twenty

o   o   o   o   o   o

OF HOBOES, BAREFOOTERS,
AND THE OPEN ROAD

I

Wainwright, *Alaska, sits on a spit* of land at the edge of the Arctic Ocean, just within the boundary of the National Petroleum Reserve. An old Inupiat map from 1853 showed that the fishing camp used to be called Olgoonik. But coal was found along this part of the Chukchi Sea coastline in the early twentieth century, and it seemed only proper to anglicize the name of the town. The first naval report from the Arctic area had been written in the 1820s by Lieutenant John Wainwright. Later the navy honored Wainwright (if you want to call it an honor) by naming the frozen town after him. During the winter in Wainwright, temperatures regularly dropped to about fifty degrees below zero Fahrenheit, and there was very little precipitation. More than 90 percent of Wainwright remained Inupiat, hunting bowhead whales and caribou to survive. But the U.S. Navy kept a lookout station in Wainwright: you never knew when, instead of beluga whales, you might see a Soviet submarine or an oil seep or a UFO.

If one were to pick a place on the globe where one wouldn't expect to find the poet Allen Ginsberg in the summer of 1956, it could have been Wainwright. But Ginsberg, depressed because his mother, Naomi, had died in June, signed up as a deckhand and boarded the USNS *Sgt. Jack J. Pendleton* (T-AKV-5)—a cargo ship constructed during World War II—for the summer months while City Lights Books was preparing *Howl and Other Poems* for publication. His employer was the U.S. Merchant Marine. The *Pendleton* had been refitted with radar and enlarged hatches in 1948 and usually worked the central Pacific Ocean, visiting ports in Japan, Korea, Okinawa, Taiwan, and the Philippine Islands and restocking U.S. radar stations along the Distant Early Warning (DEW) line with foodstuffs and supplies for the coming winter. Ginsberg was desperate for money and also hoped that the stark Arctic scenery might help him shake off the blues; it had worked for Rockwell Kent. Ginsberg earned $450 a month, the equivalent of $3,500 a month in 2010 dollars. But, far from finding enlightenment in the Chukchi Sea, as Muir had, he grew even more depressed at the sight of the bruised skies, coal storage tanks, Eskimo skid rows, wharf shacks, rocks, and general bleakness of Wainwright. "Settled down in trip more, now up at a place in Arctic Circle called Wainwright, Alaska—so far no ice, snow, icebergs, aurora, whales, dolphins, seals, fish," he complained in a letter to the painter Robert LaVigne. "Nothing but grey sea and occasional bright day, and day which truly does last all night. The light if you're interested in these northern lights has a kind of teablush-grey immanence, as if not out of sun (usually hidden behind solid cover of clouds also dead grey color past midnight) but lunar reflected out of the water."[1]

What made Wainwright even worse for Ginsberg was the fact that the *Pendleton* had been quarantined half a mile offshore by the merchant marine, because of a measles epidemic in the Native villages. (The memory was still raw, in the Bering Sea region, of the "great sickness" of June 1900, when a vicious strain of influenza wiped out half the population of Alaskan villages with "lightning force."[2] Also, all around Arctic Alaska tuberculosis—which accounted for one-third of all Native deaths in the territory—was always a threat.) Using field glasses, he could see the village crouched on the cliff: a cluster of about seventy dreary, ramshackle edifices. Jack Kerouac had made a steamer voyage to Greenland in 1942 aboard the SS *Dorchester* and described it romantically in his first (unpublished) novel, "The Sea Is My Brother." To Ginsberg, however, the landscape was profoundly desolate. Adding to the grim bleakness, the *Pendleton*'s captain had a persistent fear that a flood tide or a northerly wind would sink the ship, and all the cargo would be lost. "Northern latitudes look flat and the land of Alaska a pencil line on the edge of horizon from where we are," Ginsberg wrote to his friend, "and the further Northward stretches up another thousand miles to the pole in the daylight streaked with clouds."[3]

As the *Pendleton* steamed farther north up the Chukchi Sea, Ginsberg's mood grew darker. Nothing noteworthy happened, there was just the ache of boredom. According to the merchant marine's plans, the *Pendleton* would moor off Point Barrow, not far from where the humorist Will Rogers's plane—the *Aurora Borealis*—had gone down in 1935 during a violent gale. Meanwhile, Ginsberg would peer out over the wet railing into the cold summer dusk, too often asking himself *why*. There is no record that he saw any other ships on the horizon. A sharp pang of regret penetrated him as the *Pendleton* headed toward the north pole. The Chukchi Sea shoreline changed almost minute by minute but became no less desolate. Large scattered masses of blue, green, and white ice drifted forlornly. "I am on the sea north of Alaska 1000 miles from the Pole," Ginsberg wrote to his grandmother Buba. "The sun is up all night, and ice flows by on the edge of the ocean day after day. I spend my evenings reading through the books of the Old Testament."[4]

Point Barrow, frozen and windswept, was the most northerly outpost in Alaska. A thick fog suddenly swallowed the *Pendleton*. Sea ice encircled it as it steamed ahead to port, with the crew hoping for a safe anchorage. Ginsberg, carrying a clipboard that held numerous cargo release papers, was to oversee the unloading of supplies at the U.S. Navy station. In the Inupiat tongue, the geographical location of Point Barrow was *Ukpeagvik*, "the place where we hunt snowy owl"; in the requisition office two stuffed owls were on display. The midnight sun caused the sleep-deprived Ginsberg to wander about Point Barrow like a zombie. Darkness is the natural signal for human glands to produce melatonin—the hormone that most affects sleep. Body clocks get scrambled in the Arctic. Ginsberg was among the victims.

Before Ginsberg left San Francisco, he had heard sailors describe Point Barrow as the Arctic transportation hub. Now he could see, with his own eyes, that besides a few weather station buildings and conical Native huts, Point Barrow was nothing much. Working to counter his despair, however, was a gladdening thought: before setting sail he had optimistically mailed prepublication copies of *Howl and Other Poems* to T. S. Eliot, Ezra Pound, and William Faulkner, though he didn't know any of them. City Lights Books was bringing out *Howl* on November 1, 1956, as the fourth volume in its Pocket Poets series. The San Francisco poet Lawrence Ferlinghetti, owner of City Lights Books, had begun the series with his own collection *The Gone World*. Ginsberg, like any author, was bursting with anticipation and longing to actually touch his own finished book. "So have been up and down north coast of Alaska for a month, now at northernmost Point Barrow," Ginsberg wrote to Jack Kerouac in mid-August (Kerouac was at Desolation Peak in the North Cascades, working as a fire lookout, deep in solitude for sixty-three days). "Sun is out all night or was in midsummer last week, dread ghastly pallor all night through clouds, and this week fantastic burning iron sun going down at edge of horizon for a few hours, clear weather. The water always moving clouds, always moving, birds same clouds and me same like a transparent shifting haze everywhere changing."[5]

Ginsberg was unlike John Muir in that Alaska didn't inspire his creative muse very much; although on August 10 he wrote the poem "Many Loves" from the Arctic. The primary intellectual lesson he squeezed out of his job with the merchant marine was how viciously the Chukchi Sea current attacked ships. Whalers considered the waters between Icy Cape and Point Barrow the most treacherous north of New Zealand. The Arctic sea-lanes were in the field of a strong northward magnetic pull that made timepieces run backward. Frequent fog could turn dangerously heavy within seconds in an unexpected rain shower. Nobody was really ever prepared for the strange turbulence that could suddenly appear with no meteorological rhyme or reason.

Once, in poor visibility, the *Pendleton*'s navigator accidentally rammed the ship into a huge ice floe, causing serious damage to the hull and cracking the fantail. The captain was then forced to make a two-day detour around the floe. Saltwater seeped aboard the ship. Vacuums were brought out. Divers in what Ginsberg described as "Mars suits underwater" tried to fix the damage while the ship was at dockside in Point Barrow. At least no one had to worry about working the graveyard shift: Barrow had eighty-five days of continuous daylight from May 10 to August 2.

Ginsberg, wandering around in the thick weather, did a little paperwork in the village center, thankful for the chance to stretch his legs, and thought about the fame *Howl* might soon bring him. While supplying a storage shed onshore, he contemplated the U.S. Air Force radar station stuck here on top of the continent. This was the cold war era, and some Democrats—saying that the Soviet Union had the "missile edge"—wanted Alaska to become a launch area. Ginsberg wondered if there weren't already enough atomic missiles that could be fired, from underground bunkers, over the North Pole to destroy the Soviet Union. Rumors of polar bears on the ice, always bandied about in Point Barrow, were also troubling to him. Feeling unsafe, he went back aboard the *Pendleton* and continued reading the Bible.

While Ginsberg was at Point Barrow, Kerouac was working on his *Scripture of the Golden Eternity*: sixty-six easy-to-contemplate nuggets of personal wisdom from the Buddha. Corinth Books would publish it as a pamphlet in 1960. Scripture 63, written while Kerouac was at Desolation Peak as a forest lookout, dealt with Coyote; Scripture 62 echoed passages in Robert Marshall's *Alaska Wilderness*. "The world has no marks, signs, or evidence of existence, nor the noises in it, like accident of wind or voices or heehawing animals, yet listen closely as the eternal hush of silence goes on and on throughout all this, and has been going on, and will go on and on," Kerouac wrote. "This is because the world is nothing but a dream and is just thought of and the everlasting eternity pays no attention to it. At night under the moon, or in a quiet room, hush now, the secret music of the Unborn goes on and on, beyond conception, awake beyond existence. Properly speaking, awake is not really awake because the golden eternity never went to sleep: you can tell by the constant sound of Silence which cuts through this world like a magic diamond through the trick of your not realizing that your mind caused the world." [6]

Just two weeks after the reading at the Six Gallery, Kerouac and Snyder took off for Yosemite National Park to climb the 12,000-foot Matterhorn Mountain. (The outing, complete with raisins, haiku sessions, and homemade chocolate pudding, became the anchor for *The Dharma Bums*.) When Ginsberg returned to San Francisco from Point Barrow, happy to be in a softer climate, he learned that the *Nation* was going to run an explanatory article about the "San Francisco Poetry Renaissance," which had been launched at the Six Gallery reading of October 7, 1955. Since he was considered the publicist for the beat generation, he wrote to the journalist Carolyn Kizer of the *Nation*, saying that Kerouac, Snyder, Whalen, and McClure were poetic geniuses. Ginsberg pleaded with

Kizer not to write her article in a "condescending tone," adding, "that's first paramount." Kerouac was just returning to civilization from Desolation Peak in Baker National Forest in Washington; Snyder was off to study Buddhism in Japan; Whalen was wandering around the Sierras; and McClure was married and with young children, reading Haeckel, and busy trying to protect marine life as John Steinbeck and Ed Ricketts were doing around Monterey Bay (on California's central coast)—so the burden fell upon Ginsberg to articulate what all the hullabaloo in San Francisco was about. "Generally the method is as in Buddhist Zen Archery or Koan Response," Ginsberg wrote to Kizer, trying to explain the ethos of the "dharma bums," "long continued practice at spontaneous exactness of expression requiring years of 10–16 hours a day practicing uninterrupted transcription of the droppings of the mind upon a page—until form, deep form, begins to appear, emerge out of the sea."[7]

II

With the reading at the Six Gallery in 1955 serving as an impetus, Alaska opened up to spiritual wanderers, seekers of the northern lights, tripsters, permaculturists, wildcrafters, greenhousers, seedsmen, backpackers, quartz collectors, kayakers, misfits, highway bums, seasonal workers, dropouts, malcontents, and survivalists. To longtime Alaskan boomers and sourdoughs, it was as if all of San Francisco's mystics were arriving in their territory in search of bliss. If Kerouac was right in saying that in the Lower Forty-Eight "the woods" were "full of wardens," then Alaska was a land where a free-spirited drifter could still "cook a little meal over some burning sticks in the tule brake or the hidden valley."[8] Land was still very cheap: you could easily purchase ten acres for $1,000. Motor homes were welcome in public domain lands. Squatting wasn't frowned upon. Instead of seeking gold, the young people now coming to unconventional Alaskan enclaves like Haines and Sitka were seeking *self*. Unlike Rotary Club types, whose belief in America's future was limitless, these self-seekers were turning toward Buddhism, Hindu reincarnation, vegetarianism, groovy drugs, social consciousness, and yoga—away from the flag and toward the prayer mat—and were fearful of an atomic or chemical holocaust.

In Homer, Alaska—at the tip of the Kenai Peninsula—a new wave of young seekers found the deeply forested region a spiritual haven, far from the mad rush of consumerism and conformity. Seeking solace in the sea, sky, and mountains of the Kenai Peninsula, particularly the temperate zones, they hoped that subsistence farming and fishing were the way off the treadmill of making money. As Gary Snyder had said, these seekers all wanted to "create wilderness out of empire." [9] Homer's slogans became "The Cosmic Hamlet by the Sea" and "Living on the Edge," and were a way of giving the finger to Main Street. To the professional fishermen in Homer, who thought of their village as the "Halibut capital of the world," all these cosmic-minded kids with no money were a disturbing trend. "Humans don't own the earth," Lady Greensleeves and Spoonguy have said about the ethos in Homer. "Pacha Mama, Mother Earth, la madre tierra, she bears us on our destiny, and herstory is a vast saga." [10]

Curved around Kachemak Bay, Homer—the magnet for the beats in Alaska—was a clannish fishing village centered on a low, treeless spit (a long, thin gravel bar jutting out into the water). Muir had found the Homer Spit—where fishermen caught thousands of Pacific halibut and Pacific lampreys—enchanting when he sketched Kachemak Bay in 1899. The spit was surrounded on both sides by the exchanging tidal flows of Cook Inlet and the Gulf of Alaska. Low, wooded mountains rose on one side of the spit; on the other side was a rolling ridge of glistening glaciers. At twilight, Homer glowed in what some people described as a blanketing halo. "Light," Muir had written. "I know not a single word fine enough for Light . . . holy, beamless, bodiless, inaudible floods of light." [11]

Homer, a magnet for vegetarians, may be where seaweed became a popular health food in the 1950s. Bands of seaweed—such as *porphyra* (black seaweed), *palmaria* (ribbon seaweed), and *macrocystis* (giant kelp)— became a subsistence food for hitchhikers along Kachemak Bay. Such seaweeds were rich in minerals, vitamins, and carbohydrates. [12] Likewise, the clams and mussels of Kachemak Bay, whose beds were in the mud- flats, were also an attraction; small but succulent, they were among the best-tasting in the world. Nestled along Kachemak Bay was a huge raft of sea otters. Daily they swam about in these highly productive waters, gorg- ing on shellfish.

When the Harriman Expedition visited Kachemak Bay in 1899, Charles

Palache of Harvard University, the mineralogist aboard the *Elder*, noted the "interesting geology" around Homer; the area gave him a newfound interest in crystallography.[13] John Burroughs, however, found "nothing Homeric about the look of the place." [14] But he loved seeing the volcanic peaks, Iliamna and Redoubt, sixty miles across Cook Inlet to the west.

The Kenai Peninsula was ripe for the beat generation ethos after Ginsberg's *Howl* was published in 1956, followed by Kerouac's *On the Road* in 1957. As Muir had told an earlier generation, "Go to Nature's School—the one true University." [15] Homer was a natural place for the beat philosophy to take root because the village spit—not San Francisco—was truly the end of the road in America.[16] While Snyder and Whalen were injecting Zen Buddhism into the poetry of California, the Barefooters, a group based in the Los Angeles area, were melding Hare Krishna, reincarnation, and Henry David Thoreau's "Simplify, simplify" into a heady cocktail. The back-to-nature cult had started in 1948 as the Wisdom, Knowledge, Faith, and Love Community (WKFL). A subgroup with theatrical ambitions in the Los Angeles area performed a Christmas play in which none of the performers wore shoes: hence the name Barefooters.

The WKFL was led by Krishna Venta, who sought martyrdom. The members permanently shunned shoes; the men refused to get a haircut until world peace was achieved; the women dressed in long, flowing white gowns and liked to serve apple butter on homemade wheat bread. The Barefooters intended to wear their holy robes until universal love rained down. Love and service were the goals of WKFL. These cultists, forerunners of the San Francisco hippies, devoted their varied talents to humanitarian endeavors such as helping the poor and homeless and extinguishing forest fires. "Bare feet keep one connected to the earth," Brother Asaiah, a WKFL leader, explained. "One doesn't need blinders on one's feet any more than one's eyes. We learn about the earth through our feet. We learn to tread lightly on earth and not dally too long in one place." [17]

During the summer of 1956, six Barefooters—known as the Fountain of the World contingent—left Canoga Park, California, for Homer, Alaska. They had been practicing the "beat" life years before the term was used. Word spread throughout the Kenai Peninsula that beatniks (a term coined in late 1957) were arriving en masse, hitchhiking along Highway 1 but looking too bizarre to get rides. (Alaskan lumbermen prided themselves on their own libertarian values. But what could they make of long-haired people in biblical garb walking barefoot in the snow without guns?) Krishna Venta had a vision of colonizing the Kachemak Bay area as Brigham Young had once settled Mormons around the Great Salt Lake of Utah. The Barefooters would protect the natural world of Kachemak Bay not as a possession but as a responsibility. All religions would be embraced; they weren't dogmatic about reincarnation as *the* way, although the Indian religious traditions of Hinduism, Jainism, and Sikhism, and particularly the transmigration of the soul, seemed to be their prevailing ethos. Feet, however, were their fetish. They held ceremonies in which they marveled at the evolution of the foot's anatomy: its thirty-three joints, twenty-six bones, and twenty muscles.

Acquiring three homesteads in the Fox River valley, about a half-hour drive from Homer, the Barefooters established a commune in 1956—when Ginsberg was in Point Barrow on the *Pendleton*. They named their land Venta. Although the Barefooters didn't make elegant handcrafted chairs, there was something of the Shakers in them. What Krishna was trying to teach was avoidance of *avidya* (one's true self), since the self led to ignorance and militarism. The Barefooters were more freakish than Ginsberg was in the late 1960s, when he wanted to levitate the Pentagon.

Krishna Venta (originally Francis Herman Pencovic, born in San Francisco on March 29, 1911) was a very popular leader. By the mid-1950s he had tens of thousands of followers. Venta had messianic blue eyes and a tangled beard like Charles Manson's, and his favorite subject was himself. He stated matter-of-factly in April 1948: "I may as well say it: I am Christ. I am the new Messiah." Angry that newspapers kept calling him Francis Pencovic instead of Krishna Venta, he had his name legally changed in 1951. When asked why he was the new Christ, Krishna Venta claimed that he had led a convoy of rocket ships from the burning planet of Neophrates to save Earth; even L. Ron Hubbard, whose first scientology writings appeared in 1951, thought he was weird.

Dormitories were built at Venta, along the Kachemak Bay mudflat on the far outskirts of Homer. The Barefooters became children of the Kenai Peninsula tides. Outside their front doors, glacial erratics dotted the flats. Driftwood and detritus hourly washed up on their beach. Mushrooms grew along the horsetail-fringed shore—not psychedelic ones, for the

Barefooters were opposed to using drugs. Moose browsed around their acreage eating dwarf birch and willows. But being a Barefooter during the winter months was of course dangerous and nonsensical. "It wasn't even the Alaska icy roads that stopped the Barefooters from going barefoot," recalled a former state senator, Clem Tillion. "As long as they kept walking, when on the ice, they were fine. If they stopped, the heat from their feet melted the ice, and they stuck to the ice, so they didn't just stand when they were barefoot. It was actually the sparks from welding in their shop that brought about the decision to clothe their feet." [18] Alaska's weather had, alas, forced the Barefooters to wear thick leather boots.

Krishna Venta's principal surrogate in Homer was Brother Asaiah (originally Claude Bates, raised fatherless in Pilot Mountain, North Carolina). Shuttling between Ventura County and the Kenai Peninsula, keeping the Homer contingent well supplied for the hard winter months, Brother Asaiah was treasured by all the Barefooters. "He became our beloved Brother Asaiah," Martha Ellen Anderson recalled. "His consciousness propels our lives in directions we often know not where but the path is not unknown. He expressed his truth, international, intercultural, and universal, in the last frontier on this earth, in our little town at the end of the road, Homer, Alaska, our cosmic hamlet by the sea." [19]

For all their peaceable words, however, the Barefooters had a darker side. On December 10, 1958, Krishna Venta was murdered in Chatsworth, California, by two disgruntled followers. Claiming that he was embezzling funds and seducing their wives, they strapped on twenty sticks of dynamite and blew up themselves, Krishna Venta, and seven other Barefooters. The explosion also burned more than 200 acres in California. A shock wave touched youth communities such as Santa Monica and Venice Beach: How could such destruction emanate from the seemingly benign Barefooters? Hadn't members volunteered in soup kitchens, wildlife reserves, organic farms, and orphanages? The victims of the explosion had included a seven-year-old girl and a baby; how could this be explained? Brother Asaiah was left holding the torch for Krishna Venta's followers, trying to make sense of what had happened. Alaskan newspapers naturally reported the tragedy, pointing out that the cult had a presence in Homer. Shaken, Brother Asaiah nevertheless came north, preceded by a taciturn message: "Heading to Homer."

Driving up the Richardson Highway to the Wrangell Mountains, then heading west to Anchorage, Brother Asaiah may have felt optimistic. Krishna Venta had been his spiritual teacher—and he would continue to convey Venta's philosophy of love in Homer. After a few days in Anchorage, Brother Asaiah headed into the Chugach Mountains along Highway 1. After a night of camping, he headed down the western side of the Kenai Peninsula and saw Redoubt Volcano looming across Cook Inlet like a watchtower. He pulled into Homer and bought a trailer-like home from a local realtor on Lucky Shot Street. To make ends meet, he got a janitorial job. So suspicious were his outsider manner and world view that the police wondered about his sanity. But after a while, the community of Homer got used to Brother Asaiah's eccentricities. Slowly but surely the inhabitants adopted him as one of their own.

Warmhearted, deeply mystical, convinced that the world needed to be rid of nuclear weapons, Brother Asaiah became the spiritual leader of nonconformist Homer. He probably did more than anybody else to inject the word *cosmic* into the American parlance of the late 1950s. "When the Barefooters arrived, there were a lot of John Birchers living in Homer," Martha Ellen Anderson recalled. "They wouldn't so much as talk to Brother Asaiah. The Birchers were about the conquering spirit of Alaskan lands. The Barefooters were living a whole-earth philosophy. But their kids all got to know one another. Eventually the Barefooters were accepted. What everybody in town had in common was this strange draw to how the land met the sea in Homer." [20]

Brother Asaiah brought an old-time homesteader ethic to Homer. As a community leader he encouraged Barefooters to grow their own food, construct spruce-log buildings, and cook communal meals. He promoted social services in Homer when there weren't any. Owing in large part to Brother Asaiah's leadership, Homer offered social services such as Alcoholics Anonymous, a women's clinic, an abuse shelter, meals on wheels, and an elder hostel. The Barefooters also donated land to the city of 5,500 to make the WKFL Peace Park. A hospital was built, and the Family Theatre opened. Long before Whole Foods got started in Austin, Texas, the Barefooters, led by Brother Asaiah, promoted organic foods. Brother Asaiah was like a one-man Great Society, applying the principles of social work to Homer, earning praise even from right-wing townsfolk who were initially skeptical about him. Until his death in March 2000 he was the heart and soul of Homer. "Attempting to capture the essence of Brother Asaiah seems akin to trying to catch a moonbeam in a mason jar," Governor Jay Hammond of Alaska (in office from 1974 to 1982), explained.[21]

Another presence in unconventional Alaska was the sea goddess Sedna. Long a part of Inupiat mythology, popular in shaman art along coastal areas north of the Arctic Divide, Sedna was supposedly a mermaid-like woman who lived in a huge mansion on the seafloor. In some renderings, Sedna had a fishtail and caribou antlers. So strong was Sedna's appeal that when NASA discovered a new planet in 2003, it was named VB 12 "Sedna." In one enduring story, Sedna refused to marry a man. Her angry father threw her into the sea, chopping off her fingers for good measure. Her fingers turned into sea mammals such as seals and walrus. Sedna stayed on the ocean bottom, deciding, according to her all-powerful whim, whether marine game should be withheld from Eskimo men. Without this food, the men would perish.[22]

For liberated women of the 1950s who were moving to Alaska, Sedna's story involved turning abuse into empowerment. There was a rejection of marriage, a cruel father, societal ostracism, and finally Sedna herself—holding all the power, making men beg for sustenance. Sedna and mermaids became popular during the 1950s among avant-garde artists in Anchorage and the Kenai Peninsula—perhaps not surprisingly, in a state with 33,000 miles of coastline. The strength and persistence of Sedna's legend spoke to a confident belief that in the male-female exchange, the woman held sway. Interestingly, in the biological sciences, once a male domain, women were becoming the top marine biologists in Alaska, Hawaii, and the Lower Forty-Eight by the 1950s. Also, national wildlife refuges began to be named after women: Elizabeth A. Morton in New York, Rachel Carson in Maine, and Julia Butler Hansen in Washington.

III

Jack Kerouac never came to Homer, never met Brother Asaiah, and evidently never learned about the legend of Sedna. But the commune at Homer was in existence a year before *On the Road* was published and more than two years before the term "rucksack revolution" was coined in *The Dharma Bums*. Kerouac had predicted the "rucksack revolution" in *The Dharma Bums* as an imminent, consciousness-changing movement in which city dwellers would light out for places like the windswept Kenai Peninsula seeking personal renewal.[23] (The hitchhiking, communal back-to-nature movements that absorbed many baby boomers of the 1960s bore out this prophecy.) In *The Dharma Bums*, Kerouac wrote about the glory of going barefoot, of feeling connected to the earth without oppressive footwear, of taking "off my shoes" and sitting in a lotus position feeling "glad."[24] Sometimes being primitive like a caveman felt superior to living above a Laundromat in New York or a restaurant in San Francisco. "If Cro-Magnon man was less subject to degenerative diseases and less prone to modern genetic and actual defects such as caries and tuberculosis," Michael McClure mused in *Lighting the Corners*, "the artist could idealize him and begin a review of history from that point."[25]

Following the success of *The Dharma Bums*, feeling footloose and fancy-free, Kerouac wrote his wilderness essay, "The Vanishing American Hobo," for *Holiday Magazine*; it was included in his omnibus of drifter essays, *Lonesome Traveler*, published in 1960 by Grove Press. (The novelist John Dos Passos would also soon write an essay about Alaska's Glacier Bay for *Holiday*.) This was Kerouac's first truly autobiographical work, comprising eight sparkling essays. Kerouac detailed his stints as a brakeman in California and as a fire lookout atop Desolation Peak in the North Cascades. He seemed to have the soul of a bedouin. "There is something strange going on," Kerouac complained; "you can't even be alone any more in the primitive wilderness." To Kerouac the Eisenhower era was a police state and was killing the noble traditions of camping, tramping, and trailblazing in favor of a homogenized monoculture of groupthink. Individuality and authenticity were being stamped out. The international economy was on the rise. If you wanted to sleep out under the Milky Way along a roadside, policemen would demand identification and treat you as a vagrant.[26]

By 1959 Kerouac had become a hero of the nonconformists. Groups like the Barefooters were an early version of the hippies who hitchhiked to Alaska throughout the 1960s, searching for revelations in nameless woods. Feeling blessed, they wanted to escape the confines of the Lower Forty-Eight. Kerouac spoke to later young people disenchanted with postwar abundance, thirsting for a deeper truth than air-conditioning and missile technology. The neoconservative critic Norman Podhoretz, in the *Partisan Review*, dismissed *On the Road* as anti-American, and as promoting drug use, free sex, and joblessness over the Protestant work ethic.[27] What Podhoretz didn't say was that the "know-nothing" beats, as he called them, were bravely asking questions about an accident at a nuclear power plant in Windscale, England—and about Minamata disease, a neurological syndrome caused by poisonous mercury in waters.

In *The Dharma Bums*, the poet Gary Snyder (Japhy Ryder) represented the open road: a lineage that could be traced through American literature from Thoreau to Whitman to Muir. Despite all the commentary about the novel's overt sexuality ("yabyum"—two men with one woman—adds spice to the story), *The Dharma Bums* was, in truth, an intersection of Christianity and Buddhism. Kerouac's overriding message was, "Charity shall cover the multitude of sins." His mountaintop exhortations represented a great original American artist at his absolute prime; the descriptive writing equals the best of Thomas Wolfe and John Muir. "I'll tramp with a rucksack," Kerouac wrote, "and make it the pure way."[28]

Perhaps more than any other novel, *The Dharma Bums* conveyed the value of wilderness to young audiences in the 1960s. Kerouac's words pulled readers toward a craving for outdoors experiences, for almost mystical reasons. "Logs and snags came floating down at twenty-five miles an hour," Kerouac wrote. "I figured if I should try to swim across the narrow river I'd be a half-mile downstream before I kicked to the other shore. It was a river wonderland, the emptiness of the golden eternity, odors of moss and bark and twigs and mud, all ululating mysterious visionstuff before my eyes, tranquil and everlasting nevertheless, the hillhairing trees, the dancing sunlight. As I looked up the clouds assumed, as I assumed, faces of hermits."[29]

Understandably, Kerouac deeply resented any belittling of his romantic yearnings for Walt Whitman, Huck Finn, and Herman Melville. Combining Bob Marshall's wilderness philosophy with Gary Snyder's belief in nature as a healer of the soul, Kerouac defended the hobo tradition in a torrent of heartfelt, first-rate prose. Writing from a cabin at Big Sur, where the rugged Santa Lucia mountains dropped straight into the Pacific Ocean and huge waves slapped in rhythmic fury against towering sea rocks, Kerouac lamented the mainstream culture and its need to commodify *everything*, even its national parks. In *Big Sur* Kerouac, with a charitable heart, objected to the end of "barefoot kids" with "a string of fish," warming themselves by wood fires while camping out in secret coves along the Pacific coast. The Barefooters were doing that in Homer, but most American families were now driving station wagons into sacred landscapes like the Painted Desert or Mount McKinley, "sneering" over a "printed blue-lined roadmap" and worried silly about getting "the car washed before the return trip."[30] Or, perhaps, these families headed to Alaska on a cruise ship, listening to music and eating buffets of chemically enriched foods five times a day.

Lonesome Traveler was filled with impressionistic prose riffs; its central premise was the enduring virtues of hoboing in the wilderness. In the jargon of Broadway theater, Kerouac "believed his own show": spontaneous prose enriched by Buddhist philosophy, transcendental yearnings, and American outdoors romanticism. When Grove Press published his essays, Sputnik had been launched, Americans were worried about a supposed "missile gap" relative to the Soviet Union, and NASA's space programs were headline news, so Kerouac's meditations about the open road seemed antiquated. But the essays took on relevance when the U.S. Atomic Energy Commission drew up plans to test nuclear weapons on the Aleutians. "The Jet Age is crucifying the hobo," Kerouac wrote, "because how can he hop a freight jet?"[31]

Kerouac was concerned that in mature America "camping" was deemed a "healthy sport" for the Boy Scouts but "a crime for nature men who have made it their vocation." With a rucksack on his back, Kerouac had wandered through America from 1948 to 1956. But he abruptly halted his hitchhiking because of ugly television news stories about the "abominableness of strangers with packs passing through by themselves independently" in frightened suburbia. Beatniks were considered dangerous perverts to be avoided at all costs. To Kerouac, untrammeled places like the North Cascades, the Brooks Range, and the Kenai Peninsula offered the last best hope for disappearing into the wilderness to find oneself, as Rockwell Kent had done at Fox Island in 1920. In *Lonesome Traveler*, Kerouac noted the great hoboes in American history, from Ben Franklin to William O. Douglas. Lovingly he declared John Muir a hobo who "went off into the mountains with a pocketful of dried bread, which he soaked in creeks."

To Kerouac the great Teddy Roosevelt was a "political hobo" of the first or-der. Hadn't the poet Vachel Lindsay enriched America by his "troubadour" hobo wanderings, giving farmers verses in exchange for homemade pies?

Kerouac was frustrated that the open road was under assault by a police state mentality. Douglas had complained about railroad cops beating ho-boes outside Chicago during the Great Depression; now, Kerouac voiced a similar complaint in *Lonesome Traveler* about "great sinister tax-paid police cars (1960 models with humorless searchlights)" bearing down on The Wilderness Society types who were only looking for "hills of holy silence and holy privacy." To Kerouac, the celestial seeker, there was "nothing no-bler" than to "put up with a few inconveniences like snakes and dust for the sake of absolute freedom."[32] Kerouac was insisting, in 1960, that vagrancy wasn't merely legal; it was part of the patriotic Thoreauvian tradition that made America unique. To Kerouac, Johnny Appleseed (whose real name was John Chapman), the Swedenborgian orchidist who dropped seeds of fruit-bearing trees from the Berkshires to the Ohio Valley chanting "the Lord is good to me," a wanderer who was even kind to skunks, should be celebrated as an American counterpart of Saint Francis of Assisi. Nothing was more liberating to Kerouac than to live in a wilderness where, over-night, you could be reborn, choosing your own new name and identity: Au-rora Borealis, Brother Asaiah, Sedna, Japhy Ryder, or Johnny Appleseed.

But the Anchorage Museum of History and Art circa 2010—funded in part by BP and Shell—doesn't consider beats, Buddhists, or Barefooters* a part of state history worth remembering. Instead it prominently displays under glass the bronze boots worn by William G. Bishop when he ordered the Richfield Oil crews to drill the discovery well for the Swanson River oil field on the Kenai Peninsula in 1957. This was a gift from the Anchor-age Chamber of Commerce, which was promoting the slogan "Drill today, drill tomorrow." Down south in Homer, however, every summer a "Howl" camp was held, at which naturalist instructors took children backpack-ing, rock climbing, and bird-watching.

---

* Nike and New Balance, perhaps influenced by Brother Asaiah, did design a "barefoot shoe" post-Y2K with a special Vibram Fivefingers sole; it was like a latex glove for the foot.

Chapter Twenty-one

∘ ∘ ∘ ∘ ∘ ∘

SEA OTTER JONES AND MUSK-OX MATTHIESSEN

I

If contestants on a quiz show were asked to identify the most valuable market fur in the world, "sea otter" would be the right answer. Sea otters have a close-packed dark coat with silvery streaks. It's lush to touch and gorgeous to look at. By the time the team of Theodore Roosevelt and William Temple Hornaday came to the rescue of sea otters in 1911, a single pelt was worth about 2,000 pounds in London. In documentaries, Walt Disney portrayed the species as cuddly and doglike, but in truth sea otters aren't to be messed with: they are on average four to six feet long, weigh ninety pounds or so, and have razor-sharp teeth that can rip into flesh. In Alaska, they have long been pursued for their fur by Native tribes, and in the early nineteenth century these otters could be seen floating on their backs eating shellfish, and frolicking in kelp beds and around offshore rocks. "A sea breaks, the gull lifts, and the otters slide beneath the surface," Peter Matthiessen wrote in *Wildlife in America*, "to rise again like black shadows in the semitransparent water beyond the foam."[1]

Banning the sale of sea otter fur helped the species survive in the Aleutian Islands. But during World War II, troops stationed in Alaska often shot at them recklessly, forcing them to the brink of extinction. At last, however, the species found a steadfast ally in an employee of the U.S. Fish and Wildlife Service who was known in the Alaska circuit as Sea Otter Jones. He was the happiest in the spring, when floods of migratory birds returned to the Aleutians, following the primordial "river in the sky"—flyways that lured birds to Alaska from five continents.

Bob "Sea Otter" Jones, who worked for U.S. Fish and Wildlife from 1947 to 1980, was born on August 3, 1916, in Millbank, South Dakota, a farm town of 2,500 people along the South Fork of the Whetstone River. Jones was an animal lover from a very early age. By the time he was ten, he wanted to be a field biologist. A beneficiary of a conventional midwestern upbringing, Jones excelled in high school. His father, an attorney, encouraged him to learn how to operate a ham radio. Skilled in electronics, Jones graduated from South Dakota State College in Brookings.[2] "After college all he wanted to do was go to Alaska," his wife, Dorothy, recalled. "People think he was an outdoorsman wearing a flannel shirt, but he liked ruffled shirts and opera."[3]

When the Japanese bombed Pearl Harbor in December 7, 1941, the twenty-five-year-old Jones enlisted in the U.S. Army's Signal Corps. He became a licensed radio operator. Although the army initially assigned him to New Jersey, he had a persistent vision of himself working in Alaska. The colossal territory appealed to his intense interest in wildlife resource management and in long-distance radio. "I had conceived a desire to come to Alaska," Jones recalled, "so I made that known. There were no assignments in Alaska at the time, so I was sent to Fort Lawton in Seattle, that was the nearest opening."[4]

Ruggedly handsome, with sandy hair and aristocratic features, Jones—a welterweight—wasn't tall. But his personality was so huge that he seemed as big as a linebacker. Fascinated by meteorology, Jones dreamed of someday living in the Aleutian chain, considered the stormiest area in North America. In August 1942 he got his wish. He was assigned to Adak Island (part of an island arc between Alaska and Asia), to radio-monitor Japanese air traffic and use high-powered telescopes to watch for Yokohama's naval fleet. Even for a Dakota boy, used to blue winters, Adak took some getting used to. The winter months on the forlorn island were unbearable. Living in a tent, First Lieutenant Jones and his colleagues cobbled together a diesel-burning stove to stay warm. The wind came at them like needles; its baritone howl was deafening. South Dakota seemed like a tropical rain forest by comparison. Jones developed a new appreciation for Aleuts who had made clothing from bird skin to stay warm. "It isn't an extremely wet climate but it's damp all the time and a little moisture in a 40 to 50 knot wind is enough to go a long ways," Jones recalled. "Some of the guys couldn't hack it and they'd lose their marbles. Well, sometimes in the middle of the night the tent would collapse, a pole would break, or the whole damn thing would blow away. And this didn't happen just now and then, it was a routine sort of business."[5]

Wearing blue jeans, three layers of long underwear, a wool hat, Canadian work boots, and a bulky green parka, Jones learned the art of survival in the Aleutians. He felt like a marooned pilot shot down over enemy lines in Amchitka and Adak, monitoring Japanese aircraft. He was responsible for installing radar on Adak (the frequency was about 100 kilocycles). "There were seven or eight of us, a small detachment," he recalled. "We were sent out with portable radar, the first portable radar the U.S. Army ever put in the field. We went to Bird Cape, at the west end of Amchitka, where we could look into Kiska Harbor (fifty miles away)."

The Aleutians were a huge bow-shaped chain of seventy islands extending for 1,000 miles from the Alaska Peninsula to Kamchatka. These treeless islands had survived the upsurge of mountains and their erosion through the ascent and descent of ocean waters. Amchitka was therefore a tectonically dangerous place for Army Air Corps troops to be stationed. If the freezing temperatures didn't get them, hot lava could. Since 1832 no Aleuts had lived on the island. But rafts of sea otters populated isolated coves and bays; this kelp-rich ecosystem was their last stand. The army used the southernmost of the Rat Islands group as an airstrip throughout World War II. Planes regularly flew in and out. Huge maneuvers were sometimes held as a decoy to distract the Japanese. The U.S. government ordered forced evacuations of Aleut villages—an unfair imposition on blameless citizens. The waters around Amchitka were extremely rough; the destroyer *Warden* (DD-352), for example, was grounded and sank in 1943, drowning fourteen men. But for a South Dakotan outdoorsman like Jones, Amchitka was a wild and wonderful place. He kept regular field diaries of the spectacular waters teeming with exotic waterfowl. "I was especially interested in the emperor goose," he recalled. "I had never seen that goose before, coming from the Great Plains."

At Amchitka, Jones started studying sea otters in earnest, learning how to scuba dive in the coldest waters in the world. He criticized army troops who used the sea otters, a fairly depleted species in 1943, for target practice, as if otters were filthy vermin. Incensed, Jones later reminded his trigger-happy colleagues that part of the U.S. mission was to protect the wildlife in the Aleutian chain—not devastate a charming endangered species. The legend of Sea Otter Jones was born. "We knew the presence of sea otters there was important," Jones recalled. "The military command was aware of that and wanted to protect them to the degree possible, so that those of us who were interested found our way into that extra activity."

Whereas other servicemen in the Aleutians couldn't wait to get back to Nebraska or New York, Jones wanted to be a wildlife biologist there. Bouncing around Amchitka, Adak, Ogliuga, and Little Sitkin islands, he marveled at the high density of wildlife. Sweeping in a curve more than 1,000 miles long from the end of the Alaska Peninsula toward Kamchatka, the Aleutians connected North America with Japan, China, Korea, and all the Asian nations of the far east. "To the traveler from the south, approaching any portion of the chain during the winter or spring months, the view presented is exceedingly desolate and forbidding," Muir had written in *The Cruise of the Corwin* about the Aleutians. "The snow comes down to the water's edge, the solid winter-white being interrupted only by black outstanding bluffs with faces too steep for snow to lie upon, and by the backs of clustering rocks and long rugged reefs beaten and overswept by heavy breakers rolling in from the Pacific Ocean or Bering Sea, while for ten or eleven months in the year all the mountains are wrapped in gloomy, ragged storm clouds."[6]

When World War II ended, Jones decided to live in Alaska permanently, and his dream was to make a career as *the* biologist of the Aleutians. He was one of the rare breed who enjoy calamitous weather. Generous homesteading provisions were offered to veterans like Jones by the U.S. government. Adding to the appeal of Alaska were enhanced communication systems, highways, and a road connection to the Lower Forty-Eight. As a first step Jones moved to Kodiak Island, bought a skiff so that he could go shopping, and started studying sea otters on his own. Working on the salmon boats for day wages—a truly hard way to earn a dollar—Jones decided that he preferred roadless areas to roads. It was a hand-to-mouth existence, but his experiences studying otters continued. If Jones could have promoted himself as well as Crisler, Hollywood might have made a movie about his life.

No single person did more than Jones to help sea otters become a protected species on the remote Aleutian island communities between the North Pacific and the Bering Sea. The Aleutian fishermen despised otters because they raided oyster beds, but Jones educated the geographically scattered people on Akutan, on Unalaska, and at the port of Dutch Harbor to leave the sea otters alone. His own headquarters and home were at Cold Bay, a main commercial center on the Alaska Peninsula. His combined base of operations there was a tiny structure in which he kept his few belongings: binoculars, framed pictures, a shaving mug, and shotguns. On his iron bed was a quilt from South Dakota. And on his record player, often at odd hours, there was typically something by Mozart or Bach.

In 1948 the U.S. Fish and Wildlife Service hired Jones to oversee the entire Aleutian chain for the Department of the Interior. Paid meager wages, Jones at least was his own boss for about 340 days a year; his immediate supervisors didn't like flying much farther south than Homer. Although Theodore Roosevelt had started protecting Aleutian mammals in 1908, Jones was the first college-trained warden-manager appointed.* The gateway town to the Aleutians' East Borough, Cold Bay, was only a block long. The deprivations there were considerable. Fewer than 100 people lived in the town. Every week, it seemed, the land trembled with an earthquake.

During World War II, Fort Randall was created as a base camp for the 11th Air Force at Cold Bay. Quonset huts housing nearly 20,000 U.S. troops were built near the shore. Japanese bombs fell on the nearby village of Unalaska in 1942, but Cold Bay was unscathed. After the war the soldiers left the aptly named Cold Bay. But Jones stayed, living with a couple of weather-service specialists, a few fishermen, and occasionally some stopover wildlife tourists—Audubon Society types—who lodged at the wind-chafed World Famous Weathered Inn. The U.S. Fish and Wildlife

---

* In the 1920s five federal game wardens had been appointed to the Aleutians: Doug Gray, Frank Beals, Donald Stevenson, C. C. Loy, and D. A. Friden. None had a college degree.

Service hoped Jones could help create the 300,000-acre Izembek NWR, a rocky outcropping for 130,000 Pacific black brant (*Branta bernicla nigricans*), 62,000 emperor geese (*Chen canagica*), 50,000 Taverner's Canada geese (*Branta canadensis taverni*), 300,000 ducks, and 80,000 shorebirds. During the windy months, the Steller's eider (*Polysticta stelleri*)—one of the most beautiful birds in the world—wintered along the thirty-mile Izembek Lagoon, which had the world's largest eel grass beds. No wetlands in all of America held an abundance of wildlife that could rival the Izembek. Its panorama of a U-shaped valley, ancient glaciers, and hot springs made it the best-kept secret in America. "In my opinion, it was the finest assignment the [U.S. Fish and Wildlife] Service had," Jones said of Cold Bay. "I wanted to be where there were animals and not many people, and it fulfilled both categories."[7]

In 1953 Jones married Dorothy, a native of California. What made Dorothy unique as a bride was that she tolerated Bob's pet sea otter, Harriet. Dorothy quickly learned that the future of sea otters was bleak, and that conservation biologists had to come to their rescue—fast. What most worried her husband was the scarcity of sea otters in the Aleutians. The Aleutians, in fact, were long the home of the greatest concentration of sea otters in the world.[8] Every day he dutifully studied their goings-on. The 1911 Hornaday-Roosevelt Treaty had temporarily saved the sea otters from extinction. But the illegal black-market slaughter of sea mammals by Japanese and Russian pirates continued. Jones was determined to bring the otters back to their full glory. In California, a historic home of sea otters, only a small band of 638 were alive.[9]

As Jones worked for the U.S. Fish and Wildlife Service, dividing his residency between Anchorage and Cold Bay, he was intrigued by the thickness of the sea otters' fur—the thickest mammal fur in the world. Pushing a finger through the dense fur was futile; you couldn't get to the skin. Children, in particular, adored the sea otter because it seemed so frisky and joyful, showing off in the kelp, lying on its back eating oysters and clams. Most otters slept on their backs in the water, usually amid the tangled kelp and seaweed. At the Cold Bay station Jones had another pet otter, named Hortiser. Nothing seemed to exhaust this pet. What Jones learned about sea otters (as a species) from the ones he befriended on the Aleutians was their apparent joie de vivre. As in a scene from Gavin

Maxwell's *Ring of Bright Water*, these otters would dive 300 feet down, then would surface, seemingly full of glee and laughing among themselves about their underwater antics.

Frolicking with the sea otters around Cold Bay and counting the birds that congregated around the Izembek Lagoon constituted the entertaining part of Jones's U.S. government job. Far more menacing was taking his dory out in rough Bering Sea weather to patrol the other islands in the chain. The giant surf on Buldir Island was particularly rough for landing a small boat with a small outboard motor. Rain . . . hail . . . sleet . . . snow . . . storm . . . surf. No matter what the weather was like, Jones would make the rounds along the rugged islands of today's Alaska Maritime National Wildlife Refuge. Over the years he had a lot of colorful names for his boats: *Water Ouzel, Phalarope, Dipper,* and *Wandering Tattler.* Daily he took field notes about seabirds, invertebrates, transplant geese, and foxes in need of removal. "On bright clear days the approach of the dory by a cascade of cormorants and puffins from the cliffs and then we traveled under a veritable canopy of wings," Jones jotted in a notebook in 1959. "When a sea fog lay close and we ran on compass courses, quite out of sight on land though we knew it to be near, the smell of whitewashed cliffs was a beacon guiding us and a sudden avalanche of birds, bursting out of the murk, pinpointed our location."[10]

Safeguarding the Aleutians brought Jones a lot of satisfaction. The biologist Olaus Murie, always circulating around Alaska, shared with Jones much of his research pertaining to birdlife in the Aleutians. Murie was overseeing a project to rid the islands of the foxes introduced in the 1920s, 1930s, and 1940s: these foxes were leading to the demise of the Aleutian Canada goose. Wanting to experience the underwater life of sea otters, Jones got a wet suit and started scuba diving in the kelp beds, nearly living with the otters (an unheard-of practice before Jones). Conditioning allowed his body to become almost immune to the powerful numbing effect of the water. "It became apparent that if we were to get information about what happened below the water surface," he said, "we'd better go down to take a look."[11]

Jones was perplexed about why the sea otters were dying off in record numbers during the winter months. His scuba dives off fogbound Amchitka and Adak produced an answer: the otters' food source was depleted. Rock oysters weren't alive on the bottom; the shells were without much muscle. Every time the otters *seemed* to be feasting, cracking open shells, they weren't getting much meat. "The otters kept on eating sea urchins, but they weren't getting any nutritional value," Jones recorded, "and downhill they went."

Owing to the sea otter rehabilitation project undertaken by U.S. Fish and Wildlife in the Aleutians, a recovery took place. Under Jones's leadership, sea otters were reintroduced to Attu in the mid-1950s. Starting in 1954, the otter population increased at a healthy rate of at least 5 percent annually. A comeback was happening, one oyster bed at a time. "The number of animals we released at Attu was well below the level where the population could sustain itself," Jones wrote. "I concluded it was for the better to expand the necessary protection to otters and let them expand than to try to introduce them. When a sea otter population really begins to grow, it will swamp the survival of any artificial introduction."

Another of Jones's jobs was to make sure that people in boats or other trespassers didn't hunt sea otters, which were like sitting ducks. Sometimes he would trap otters by using tranquilizer guns. "Unlike a seal that often sinks," Jones explained in an official oral history of the U.S. Fish and Wildlife Service, "sea otters float." From 1958 to 1959 he introduced caribou to Adak Island. On Amchitka, working with the newest science, Jones helped reestablish the Canada goose by trying to get rid of island rats. Agattu, in particular, was plagued by these rodents. Sometimes Jones would use a slide-action 12-gauge Winchester Model 12 to shoot them. "All we had access to was poison and that was not good enough," Jones reflected, "and besides, you have to watch where the poison goes. You don't want it to go into the eagle population." [12]

Although Jones felt good about his work for U.S. Fish and Wildlife, including the introduction of caribou on Adak Island, he was apprehensive when he heard rumors that the Atomic Energy Commission (AEC) wanted to conduct a series of underground nuclear explosions on Amchitka Island. Ironically, the Aleutians' greatest strength as a wildlife incubator—their remoteness—was now their most dangerous liability. To Jones, it was strange to think that the U.S. government wanted to detonate an atomic bomb in the "ring of fire," also called the "volcano belt"). "He blasted the

AEC," his wife recalled in an interview in 2010. "He went to Fairbanks to protest. His outspokenness came from a disbelief that a U.S. government agency could be so reckless." [13]

Within an afternoon's motorboat ride from Homer on the Kenai Peninsula were four major volcanoes: Redoubt, Douglas, Iliamna, and Augustine. Sometimes they looked like picture-postcard peaks, particularly when blanketed in snow. But eruptions were—from a geologic time perspective—commonplace. Mount Douglas, which guarded the entrance to Cook Inlet just north of the Shelikof Strait, had a highly acidic crater about 525 feet wide. To set off an atom bomb on an Aleutian island could very easily trigger earthquakes, causing smoke plumes to rise 50,000 feet in the sky. Then again, the Soviet Union had conducted more than 2,000 nuclear tests on the island of Novaya Zemlya (including a fifty-eight megaton, which is considered the biggest explosion in world history) and nobody at the United Nations was chastising the Kremlin.[14] Why couldn't Americans understand that wildlife and nuclear explosions didn't mix?

When the AEC did detonate nuclear bombs on Adak in 1961, all Jones could do was weep. As a government employee, he felt trapped in a corridor with no exit, or a tunnel with no opening. If he quit U.S. Fish and Wildlife because of the detonation on Adak, who would look after the sea otters and emperor geese? The U.S. government also detonated nuclear bombs on Amchitka Island in 1965, 1969, and 1971.[15] All Sea Otter Jones could do was try to protect the wildlife in the Aleutian chain—and outfit his home for solar energy, a small first step in the green movement. To reassure himself that humans weren't invariably monsters, he'd play Bach's Toccata and Fugue in D Minor over and over again on the turntable.

II

Coinciding with Jones's work in the Aleutians during the 1950s was Peter Matthiessen, a New York–based writer determined to document American wildlife in peril. Born on May 22, 1927, in a Manhattan hospital, Matthiessen was raised in Connecticut near Bedford Village, New York. The Matthiessen family was full of nature lovers. Peter's father, Erard Adolph Matthiessen, became a spokesman for the Audubon Society and The Nature Conservancy; a hunter, he had nevertheless been inculcated with ecology by his two boys, Peter and George. The Matthiessens loved Connecticut, where they collected reptiles, bird-watched, and hunted. They lived in Long Ridge, an idyllic rural Connecticut town northwest of Stamford. "My brother and I were always finding animals," Peter Matthiessen recalled. "We played near the Mianus River. One afternoon we found a copperhead den on our property. We took seven as pets. Mother made us get rid of them." [16]

Erard Matthiessen joined the U.S. Navy during World War II to help design gunnery training devices. Young Peter followed in his father's footsteps, joining the U.S. Navy during the Truman years. But his true love was bird-watching. The shifting patterns of nature fascinated him. As a student at Yale University, Matthiessen studied biology and ornithology. Nature writing became another of his passions, fanned by his reading of Thoreau and Muir. In his early short story "Sadie"—which won the Atlantic Prize—Matthiessen demonstrated a flair for descriptive writing. Upon graduating from Yale in 1950, he married Patricia Southgate. Bold, daring, filled with artistic inspiration, Matthiessen moved with his bride to Paris, where they took classes at the Sorbonne. To promote literature, Matthiessen cofounded the *Paris Review* in late 1951, along with Harold Humes and George Plimpton. By 1953, this handsome monthly English-language journal offered its first issue. (The same year, the Matthiessens had a son, Luke, and Peter finished his first novel, *Race Rock*.)

Matthiessen approved an American bald eagle donning a Phrygian cap for the journal's logo. The *Paris Review*, based in New York City, ran a long interview with authors in every issue. With Jack Kerouac publishing "The Mexican Girl" and Samuel Beckett contributing a selection from *Molloy* to the *Paris Review* in the 1950s, one could reasonably assume that the journal was simply an antiestablishment publication promoting avant-garde arts and letters. But as Matthiessen divulged in 1978 to the *New York Times*, he had "invented" the *Paris Review* as a cover for his spying for the CIA.[17] "I was only in the Agency for two years—1951 to 1953," Matthiessen recalled. "Trending left, I quit over a disagreement on my Paris assignment. Plimpton, who had been in Cambridge, took over the *Review*. My interest was in writing fiction. The *Atlantic Monthly* had published two of my pieces. But fiction paid poorly. So I started writing nonfiction essays for magazines to live."[18]

Plimpton became the *Paris Review* editor in chief in 1953, in New York City. The journal had nothing to do with the CIA. Meanwhile, Matthiessen worked on both charter and commercial fishing boats (flatfish, blowfish, tarpon, and tuna) out of Montauk, Long Island, earning extra money from the blue-green depths of the Atlantic. He also captained various shark-watching excursions. "I'd see eighty or ninety sharks in a day as a boy," he recalled. "Now they're scarce everywhere." Matthiessen was collecting good material on sharks, possibly to use in a book or article. Worried that huge corporations were destroying the planet, concerned that the U.S. government was doing too much nuclear testing, Matthiessen decided to become a generalist biologist in the Hornaday vein. During the cold war, the Bering Sea seemed like a moat protecting Alaska from invasion. But Matthiessen was concerned that the Aleutian chain and the Pribilofs—with their wildlife—were being killed by the Atomic Energy Commission.

Besides his rambles in Connecticut and his charter fishing off Montauk, another influence on Matthiessen, in biotic terms, was his brother, George, who studied marine biology at Princeton University and conducted research at Woods Hole Laboratory. While writing wildlife articles for *Sports Illustrated*, Matthiessen was accumulating reams of information about North American birds and animals. He had a brother who willingly served as a marine consultant. "I wasn't planning on writing *Wildlife in America*," Matthiessen remembered. "I didn't *want* to write the book. But I had done all this research. Back then magazine editors tended to treat young writers very badly. I had all this material. So why not a book?" Nobody since Hornaday had written comprehensively about endangered species, wildlife in crisis. "So I went to discover wild America," he recalled. "U.S. Fish and Wildlife in the 1950s was always very helpful. They had Rachel Carson, whom I unfortunately didn't meet, writing enormously powerful pamphlets and papers, all quite lyrical and influential."[19]

Matthiessen, by now separated from his wife, loaded his forest green Ford convertible with books by naturalists such as Spencer F. Baird, A. C. Bent, and Roger Tory Peterson; tossed in a .20-gauge shotgun and a down sleeping bag; and headed west. "The gun was for protection," he recalls. "Perhaps I thought I might need to shoot a bird to eat in the Mojave or Sonora. But I never once used it." What he did use was the booze bottles that were also among his essentials in lighting out for the territory. His mission "on the road" was to document the history of wildlife struggling to coexist with humankind during the atomic age.

As a scholar, Matthiessen knew the history of wildlife extinctions—such as the Carolina parakeet, Steller's sea cow, and Merriam's elk—and wrote high-minded eulogies, including his elegant lament for the Labrador duck. (One of the last of this species was shot off Martha's Vineyard in 1872 by Daniel Webster for the Smithsonian Institution.) What made Matthiessen different from Hornaday in *Our Vanishing Wild Life* was his novelistic trick of imagining that someday flocks of Labrador ducks would be back in the marshlands of the Atlantic coast. "Today, off Long Island's beaches, on a still day of winter the great rafts of black and white pied sea ducks are a fine sight," Matthiessen wrote, "the trim old-squaw and neat bufflehead, the mergansers and goldeneye and dark, heavy-bodied scoters. The sharp air is clean, virtually odorless, and only the strange gabble of the old-squaws breaks the vague murmur of the tide along the shore. Alone on the beach, one can readily imagine that, momentarily, the loveliest pied duck of them all might surface, startled, near a sand spit, the white of it bright in the cold January sun, as it did winter after winter long ago."[20]

From 1956 to 1958, Matthiessen was quite a sight driving around the United States in his beat-up convertible, visiting national wildlife refuges such as Aransas in Texas and the Lower Klamath in Oregon. Books were piled up on his backseat. While camping in the Great Plains, he became fond of George Catlin's depictions of animals; he used them as endpapers for *Wildlife in America*. For the first time Matthiessen also discovered the narrative panache of Francis Parkman's *The California and Oregon Trail*. He was a sponge for information and found the species reports of Dr. C. Hart Merriam especially valuable. Circulating among the U.S. Fish and Wildlife biologists of the upper Midwest, who were working on various federal refuges, Matthiessen started drafting chapters for his first nonfiction book, *Wildlife in America*. "Forests, soil, water, and wildlife are mutually interdependent," Matthiessen wrote, "and the ruin of one element will mean, in the end, the ruin of them all."[21]

Dutifully keeping journals, reading everything possible about the wildlife protection movement of Roosevelt, Grinnell, and Hornaday when he stayed at campgrounds, parking lots, and motels, Matthiessen, who in the late 1960s would convert to Zen Buddhism, eventually flew to Alaska in May 1958 for research. "I started off in the Kenai Peninsula and flew everywhere in Alaska," he recalled. "U.S. Fish and Wildlife took me all around. There weren't roads back then to get around. You had to fly. Every big gravel bar we saw had a wrecked plane. Pilots constantly crashed. They'd break a shoulder blade but walk off relatively unscathed. It was very foggy, I remember, and navigation was a worry."[22]

Seeing the brown bears on Kodiak Island was a must for Matthiessen. But the far north was the magnet that tugged on his psyche. Roger Tory Peterson had written in 1955 that Arctic Alaska was, without question, the "wildest" remaining part of "wild North America." Matthiessen heard that call. He also started observing the caribou and Dall sheep on the North Slope. And everything about Alaska's bear population grabbed his attention. Matthiessen had read the reports of the Harriman Expedition about its successful hunt for Kodiak bears in 1899. Now, in the late 1950s, the Kodiaks, like the grizzlies, were becoming endangered race. Polar bears had become even more scarce. At the beginning of *Wildlife in America*—which Viking published in 1959—Matthiessen included a color plate of a polar bear drawn by John W. Audubon (son of the great ornithologist).

In a chapter titled "Land of the North Wind," Matthiessen included black-ink drawings by the illustrator Bob Hines of a polar bear, Alaska worm salamander, Aleutian tern, whiskered auklet (*Aethia pygmaea*), northern right whale (*Eubalaena glacialis*), ribbon seal, Kodiak bear, barren-ground caribou, grayling, and woodland caribou, and a wolf pack. In 2010, in an interview, Matthiessen said he was pleased that Hines—his illustrator—also provided the drawings for *Silent Spring*. "In the state of Alaska," Matthiessen wrote, "America has a splendid chance to demonstrate that the hard lessons of conservation have been learned, for the great part of it is still under federal jurisdiction and, protected from the excesses of private exploitation, remains unspoiled. The effects of statehood on this unique wilderness should not be the responsibility of its inhabitants alone, for the future of Alaska is crucial to the nation."[23]

Matthiessen investigated the great fisheries throughout Alaska. With awe he visited Native villages, with salmon drying in rows. Point

Barrow—Allen Ginsberg's forlorn radar-station outpost in the summer of 1956—was to Matthiessen a magical place where the rose-tinted Ross's gull (*Rhodostethia rosea*) sometimes appeared after wandering all the way from the Asian Arctic. His prose meditation on the scarcity of Ross's gull predated *The Snow Leopard* (which won the National Book Award), about his search in the 1970s for these rare Central Asian cats, which adapted to cold mountainous environments. Matthiessen knew of Edward Curtis's 1899 photographs of Inuit village life, which appeared in the report of the Harriman Expedition. How had life on the isolated North Slope changed in more than half a century? Not much, it turned out. What really surprised Matthiessen about Point Barrow was that *he* was the strange roadside attraction. "When we got out of the plane, Inuit people were snapping photos," he said. "We were the odd visitors. But they had Kodaks" [24]

It becomes apparent in *Wildlife in America* that Matthiessen wanted to keep Alaska wild—free of fish propagation, fur farms, and reindeer ranges. On the eve of statehood all anybody would talk about was North Slope oil concessions. Saloons in Fairbanks were abuzz with stories of new fields. Just as Ansel Adams had visited Alaska's national parks, the twenty-nine-year-old Matthiessen focused on the national wildlife refuges: the Pribilof Islands for seals; Kodiak Island for bears; and the Kenai Peninsula for moose. Matthiessen believed all three of these species would survive the onslaught of timber agents and oil geologists. And his bush pilot flew him in a Cessna over the National Petroleum Reserve near what was about to become the flagship Arctic NWR.

The highlight of Matthiessen's travels for *Wildlife in America* was touring Arctic Alaska in May 1957. Two U.S. Fish and Wildlife pilots—Jim Branson and Ray Tremblay—took him around on an animal survey east and west of Point Barrow. Just to see all those caribou thronging across the tundra, and Arctic primroses popping up on the pebble-strewn beaches, was life-changing. They flew along the Arctic Ocean toward Canada, feeling minuscule. "From the sky I could see the National Petroleum Reserve, where wildlife was thick," Matthiessen recalled. "And the caribou herd was unreal. I was determined to come back someday."

But Arctic Alaska, the fragile tundra that the Muries had fought to protect, worried Matthiessen. A warning prayer from a Togiak elder stuck with him: "If we fail to save the land, God may forgive us, but our children won't." [25] Much like Bob Marshall, Matthiessen used a book—in his case *Wildlife in America*—to promote the "sequestration of inviolate primeval wilderness for posterity" in the Arctic. To Matthiessen the Wilderness Bill—which was pending in Congress during the late 1950s—needed to be passed. It was senseless, he argued, for America to rethink land policy every four years to deal with special interests or political expedience. Matthiessen understood Alaskans' need to timber, drill, and mine. His chief concern was that once a place earned the designation of a wildlife refuge, it should be left alone. Matthiessen believed that the Arctic land needed to be protected in perpetuity. Gold and oil might be found decades later, he argued, but this possibility didn't mean that the Arctic Game Reserve should be reopened for oil derricks or mine shafts. Otherwise, Yellowstone or the Tetons could become a natural gas reserve. "Glimpsed from the air between banks of cold rolling fogs, the region is beautiful and forbidding," Matthiessen wrote of the Arctic. "Its tundra is desert of a kind, but the great beauty of Alaska lies in its bleakest areas—tundra, ice pack, glacier, and bare mountain, with their unique and precious complement of life." [26]

III

Wildlife in America *was a nonfiction work* set during the late Eisenhower era. Half of the book was a eulogy for extinct species. At times the prose read like a lyrical forensic report, a long-distance gaze backward in time to the sad legacy of mankind's inhumanity toward animals. Matthiessen sadly documented how reckless Americans had been toward the bison, manatee, flamingo, and sea otter. The grimmest Greek tragedies were mild compared with stories of Alaskan wolf exterminators, trophy hunters who sought Dall sheep, and reckless fishermen who overfished and then blamed bald eagles for poor harvest seasons. Matthiessen didn't report on American highway life like Kerouac in *On the Road* (or like John Steinbeck in his 1962 memoir *Travels with Charley*). There was no ego on display in *Wildlife in America*. Matthiessen wasn't a preacher, a braggart, or a show-off about his Alaskan literary expeditions. But Matthiessen was prescient in warning about the toxicity of DDT, ammonium phosphate, and organophosphates and their effect on wildlife. His first book is both a throwback to the meticulous zoological research of Hornaday and an environmental manifesto giving—from the eastern establishment—credence to McClure's bitter "For the Death of 100 Whales" and Snyder's more hopeful "A Berry Feast."

Although the *Paris Review* had no conservationist agenda, both Matthiessen and Plimpton were ardent Auduboners. Matthiessen took from his odyssey on the road a newfound sense of himself as a world traveler. Like the humpback whales, sea otters, and Canada geese he saw in Alaska, Matthiessen recognized himself as a migrant. So when he spied on a Northern wheatear (*Oenanthe oenanthe*) along the Bering Sea he knew it would join its Siberian counterpart on a trek across Russia before heading south through Turkey and Syria into eastern Africa. The bird spends its winter in the sub-Saharan grasslands of Africa after traveling more than 7,000 miles from its breeding grounds. Traveling the world to find great white sharks, snow leopards, and caribou herds became his specialty. There was a lot of Roosevelt and Hornaday in Matthiessen's approach. But there was also a spiritual awakening about protecting wild places that reflected the beat writers Snyder, Whalen, McClure, and Kerouac.

One thing that differentiated Matthiessen from the "dharma writers" on the Pacific coast was his establishment credentials. The *Paris Review*'s contributors also wrote for the *New Yorker* and attended Warhol's Factory happenings. They attended Truman Capote's parties at the Waldorf-Astoria and befriended the Kennedys. More than anybody else in the 1950s, 1960s, and 1970s, Matthiessen was connecting the beat's energy into the main consciousness of his time. Matthiessen regularly ingested LSD, getting it from a renegade Ivy League psychiatrist known as Dr. John the Night Tripper. Among the major writers of the era, perhaps only Ken Kesey and Hunter S. Thompson dropped more acid than Matthiessen. But his reason—as with McClure and peyote—was a longing to feel man's relationship with nature. "On acid I felt the unity of all nature," he recalled. "It was thrilling to feel yourself as part of the whole planet. But I stopped that at some point. We learned that you can achieve the exact sensation through Zen. It's slower, but purer and healthier all around." [27]

Besides gravitating to psychedelics, Zen Buddhism, and remote wilderness areas (like the beats), Matthiessen also championed Native Americans' rights. When the Sioux leader Leonard Peltier was arrested for the Wounded Knee massacre and convicted in 1977, Matthiessen defiantly stood up for his release from Lewisburg prison; it became a long crusade. At Point Barrow in 1961 the Inupiat protested against limits set by the International Treaty in favor of unlimited hunting and wanted no nuclear tests by the AEC; Matthiessen sided with them in both causes.

Another thing that differentiated Matthiessen from the beats was his embrace of Alaska as his special landscape. Nowhere else were the mirages so profuse: on clear summer days, the tundra shimmered like a dragonfly's wings. And how could anyone not be impressed by 900,000 wild caribou? Owing to the success of *Wildlife in America*, he next decided to live on Nunivak Island and write *Oomingmak: The Expedition to the Musk Ox Island in the Bering Sea*. Nunivak was the offshore part of Yukon Delta National Wildlife Refuge, which Theodore Roosevelt had created by an executive order in 1909. A volcanic island of Cretaceous sedimentary rock, Nunivak—with the Etolin Strait separating it from the mainland—had an end-of-the-Earth feel. It was the year-round home of Cup'ik-speaking Eskimos and the summer home of cliff-nesting seabirds such as puffins, murres, and kittiwakes. The only village on the island—which was sixty-five miles long and forty-five miles wide—was Mekoryuk, with a population of fewer than 200.

Because Nunivak wasn't on the Alaskan mainland, it was extremely difficult to reach; also, it was protected by the U.S. Fish and Wildlife Service. Thus a herd of shaggy, surreal musk oxen were released on Nunivak in an effort to restore the species to Alaska. To Matthiessen, the musk oxen were the "last of a great Ice Age family of goat-antelopes that includes the European chamois."[28] They had a wonderful aura about them. About 480 of the musk oxen were still alive on the island when Matthiessen started tracking them on the boggy tundra like a field biologist. All of these Greenland musk oxen were descended from a group of thirty-three calves, which had been imported to Fairbanks in 1930 as breeding stock to help restore the herds to their former range in Arctic Alaska. In 1935 and 1936 eighteen males and nineteen females from the University of Alaska Experiment Station in Fairbanks were taken to Nunivak Island and released. In the 1960s, working with researchers from the University of Alaska and the Institute of Northern Agricultural Research, Matthiessen helped relocate a herd of Nunivak musk oxen back to Fairbanks.

Matthiessen brought out *Oomingmak* (Eskimo for musk ox), in 1967; an excerpt ran in the *New Yorker*. Two years later, Matthiessen was part of the team that relocated fifty-three musk oxen from Nunivak Island to the Arctic National Wildlife Refuge (ANWR).

A westerner who considered himself a kinsman of the far east, Matthiessen traveled to some of the most remote places on earth during his impressive literary career. Believing in the old maxim that the most dangerous thing to do with one's life is to stand still, Matthiessen carefully studied wildlife in a dazzling array of remote habitats. The titles of his nonfiction books speak for themselves: *The Birds of Heaven, Travels with Cranes, Tigers in the Snow, The Tree Where Man Was Born* (which was nominated for a National Book Award), and *The Snow Leopard* (which won that prestigious award). He also helped Secretary of the Interior Stewart Udall write *The Quiet Crisis*—a landmark environmental manifesto—in 1963.[29]

Among all the places Matthiessen visited, however, Arctic Alaska remained foremost in his memory. Whenever the chance presented itself, Matthiessen would tour the region north of the Arctic Divide, meeting with Gwich'in people, observing the musk oxen, and studying the drift timbers at the eastern end of Icy Reef. As a correspondent for the *New York Review of Books*, he wrote marvelous essays on beluga whales and even a white wolf.[30] Whenever he was asked by a conservation group, such as the Audubon Society or the Alaska Wilderness League, to help preserve the Arctic, Matthiessen obliged. His 2003 essay "In the Great Country" (published in the photographic book *Seasons of Life and Land*) remains the most poignant essay ever written about the Arctic NWR. "I am outraged," he wrote, "that the last pristine places on our looted earth are being sullied without mercy, vision, or good sense by greedy people who are robbing their fellow citizens of the last natural bounty and profusion that Americans once took for granted."[31]

Chapter Twenty-two

· · · · · ·

RACHEL CARSON'S ALARM

I

S trange to think that *Walt Disney*—who had done so much to help protect wildlife with *Bambi, Seal Island*, and *White Wilderness*—gave popular credence to the romantic thrust of what William O. Douglas and
Jack Kerouac were arguing in the 1950s about ramblers' rights. Among the
most popular movie shorts of the late 1940s had been Disney's film about
Johnny Appleseed (starring Dennis Day). Suddenly Appleseed's grave in
Fort Wayne, Indiana, became a pilgrimage site for environmentalists.
Everything about the historical Johnny Appleseed spoke of forest protection in an era of logging. Historical texts, often infused with folklore,
reported that the footloose and fancy-free Appleseed always dressed in
ragged clothes, wore ill-fitting shoes without socks, and willingly ate table scraps. In 1871 *Harper's New Monthly Magazine* had portrayed Appleseed
as a wandering mystic. A ninety-nine-year-old friend of Appleseed in
Wells County, Indiana, remembered Johnny's tramps along the Maumee
River of Ohio-Indiana. Appleseed, he said, was "crazy as a loon," always
with "an apple in his hands, turning it over and over, wiping it off, and
then picking out the seeds, and putting them in his pocket."[1]

Disney also promoted the tramping tradition in the 1955 film *Davy
Crockett: King of the Wild Frontier*. The movie's theme song—"The Ballad of
Davy Crockett"—was known to virtually every kid in America. The words
began: "Born on a mountaintop in Tennessee/Greenest state in the land
of the free." Kerouac, Whalen, Snyder, and Douglas also knew some-

thing about mountaintops—and wilderness lovers like themselves were suddenly in vogue among adolescents along with Disney's Davy Crockett. When the film's star, Fess Parker, went to Washington, D.C., he was mobbed by 18,000 to 20,000 fans at the National Airport. Speaker of the House Sam Rayburn and senators Estes Kefauver of Tennessee and Lyndon Johnson of Texas took Parker to lunch and were overwhelmed by joyful cries of "We love you, Davy."[2]

When it came to the promotion of Arctic Alaska, the last frontier, Disney also delivered for the wilderness movement. In 1956 Lois Crisler had published her popular memoir, *Arctic Wild*, about spending the winter and spring photographing wolves and caribou in the Brooks Range. Now it was time for the movie.[3] "Disney's focus on the 'timeless' frontier region of the Pacific Northwest, and particularly Alaska, as the setting for many True-Life Adventures coincided with public campaign efforts to preserve wilderness areas in the far north led by such organizations as the Conservation Foundation and The Wilderness Society," Gregg Mitman writes in *Reel Nature*. "The completion of the Alaska Highway in 1948 threatened what many conservationists like Robert Marshall, founder of The Wilderness Society, had hoped in 1938 would become a permanent place to relive 'pioneer conditions' and the 'emotional value of the frontier.'"[4]

To promote the forthcoming Disney documentary *White Wilderness*—a groundbreaking precursor of today's *Deadliest Catch* and *Man vs. Wild*, Alaskan adventures on the Discovery Channel—photographs were circulated of Herb and Lois Crisler eating roast frog in Oregon and hand-feeding wolves in Alaska. At a time when the cold war pervaded American life and newspapers were filled with grim reports about Khrushchev, Mao, and the hydrogen bomb, the back-to-nature movement found a place in pop culture and was a huge success at the box office. When Disney released *White Wilderness*, about the Crislers, in 1958, the critics praised the wildlife photography. Never before had the migration of caribou, the howls of wolf packs, and the antics of grizzly bears been experienced by so many people. Disney's nine cameramen caught all the inherent drama of Alaska's spring thaw and winter freeze. Moviegoers' hearts raced as lemmings "committed suicide" by jumping off cliffs, a wolverine attacked a fleeing rabbit, and polar bears swam in the Arctic Ocean in search of seals. For use in schools Disney had *White Wilderness* cut into fifteen-minute capsule specialty films such as "Large Animals of the Arctic" and "The Lemmings and Arctic Birdlife."

In the 1980s the Canadian Broadcasting Corporation newsmagazine *The Fifth Estate* (a counterpart of CBS's *60 Minutes*) attacked *White Wilderness* as nature faking and as having involved cruelty to animals. Disney's cameramen were accused of forcing lemmings off a cliff into the Arctic Sea. And the cute scene of a polar bear cub tumbling down a snowy embankment had been shot, allegedly, in a film studio in Calgary.[5] These were serious charges, but in 1958, when *White Wilderness* won an Academy Award for best documentary, they didn't seem to matter to the "Save Arctic Alaska" movement.

The Crislers had done an impressive job of promoting the enduring beauty of Alaska in both *Arctic Wild* and *White Wilderness*, and, building on the status established with their cult work *A True-Life Adventure: The Olympic Elk* in 1952, they became television celebrities and were sought for speaking engagements from Los Angeles to New York. Somehow, in the era of the cold war and containment, Disney's Arctic was therapeutic, a reminder that parts of America were still wild. Retreating to Crag Cabin, their home near Lake George, Colorado (forty miles from Colorado Springs), the Crislers started raising wolves and dog-wolves in their fenced-in backyard. Nature appeared to have been domesticated, with Disney's help. After having brought public attention to the Olympic Mountains, they now made Americans aware that the United States owned part of the Arctic. The Naval Petroleum Reserve had been claimed for oil in 1923; now some adjacent acres were up for grabs. The Crislers' love of wolves far outdistanced Gary Snyder's humorous affection for coyotes or Peter Matthiessen's calm appreciation of musk oxen. "Sometimes [the female] ululated, drawing her tongue up and down her mouth like a trombone slide," Lois Crisler wrote. "Sometimes in a long note she held the tip of her tongue curled against the roof of her mouth. She shaped her notes with her cheeks, retracing them for plangency, or holding the sound within them for horn notes. She must have had pleasure and sensitiveness about her song for if I entered on her note she instantly shifted by a note or two: wolves avoid unison singing; they like chords."[6]

The *New York Times* extolled Lois Crisler for contributing to our "knowledge of animal behavior." While the untamable Ginsberg was reading "Howl" at the Six Gallery, Lois Crisler was giving slide presentations of wolves actually howling all over Alaska. While Matthiessen was studying sandhill cranes that breed in the wetlands of western, northern, and interior Alaska (and also catching up with them near Corpus Christi, Texas), Crisler was having Colorado schoolchildren pat wolves as if these animals were poodles. The field biologist David Mech—author of *The Wolf: The Ecology and Behavior of an Endangered Species*—later credited the Crislers with giving the wolf a makeover from a rangeland menace to a beguiling trickster with the heart of a dog. Without *Arctic Wild* and *White Wilderness*, in which wolves were the dignified heroes, the reintroduction of the species to former ranges such as Yellowstone National Park, Idaho, and New Mexico would have been hugely unlikely.[7]

While Justice Douglas fumed about Disney's nature faking, Olaus Murie knew that True-Life Adventures would attract a new, young audience to the cause of wildlife protection. By trying to capture the quirks of animals in his documentaries, Disney aroused sympathy for them. Disney, in fact, told Olaus Murie that he loved to watch squirrels playing games outside his window; they seemed to have "personalities just as distinct and varied as humans." [8] Murie knew that Disney was overly romantic and given to anthropomorphism—Disney saw human qualities in animals—but he was an important ally of conservation, and Murie did not want to be rude to him. "All nature," Murie once wrote to Disney, "has much in common among its various forms; certain general laws, certain general reactions, and much that can be predicted under many circumstances. But, and I hope this is not too paradoxical, there are many distinct facets that have individuality." [9]

With Douglas being the only dissenter, The Wilderness Society rallied behind Disney's True-Life Adventures filmed in Alaska. In an article in *Living Wilderness*, Murie marveled at Disney's unique ability to capture "the simple beauty of untouched woodlands and their wild inhabitants" (although Murie later criticized Disney's film *The Living Desert* as overemphasizing "tooth and claw" and as being exploitative).[10] And that was the most controlled praise for *White Wilderness*. Another reviewer for *Living Wilderness* said that Disney was the best friend conservationists had, "a sun ripening the grain for wilderness advocates to harvest!" The National Audubon Society bestowed on Disney its prestigious Audubon Medal.[11]

II

There are many wonderful stories about the friendship that developed between Lois Crisler and Rachel Carson following Disney's release of *White Wilderness*. Crisler spent time at Silver Spring with Carson, at exactly the time when America's gifted marine biologist was tormented by the problem of how to make her controversial scientific findings public and was marshaling damning evidence that DDT was poisoning animals and making humans sick. Carson's correspondence with Crisler between 1956 and 1960 shows her promoting the "wonderful book" *Arctic Wild*. Together Carson and Crisler formed a sort of iron sisterhood that eventually included Beverly Knecht, Dorothy Algire, Irston Barnes, and a few others. When Carson had radiation treatments for breast cancer, she confided in Crisler her fears and her determination to become a *survivor*. Furiously, Carson worked on *Silent Spring* (the title came from a poem by John Keats), struggling with her own ill health, determined to ring an alarm bell about the lethal effects of pesticides, DDT, and other toxic chemicals. Carson and Crisler were planning to hike trails together in the Colorado Rockies once Carson finished *Silent Spring*. They wanted to brainstorm on how to stop poisoned bait being dropped from planes by U.S. Fish and Wildlife agents all over Arctic Alaska; the agents were trying to exterminate only wolves, but the chemicals were also killing polar bears and grizzlies.[12] "I feel really over the hump now—there remain only two new chapters to do, plus of course a lot of final revision," Carson wrote to Crisler in August 1961 from Maine. "There has been good solid progress this summer and at last it moves with its own momentum."[13]

Equally worried about contaminants invading water, air, soil, and vegetation in the late 1950s was Justice Douglas. From his imposing office at the Supreme Court he sought cutting-edge data on radioactive waste, fallout from nuclear explosions, detergents used in homes, and chemical wastes from factories. The Bureau of Land Management (BLM) was spraying public lands, Douglas believed, with toxins. Couldn't people understand that when cattle ate grass sprayed with DDT, the milk would be contaminated? Didn't BLM comprehend that, say, spraying DDT on sagebrush killed the willows, too? At his small ranch in Goose Prairie,

Washington, Douglas—after horseback riding in the Snoqualmie National Forest, which adjoined his eight-acre spread—would write flawless prose about the degradation of the planet by big corporations.[14] Nobody, in fact, cheered Carson on in her writing of *Silent Spring* with more fervor than Douglas. She reciprocated by quoting, in *Silent Spring*, from Douglas's dissent in *Murphy v. Butler* (1960)—a landmark environmental case involving people on Long Island who wanted to ban the use of DDT to arrest Dutch elm disease.[15] "The Great God the Dollar has sent us recklessly into chemical controls that have upset the biotic community," Douglas scolded. "Some controls of insects are necessary, but they must be carefully designed and applied."[16]

III

Peter Matthiessen called Silent Spring the "cornerstone of the new environmentalism." The main, stunning thrust of Carson's book was that Americans were poisoning themselves by misusing synthetic pesticides. Every farmer or outdoors worker was affected. Bringing into her narrative the ecological history of the world, plus her own bona fides as a longtime marine biologist with the U.S. Fish and Wildlife Service and as the acclaimed author of *The Sea Around Us*, which was published in 1952 and won a National Book Award, Carson was a scientist with an abiding social conscience. Over the years she had accumulated powerful allies, had become a highly respected science writer, and had become a member of the American Academy of Arts and Letters. Known for her fierce spirit, and for her abruptness, Carson had shrugged off admonishments from her peers who accused her of courting popularity. Somehow she intuited that during the cold war the reading public needed trustworthy voices to speak about the natural world. *The Sea Around Us* was on the *New York Times* list of best sellers for eighty-six weeks. By writing complex marine biology in such an accomplished way, with integrity shining forth from every line, Carson had a kind of power that transcended Lois Crisler's livelier Arctic exploits. At the *Atlantic Monthly, New Yorker, New York Review of Books*, and *New York Times*, Carson could do no wrong.

Chemical corruption of earth was a big topic for many marine biologists to get their hands around. Carson had learned about DDT—the *insect bomb*—shortly after World War II. Although she didn't write about DDT until *Silent Spring*, she was accumulating disturbing scientific information about its deleterious effects throughout the 1950s. Success tends to breed intense jealousy in America, particularly for a woman in what was then the male-influenced world of laboratory science. There was gossip that the mild-mannered Carson was the lesbian lover of Dorothy Freeman of Maine, that she was merely a stalking horse for the Audubon Society, and that her name was mud at the USDA. Perhaps it was all true. But who cared? The U.S. government's reckless spraying of fire ants with toxic pesticides and its poisoning of rivers and lakes needed a whistle-blower; Carson stepped into the role with true courage. Boldly, she claimed that the pesticides were biocides and caused cancer in humans. With the stakes so high, Carson's personal life was irrelevant. But to set the record straight, in the late 1950s Carson was taking care of her sick mother, helping to raise an orphaned five-year-old nephew, and combating a duodenal ulcer. It was the support supplied by her woman friends, Lois Crisler among them, that helped Carson persevere in writing the "galvanic jolt," as the naturalist E. O. Wilson of Harvard called *Silent Spring*, when the entire U.S. chemical industry maligned her character.[17]

Long before Carson and Crisler, women had been important in the U.S. conservation movement. There was the indomitable Isabella Bird, whose explorations of the Rocky Mountains in 1873 had a distinctly feminist goal: "simply to experience the place the same as any male nature lover."[18] Her memoir *A Lady's Life in the Rocky Mountains*, based on the letters she sent from Colorado to her sister, remains a classic evocation of the Rockies' wilderness as a "place of freedom from civilization."[19] Even more significantly, Mary Hunter Austin came onto the literary scene in 1903, writing *Land of Little Rain*, an elegiac memoir promoting conservation of the American Southwest. Every page had the feel of hand-polished turquoise. Death Valley and the Mojave Desert were, finally, not dismissed as wastelands but celebrated as bountiful ecosystems. Bird and Austin are taught in courses in environmental history, but other activists haven't been given their due. Whether it was saving the Palisades along the Hudson River or the ancient ruins at Mesa Verde or stopping saw gangs from clear-cutting California's sequoias, women's organizations were often in the front ranks of the preservation movement. Pick your state and you'll find heroines. In Minnesota there was Lydia Phillips Williams, who protected the Chippewa National Forest from becoming board feet. In Calaveras County, California, Harriet West Jackson prevented timber barons from devastating Calaveras Groves. By 1915, more than 50 percent of the members of the National Audubon Society were women. By the late 1920s, when Herbert Hoover was in the White House, the same was true of the National Parks Association.[20]

The novelist Edna Ferber, author of *So Big, Show Boat, Cimarron,* and *Giant,* also entered the "wild Alaska" movement in the 1950s. To gather material for her 1957 novel, *Ice Palace,* Ferber made five trips to Alaska. There is a wonderful photograph of Ferber bundled up in winter clothes, hood covering her ears, hands deep in coat pockets, taken in the Arctic village of Kotzebue. Ferber thought Alaska was pure magic. A love letter to Alaska, *Ice Palace* was sometimes called the *Uncle Tom's Cabin* of the movement for statehood. With an unerring eye for detail, Ferber wrote about parkas, salmon fisheries, and mining-camp prostitutes; her portrait of Alaska as it was transformed from a territory to a state remains timeless. "Alaska," she said, "is two times the size of that little bitty Texas they're always yawping about."[21]

In Arctic Alaska, Rachel Carson, Lois Crisler, and Mardy Murie were at the forefront of the conservation movement. They were in 1960 what Roosevelt, Muir, and Burroughs had been in the first decade of the twentieth century. Carson's *Edge of the Sea* offered essential scientific arguments for protecting Alaska's unparalleled marine life. Crisler's *Arctic Wild* brought wolves and caribou into the category of spectacular North American animals worthy of federal protection. And in 1960 Murie, disseminating her detailed diaries of the Sheenjek River Expedition of 1956 among friends in The Wilderness Society, helped persuade the Eisenhower administration to protect more than 8.9 million acres (increased to more than 19 million acres in 1980) of the Arctic Range. Murie's film *Letters from the Brooks Range* shows her washing clothes in Arctic waters, a modern-day embodiment of the pioneer woman.[22] All three women were effective conservationist crusaders in 1959, for they placed the ideas of ecology within the broader context of the cold war and frowned on nuclear testing in far-flung ecosystems such as the Aleutians.

While schoolchildren were watching Disney's *White Wilderness* in biology classes and theaters in 1959, Carson sent a letter to the *Washington Post* warning that the pesticides had arrived and were destroying birdlife. This letter awoke Americans to the toxic perils in their own backyards. Some of Carson's biological research had been reinforced by Christine Stevens of the Animal Welfare Institute.[23] The National Audubon Society gave further credence to Carson's brave research, documenting the declining populations of bald eagles as a result of DDT.[24] In Alaska, as Matthiessen noted in *Wildlife in America*, there was a chance to save the last great wilderness. "To many of us this sudden silencing of the song of birds," Carson wrote, "this obliteration of the color and beauty and interest of birdlife, is sufficient cause for sharp regret."[25] To Mardy Murie, the combination of Project Chariot and DDT was too much to bear. With Alaska's statehood looming, a quid pro quo to save the Arctic Range had to be worked out quickly. The "save the Arctic" movement needed to quickly gather a head of steam.

Chapter Twenty-three

o   o   o   o   o   o

SELLING THE ARCTIC REFUGE

I

W hether travelers approached Arctic Alaska by plane, boat, or dogsled, a hush fell over most of them. They seemed to be entering God's no-trespassing zone. For much of the year, the Arctic was frozen off from outsiders, though the Gwich'in and Inupiat traveled the North Slope year-round. Visitors lucky enough to come in the summer months, particularly those trained to understand the flora and fauna seen on a day's hike, were likely to return to civilization as prophets of the wilderness, reverent disciples of the quiet world. Arctic Alaska was God's own altar on Earth, an undatable place so obviously hallowed that no human footprint should ever be too deeply imprinted in the frozen tundra or sea ice. In the delicate northeastern corner of Arctic Alaska that the Muries were trying to save, horrible ruts produced by U.S. Navy vehicles retained their depth for decades, slashing the permafrost as boldly as if they were freshly made. From above—from a bird's-eye view—a traveler could see ancient caribou trails etched into the tundra. Those witnessing the actual migration were often overcome with a stabbing wave of exaltation. Other game trails followed stream corridors and hoof-beaten switchback paths up limestone hillsides. The question that American environmentalists of the mid-1950s were asking was: could the industrial order leave much of a treasured landscape free from development? Or, as Mardy Murie asked, "Will our society be wise enough to keep some of 'The Great Country' empty of technology and full of life?"[1]

Ever since the Sheenjek Expedition of 1956, Olaus and Mardy Murie had lobbied for an inviolate 8.9 million-acre Arctic National Wildlife Refuge from sixty miles east of Prudhoe Bay all the way to the Canadian border.* The proposed site was bounded on the north by the Arctic Ocean (Beaufort Sea), on the east by Canada, and on the west by the Canning River, and led south to a point beyond the lovely crest of the Brooks Range. When discussing Arctic ecosystems, the Muries often used the word *fragile* to help laypersons understand the interconnectedness of the far north wilderness. The elimination of one species could cause a chain reaction affecting others. Lemmings and sparrows were as important to the Muries as polar bears. They had also studied twenty-three types of spiders found in the Arctic.[2]

Bursting with enthusiasm, convinced that Arctic Alaska could be saved, the Muries launched a comprehensive plan to convince Alaskans that the time for preservation was *now*. This seven-year push for the Arctic Refuge coincided exactly with the movement for Alaska's statehood, which was under way following a 1955 constitutional convention in Fairbanks.[3] To the Muries, the land forming the Arctic Alaska refuge was the most majestic panorama of wilderness in North America. It presented life in consummate ecological harmony. Winning the fight against the proposed dam in Dinosaur National Monument emboldened the Muries to seek another victory in Arctic Alaska.

Bringing dozens of photographs they had taken with Justice Douglas while camping in the Brooks Range along the Sheenjek River, Olaus and Mardy Murie spent more than two weeks in Alaskan cities in the fall of 1956, talking about the Arctic with the Territorial Land Commission and local news organizations. Olaus's *Elk of North America* was a classic study of the Jackson Hole elk herd, and many Alaskan outdoorsmen hoped

---

* What would become the Arctic National Wildlife Range in 1960 was later enlarged from 8.9 million acres to 19.3 million acres and redesignated the Arctic National Wildlife Refuge by the Alaska National Interest Lands Conservation Act of 1980 (ANILCA). As Roger Kaye points out in *Last Great Wilderness*, throughout the 1950s the designations *range* and *refuge* were essentially synonyms. Oil-gas companies call the area ANWR. Environmentalists call it the Arctic Refuge. I prefer Arctic NWR.

he'd now fight for the preservation of caribou. In Juneau the Muries met with U.S. Fish and Wildlife officers, garden clubs, and Alaskan politicians. Their lobbying culminated when Olaus Murie showed slides of Arctic Alaska to the Tanana Valley Sportsmen's Association (TVSA) at a stag dinner in Fairbanks. Besides the great caribou herds and Dall sheep groupings, more than 300,000 snow geese (*Chen caerulescens*) fed on the Arctic tundra in autumn before migrating to their wintering grounds in California. The TVSA bird hunters wanted to be sure that this migration would continue for their children's children to enjoy. "Afterward several came to me," Murie wrote to George L. Collins, "and fervently promised their support, and greatly surprised me by giving me an honorary life membership in their organization."[4]

Even though Olaus and Mardy Murie were ecologically-minded, they had no serious qualms about genuine hunters. Unlike some "faux hunters" who guzzled beer and then stomped into the autumn woods to kill deer for a trophy, many serious Alaskan hunters (both Native and Euro American) had an almost Paleolithic reverence for animals. These *real* hunters used their body and senses with a trained acuteness, actually getting into the thought processes of the stalked animals. Where would a grizzly be catching salmon today? What bog would a moose prefer in a cold drizzle? The poet Gary Snyder wrote about this kind of genuine hunter in *Earth House Hold*: "Hunting magic is designed to bring the game to you—the creature who has heard your song, witnessed your sincerity, and out of compassion comes within your range. Hunting magic is not only aimed at bringing beasts to their death, but to assist in their birth—to promote their fertility."[5]

The Muries were convinced that there were members of TVSA who, like the Gwich'in, knew the magic of the animals they killed. Not that the sportsmen's association didn't also include "slob hunters" and "gun nuts" among its members. But the Muries were betting that a number of TVSA leaders—whom they knew as friends for decades—would join the Arctic preservation cause because they intuitively understood Rousseau's theory of the noble savage: the ancient notion that humans still had a lot to learn from the primitive world. Congress had granted TVSA twenty acres of land along the Chena River (an unusual allocation for any sportsmen's club) for two reasons: to teach Alaskan children how to safely use firearms, and to promote the "fair chase" ethics of Theodore Roosevelt's wildlife conservation policies.

Almost like a theologian, Olaus Murie spoke to the TVSA about the spirituality of the Brooks Range and the coastal plain of the Beaufort Sea. The 120,000-head Porcupine caribou herd was his best selling point. Pregnant female caribou came to the coastal plain to give birth in May and June. Since the Pleistocene age, the Muries' proposed Arctic range was also home to the northernmost population of Dall sheep, whose curled horns TVSA hunters coveted. And Murie had preservationist selling points for anglers. America's largest and most northerly alpine lakes— Peters and Schrader—were also located in the proposed 8.9 million-acre Arctic Refuge. As Justice Douglas had found out, the braided rivers were rife with grayling in the summer. Most important, northeastern Alaska was the home of the Gwich'in people, who considered themselves one with the caribou herds. Murie made it clear that the proposed Arctic Refuge was, as Rick Bass put it in *Caribou Rising*, "as wild as when it was first created." [6]

Convincing the antigovernment types in the TVSA that withdrawing 8.9 million acres of Arctic tundra for either U.S. Fish and Wildlife or the National Park Service wasn't easy, even for Olaus Murie. The U.S. Geological Survey had barely mapped Arctic Alaska. Who knew what riches lay under the permafrost? Oil seeps had been spotted between Point Barrow and Prudhoe Bay along the Beaufort Sea. Ore deposits were considered probable on the tundra. In fact, the Alaskan mineral extraction industries—both local and national—abounded with rumors that zinc, copper, nickel, and platinum were to be found in the Muries' proposed Arctic Refuge. Naturalists like Olaus and Mardy were opposed to coal mining, oil drilling, and wolf hunting—activities that many TVSA members thought made Alaska great. "He was a mild-mannered fella," Charles Gray, an unrepentant aerial wolf hunter, recalled of Murie's attempts in the fall of 1956 to lobby the TVSA. "He was sincere and had facts." [7]

A few days after lobbying the TVSA, Murie wrote to Howard Zahniser of The Wilderness Society explaining his firm conviction that to persuade fiercely antigovernment Alaskan residents to protect the Arctic for recreational and aesthetic reasons took patience: "a lot of psychological progress will have to be made before enough Alaskans favor further federal reserves, that is a phobia in Alaska." Zahniser, operating from Washington, D.C., frustrated by the five-hour time difference with Fairbanks, didn't have much patience for the hand-holding style of Olaus Murie. He didn't believe in Snyder's "hunting magic." To Zahniser, who had suffered a heart attack in 1951 and understood the meaning of *borrowed time*, most Alaskans were shoot-em-up types, uneducated in modern principles of conservation and ecology, reckless stewards of the land whose own front yards resembled town dumps with rusted Chevrolets and broken bottles littering the unmowed lawns. Left to their own devices these north country fools, ready to do anything for a fast dollar, would foul the Arctic. Extolling the virtues of the Alaskan tundra, Zahniser, who had started drafting a wilderness bill, wanted the Arctic Refuge rammed down the territory's throat while statehood was a pending issue. The time for the federal government to strike, Zahniser believed, was *now*. "Will the wilderness disappear," Zahniser, who had never visited Alaska, asked Murie, "while we are waiting to be good psychologists?"[8]

Such exchanges between Murie and Zahniser were commonplace in the late 1950s. As a member of The Wilderness Society's governing council, Murie worried that Zahniser's in-your-face style was alienating congressmen and threatening the society's tax-exempt status.[9] Unlike Zahniser, Olaus and Mardy Murie were beloved in Alaska. Powerful friendships had been built up by the couple over the decades. Even though the Muries' primary home was Moose, Wyoming—which had grown into a campus of seventeen ranch structures—they were embraced by the Fairbanks community, and Mardy had many childhood friends in town. Olaus had proudly received an honorary membership in the Pioneers of Alaska—the venerable sourdough club. "Some years ago I received in Alaska one of my most valuable treasures," Olaus Murie wrote, "It was not a gold nugget. It was an honorary membership in the Pioneers of Alaska."[10]

Deeply respectful of outback types who made a living in the far north, Olaus and Mardy were friendly toward Alaskan miners, hunters, and homesteaders; and this attitude made environmentalist fund-raisers in the Lower Forty-Eight uneasy. Olaus and Mardy Murie's consensus-building style with Alaska's NRA types took up a lot of precious time. But the Muries insisted that the Arctic movement needed Alaskan sportsmen as partners. Furthermore, they also wanted the Gwich'in who lived just outside the proposed Arctic Refuge to become allies. "While we were camped on the Sheenjeck River, a group of Indians came up and camped across the river on a hunting expedition," Olaus Murie wrote in *Living Wilderness*. "We had some good visits with them. These represented the first human settlers of Alaska; they fit in with wilderness living, and our system of wilderness areas does not intend to interfere with hunting and trapping by such people."

Searching for influential allies, Olaus turned to George L. Collins of NPS to explain why Brower's confrontational activism wouldn't work in Alaska. "George," Murie explained to Collins in late 1956, "in this whole project I have adopted a go-easy method. As an old-timer up north said to me once: 'Easy does it.' I met with many people, from Fort Yukon to Juneau and I can't remember a time when I came right out and said: 'Support this wilderness proposal.' I told them what our experience was, and I sincerely wanted them to make up their own minds. Without the sincere backing of people who have thought the thing through, I feel we can get nowhere."[11]

Fairfield Osborn Jr.—who was president of the New York Zoological Society and whose 1948 book *Our Plundered Planet* was an eye-opening critique of humans' reckless stewardship of Earth's natural resources—was carefully monitoring the Muries' advocacy of the Arctic Refuge. Osborn worried because the proposal to withdraw more than 8.9 million acres had no proper name, such as Yellowstone or Mount McKinley. "The Arctic Range" sounded like the entire north pole. Perhaps if the proposal was signed into law by Eisenhower, the land could be called the "Pioneers of Alaska Range," maybe the "Theodore Roosevelt Refuge," or the "William O. Douglas Reserve." The problem with the name Arctic National Wildlife Range, it seemed, was that the acronym, ANWR, sounded like a Saudi oil field. Osborn, however, agreed with Olaus Murie that wilderness hunting be allowed on the Arctic Refuge, or Arctic NWR (whatever name was chosen), and that getting the 500-member TVSA on board was essential.

When Lois Crisler discovered that Olaus Murie (of The Wilderness Society) and Fairfield Osborn Jr. (of the New York Zoological Society) were promoting hunting—hunting of her beloved wolves!—in the proposed Arctic NWR, she felt betrayed. She wrote a searing letter to Murie denouncing the "hunting syndrome" as a manifestation of males' cruelty to animals that shouldn't be perpetuated in the modern era. Crisler was most disturbed by the U.S. Fish and Wildlife Service's predator control program as it affected wolves; it involved carnage unacceptable in the postmodern world. Crisler reminded Murie that he himself had written an article in *Audubon* magazine calling for a "wholesome impulse of generosity toward our fellow creature." Using recent ecological studies to make her point, Crisler described hunting as "neurotic behavior," which was "no longer rooted in the demands of reality."[12] Like Zahniser, she wasn't impressed with the concept of "hunting magic" as an argument for killing wolves; in fact, the Alaskans she encountered in the Brooks Range when she was writing *Arctic Wild* were cold-blooded killers.

Because Olaus Murie had defended the Crislers' and Disney's *Winter Wonderland* from accusations of nature faking, Lois's rebuke stung. Murie, who had devoted much of his life to helping Alaskan wildlife prosper, was now being painted by the Crislers and by Rachel Carson as having sold out to the hunting lobby. Frustrated, Murie wrote to Osborn, who had remained above the fray as a mediator in the dispute, that— unequivocally—environmentalists "should not bring into this wilderness project the controversial wolf question."[13] Killing *Canis lupus* was a traditional Alaskan ritual that would be stopped only by endangered species laws. By contrast, Alaskans who loved the land wanted the caribou herds to be permanently protected: the caribou were an embodiment of wild Alaska itself. Both Olaus and Mardy wanted to promote the Arctic NWR with Robert Service's poetry, memorized in grade schools from Ketchikan to Nome—not scold Alaskans for believing that wolves were menacing predators.

Olaus Murie, whose views used to be in the avant-garde of wildlife biology, was now being denigrated as passé. Crisler and Carson represented the new, uncompromising voice of the environmental movement of the late 1950s. No longer were activists interested in making trade-offs with hunters about slaughtering animals for sport. The Crisler-Carson forces considered the Boone and Crockett Club, the Camp Fire Club of America, and the TVSA antiquated and the enemy, not much better than, say, Humble Oil. Destroying wildlife didn't make any sense in the era of toxic chemicals, plastic, and DDT. They thought that debating the intricate rules of hunting licenses was an exercise for numbskulls. A new ecological consciousness had arrived. If the Gwich'in hunted caribou for subsistence and as a spiritual quest, that was one thing; the Crislers and Carson thought their traditions should be honored. But for sport hunters to shoot animals, and to pay NRA membership fees, was degenerate, murderous behavior.

In the essay "Where Wilderness Is Complete" in *Living Wilderness* magazine, Crisler—now a member of The Wilderness Society's governing board—wrote poignantly about the immense complex called the Brooks Range. From Crisler's perspective Arctic wonders such as Mount Michelson, Mount Chamberlin, and Togak Peak needed *full* protection: hunters should not be allowed to slaughter wolves indiscriminately or to kill migrating caribou for the antlers. The Brooks Range, Crisler wrote, was the "only authentic living wilderness left for humans to learn from—to learn something more important than scientific knowledge; to learn the feel of a full response to a total situation involving other lives." [14] Crisler said that Alaskan roughnecks were actually sick-minded cowards who would derive "great fun" from flying a plane in circles to terrify a "small furred animal veering and running beyond what the heart of flesh and blood can endure." In Fairbanks when hunting season started, these "slob hunters" would celebrate by getting drunk in bars such as the Big-I Pub and Lounge on Turner Street. Sounding like Cassandra, Crisler warned that "tomorrow" would bring "that final sportsmen's weapon the jet helicopter with silencer." [15]

Crisler set up the debate over the Arctic Refuge in terms of evil versus good. God was telling businesses to weave their commercial webs elsewhere; here at the top of the world the environment should be left alone. Only a gambler infected with boom fever and willing to defy the odds would believe that oil could be safely drilled in the Arctic. The environmentalists and the Gwich'in and Inupiat (who relied on sea mammals and caribou for fundamental subsistence and socioreligious values) were David, while extraction corporations and hunters were Goliath. The Gwich'in lived in villages to the south of the Brooks Range and believed the coastal plain had to be protected because it was where the sacred Porcupine caribou thrived. These people needed caribou to make boots, sleeping robes, mittens, shirts, and tents. The Gwich'in used every part of the caribou: for example, rawhide (to make tambourine drums), antlers (to make knives), and skin bladders (to haul water). [16] Crisler feared that in the long term, "big oil" would come to the calving grounds. Short-sighted, dollar-obsessed oil companies, she believed, would lay waste to the caribou and the landscape with rigs, roads, drills, and spills. "Here in the Brooks Range the biggest of all historical moments, man against nature, meets actual living wilderness making its last stand," she wrote. "So far man has always won; living wilderness has always perished into desert or mere scenery." [17]

II

Alaska's North Slope in the mid-1950s was still a land of life that had not yet been depredated. Not much had changed in the Brooks Range since the first Alaskans arrived somewhere between 33,000 and 13,000 years ago across the Bering Strait from Siberia during the second stage of the Wisconsin glaciation. Native village elders in Point Hope and Wainwright, it seemed, had little interest in turning the serene Arctic tundra into oil fields like those in Texas or nuclear testing grounds like those in New Mexico. All the indigenous tribes had learned to survive in extremely low temperatures and to live in "peaceful intimacy with all the animals." [18] These humans had found ways to use everything from whale blubber to polar bear fur to stay warm. Some parts of Arctic Alaska (the Brooks Range, in particular) had experienced at least twenty periods of glaciation during the past 2.5 million years, and here as everywhere, the fittest species survived. In summer, hundreds of thousands of birds hatched on the Arctic tundra. Tens of thousands of caribou congregated, calved, and migrated along the coastal plain of the Beaufort Sea, as they had done during the Pleistocene epoch. As Ernest Thompson Seton wrote in his 1911 book *The Arctic Prairies*: "The Caribou is a travelsome beast, always in a hurry going against the wind. When the wind is west all travel west, when it veers they veer . . . but they are ever on the move." [19]

Olaus Murie, struggling against cancer (a melanoma), understood that there is no peace unto the wicked. Today's wilderness could be a garbage dump a year later. An oil company, for example, would not hesitate to flout any scruple or ignore any communal value, for profit. The forces of light, the ecologically conscious people who were stewards of God's land, had to make a public stand over Arctic Alaska. When you're sick, as Murie

was, surviving felt pointless unless there was a last act aimed at helping preserve beauty for tomorrow's children. Murie knew life was transitory. All biologists understood this unalterable Darwinian fact. If the
wilderness movement could establish a huge Arctic NWR, with no roads
for hundreds of miles in all directions, where the evolutionary processes
were left to continue their natural ebb and flow, then the 1950s generation of conservationists would be able to claim that they had stood up to
the postwar industrial beast. By saving Arctic Alaska—or at least a swath
of the Beaufort Sea coast, the tundra plain, the glacier-capped peaks of
the Brooks Range, and the spruce and birch forests of the Yukon basin—
Murie could die content.

Throughout 1957–1958 the proposed Arctic NWR was a bureaucratic
conundrum, which the Muries wanted solved. Nobody knew for certain
whether to push for withdrawal under the Antiquities Act, as Theodore
Roosevelt had done with landscapes such as the Grand Canyon and Devils
Tower from 1906 to 1909. Roosevelt's approach tended to infuriate Congress. This mechanism of executive orders had helped the Alaskan wilderness movement establish Katmai National Monument in 1918, Glacier
Bay National Monument in 1925, and Kenai Moose Range in 1941. Bypassing Congress had the virtue of avoiding brouhahas and filibusters. Obviously, this approach also had the appeal of quickness.[20] But in the long
term, working through Congress also had virtues. "The area will be safer
for all time if Alaskans themselves are behind it," Olaus Murie wrote to
Osborn about the Arctic campaign. "That's why I am so concerned over
developing this general Alaskan attitude."[21]

The Sierra Club entered the effort in March 1957 during the club's Fifth
Biennial Wilderness Conference, held at the Fairmont Hotel in San Francisco. George L. Collins of the NPS, serving as chair, spoke eloquently about
the Arctic's seemingly infinite space with his usual fullness and strength:
its exultant grandeur, solemnity, forlornness, abundant wildlife, and
dancing northern lights. To most of the conservationists in San Francisco,
the Brooks Range was the last great wilderness. The conference served as a
clearinghouse for all the best proposals for saving the Arctic. Lowell Sumner spoke about the Malthusian population explosion. Starker Leopold—
who had written the fine introduction to Lois Crisler's *Arctic Wild*—dealt
with the *morality* of saying no to "big coal" and "big oil." Howard Zahniser

pushed forward his wilderness bill (which had just been introduced in the House and Senate). How amazed Bob Marshall would have been that his wilderness ethos—vast tracts of pristine land with no roads—had gathered so much momentum in the nearly twenty years since his death.

Because the gathering at San Francisco totaled more than 500 people, a small group of Marshall's admirers reconvened after the event in a boardroom to *definitively* determine whether to promote the 8.9 million-acre Arctic reserve as ANP (Arctic National Park) or Arctic NWR (Arctic National Wildlife Refuge/Range). Time was precious. An appropriate name could make a huge difference in the long run. This smaller meeting brought insiders together—high-profile outdoors enthusiasts whose careers had taken them into national conservation politics. Representing the NPS were Collins, Sumner, and Conrad. The U.S. Fish and Wildlife Service had Dan Janzen and Clarence Rhode (Rhode was an advocate of protecting Alaskan wildlife but nevertheless wanted wolves shot on sight). The Bureau of Land Management was represented by its director, Edward Woozley. The Sierra Club had the husband-and-wife team of Richard and Doris Leonard (best known in San Francisco for running the Cragmont Climbing Club, which promoted modern rappelling around Berkeley). But it was The Wilderness Society—represented by the Muries and Zahniser— that delivered the facts and the firepower to the discussions. In Alaska the Sierra Club deserved a lot of credit for pushing Glacier Bay forward to eventual national park status. But the Arctic NWR was the pet cause of The Wilderness Society. When, a couple years later, Zahniser testified before Congress about establishing the Arctic NWR, he claimed that in the Lower Forty-Eight true wilderness had been displaced by development.[22]

Oddly, the consensus at San Francisco was that collaborating with Secretary of the Interior Seaton made the most sense, and that the 8.9 million acres should be called the Arctic National Wildlife Range. *Range* was a designation given to areas with big-game animals. The words *sanctuary* and *refuge* were rejected as sounding too "environmental," in Crisler and Carson's sense. Later, in 1980, *range* was replaced with *refuge*. Everybody present at San Francisco thought the Arctic Range (what became the Arctic NWR) would have made an ideal national park. But it was seriously doubted whether Secretary of the Interior Seaton would ever sign off on such a grand preservationist scheme. "The majority favored wildlife range designation, so we made it unanimous," Collins explained. "The main thing was to get agreement on *something*."[23]

Another important activist had also signed on for the effort to create the Arctic NWR; at the San Francisco meeting, Sigurd Olson entered the Arctic movement for the first time. His book *The Singing Wilderness*, published in April 1956, had found a cult readership for its promotion of the joy and wonder of the outdoors. There was much to recommend Olson, a Minnesotan environmentalist and canoeist who would later be credited with creating Voyageurs National Park in 1975. Never did Olson know a day of leisure when it came to protecting the American wilderness. The handsome, silver-haired Olson felt such oneness with the "boundary waters" of Minnesota that he called his state of mind a "wilderness theology." A "great peace" engulfed Olson when he was in the outdoors. As the naturalist Roger Tory Peterson noted in the *New York Herald Tribune*, Olson wrote the best prose ever about the "northwoods country."[24] Unusually for a book about conservation, *The Singing Wilderness* appeared on the *New York Times*' best-seller list (as number sixteen).

As president of the National Park Association, Olson, who also taught biology at Ely Junior College (now Vermilion Community College), made an appointment to see Seaton within weeks after Seaton's confirmation as secretary of the interior. A fast friendship ensued. Olson, in fact, served as a consultant to the Department of the Interior and to the NPS from 1956 to 1961. If any pro-wilderness activist could be said to have Seaton's ear, it was this deeply honest, soft-spoken Minnesotan, who epitomized generosity of spirit. Conservationists of all stripes, enthralled by *The Singing Wilderness*, were heartened when Olson pronounced Seaton a "fine chap" who "wouldn't repeat the mistakes of McKay."[25]

Fred Seaton was born in Washington, D.C., on December 11, 1909. His father, Fay Seaton, served as assistant to a progressive Kansas Republican, Senator Joseph Bristow (a Bull Moose in 1912). When Fred was a child, the Seatons moved to Manhattan, Kansas, where his father owned the *Manhattan Mercury* (later the *Manhattan Chronicle*). Eventually the family had a financial stake in newspapers in Alliance, Nebraska; Sheridan, Wyoming; and Deadwood, South Dakota. Outgoing, friendly, and a solid B student, young Fred attended Kansas State Agricultural College (which later became Kansas State University) from 1927 to 1931; there, he held the post of director of sports publicity. But because he was nine science credits short, Seaton never officially graduated from college—although this situation was rectified when Kansas State University awarded him an honorary doctorate in 1955. Fred's father purchased the financially troubled *Daily Tribune* of Hastings, Nebraska. It was another newspaper trophy. In 1937 Fred moved to Hastings to run acquisitions. As the elder of two boys and the first out of college, Fred took over publishing responsibilities at the *Daily Tribune*. Seemingly overnight, he turned it into a profitable business. He went on to become city editor of the *Manhattan Mercury*. As a pioneering publisher, Seaton figured out how to develop the newspapers' stories from the wire services.[26]

Of medium build, with grayish blond hair and sharp blue eyes, Seaton was a real white-shirt downtown Republican, proud to be in the party of Lincoln and TR. He was always meticulously groomed. During the 1936 presidential election Seaton served as the personal secretary of the Republican nominee Alfred Landon. Always a great team player, Seaton was a consistent Republican, never once casting a vote for a Democrat. As a political consultant, Seaton accumulated Republican jobs in both Kansas and Nebraska. While brash in temperament, he had fine manners. When he died in 1974, the *New York Times* noted: "It was said of him that no one in politics was wiser in the ways of not giving unnecessary offense."[27]

When a legendary U.S. senator, Kenneth S. Wherry of Nebraska, died in December 1951, Seaton was selected by Governor Val Peterson to fill the sudden vacancy. Earnest, unflagging, and more cerebral than ideological, Seaton served only a little over a year in the Senate, just enough time to be called "senator" by constituents. But Eisenhower liked Seaton, considering him a fellow Midwesterner full of modest intensity. That wasn't unusual. Everybody liked Fred because Fred liked everybody. In a long public career in the Great Plains, he never really received bad press. Now, Seaton's career took off. When Seaton got married, Alf Landon attended the wedding. Seaton and his wife, Gladys, adopted four children. From 1945 to 1949 he was elected to the Nebraska unicameral legislature. His first real political hero was Harold Stassen, the boy wonder who had been elected governor of Minnesota at the age of thirty-one. In 1948, Seaton managed Stassen's unsuccesful bid for the U.S. presidency.[28]

When Eisenhower became president in 1953, he appointed Seaton as assistant secretary of defense for legislative affairs (1953–1955), then as administrative assistant for congressional liaison (1955), and then as deputy assistant to the president (1955–1956). Like a utility infielder in baseball, he could fill various slots. When Secretary of the Interior McKay fell ill in 1957, Eisenhower asked the forty-six-year-old Seaton to be McKay's successor. McKay's tenure had been rocky; he had been accused of making sweetheart public land deals with industries. The deeply ethical Eisenhower didn't like having a new Albert B. Fall on his hands. Seaton, who served as secretary of the interior from June 8, 1956, to January 11, 1961, proved to be an inspired choice.

Fair-minded, and not wanting to see America's natural resources mismanaged, Seaton also had the all-important advantage of being extremely close friends with L. W. Snedden, publisher of the *Fairbanks Daily News-Miner*, a fierce lobbyist for statehood. From 1956 to 1961, whatever Snedden thought needed to occur on the North Slope, Seaton concurred with him.[29] And from the outset, Seaton was determined to take a fair and balanced approach to both industry in Alaska and conservation of natural resources. Whereas McKay had tried to avoid traveling around America, preferring to operate from his desk, Seaton did travel, and he delivered about sixty speeches per year.[30]

During Eisenhower's second term, however, Herb and Lois Crisler were far greater celebrities than the Muries or Seaton, thanks to the Walt Disney Company's magic. Expressing their belief in the value of keeping the Arctic Refuge undeveloped forever, the Crislers wrote to Seaton to urge saving the "only place left on the continent where great authentic wilderness can be reserved."[31] Their letter was filled with ecological buzzwords such as *otherness, vanishing,* and *technical environment*—rather pretentiously for a couple who had stolen wolf pups from a den for a Disney movie. Instead of answering the Crislers directly, Seaton had Assistant Secretary Ross Leffler write them a courtesy reply, informing them that *if* the Arctic NWR was created, then in all likelihood mining, hunting, and trapping would be permitted.[32]

If the Arctic NWR movement had an unsung hero, it was Snedden. He admired the "fair chase" ethics of the TVSA and loathed the new generation of aerial hunters and guides, whose activities were becoming a trend in the late 1950s. Using Super Cub bush planes, pilots would land in the middle of a caribou herd, and then the hunters would fire at the frightened animals. If this deplorable kind of hunting was allowed to continue unabated, the Arctic would be depopulated of game animals. "With American population—and world population—growing at an explosive rate," Snedden warned, sounding like TR, "the natural pattern of life which existed in the area since the dawn of time . . . its game and primitive scenic beauty—could cease to exist." With statehood pending, Snedden urged Alaskans to act "now" to "prevent the destruction and slaughter of game animals tomorrow." [33]

Olaus and Mardy Murie were proud of the *Fairbanks Daily News-Miner* for taking a pro-conservation stand. And they had another unexpected ally at U.S. Fish and Wildlife: Clarence Rhode, the aerial wolf hunter who nevertheless thought the Arctic should be a wilderness preserve. Rhode, an employee of U.S. Fish and Wildlife since 1935, had learned to respect the Arctic as a wilderness like none other. God, he believed, had made the Arctic perfect. According to Collins, Rhode had an "inside track" with Seaton on all issues concerning Alaskan lands.[34]

A law-and-order type, Rhode enjoyed busting salmon canneries around Bristol Bay and the Alexander Archipelago for overfishing. But animal rights activists such as Herb and Lois Crisler, who thought that wolves were cuddly dogs, left Rhode cold. Leftists, he believed, were hypocrites. "Raising a big moose crop," he once declared, "is farming the land exactly as if [one] raised Hereford Cattle."[35]

Nevertheless, Rhode had defended the integrity of Franklin Roosevelt's Kenai National Moose Range. To Rhode, this moose range didn't impede the economic advancement of the territory. The Bureau of U.S. Fish and Wildlife managed the range, from Rhode's perspective, as if the Kenai moose (which had the biggest antlers of all the deer in the world) were a treasured species. That was a good thing. The Kenai Peninsula, however, was the best place to live in Alaska, and settlement there was being thwarted by the moose. When Richfield Oil Corporation of Los Angeles found petroleum in July 1957, the boomers in Anchorage turned against FDR's moose range. Since World War II the Alaskan economy had been sagging. Now, with this discovery of oil, boomers anticipated a profitable new rush. "I have reports," Seaton said, "that things are almost back to the gold rush days."[36]

Rhode tried to prevent the Kenai Moose Range from being dismantled, and to persuade the TVSA to take up the preservationist cause. "There is much pressure in Anchorage, backed by the Chamber of Commerce and oil interests, to convince everyone oil exploration and development will not harm moose habitat in any way and might even enhance it on the Kenai Moose Range," Rhode wrote to Olaus Murie. "Some of the proposals call for a road network in a grid fashion every quarter mile. I cannot agree that would be helpful in maintenance of the type of moose habitat, which appeals to me, but it is difficult to convince the hungry promoters. It even appeals to some moose hunters who feel they would have no difficulty with such a network or killing a moose where they could back up the car to load them."[37]

The political dispute over moose became fierce. A real-estate developer in Anchorage, Marvin R. "Muktuk" Marston, circulated the slogan "Make these moose move over and make room for people." In 1955 McKay, who was then still secretary of the interior, had granted Richfield operational drilling leases in the moose refuge. (During his tenure as secretary of the interior he had, however, created nine new wildlife refuges and had refused to let the U.S. Army take control of the Wichita Mountains Wildlife Refuge, a buffalo range created by TR in 1905, from the Department of the Interior even though the refuge was adjacent to an ever-growing Fort Sill.) The National Wildlife Federation (NWF) defended the Kenai moose population, but the Alaskan zeitgeist in general was drill-drill-drill. The Alaskan politician Walter Hickel, later to become President Richard Nixon's secretary of the interior, was furious that sentimentality regarding moose was slowing down economic development. Hickel reminded Alaskans that in 1910 the defiant Cordova "Coal Party" had organized citizens to dump crates of coal into Prince William Sound to protest against Gifford Pinchot's federal policy of tying up resources.

After congressional hearings in December 1957, Seaton sided with the oil industry. In August 1958 he opened up 50 percent of the Kenai Moose Range for oil exploration. This action directly contradicted his claim in the *New York Times* that oil and wildlife refuges didn't mix. To Seaton, in the end, it made little sense to allow every moose its own 500 acres of prime Kenai real estate to browse.

III

W hy Rhode allied himself with the Muries so zealously with regard to the Arctic NWR is a mystery. Keep in mind, however, that in 1957 oil hadn't yet been discovered there. (Richfield's discovery was at Swanson River on Alaska's Kenai Peninsula.) Also, despite his tough pose, Rhode knew that the migratory caribou of the Arctic (unlike Kenai moose) indeed *did* need thousands of miles of rangeland to survive: the conservationists weren't making that up. In addition, Rhode, whose views on conservation were like those of the old-style homesteaders, believed that the Brooks Range, as it unfurled closest to the Beaufort Sea, was, along with Bristol Bay and Kachemak, perhaps the most beautiful part of Alaska.

The Muries, now working with Rhode, Snedden, and Seaton, set about the task at hand: the Arctic NWR. The *Fairbanks Daily News-Miner* wrote a powerful endorsement of the refuge in the fall of 1957: "We favor the proposal for the Arctic Wildlife Range," its editorial read. "We think the complaint of those opposing it is akin to that of a small boy who has just been given a pie much larger than he can eat but who cries anyway when someone tries to cut a small sliver out of it. We ask those who would raise strong protest over reserving this comparatively small sliver to stop and ponder the fact that the 20,000,000 acres now being made available for development by Secretary Seaton's action comprises an area which exceeds the total land area of five New England states combined." [38]

The attitude toward the Arctic NWR in Anchorage, however, was decidedly negative; this was considered just another federal lockup of Alaskan land. If the Arctic NWR drew tourists, Fairbanks would become the hub city. After victory in the Kenai, developers weren't inclined to forget about Arctic real estate. Rhode candidly wrote to Olaus Murie that Alaskans opposed "everything" proposed by the U.S. government except "immediate statehood"; they felt almost unanimously that "exploitation" of the land should always be the first principle. [39] In Kaktovik—the coastal village that in 1923 became a trading post for the Arctic NWR area—some Natives wanted assurances that their own tradition of caribou hunting would be preserved.

Once again, Olaus flew into Anchorage from Moose, Wyoming, to start working toward acceptance of the Arctic NWR. Osborn's New York Zoological Society, along with its affiliate, the Conservation Foundation, financed Murie's promotional and educational tour around Alaska. (Osborn also had his Conservation Foundation pay for a nine-minute film, *Letter from the Brooks Range*, which was narrated by Olaus and Mardy Murie.)[40] Meanwhile, Olaus Murie's little book *A Field Guide to Animal Tracks*, published in 1954, had become extremely popular with American outdoorsmen. That field guide—part of a series edited by Roger Tory Peterson for Houghton Mifflin—enabled Olaus to get interviews into which he slipped promotions for the Arctic NWR. Impressively, Murie had done all the intricate drawings of paw prints in *Animal Tracks* himself. From 1956 to 1960 a succession of radio interviews were set up for Olaus in Alaska so that he could discuss both his book and saving the Arctic NWR. Working in Olaus's favor was the Inuit belief that *nanook* (the polar bear) had human intelligence—this anthropomorphic notion was the inspiration for a number of children's books. Every souvenir shop in Anchorage or Fairbanks promoted the Arctic polar bears as lords of the last great wilderness. "The trip was evangelism, not adventure," Mardy wrote, "Olaus was speaking and showing slides of the north country before every possible organization."[41]

Olaus Murie struck paydirt when he lobbied the TVSA in Fairbanks for the second time. Never mentioning the issue of killing wolves, and refusing to grovel, Murie showed slides of caribou herds, white ptarmigan, and beautiful streams rich with grayling. Murie was subtly presenting The Wilderness Society's plan for the Arctic NWR ("wilderness as wilderness") with a few sportsmen's provisions for mass "recreational use" that allowed non-airplane, non-helicopter hunting. That evening TVSA members voted on supporting the Arctic NWR, and Murie won, forty-three to five. Murie had been right to fight for the Arctic NWR on this level. As Mardy Murie later noted, her husband had "a natural ability" to deal with Alaskan outdoors types.[42]

With the TVSA on board, Rhode moved quickly with the "Suggested Plan of Administration of Regulations" (the first of many U.S. Fish and Wildlife withdrawal documents). Zahniser's imprint—a lot of cunning legal work—was obvious in this initial document. Murie now met with the Alaska Federation of Women's Clubs and three garden clubs. His slides of caribou on their spring migration (photos that did not show the swarms of mosquitoes) were awe-inspiring. Many of the clubwomen were fascinated to learn that caribou were the only deer in which both sexes grew antlers.[43] They unhesitatingly signed the Arctic NWR resolution. Murie also procured the support of the Izaak Walton League's influential Anchorage chapter. Momentum was building for the Arctic NWR. Rhode started receiving supportive letters from conservation-minded clubs all over the territory. Bob Marshall's dream of a roadless Arctic was finally becoming reality.

Besides pushing for the Arctic NWR, wildlife enthusiasts, such as "Sea Otter" Jones in Cold Bay, were pushing hard for federal protection for the Kuskokwim and Izembek refuges (brackish wetlands that were extremely important for the Pacific black brant.[44] Overseeing the Aleutian district for U.S. Fish and Wildlife, Jones wanted the federal government to better protect otters and birdlife. The word was that President Eisenhower, who was negotiating with twelve other nations a complicated international treaty *not* to develop Antarctica, thought Alaska, which was seeking statehood, might be a good place to create a few additional wildlife refuges to burnish his conservation legacy. The *Fairbanks Daily News-Miner* thought so, too. All three of the major refuges—Arctic (14,000 square miles), Kuskokwim (2,924 square miles), and Izembek (680 square miles)—made sense to the *News-Miner*. Unlike those on the Kenai Peninsula, none of these proposed lands were thought to be rich in timber, oil, or coal. As Eisenhower reluctantly moved toward admitting Alaska as the forty-ninth state, it made sense for the Department of the Interior to have these national wildlife refuge proposals drawn up, detailed, ironed out, perhaps ready for Congress to debate, and—it was hoped—signed into law.

With most Alaskans wanting statehood, arguments about letting U.S. Fish and Wildlife save 8.9 million acres of the Arctic had a low priority. If Alaska had already been a state in the spring of 1957, the opposition to the Arctic NWR would probably have carried far greater weight. Now, even the Fairbanks Chamber of Commerce supported the Arctic NWR. The assistant secretary of the interior for fish and wildlife, Ross Leffler, toured the proposed Arctic Range site in July 1957. Rhode piloted Leffler all around the Brooks Range, exploring the immense world of extremes, contrasts, enlightenment, and wonder as best he could. Leffler was awed by the presentation. The Department of the Interior issued a press release announcing its hope of establishing the Arctic National Wildlife Range. On July 13 the *Fairbanks Daily News-Miner* ran the headline "Arctic Wildlife Area Is Proposed."*

For the Muries, Leffler's announcement was a godsend (as was the *Times*' story). The federal government was now fighting on their side to protect hallowed ground. In the dispute over the dam at Dinosaur National Monument, some 175 organizations had worked against a U.S. government project that threatened to destroy the environment. By contrast, the Arctic NWR had the Department of the Interior on *its* side.[45] Still, the department wanted Alaskans to see the project. And the terms of engagement were now clear: congressional authority instead of executive order. The Muries didn't get a pure wilderness: there was a provision that allowed "limited mineral entry" at the "secretary's discretion,"[46] and this clause worried Olaus and Mardy. But a deal had to be made. By approving the Arctic NWR, Eisenhower had suddenly become a friend of The Wilderness Society (at least for the duration of this particular fight). Having Fred Seaton as secretary of the interior was proving to be a boon to conservationists, as Sigurd Olson had promised. The Muries were acutely aware that there could be many more plot twists, but victory was in sight.

Working for Seaton at the time was Ted Stevens, a former Fairbanks district attorney and legal consultant to the *News-Miner*. Stevens, in his mid-thirties, had earned the Distinguished Flying Cross during World War II with the Army Air Corps, for heroism in the China-Burma-India theater. There was no limit to his enterprise. He was known as Mr. Alaska. Now, in 1957, wanting to rise quickly in the bureaucracy, Stevens was responsible for tweaking the legal intricacies of the Arctic NWR agreement. Ironically, Stevens, when he was a U.S. senator from 1968 to 2009 (at forty-one years, the longest Senate stint by a Republican in U.S. history), fought hard to open the Arctic NWR for drilling. But in 1957 nobody knew that there might be a lot of oil in the northeastern part of Arctic Alaska. And

---

* On September 29, 1957, the *New York Times* ran a story saying that Secretary of the Interior Fred A. Seaton planned to virtually disallow oil and gas drilling in wildlife refuges.

Stevens, an up-and-coming Republican, was glad to be working closely with President Eisenhower, creating alliances aimed at withdrawing lands for the Arctic NWR.

The U.S. government paper "Establishment of Arctic Wildlife Range" (released in November 1957) included the language of Olaus Murie, Zahniser, Collins, and Sumner, and also a lot of paraphrasing. The most significant statement was that the Arctic NWR offered the "ideal opportunity" for the United States to save an "undisturbed portion of the Arctic large enough to be biologically self-sufficient." [47] For the holiday season, Murie returned to Washington, D.C., with slides from his Sheenjek River Expedition of 1956 (including photos of Supreme Court Justice William O. Douglas, who was busy writing up his stories about the Brooks Range for a memoir to be titled *My Wilderness*). Douglas, decidedly skeptical about technology, became a promoter of the Arctic NWR in the corridors of power in Washington, D.C. "Here were pools never touched by man," he wrote of the Arctic, "except perhaps by the awful fall-out from the atomic bombs that slowly disseminate [over] the whole earth." [48]

As a clever strategy, a group of fifty-five Alaskan leaders—forty-nine men and six women—assembled at Constitution Hall on the campus of the University of Alaska near Fairbanks to draft a constitution. The event was modeled on the 1787 convention in Philadelphia where the Constitution of the United States was written. Tired of waiting for statehood, Alaskans were taking matters into their own hands. The constitution drafted at these sessions demonstrated that Alaskans were more than ready to become the forty-ninth state. [49]

But problems were brewing for Arctic Alaska. In the spirit of a quid pro quo, Seaton announced that Public Land Order (PLO) 82 of 1943 (FDR's executive withdrawal of 48 million acres north of the Brooks Range from civilian exploitation or development) would be modified. The land withdrawn by this order had included Harding's 23 million-acre Naval Petroleum Reserve plus about 26 million acres more. Seaton was in effect saying: allow the Arctic NWR to be saved for conservation and we'll open other federal Arctic lands up for mining or drilling. The *Fairbanks Daily News-Miner* applauded this, and on November 20, 1957, 20 million acres of PLO 82 land were opened for Alaskans to develop. Snedden, who was allied with Seaton, ran a 144-page edition of his *Fairbanks Daily News-Miner* extolling the decision: "Seaton Opens Arctic Gas Oil." The 8.9 million-acre Arctic NWR was buried deep in the story as a secondary event.

Olaus and Mardy Murie were worried about PLO 82. They now understood that when the Department of the Interior endorsed the Arctic NWR, this was merely the first step along a tortuous road toward making it permanent. The whole effort could still be obstructed. They warned Osborn and Zahniser to be realistic and keep the champagne corked: premature celebration was a curse of political novices. Charles Sheldon, for example, thought he had saved Mount McKinley in 1906, but it took him until 1916 to get the job done in Congress—a full decade of nonstop lobbying. Alaska's Territorial Department of Mines wasn't going to allow the Arctic NWR without a hellacious fight. Vague language about allowing mining in the Arctic NWR wouldn't placate developers and speculators. Once the Arctic NWR became America's largest national wildlife refuge, they understood, drilling, trenching, and dynamiting wouldn't ever be allowed. The miner Douglas Colp spoke for many when he described the Arctic NWR as a "preposterous fantasy" of New Dealers and wilderness fanatics of the 1930s, now suddenly being embraced by the Eisenhower administration in the 1950s. The Alaska Miners Association flatly rejected the idea of giving caribou herds and seagulls priority over people's jobs.

Sensing that public opinion in Alaska was turning against the Arctic NWR, Snedden once again rallied to the side of the Department of the Interior. On January 29, 1958, his *Fairbanks Daily News-Miner* published another editorial in favor of the Arctic NWR. The newspaper said that the Arctic Range was "one of the most magnificent wildlife and wilderness areas in North America . . . undisturbed as God made it," and that in coming decades tourists from all over the world would come to see the caribou herds, polar bears, and snow-white owls: "Thousands of tourists with cameras and fishing gear will leave many millions of dollars in Alaska, on trips to visit the Arctic Wildlife Range, the only one of its kind in the world." [50]

While the Arctic NWR was hotly debated in Fairbanks, the big story in Alaska was statehood. President Eisenhower, it seemed, was lukewarm about admitting Alaska into the union as the forty-ninth state. A lot of Republican donors—particularly in the canned salmon industry—worried that statehood would mean higher taxes and stricter regulation of fishing. Austin E. "Cap" Lathrop, Alaska's only business tycoon, threatened to shut down operations in the 1940s if statehood came about. Lathrop was paying few taxes on his coal mine, bank, theater, and other operations.[51] "To my mind," Eisenhower said in 1953 about statehood, "not yet has the Alaskan case been completely proved." In his 1954 State of the Union address, Eisenhower championed statehood for Hawaii but *not* for Alaska. With the cold war on, Eisenhower thought Alaska should be fortified as a national defense headquarters. Why cede federal land to create a state after the U.S. government had poured so much money for infrastructure into Alaska during World War II? Politically, Eisenhower feared that admitting Alaska as a state would mean two new Democratic senators.[52]

Eventually, on July 7, 1958, Eisenhower reluctantly and unenthusiastically signed the statehood bill. The deed was done in the privacy of the White House; no Democrats were in sight, and only a couple of reporters were allowed to witness the historic event. "OK," Eisenhower said, sounding almost disgusted, "now that's forty-nine." Alaskans threw a Statehood Day party. The *Anchorage Daily Times* ran a huge headline: "We're In." Suddenly, Alaska was in the glare of the media. A lot of upbeat stories were published under headings such as "Visit Wild Alaska." There were also upbeat stories about Alaska's four producing oil wells and the further exploration that was under way. And Japanese companies were now interested in procuring Alaska's raw minerals.[53] Much was made of all the roads and infrastructure that had been built during World War II and had opened Alaska for commerce.

In late August 1958, with statehood being finalized, the proposed Arctic NWR was jarring front-page news throughout Alaska because of an aviation disaster. Clarence Rhode, his twenty-two-year-old son Jack, and the federal wildlife enforcement agent Stanley Frederickson flew their twin-engine "Grumman Goose" on a roundtrip mission around the Brooks Range on a law-enforcement patrol, in part to locate caribou herds exactly so that these herds could be shown to a group of conservationists in the coming days. The Rhodes and Frederickson were also going to check up on Dall sheep in the Porcupine Lake area.[54]

But then tragedy struck Rhode. The plane crashed somewhere in the vast Brooks Range. For weeks search-and-rescue missions were ordered, but nobody could find the wreckage. The search involved 260 people in almost thirty geographic zones. Rescuers traveled up and down the Koyukuk, Alatna, Chandalar, Porcupine, and Old Crow rivers by plane, all to no avail.[55] Plane wreckage was almost impossible to find in the forbidding Brooks Range in 1958, without modern radio links, flight black boxes, or downed-plane tracking devices. After months of failure, the men were at last pronounced dead. The wreckage was not found until 1979. "He died on the divide of his beloved mountains on the eve of what would become the national environmental movement of the 1960s," Debbie S. Miller wrote in *Midnight Wilderness* after personally seeing the wreckage. "His life ended at the very time the battle began to establish his northeastern corner of Alaska as a wildlife range."[56]

What concerned conservationists like the photographer Ansel Adams about the movement for statehood was that the Department of the Interior was willing to make deals with big oil-gas and mining concerns. Instead of trying to cultivate a cozy relationship with Seaton, as Sigurd Olson had done, Adams thought the Sierra Club should hold out until after the 1960 presidential election, in which Lyndon Johnson or John F. Kennedy—both Democrats, and both far more in favor of national parks than Eisenhower was—had a good chance of beating Vice President Richard Nixon (the likely Republican nominee). "I think," Adams wrote to the environmentalist J. F. Carithers on December 19, 1959, "the conservation organizations are too scared of Uncle Sammy's briefcase men for their own good." Adams was sickened by the way the U.S. Forest Service, in particular, was trying to "milk wilderness for all it is worth."[57]

On January 3, 1959, Eisenhower signed the official proclamation transforming Alaska from a territory to a state. This time Eisenhower stood with a number of Alaskan dignitaries—senators-elect E. L. Bartlett and Ernest Gruening; representative-elect Ralph Rivers; the former territorial governor Mike Stepovich; the acting governor, Waino Hendrickson; and Bob Atwood, publisher of the *Anchorage Daily Times*. Also present, and beaming with joy, was Fred Seaton. Signing pens were handed out by the handful. An American flag with forty-nine stars was unfurled—now a collector's item because Hawaii became the fiftieth state on August 21, 1959. As Eisenhower had feared, Alaska's first two U.S. senators were indeed Democrats. The first two senators from Hawaii were Oren E. Long (a Democrat) and Hiram Fong (a Republican), so the addition of the two new states brought three Democrats and only one Republican to the Senate.

What nobody knew for certain throughout 1959 was what Alaskan statehood meant for the wilderness movement. But Howard Zahniser of The Wilderness Society and Ira Gabrielson of the Wildlife Management Institute kept up the intense lobbying effort. In May they got a big break. Senator Warren G. Magnuson of Washington, a Democrat, introduced legislation to create the Arctic NWR. The Department of the Interior would be the administrator of the refuge, through the U.S. Fish and Wildlife Service. Predictably, Senator E. L. Bartlett denounced the legislation as a federal land grab in the new state. A fight was under way.

To ecologists of the late 1950s, something larger was at stake in the debate over the Arctic NWR: the planet Earth. If the last great wilderness was wrecked by humans, exploited for profit, what did that say about the future of the Amazon, Serengeti, or the Yangtze River? Shouldn't some places remain inviolate? The politics of the Arctic NWR fight coalesced in such a way that if the environmental movement suffered a loss, a dozen growing wilderness nonprofits would lose the momentum they had achieved in the controversy over Dinosaur National Monument. With the world population predicted to be 7 billion by 2010, wouldn't some truly wild places be needed as ecological buffers? Hadn't Eisenhower done the right thing by declaring Antarctica a free zone? Shouldn't the same type of global preservation take place in the Arctic?

Was the Atomic Energy Commisson's Project Chariot really going to explode approximately 2.3 megatons of nuclear bombs and other nuclear devices—equivalent to about half of all the explosives of World War II—to construct an artificial harbor at Cape Thompson on the North Slope? Was Edward Teller so committed to nuclear weapons that he didn't care about radioactive contamination from the blast? It was the threat of Project Chariot that impelled Ginny Wood and Celia Hunter—the two WASP pilots from Washington State—into grassroots conservationism in 1960. If Seaton needed petitions for the Arctic NWR signed by Alaskans to deflect criticism that the Eisenhower administration had turned as soft as the Sierra Club, they could gather the signatures. They would do anything to prevent Arctic Alaska from becoming an atomic test range or an American version of a Saudi oil field.

EPILOGUE:
ARCTIC FOREVER

*This is the place for man turned scientist and explorer, poet
and artist. Here he can experience a new reverence for life
that is outside his own and yet a vital and joyous part of it.*
—WILLIAM O. DOUGLAS

I

For anybody planning a trip to what became the Arctic NWR, William O. Douglas's engrossing *My Wilderness*, published in early 1960, should be mandatory reading. When he was north of the Brooks Range—the great watershed dividing the Arctic from the Alaskan interior region—Douglas felt as if a time machine had taken him back to the beginning of the world. Everything was primordial, uncontaminated, and fresh. Ralph Waldo Emerson once wrote that the "world laughs in flowers." Nowhere was this metaphor truer than in the Arctic, where primroses and forget-me-nots bloom in the summer along the Sheenjek River, suffusing its banks with pink and purple. Botany and animal life fill every page of *My Wilderness*. In the chapter on the Brooks Range, Douglas wrote about seeing caribou hooves crush grass, befriending an arctic ground squirrel (*Spermophilus parryii*), and watching grizzlies dig hummocks. There are scenes of golden eagles nesting near his base camp, and happy-go-lucky pintail ducks scouring for food in the velvety hummocks of the range. To Douglas, Arctic Alaska—like Antarctica—was too precious to permit destructive oil-gas and mining activity, particularly since the future would bring clean energy.

Douglas made clear in *My Wilderness* not only that the Eisenhower administration should create the Arctic NWR, but that its 8.9 million acres should remain untouched by civilization. It would be a laboratory for biologists intent on discovering the natural order before man changed the rhythm of creation. Douglas had done the math in 1960 and had learned that only 2 percent of American land was roadless or a wilderness. Fuming at utility corporations, federal agencies, stockmen, timber barons, and oil-gas executives—"the modern Ahabs" who saw a cliff and thought in "terms of gravel"—Douglas insisted that the Arctic must remain a living wilderness for both scientific observation and aesthetic wonderment.[1] "Potbellied men smoking black cigars, who never could climb a hundred feet," Douglas said, referring to the intrusion of corporate developers into the Pacific Northwest and Alaska, "were now in the sacred precincts of a great mountain."[2]

My Wilderness was illuminating about Arctic life and, considered simply as literature, elegantly written. Douglas wrote about 300-year-old white spruces, about wild cranberry, and about measuring a wolf's paw print (six inches by 5.1 inches). At an Arctic campsite in the upper reaches of the Sheenjek River alongside Last Lake (the latter designation credited to the Muries), Douglas went fly-fishing and recorded the experience. The reader could almost feel the grayling tug at the line. As camp chef for a few nights, Douglas cooked grayling for dinner on the creek-side grill for fellow members of the Sheenjek Expedition of 1956. There was also a sense of urgency in *My Wilderness* regarding alternative sources of energy. Fossil fuels, he worried, were choking the planet to death. *My Wilderness* was also clearly the work of an erudite globetrotter. Without showing off, Douglas compared the wolves of Sheenjek Valley to wolves he had previously studied in Afghanistan and Persia. Alaskan wolves, in fact, found a very effective defender in Douglas. "The sight of a wolf loping across a hillside," Douglas wrote, "is as moving as a symphony."[3]

Ethel Kennedy—whose husband, Robert F. Kennedy, was murdered in 1968 while running for U.S. president—fondly remembered Douglas's nonstop promotion of the Pacific Northwest and Alaska. To Douglas, the region from Big Sur in California to Homer, Alaska, 3,000 miles away, was an ecotopia. When he talked about the lush green zones along the Pacific coast, he would also promote the notion of a "wilderness bill of rights" to protect "the region's rivers and lakes, the valleys and ridges," from "mechanized society."[4] In 1962 Robert and Ethel Kennedy had joined Douglas and his wife Joanie for a week of camping along the Olympic beaches. Douglas loved cooking rainbow trout for the Kennedy party and gave his recipe as follows: "Set rock at 45 degree angle, and heat upper side with fire; salt and pepper trout and roll in flour and place on heated face on rock; do not turn; rock will cook underside and campfire will cook topside; serve when trout is deep brown."[5]

"Bill had an enormously open mind around the campfire, talking about the world," Ethel recalled. "He didn't pontificate. He was refreshing. He took us to the rain forests—which, I might add, are appropriately named. We all got soaked on the trail, day in and day out, but Bill didn't seem to notice. He was serious about us seeing *his* wilderness. A lot was made of the fact that Bill had gotten Mercedes a special gift for their anniversary. Our group, a few couples, kept speculating what it was: a diamond brooch or necklace. When the big moment came, Bill presented her with her own ax for chopping wood. That was the big romantic gift."[6]

The Kennedys learned on the trail just how devoted Douglas was to deep silence and utter seclusion. To Douglas, the great American outdoors was quiet medicine for the shattered urban soul. Douglas, in fact, knew a lot of U.S. veterans who ended up staying in Alaska because the open land offered healing and solace. In a marvelous extended essay published by the Orion Society, the poet Terry Tempest Williams called Alaska's wildlife refuges, with their liberating effects, the "open space of democracy"; Douglas would have liked that phrase.[7] Men who had seen combat in World War II—such as Morton Wood, who ran the Denali Lodge with his wife, Ginny Wood—needed the Alaskan wilderness to spiritually heal after seeing so much blood spilled. Wild areas such as the proposed Arctic NWR, Douglas believed, could bring God back into the lives of disillusioned ex-soldiers like Wood. These war veterans would backpack for days, weeks, or even months. Fresh air was the real curative for a soldier. The clean air off the Arctic Ocean, for example, was far more healthful than the psychotherapeutic drugs or morphine distributed at a dozen facilities similar to Walter Reed Hospital. A profound sense of humility fell over people on the tundra. The soul became *whole* again. Many veterans of World War II and the Korean War were proud that so much of Alaska was public land—it was wild America for the *people*.

With regard to the politics of wilderness, however, Douglas was a pragmatist, not a dreamer. He understood that with regard to conservation, no important cause was ever permanently won or lost. The combat always had to be renewed and the rationale for preservation reiterated. Every time America went to war, opportunistic companies, capitalizing on national fears and anxieties, claimed that the Tongass should be clear-cut or that Cook Inlet should become an oil field. Executive orders and legislation, once so potent, would over decades become dim and faded documents with none of their original preservationist passion. Thus Sitka National Historic Park—America's great totem pole field—was seized in 1942 by the U.S. military, which removed huge quantities of gravel from the park's shoreline, devastating the environment. No part of wild America was safe when an economic crisis arose. Every new generation would have to fight for the integrity of the Denali wilderness or Glacier Bay. The money-grubbers, Douglas believed—those who couldn't recognize God's artistry—were always going to swarm like a plague of locusts onto the land, destroying its splendor. The mistake conservationists made was believing in total victory. No wild place was ever safe from Moloch.

To the Muries the fight for the Arctic NWR was about the Brooks Range and coastal plain, caribou calving areas, and polar bears' denning. Douglas concurred with these sentiments. But he also saw the preservation of those 8.9 million acres as a victory of the quiet world over the sonic boom. He wanted corporate noise polluters regulated, fined for selfishly stealing people's right to quiet so that their boards of directors could become multimillionaires. In his opinion in *United States v. Causby*, Douglas agreed with a chicken farmer who claimed that noise from U.S. military airplanes had caused his poultry to die of panic. Douglas also felt that he personally had a God-given right to ride horseback on a "precarious mountain trail" without a sonic boom or the roar of jet engines frightening his mount and putting himself in danger of being tossed.[8]

II

O n February 26, 1960, just a few weeks after *My Wilderness* was published, the Alaska Conservation Society (ACS) was founded. Realizing that Olaus and Mardy Murie needed *local* help with their campaign for the Arctic NWR, a group of activists in Fairbanks began a policy assault that continued throughout 1960—and worked. The goal of the ACS was to marshal local opinion for the Arctic NWR and thereby help Secretary of the Interior Seaton get the job done in Washington, D.C. The driving forces were Celia Hunter and Ginny Wood, the women who had been WASP pilots during World War II and who were now committed to what would come to be called ecotourism. If Costa Rica could attract tourists to its tropical rain forests, then, logically, Alaska could promote temperate rain forests. Spiritual reward, however, not profit from tourism, was the primary motivation for creating the ACS.

To Wood and Hunter, the Arctic was unlike any other place they had flown over in Alaska. The light, the sedge, and even the soil were *different*. When Hunter flew from Fairbanks to Kotzebue in late August and early September, the flaming yellow birch and aspen combined with reddish brown meadows and blue waterways to form a patchwork of dramatically mixed Arctic habitats. She would see hawks circling overhead, identifiable by the multibanded tail with a broad, blackish subterminal stripe. Ice fog would roll in for hours, causing strand bands.

The harsh country outside Fairbanks had always attracted women of fortitude, with an appreciative eye for the land's expansiveness and courage enough to heed its summons, in sync with the power of the Alaskan wilderness. Both Wood and Hunter were part of this frontier tradition. On clear days, toiling at her desk in Fairbanks during the first months of 1960, Wood could see the distant mountains outside her kitchen window, through the towering birches. Since World War II, she had flown all over that range; she knew every peak like the palm of her hand. She had landed on runways and gravel bars. Along the way she had made a lot of friends in the North Slope.

The Alaska of the pioneer days was always part of Wood and Hunter's consciousness—the Klondike gold rush, aviation in the 1920s, Mount McKinley and Gates of the Arctic, the salmon runs of Bristol Bay, and, stretched out north of Fairbanks, beyond the Arctic Divide, the Brooks Range, which Robert Marshall had written about in *Alaskan Wilderness*. Having organized tours from Camp Denali from 1954 to 1959, Hunter and Wood were determined to help create the Arctic NWR before President Eisenhower left the White House. Closing Denali Lodge for the winter season from October to May, Hunter and Wood, taking advantage of their freedom during the off-season, started to organize from Fairbanks on behalf of their beloved Arctic Range. Their headquarters was a birch log home in the Dogpatch area of Fairbanks (not far from the university), and the ACS was from the beginning a typical small, personal nonprofit organization. Aspens surrounded the handsome cabin; at their Dogpatch headquarters, Hunter and Wood felt at one with nature. An owl nesting box was hung in a nearby tree, to attract wisdom.

Because Camp Denali was a seasonal business, taking people to see Wonder Lake only from April to November, Hunter relocated the office mimeograph machine to Dogpatch, and installed it on the cabin's second floor. At that time, the low-cost mimeograph, which worked by squirting ink through a stencil onto paper, was a common way to disseminate gossip and news. Ginny Wood, in fact, lived at the Dogpatch headquarters with her husband; she was always on call. Just one house over, down the dirt road, resided Celia Hunter. Both women were beloved in Fairbanks. Ginny emerged as the dauntless workhorse of the ACS, forming alliances and recruiting an impressive mélange of volunteers, networking all over the state to knit the conservation community together so that the Eisenhower administration would be forced to take the Arctic NWR seriously. Hunting guides, fishing charters, glacier tours, kayak retailers, outdoor gear shops, organic food stores—all joined the cause of the Arctic NWR because it promoted wild Alaska, the business they were all in. With regard to promoting state tourism, Hunter had an address file filled with all the right people, lovers of the wilderness who gladly signed petitions to save the Arctic NWR.

Celia Hunter testified on October 20, 1959, before the Committee on Interstate and Foreign Commerce in Ketchikan, and had made a series of arguments that deeply influenced the acceptance of the Arctic NWR by ordinary Alaskans. Quite convincingly, she showed how tourism had supplanted mining as Alaska's second-biggest revenue-generating industry. (Military construction was still first.) There was more long-term economic benefit to be gained from tourism than from hiring, say, 100 temporary tie pickers or timber testers. "The years 1958 and 1959 have seen tourist income at least double," she said, "and estimates as high as triple the figures have been given by the tourist industry. And, yet, in

spite of the decline in importance of mining, and the increasing emphasis on tourism, the whole tone of our state administration is set by the
mining interests." [9]

As with all successful new nonprofits, a hierarchy was quickly established at the ACS. Ginny Wood collected dues and wrote hundreds of
recruitment letters. Her work ethic meant a lot of envelope licking and
a lot of work through the night and into the morning hours. Conservation politics, she soon learned, involved nonstop paperwork. Throughout 1960 Wood corresponded daily with state senators, college students,
restaurateurs, small business owners, outfitters, and travel agents, and,
most important, kept the mimeograph machine humming. She pored
over territorial records, land deeds, and loads of newspapers to extract
information about the Arctic. Wood's motto, printed on the first newsletter bulletin, was "Alaskans Organize." And at the Dogpatch headquarters,
caulked against winter weather, various funny, quirky aphorisms were
taped to the wall: "For God's sake don't let them make any more progress!"
and "Next week we gotta get organized!"

Wood preferred typing letters to calling people on the telephone; for
one thing, letters were cheaper. There were hardly any exceptions to
this preference, but whenever Lowell Sumner of the Department of the
Interior called, Ginny Wood felt cheerful. She liked robust men who appreciated life to the fullest. Along with the Muries, he always offered
the soundest counsel on how to make the principal issues—like saving 8.9 million acres of the Arctic—heard in the right way by the powers
that be in Washington, D.C. "We both loved our airplanes as much as the
Arctic," Wood explained about her friendship with Sumner. "Whenever
I'd be in the most remote Arctic places like Nome or Barrow or Coldfoot, I'd invariably bump into Lowell. Olaus was very mellow, always
taking his biology seriously. Lowell liked to see things from the sky . . .
like me." [10]

Wood called the ACS newsletter, which began getting mimeographed
in March 1960, the by-product of a "subversive" press. [11] Because Hunter
also did serious fund-raising for The Wilderness Society, she added a
wider conservationist net to the homemade newsletter from Dogpatch.
By contrast, Wood tended to fill the ACS newsletter with folksy woodlore.
Visually, the five-page newsletter was like a church bulletin. People in

Fairbanks committed to Ginny Wood were known as Friends of Ginny, or FOGs. The acronym was a perfect fit because Wood flew her Cessna even in the worst weather imaginable, feeling responsible for linking North Slope bush communities to Fairbanks when an emergency occurred. If, say, a physician or funeral director was needed in an Arctic town, Wood always volunteered her pilot services pro bono to fly the person out from Fairbanks. Locally, she was known as an ace bush pilot, an all-around good Samaritan, and an Arctic activist. Nobody ever accused Ginny of harboring any confusion on issues related to conservation.

Hunter and Wood did a few clever things when creating the ACS. Like Edna Ferber in *Ice Palace*, they boisterously touted Alaska's unequaled greatness. In particular, they bragged about how abundant Alaskan wildlife was, compared with the depleted wildlife of Oregon; how superior their air quality was to that of smoggy California; and how many more vodka-clear lakes Alaska had, compared with those in Minnesota. "Fortunately, we came into statehood with our natural resources relatively intact and we have the chance to profit by the mistakes made by other states," the first newsletter read. "Whether we choose to learn by the mistakes of others, or to learn by making them over again ourselves was up to the individual citizens as well as our representatives in government and the professionals in public service. In most other fields of endeavor, mistakes may cost time or money, but they can be corrected. With wilderness and with wildlife resources, you don't get a second chance. When they are gone, they are gone."[12]

Nobody ever sold the idea of saving 8.9 million acres with quite the gusto of Ginny Wood circa 1959–1960. Whether she was writing in the ACS newsletter or testifying before a congressional committee, Wood insisted that saving the Arctic Refuge was in the tradition of Daniel Boone. Both Hunter and Wood knew the right buzzwords to use for Alaska: *individual* and *wilderness*. The *Fairbanks Daily News-Miner*, established in 1903, set the tone with its motto: "Independent in All Things . . . Neutral in None." The ACS appealed to Alaska's chauvinistic sense of being the *last frontier*. Harking back to the days of 1898, when the Klondike gold rush transformed Alaska into a boom land, Wood claimed that the descendants of the early pioneers now had a sacred preservationist obligation to uphold the traditions:

We Alaskans must reconcile our pioneering philosophy and move on to the realization that the wild country that lies now in Alaska is all there is left under our flag. Those who see the wildlife range as a threat to their individual rights refuse to face the fact that unless we preserve some of our wild land and wild animals now, the Alaska of the tundra expanses, silent forests, and nameless peaks inhabited only by caribou, moose, bear, sheep, wolf, and other wilderness creatures can become a myth found only in books, movies, and small boys' imaginations as the Wild West is now. And I regret as much as anyone that the frontier, by its very definition, can only be a transitory thing. The wilderness that we have conquered and squandered in our conquest of new lands has produced the traditions of the pioneer that we want to think still prevail: freedom, opportunity, adventure, and resourceful, rugged individuals. These qualities can still be nurtured in generations of the future if we are farsighted and wise enough to set aside this wild country immediately and spare it from the exploitations of a few for the lasting benefit of the many.[13]

There was another factor in the debate of 1960 over the Arctic NWR. In Alaska—with a population of only 250,000—politics were personal. For more than a decade Wood and Hunter had done neighborly favors for people living in Nome, Cold Bay, and all points between. Few Alaskans trusted the federal government much—with the notable exception of the armed forces. Wood and Hunter's notion of having the U.S. Department of the Interior control the 8.9 million acres of Arctic Alaska wasn't something the average citizen of Juneau, Anchorage, or Fairbanks would automatically approve of. But doing a favor for Ginny Wood or Celia Hunter—that was a different matter entirely. Cashing in all their chips, recruiting friends to join the ACS, Wood and Hunter started circulating the pro–Arctic NWR newsletter all around Alaska.

Another obvious step for ACS was lobbying in tandem with Alaska's premier conservationist groups—the Alaska Sportsman's Council, Tanana Valley Sportsmen's Association, Fairbanks Garden Club, and others—to keep the movement for the Arctic NWR going. There was power in unity. Everything was so hurried for the ACS during the first months of scurrying to line up allies during 1960 that there wasn't a minute to be bored. The next step for the "Arctic Forever" cause involved ensuring that the national conservation societies, such as the Sierra Club and the National Wildlife Federation, would not feel poached upon by an upstart outfit like the ACS. All these organizations had worked for the Arctic for years. Hunter and Wood reassured their allies that the ACS wasn't going to eclipse them or compete with them. The ACS never urged anyone to defect from other groups; but additional financial support for their Dogpatch operation was welcomed.

Tourists from other states, particularly those who had been at Camp Denali with Wood and Hunter, would be tapped for both moral and financial support. To give the ACS immediate credibility, Les Viereck (a veteran of World War II who had become a biology teacher at the University of Alaska) was the unanimous choice for president. The treasurer was John Thomson, an information specialist with the Agricultural Extension Service at the University of Alaska–Fairbanks, who had climbed Mount Michelson in the Brooks Range in April 1957. He would be responsible for paying the bills. In truth, the ACS was a shoestring operation, tasked with getting the disagreeable business of haggling over the Arctic NWR finished and done with.

But it was Sigurd Olson's visit to the Arctic Refuge over the summer of 1960 that seemed to influence Seaton the most. Seldom has a reconnaissance trip by a conservationist produced such fruitful results as Olson's whirlwind trip to Alaska, at the behest of the Department of the Interior. Olson was awed by the idea of the proposed Arctic NWR. He quickly understood that as with Antarctica, saving this living wilderness would make the *world* happy forever; if you lived in crowded Beijing or overpopulated Mexico City, you would want to be assured that the polar cap regions were flourishing. "I stood on one plateau one morning and could see 75 to 100 miles in all directions to four immense mountain ranges with snow-capped peaks," he wrote to friends. "Such a sense of immensity and distance, I had never known before." [14]

Olson—"Captain Wilderness"—reported on Mount McKinley and Glacier Bay national parks and recommended that the Mission 66 road plans be downsized. After counting 161 Dall sheep and reaching a better understanding of Charles Sheldon's rigorous legacy, Olson promoted the Wrangell Mountains of south-central Alaska as a potential new national park. (They became one in 1980.) At the Valley of Ten Thousand

Smokes he experienced the immediate aftermath of a volcanic eruption: the stench of acrid sulfur nearly suffocated him, and gray ash blew in the air like snow.[15] As the author of *The Singing Wilderness*, Olson raved about "big, bold, beautiful" Alaska. "I've been traveling for three or four days," he wrote to his son, "and it's just been one national park after another."[16]

Olson hadn't been as important as the Muries in getting the movement for the Arctic NWR started, but the fact that Seaton trusted him mattered tremendously in 1960. Olson came back to Washington, D.C., that summer with three policy recommendations: sign executive orders creating the Arctic, Izembek, and Kuskokwim wildlife refuges. If Congress did not take up these crucial proposals, Olson recommended that Seaton implement them by an executive order.

Also helping with the ACS lobbying was Mardy Murie. Many women would have wanted the glory of being credited in history with saving a treasured landscape like the Arctic NWR. But Mardy was different. She considered Olson, Hunter, and Wood heroes of conservation. Ever since her honeymoon in 1926, when a dogsled had pulled her over the tundra once gouged by glaciers, she had dreamed of a Brooks Range wilderness park including the coastal areas. Sharing credit with Wood and Hunter wasn't an issue for her. In 1958 Mardy had sailed with Olaus across the Atlantic Ocean to attend Finland's International Ornithological Conference. Besides marveling at how much better Scandinavians treated their landscapes than Americans, the Muries recalled the old days when Alaska didn't even have a major road.

Another shrewd organizational maneuver by the ACS was getting accredited as a nonprofit only thirteen months after Alaska achieved statehood. That single strategic decision, which took a lot of hustle to accomplish, proved crucial as the ACS sought federal protection for the Arctic NWR. On February 15, 1960, after Congressman Ralph Rivers—Alaska's only representative—withdrew his opposition, the House passed HR 7045. Rivers had proved to be a tricky ally, changing his vote continually, depending on who he was talking to. However, the Arctic Range bill—S 1899—was now in the hands of the Senate. And this was problematic. Both Democratic Alaskan senators—Bob Bartlett and Ernest Gruening—seriously objected to the establishment of the Arctic NWR. A battle was developing. It was doubtful that the Senate would pass S 1899. Therefore, the ACS knew it needed to have its ducks in a row before the arrival of Seaton, who was scheduled to speak at the University of Alaska–Fairbanks on March 3.

The audience inside the New Bunnell Building auditorium that evening included members of the ACS and supporters of the Arctic NWR who would inveigh against oil, gas, and coal development on the proposed refuge land. It was like walking into a trap. These were the Alaska intelligentsia, able to quote from *A Sand County Almanac* and identify a bird species from a distance with just a single glance. If Seaton believed that *real* everyday Alaskans like Wood and Hunter were pro-refuge, this would influence him mightily. "I had voted for Eisenhower in 1956," Wood recalled. "I thought his new Interior Secretary [Seaton] would do the right thing for the Arctic."[17] Seaton was impressed that the ACS had led Alaskans in backing the Arctic NWR even though the U.S. Chamber of Commerce thought it was a terrible idea. When asked in Fairbanks what would happen if the Senate didn't pass S 1899, Seaton snapped, "I could withdraw the wildlife range this afternoon if I choose to do so."[18]*

Throughout the fall of 1960, the fate of the Arctic NWR remained undecided. The presidential election—Kennedy versus Nixon—was preoccupying the nation. Seaton, campaigning for Nixon, postponed the decision on the Alaskan lands until after November 4. Governor William A. Eagan of Alaska made a last-ditch effort to have the Eisenhower administration turn the proposed Arctic NWR over to the state. Eagan believed that the 8.9 million acres could easily be opened to both mining and nature preservation. That meant the "big three" politicians in Alaska—all Democratic—were against it. "It is my conviction," Eagan said on September 26, 1960, "that conservation needs of the Nation and the State for an unspoiled Arctic Wildlife management area can only be achieved under State Management."[19] In a misleading letter to Seaton the governor threatened that the Arctic NWR, if established by the Eisenhower administration, would be a gross violation of state law.[20]

Sensing a threat from Eagan, Gruening, and Bartlett, the ACS attacked

---

* The Eisenhower Presidential Library provided me with a batch of Eisenhower-Arctic NWR articles that inform this chapter.

the governor in a press release. The ACS mocked the governor's notion that mining conglomerates had wilderness values. Seaton refused to respond to Eagan's plea. This snub infuriated Eagan. When Kennedy won the presidential election, Seaton knew his days were numbered. After Thanksgiving, he started clearing out his desk and preparing to move back to Nebraska.

Luckily for the wilderness movement, Sigurd Olson was invited to visit Seaton at the Department of the Interior on C Street one afternoon in early December to say hello. Olson brought up the Arctic NWR. Was Seaton at long last ready to sign off on the 8.9 million acres? To Olson's astonishment, Seaton was still of two minds. He was preparing to head back to Nebraska to run for governor in 1962. He didn't want to be vilified by the mining industry. But his heart was with Olson and the ACS. Owing to their smaller acreage, Seaton was ready to establish national wildlife refuges at Izembek and Kuskokwim. But Seaton was up in the air about the Arctic NWR, asking, "What will the Alaskans think?" Mustering all the conviction he could, looking straight at the apprehensive outgoing secretary, Olson assured his trusted friend that the smart folks in Alaska "would fall into line." [21]

Olson wasn't alone in gently pushing Seaton to do the right thing regarding the Arctic NWR in November–December 1960. Although Douglas didn't write a letter to Seaton about the Arctic NWR, two of his former wives—Mercedes Eicholz and Cathy Stone—both thought it was "highly likely" that he had lobbied the secretary of the interior. Everybody in Washington officialdom had heard Douglas hold forth on the Brooks Range, insisting that U.S. Fish and Wildlife had to protect the Serengeti of America, including its profusion of wildflowers. "The vast, open spaces of the Arctic are special risks to grizzlies, moose, caribou and wolves," Douglas would tell anybody who would listen. "Men with field glasses and high-powered rifles, hunting from planes, can well-nigh wipe them out. In this land of tundra, big game has few places to hide. That is another reason why this last American living wilderness must remain sacrosanct." [22]

One legitimate concern Seaton had was the fact that Alaska's leading Democrats—Senator Bartlett and Senator Gruening, in particular— weren't enthusiastic about the Arctic NWR. What if the Kennedy administration overturned it? Why should Eisenhower establish it with an executive order only to have the Democrats reverse him? That was a paralyzing thought for Seaton, but this is where Douglas reentered the drama. Extremely close to the Kennedys, Douglas hoped he might be chosen as Kennedy's secretary of state. Nobody knew whether Douglas would step down from the Supreme Court to take over the State Department, but it was a persistent rumor circulating around Washington that December. When Robert Kennedy asked Douglas whom his brother should nominate as secretary of the interior in November 1960, the justice had an immediate answer: Stewart Udall. "Douglas was one of my biggest promoters," Udall recalled. "We didn't see each other much, but we were clearly on the same conservationist team." [23]

Douglas was wise to recommend Udall. Raised on an Arizona ranch, a Mormon, Udall was a civil rights activist with a deep love for wild America. Udall was elected to Congress from Arizona's second district in November 1956. A gifted raconteur and a true outdoors enthusiast, soon to be an indispensable member of Kennedy's cabinet, Udall was both poet and politician. One of his closest friends was Robert Frost. The Alaskan wilderness movement was lucky to get him involved to start off the new decade of the 1960s. During his years as secretary of the interior, from 1961 to 1969, Udall would lead the heroic effort to get four national parks, six national monuments, seventeen seashores and lakeshores, and scores of new recreation areas established. His book *The Quiet Crisis* (1962) galvanized opposition against the desultory stewardship of land, sea, and air by irresponsible corporations and uncaring consumers.

III

O laus Murie was not well in early December 1960; he was still recovering from a recent lymph gland operation that he had undergone in Denver. Sometimes it seemed that his urgent work for The Wilderness Society was keeping him alive. Mardy had prayed that Olaus would live long enough for them to experience the Sheenjek River together one last time; and in 1961, just before he died, they did. Olaus, however, didn't live long enough to see his dream of a Wilderness Act—born out of Bob Marshall's Gates of the Arctic explorations of the 1920s—come to fruition in 1964. That December 1960, however, Olaus was proud that the Murie Ranch—which became part of Grand Teton National Park in 1960—had become the "heart" of the wilderness movement. It was a salon where many conservation ideas had been developed. The mimeograph machine at Dogpatch and the Muries' P.O. box in Wyoming were the cables charging the battery of the Arctic NWR movement at the decade's end. Mardy Murie would take on the role of watchdog of the Arctic NWR until her death in 2003 at the age of 101.

Nobody in the Alaskan wilderness movement knew exactly when President Eisenhower would formally issue a public land order designating the Arctic NWR. All the proper paperwork had been filed. Seaton, with renewed force around Thanksgiving, had signaled to the *Fairbanks Daily News-Miner* that it would happen soon. Still, it came as something of a surprise when on December 7, bright and early, Mardy Murie walked to the Moose post office and was handed a telegram by the postmaster. It was a press release from the previous day, issued by Seaton. The Muries had been testifying in Idaho against the damming of the Snake River and missed the historic moment; their home had no telephone service, and so they had not received the news about the Arctic NWR the night before. "I floated back that half mile through the woods on a cloud, burst through the front door," she recalled in her memoir *Two in the Far North.* "Oh darling, there's wonderful news today!" [24]

Beaming like the Cheshire cat, her eyes flashing with the excitement of a glorious achievement, Mardy waved proof that their steady, protracted effort to save the Arctic had succeeded. The press release by the Department of the Interior read: "Secretary Seaton Establishes New Arctic National Wildlife Range." [25] To Mardy it was a dream come true. Also, both Izembek and Kuskokwim (later renamed for Clarence Rhode) were designated national wildlife ranges. "Olaus was at his table at the back of the room, writing," Mardy wrote. "I held out the telegram to him; he read it and stood and took me in his arms and we both wept. The day before, December 6, Secretary Seaton had by Executive Order established the Arctic National Wildlife Range!" [26]*

---

* In 1942 President Franklin D. Roosevelt delegated his authority to withdraw public lands to the secretary of the interior. In 1952 that delegation was amended

Why did President Eisenhower approve it? This can only be conjectured, since the paper trail is so thin; but for one reason, Eisenhower trusted Seaton's instinct on Alaskan land issues. And since Eisenhower had worked so gallantly to demilitarize Antarctica, his doing something for conservation in the Arctic made sense. It's impossible, however, to measure the degree of sympathy Eisenhower felt for the Arctic NWR. There is virtually no paper trail of his views. The only public mentions that Eisenhower ever made about the Arctic NWR—his administration's crowning conservationist legacy—were minor, a notice in his "Public Papers of the President," and a bureaucratic line in his 1961 budget address.[27] Yet Eisenhower, though his rationale is unrecorded, approved the establishment of what became America's largest national wildlife refuge. Saving those 8.9 million acres was perfectly consistent with his signing of the Antarctic treaty. Few individuals had done more to preserve polar environments than Eisenhower. "Seaton told me that he didn't want to make a big deal about the Arctic Refuge because it would create a backlash with the incoming Kennedy Democrats," the incoming secretary of the interior, Stewart Udall, later recalled. "Governor Egan was squawking about it being unconstitutional. Somehow because I was from the west, Eagan thought I'd side with him and turn what became known as ANWR over to the state."[28]

When Udall was asked if he had ever considered buckling under Eagan's pressure, he said, "The thought never crossed my mind. All the Arctic Refuge meant to me when I became secretary of the interior was that our [Kennedy's] administration could do big things. If Eisenhower and Seaton could create an Arctic Refuge, then we could do similar preservationist deals in California's redwoods country and the Ozarks and Utah. All those places were of real excitement to me. They hadn't yet been completely ruined."[29]

The circumstances of Eisenhower's approval of Public Land Or-

---

and Executive Order No. 10355 was issued, delegating to the secretary of the interior central authority over operation of the federal government's withdrawal process. Thus the secretary's action in a public land order (PLO) is equivalent to that of the president. Nevertheless, in a "big deal" such as the Arctic NWR, a secretary would certainly discuss it with the president before signing the order.

der 2214, during his last days in the White House, officially designating the Arctic NWR, weren't entirely unusual for an outgoing president.
(It should be noted that Eisenhower did not sign PLO 2214; Seaton did.)
Theodore Roosevelt, for example, had saved the Olympics just forty-eight
hours before leaving the White House in 1909. Such public lands acts offered an opportunity for a timely gesture. What was odd, however, was the
farewell address that Eisenhower delivered on January 17, 1961—the most
memorable since George Washington's in 1796. Clearing his throat and
shuffling his pages behind a pair of old paper clip–shaped radio micro-
phones on his desk in the Oval Office, Eisenhower said, "In the councils
of government, we must guard against the acquisition of unwarranted
influence, whether sought or unsought, by the military-industrial complex. The potential for the disastrous rise of misplaced power exists and
will persist. Yet, in holding scientific research and discovery in respect,
as we should, we must also be alert to the equal and opposite danger that
public policy could itself become the captive of a scientific-technological
elite." [30]

The journalist Carl Rowan, who carefully studied the president's second term, thought that Eisenhower had a deeply ingrained skepticism
about technology and its effects on the environment. Rowan—who interviewed members of the administration associated with Team Seaton—
believed that Eisenhower's protection of Antarctica and the Arctic NWR
was part and parcel of this speech. "Second-term Eisenhower was a surprise," Rowan says. "Just like he helped boost civil rights—sending federal protection to Little Rock, appointing anti–Jim Crow federal judges
throughout the South—he became a conservationist, too. Not enough of
one to want to push through wilderness bills and the like. Not enough
of one to stop nuclear testing. But he thought the Arctic and Antarctica
shouldn't be destroyed. They were sanctuaries for all people." [31]

Rowan had a point. Eisenhower did help save Antarctica and Arctic Alaska from potential industrial ruin. On the other hand, a truly
ecologically-minded president would never have dreamed of allowing the Atomic Energy Commission to detonate nuclear devices around
Point Hope. Perhaps the best way to understand the Arctic NWR, then, is
through Eisenhower's initial skepticism about Alaska's statehood. Eisenhower saw Alaska, in a sense, as a possession of the federal government: a

site where the Pentagon could conduct defense exercises, the USDA could experiment with harvesting seafood, and the Department of the Interior could create national parks and wildlife refuges. Eisenhower was, it seems, skeptical about big oil, coal, timber, and the antitax movement. As Eisenhower intimated in his farewell address, huge corporations like Standard Oil, Boeing, and McDonnell-Douglas served their shareholders' interests. The U.S. government shouldn't ever be bought off with corporate dollars. Also, science, as Douglas used to say, had its drawbacks. "Science has produced instruments that make man lazier and less inclined to explore woods, valleys, ridges," Douglas would complain, in a sense echoing Eisenhower's Farewell Address. "The machine is almost a leash that keeps man from adventure."[32]

Regardless of his motivation, Eisenhower's creation of the Arctic NWR for "the purpose of preserving unique wildlife, wilderness and recreational values" was a peak moment for conservationists in the tradition of Theodore Roosevelt and Aldo Leopold. In Alaskan history this was the first time a federal unit was preserved as a national heirloom by the application of ecological principles. The founding purpose of the Arctic NWR was to preserve a wilderness, so this was a legislative harbinger for the Wilderness Act of 1964 that the Muries, Zahniser, and Douglas had been diligently working on throughout the 1950s.[33] "Wilderness," Leopold had written, "is the raw material out of which man has hammered the artifact called civilization."[34]

Ginny Wood and Celia Hunter were giddy with joy. Lois Crisler said that the wolves had also won, giving "heart and hope" to lovers of wildlife. Walt Disney wondered if there was a movie in all this. Mardy Murie, remembering that Fairfield Osborn had really started the Arctic NWR movement in 1956 by sponsoring the Sheenjek Expedition, wrote him a letter of thanks: "Sometimes it's good to have a little victory, isn't it? Even though we know also that there still has to be watchfulness, thinking and persuasion to keep the area natural, not 'developed'—a treasure for the sensitive ones, the vigorous ones, the searchers for knowledge, for all the years to come. Surely there should be a few such places on this plundered planet!"[35]

When Justice Douglas heard about the Arctic NWR, he was elated. His dream of a National Wilderness Preservation System was coming to fruition. Nobody knows what he thought that December day as rain turned to snow.* After performing his duties at the Supreme Court, he retreated to his low-ceilinged study on Hutchins Place to work on his new book for young readers, *Muir of the Mountains*. If *My Wilderness* could help save the Brooks Range, imagine how the wilderness movement could flourish with John Kennedy in the White House *and* old John Muir reintroduced to a new generation of readers. Also, receiving bigger headlines than the Arctic NWR that December 7 was the news that Douglas's friend Stewart Udall had been officially chosen to replace Seaton as secretary of the interior. "Stewart and Bill were extremely close," Cathy Stone, Douglas's fourth wife, recalled. "They hiked the C&O Canal together. They'd wear old clothes and just take off down the towpath. Once they got soaked in the rain and were mistaken for hoboes." [36]

That Christmas season, while other insiders in Washington, D.C., were attending parties, Douglas sat quietly at his desk composing *Muir of the Mountains* (to be published in June 1961 by the Sierra Club). Working with the children's illustrator Daniel San Souci, Douglas reviewed Muir's life from the Scottish Highlands to his death from pneumonia in Los Angeles on Christmas eve 1914 (around the time Hetch Hetchy was turned into a reservoir). He gave great attention to Muir's memoir *Travels in Alaska*. Douglas, in fact, had broadened his own knowledge of glaciation with Muir as his teacher. Writing a chapter about Muir's "short-legged, rather houndish, and shaggy" dog, Stickeen, Douglas was comforted that his own best friend—Sandy, the border collie—was curled up by his side. "Muir learned much about glaciers on this trip with Stickeen," Douglas wrote. "What he saw of the workings of these gigantic Alaskan icefields confirmed many of his theories about glaciation in the Sierra. Yet he learned more than this. He now knew how warm and joyous the friendship between a man and a dog can be. He learned that dogs as well as men can rise to heroic heights when danger threatens. He learned that a man

---

* Little could Douglas have known that President Jimmy Carter would pay him the honor of redesignating the Arctic NWR as William O. Douglas Arctic Wildlife Range. The new name, however, stuck only for ten months in 1980. When the Alaska National Interest Lands Conservation Act became law in December 1980, Congress renamed it the Arctic National Wildlife Refuge (ANWR).

and his dog, working as a team, can sometimes make a contribution to human knowledge."[37]

If Douglas had a philosophy, it was his dauntless belief that freedom of thought and freedom of expression were unalienable rights of all Americans. He tirelessly stated that at all costs these fundamental principles of individual freedom, protected by the Constitution, had to be preserved. Against all odds, bucking huge powerful blocs like the Morgan-Guggenheim syndicate, the Harding administration, McCarthyism, and the industrial-military complex, the wilderness movement had doggedly persevered. Some battles—a lot, actually—had been lost. But in Alaska the land skinners and despoilers had been checkmated in a number of important instances. Like trickster ravens, the Muirian preservationists often outwitted big business. The enlightened pro-wilderness minority, promoting kinship with all animal life, had a knack for pulling rabbits out of hats. Groups like the ACS, Douglas believed, were essential in a democracy. "We need Committees of Correspondence to coordinate the efforts of diverse groups to keep America beautiful and to preserve the few wilderness alcoves we have left," Douglas wrote. "We used such committees in the days of our Revolution, and through them helped bolster the efforts of people everywhere in the common cause. Our common cause today is to preserve our country's natural beauty and keep our wilderness areas sacrosanct. The threats are everywhere; and the most serious ones are often made in unobtrusive beginnings under the banner of 'progress.'"[38]

Starting with Muir, a noble band of conservationist revolutionaries— TR, Hornaday, Pinchot, Leopold, Marshall, FDR, the Muries, the Crislers, and Carson among them—stood up and said *no* to the exploiters of Alaska's wilderness kingdom. Their mythos was becoming popular on college campuses in 1960. Some places, such as the coastal plain of the Beaufort Sea or Mount McKinley, were simply too awesome to molest. The illustrator Rockwell Kent; the WASPs; the forest beatniks like Snyder, Whalen, Ginsberg, Ferlinghetti, and Kerouac—Douglas was proud to be in their victorious ranks that December. Refusing to be a cloistered justice, Douglas crisscrossed America dissenting against reckless oil drilling, clear-cutting, strip-mining, and superhighways. He worried that the Arctic NWR and other tracts of wilderness were going to fall victim to legal clauses allowing mining and timbering on federal property. "After they gutted and ruined the forests, then the rest of us could use them—to find campsites among stumps, to look for fish in waters heavy with silt from erosion, to search for game on rivers pounded to dust by sheep." [39]

But because of the Arctic NWR Douglas felt a strong current of optimism in the air. With Kennedy coming into the White House, the stage seemed to be set for a new environmental movement. Ecological consciousness was becoming mainstream. Rachel Carson was near finishing *Silent Spring*, and Stewart Udall was tapping talents like the novelist Wallace Stegner to help him write the classic ecological manifesto *The Quiet Crisis*. The new "green" movement was spreading worldwide. The legacy of John Muir was still strong; his name was becoming almost as well known as that of Paul Revere or Betsy Ross in schoolrooms. "Knowing of people's love of beauty and their great need for it, Muir gave his life to help them discover beauty in the earth around them, and to arouse their desire to protect," Douglas wrote in *Muir of the Mountains*. "The Machine, Muir knew, could easily level the woods and make the land desolate. Humankind's mission on earth is not to destroy: it is to protect and conserve all living things. There is a place for trees and flowers and birds, as well as for people. Never should we try to crowd them out of the universe." [40]

Muir, who preached the gospel of glaciers, surely would have said, "Amen."

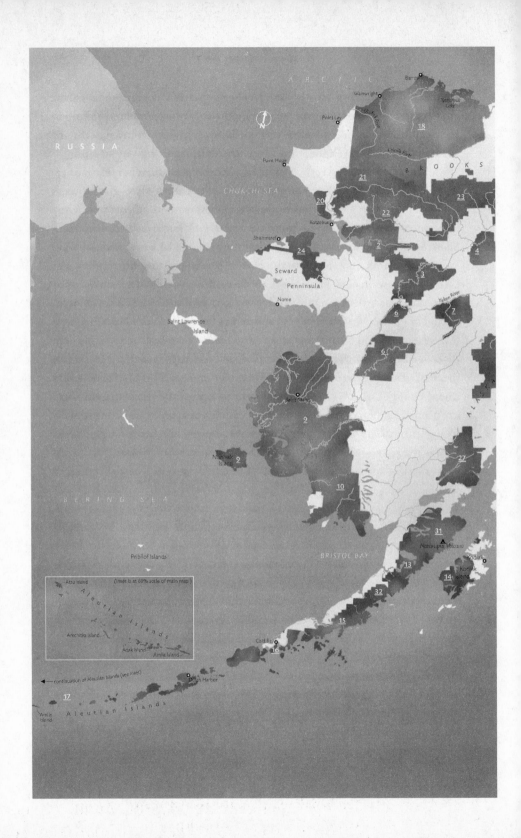

ALASKA'S REMAINING WILD PLACES

U.S. FISH AND WILDLIFE SERVICE
1 Arctic National Wildlife Refuge
2 Selawik National Wildlife Refuge
3 Koyukuk National Wildlife Refuge
4 Kanuti National Wildlife Refuge
5 Yukon Flats National Wildlife Refuge
6 Innoko National Wildlife Refuge
7 Nowitna National Wildlife Refuge
8 Tetlin National Wildlife Refuge
9 Yukon Delta National Wildlife Refuge
10 Togiak National Wildlife Refuge
11 Kenai National Wildlife Refuge
12 Kenai River Special Management Area
13 Becharof National Wildlife Refuge
14 Kodiak National Wildlife Refuge
15 Alaska Peninsula National Wildlife Refuge
16 Izembek National Wildlife Refuge
17 Alaska Maritime National Wildlife Refuge

U.S. BUREAU OF LAND MANAGEMENT
18 National Petroleum Reserve
19 Steese National Conservation Area

NATIONAL PARK SERVICE
20 Cape Krusenstern National Monument
21 Noatak National Preserve
22 Kobuk Valley National Park
23 Gates of the Arctic National Park & Preserve
24 Bering Land Bridge National Preserve
25 Denali National Park & Preserve
26 Yukon Charley Rivers National Preserve
27 Lake Clark National Park & Preserve
28 Kenai Fjords National Park
29 Wrangell-St. Elias National Park & Preserve
30 Glacier Bay National Park & Preserve
31 Katmai National Park & Preserve
32 Aniakchak National Monument & Preserve

U.S. FOREST SERVICE
33 Chugach National Forest
34 Tongass National Forest

ACKNOWLEDGMENTS

Since graduating from Ohio State University in 1982 I've traveled to Alaska many times. Kayaking around Glacier Bay National Park and hiking in the Chugach Mountains have become two of my smarter summer habits. Back in 1994, as a university history professor, I brought students on my natural-gas-fueled "Majic Bus" all the way from New Orleans to Fairbanks. We would go up the Alaska Marine Highway—which offered overnight ferry service from Prince Rupert, British Columbia, to Haines, Alaska—on the most amazing and visually impressive journey imaginable. Glaciers, bald eagles, horned puffins, blue whales, 10,000-foot snow-crested peaks, and lush forested islands were just part of the breathtaking experience of journeying through the Inside Passage described by John Muir in *Travels in Alaska*. We read classic books by Jack London, Mardy Murie, and John McPhee along the way.

I was hooked on Alaska. From this Majic Bus journey I learned that there is no more beautiful state flag than Alaska's bright gold stars on a field of blue, that the town of Homer truly is the "Cosmic Hamlet by the Sea," and that the wood bison (*Bison bison athabascae*) should be reintroduced into the Yukon Flats NWR. I've also become convinced that in the age of climate change the polar bear (*Ursus maritimus*) belongs on the endangered species list, and that Teshekpuk Lake, whose name means "the largest lake of all" (located in the National Petroleum Reserve of the western Arctic), should be designated a national park or national monument.

Out of all the books I've written, this is my favorite because it brought Alaska back into my life so fully. *The Quiet World* was conceived as the second volume of my multivolume Wilderness Cycle (the inaugural volume was *The Wilderness Warrior: Theodore Roosevelt and the Crusade for America*). Allan Nevins wrote eight volumes on the Civil War, and Dumas Malone wrote five volumes on Thomas Jefferson; my plan is to do something similar for U.S. conservation history. This present volume takes up the battles to protect wild Alaska from 1879 to 1960. The third volume of the Wilderness Cycle—*Silent Spring Revolution: Rachel Carson, John F. Kennedy, Stewart Udall, and the Modern Environmental Movement, 1961–1964*—will be published in 2013, in anticipation of the fiftieth anniversary of the Wilderness Act of 1964. Similarly, *The Quiet World* was written with the fiftieth anniversary of the Arctic NWR (created in December 1960) in mind.

HarperCollins was brave to publish *The Quiet World* (a rather hefty volume, with color-photo inserts) in our rather dreary economic times. The firm's commitment to the Wilderness Cycle—my lifework—is steadfast. The quarterback of HarperCollins is the publisher Jonathan Burnham (who embraced my multivolume concept from day one). It's a pleasure doing business with him. The editor Tim Duggan was his usual dependable self. In a hundred different ways, he helped me whip this manuscript into shape. He is a wonderful friend whom I implicitly trust. His assistant, Allison Lorentzen, is always hardworking, diligent, and kind. She helped facilitate publication with her trademark good cheer. I'm grateful to them all. Special thanks are also in order for the associate publisher Kathy Schneider and design manager Leah Carlson-Stanisic. My old buddy Kate Blum (publicist) also deserves a nod for always organizing my visits to America's esteemed independent bookstores. Lisa Bankoff, my ICM agent for almost twenty years, kept all the paperwork in order, helping me meet deadlines and commitments. At Rice University I teach a course every fall semester on U.S. conservation history; it's a terrific way to stay current in environmental history.

While writing *The Quiet World* I became engaged with numerous folks involved in the Alaska wilderness movement and U.S. conservation history. These allies include Ben Beach and Bill Meadows of The Wilderness Society (they're the greatest); Ken Rait and Mike Matz of the Pew Environment Group/Campaign for America's Wilderness; James N. Levitt of the Program on Conservation Innovation at the Harvard Forest, Harvard University; Tim Richardson of Wildlife Forever; Lauren Hierl and Emilie Surrusco at the Alaska Wilderness League; Bill Vanden Heuvel at the Franklin and Eleanor Roosevelt Institute; Cynthia Koch at the FDR Library in Hyde Park, New York; Michael Adams of the Ansel Adams Trust;

Leonard Vallender of Camp Fire Club of America; Brian Ross of Colorado Conservation Trust; Pam Miller of the Northern Alaska Environment Center; Michelle Bryant, Theodore Roosevelt IV, and Tweed Roosevelt of the Theodore Roosevelt Association; the entire staff of the Eisenhower Presidential Library in Kansas; Lowell Baier of Boone and Crockett; Ken Salazar of the U.S. Department of the Interior; Dan Ritzman of the Sierra Club; and Mike Dunham of the *Alaska Daily News*.

At U.S. Fish and Wildlife (USFW) I received immeasurable help from Warren Keogh (Alaska), Mike Boylan (Alaska), Mark Madison (West Virginia), and Paul Tritaik (Florida). The raptor ecologist Joel E. Pagel of USFW was unbelievably generous in proofreading my chapter on Adolph Murie. Ever since *The Wilderness Warrior* was embraced by USFW, and essentially considered required reading in 2010, I've been asked to speak at half a dozen different national wildlife refuges. My friend Evan Hirsche, president of the National Wildlife Refuge Association, has brought me into the loop on efforts to have Congress designate the Arctic Refuge Coastal Plain as *wilderness* and to create a Blue Fin National Marine Refuge off the mid-Atlantic.

A number of year-round Alaskans fact-checked chapters of *The Quiet World* and offered keen insights into the wilderness movement. William Reffalt, the brilliant historian of the U.S. Fish and Wildlife Service, saved me from making numerous errors. My friend John Branson, historian at Lake Clark National Park and Preserve, is a walking encyclopedia of the entire Bristol Bay area. His intellectual generosity went way beyond the call of duty. Debbie Miller of Fairbanks offered invaluable insights about Arctic Alaska. The great Kim Hearox, Nathan Borson, and Bruce Molnia—America's go-to guys on glaciers—helped me perfect my prologue on Muir. Peter Van Tuyn of Bessenyey & Van Tuyn LLC in Anchorage, the smartest environmental lawyer I've ever met, answered numerous legal questions I had pertaining to Alaskan land deed issues. Nobody, however, helped me more than Fran Mauer, a top authority on Alaskan wildlife. I owe Fran a lot of dinners for saving me from quite a few errors.

A special appreciation is in order to Braided River, a Seattle nonprofit organization working in partnership with Mountaineers Books, for bringing together the photographers who contributed color images for both the Arctic Refuge and Tongass National Forest. Braided River's executive director, Helen Cherullo, did an amazing job acquiring photo rights. Her love of the Arctic NWR and the Tongass is palpable.

During the course of writing *The Quiet World* I gave a number of public lectures in Alaska. Special thanks to U.S. Fish and Wildlife Service (USFW) and the University of Alaska–Anchorage for sponsoring them. The mayor of Homer—Jim Hornaday (a direct descendant of William Temple Hornaday, who is profiled in this book)—is the best small-city politician I know. My profile of Charles Sheldon was improved by Ken Kastens, Tom Walker, and Rose Speranza (of the University of Alaska, Polar Regions Collection at the Elmer R. Rasmuson Library). Lori McKean of Grey Towers National Historic Site answered questions about Gifford Pinchot. At the University of the Pacific in Stockton, California, the wonderful archivist Michael Wurtz aided me in going through the John Muir papers.

Nobody knows the Arctic NWR quite like Roger Kaye of Fairbanks. His classic narrative, *Last Great Wilderness*, proved invaluable. Roger is an incredible scholar, bush pilot, and conservationist. Likewise, Cindy Shogan of the Alaska Wilderness League allowed me to use her fine library as my operational base when I was in Washington, D.C. Cindy has devoted her life to grassroots activism on behalf of the Arctic NWR. Three members of the National Audubon Society—John Flicker, Thomas O'Handley, and David Seideman—helped me get my birds right. What a valuable service the Audubon Society provides to the world!

A number of characters in *The Quiet World* were able to tell me firsthand stories about the Alaskan wilderness movement. Special thanks to the poets Gary Snyder, Ed Sanders, Lawrence Ferlinghetti, and Michael McClure—all literary legends. Huge thanks to the fine scholar John Suiter, who carefully proofread my chapter on the beats. I can't wait for his full-length biography of Gary Snyder to be published. Then there is Peter Matthiessen of Long Island. Without his help I couldn't have written accurately about his first nonfiction book, *Wildlife in America*. He gladly proofread a couple of chapters. The author and photographer Dorothy Jones—wife of "Sea Otter" Jones—provided insights about her husband's career at USFW. Getting to spend an afternoon in Fairbanks with Virginia Wood, a pioneering Alaskan conservationist, was a thrill. John Suiter shared with me his unparalleled wisdom about the beat movement and ecology. Cathy Stone, the last wife of Supreme Court Justice William O.

Douglas, a supporter of the Arctic NWR since 1960, kindly told me stories about her legendary husband.

During the summer of 2010 the Brinkleys lived in Homer, Alaska, above the Jars of Clay gallery on Main Street. My entire family was with me. From our balcony we had a superb view of Kachemak Bay. The owner of the gallery, Ruby Haigh, and her husband, Tim, a home builder, adopted us. Ruby makes exquisite pottery; she sees the beauty in all objects. All the Brinkleys fell in love with the Haighs. While in Homer, I used the Pratt Museum and the Mermaid Café and Bookstore as my workspace. I always judge a town by four things: its parks, cuisine, museums, and bookstores. Homer gets an A+ in all four categories. In Homer, Alexa Graumlich, my nineteen-year-old niece, a sophomore at UCLA, helped me with research and typing. We are very proud of Alexa. My sister Leslie and her husband Jeff, a fine halibut fisherman, kept us stuffed with seafood for the Fourth of July.

The highlight of my summer was camping in the Arctic NWR with two buddies: Tom Campion (owner of Zumiez) and the Fairbanks outfitter Jim Campbell. Our Arctic trek started out in Fairbanks. Tom, Jim, and I flew first to the village of Coldfoot. This was the territory of Bob Marshall, cofounder of The Wilderness Society. After studying local history for an afternoon we joined up with Dirk Nickisch, a North Dakotan who owns and operates Coyote Air. Dirk, an amazing bush pilot, would swoop down low so I could get 360-degree views of the vast landscape. Across the Brooks Range, across the Philip Smith Mountains, on to Camden Bay and the Arctic Ocean—it was a flight that is gratifying to look back upon. A walk along the beach. A large herd of caribou nearby, their curiosity encompassing us. The memorable summer light that turns soft at midnight. Our campsite was along the serene Hulahula River. The sky was like another ocean. The nearby mountains were like ruins left over from the ice age. One afternoon in the Arctic NWR we saw a grizzly climbing up a hill, running at tremendous speed. What an awesome sight! Never had I experienced such an uplifting feeling as hiking along the Arctic Ocean in Eisenhower's great wildlife refuge. Tom, Jim, and I regularly took off our hiking boots to put on rubber waders and walk across the coastal plain streams. The tundra was always wet, even in June, because the permafrost had stopped any underground drainage.

Acknowledgments

After the trip to the Arctic NWR I was better able to understand both the geography and the politics of the area. Jim's wife, Carol Karza, helped me track down all sorts of information about the North Slope. My alter ego on this Alaskan journey, however, was the brilliant, engaging, multitalented Rachel Sibley of Alpine, Texas. A 2009 graduate of the Plan Two Honors Program at the University of Texas—Austin (with an emphasis on foreign languages and cultural studies), the twenty-four-year-old Sibley glowed with enthusiasm throughout the writing of *The Quiet World*. Working at my home office, we would blast out music from Merle Haggard to John Coltrane to Jimmy Webb and get to work. A skillful modern dancer, Rachel also taught me new songs for the guitar—such as "Wagon Wheel" by Old Crow Medicine Show and "Anchorage" by Michelle Shocked—during breaks from work. We had fun. I'm a technologically challenged person. I like my books tangible. Anything electronic rattles me to no end. Rachel, by contrast, is an Internet wizard, able to access a rare document or an elusive fact with lightning quickness. An old friend of mine, the indispensable Emma Juniper of Sedona, Arizona, first introduced me to Rachel in 2009. Emma helped us from time to time on the book, but Rachel was the driving force on this one. My wife and children consider both Rachel and Emma family.

Which brings me to Anne. It would be impossible to explain how supportive my wife is of my history projects. Our marriage is my greatest accomplishment in life. Our entire house has become filled with books, manuscripts, and historical objects. But Anne never complains. Together we're raising three splendid kids—Benton, Cassady, and Johnny—in the hills of Austin not far from Barton Springs. I feel very blessed. One reason I decided to write the Wilderness Cycle is so my children can visit all of America's great parklands before they leave for college. The summer of 2010 was the Kenai Peninsula for the Brinkleys. Our 2011 trip will be three of the national parks that Secretary of the Interior Stewart Udall had President Lyndon B. Johnson sign into law following congressional authorization: North Cascades in Washington, Canyonlands in Utah, and Redwoods in northern California. We hope to see you on the trail.

August 22, 2010
Austin, Texas

NOTES

Prologue: John Muir and the Gospel of Glaciers

1. Peggy Wayburn, "The Last True Wilderness," in Mike Miller and Peggy Wayburn, *Alaska: The Great Land* (San Francisco, CA: Sierra Club, 1974), p. 117.

2. Mark Maslin, *Global Warming* (Oxford: Oxford University Press, 2004), p. 96.

3. Adeline Knapp, "Some Hermit Homes of California Writers," *Overland*, Vol. 35, No. 205 (January 1900).

4. Michael P. Cohen, *The Pathless Way: John Muir and American Wilderness* (Madison: University of Wisconsin Press, 1984), p. 42.

5. John Muir, *My First Summer in the Sierra* (New York: Houghton Mifflin, 1911), p. 205. Also Tom Melham, *John Muir's Wild America* (Washington, DC: National Geographic Society, 1976), p. 9.

6. Bruce Molnia, *Glaciers of Alaska* (Anchorage: Alaska Geographic Society, 2001), pp. 5–7. Also see Ned Rozell, "Melting Ice," *Alaska Science Forum*, Article No. 1731 (December 30, 2004).

7. John Muir, *Travels in Alaska* (Boston, MA, and New York: Houghton Mifflin, 1915), p. 215.

8. Andromeda Romano-Lax, *Chugach National Forest: Legacy of Land, Sea, and Sky* (Anchorage: Alaska Natural History Association, 2007). (In 2008, the Alaska Natural History Association changed its name to the Alaska Geographic Association.)

9. Susan Kollin, *Nature's State: Imagining Alaska as the Last Frontier* (Chapel Hill: University of North Carolina Press, 2001), p. 28.

10. Witt Ball, "The Stickeen River and Its Glaciers," *Scribner's Monthly*, Vol. 17 (1879), pp. 805–815.

11. John Muir, "Yosemite Glaciers," *New York Tribune*, December 5, 1871.

12. Dyan Zaslowsky, Tom H. Watkins, and Wilderness Society, *These American Lands: Parks, Wilderness, and the Public Lands* (Washington, DC: Island, 1994), p. 287.

13. Kollin, *Nature's State*, p. 32.

14. Linnie Marsh Wolfe (ed.), *John of the Mountains: The Unpublished Journals of John Muir* (Madison: University of Wisconsin Press, 1979), p. 245. (Reprint of the original, Boston, MA: Houghton Mifflin, 1938.)

15. Robert Engberg and Bruce Merrell, *John Muir: Letters from Alaska* (Fairbanks: University of Alaska Press, 2009).

16. Muir, *Travels in Alaska*, pp. 314–315.

17. John Muir, "Notes of a Naturalist: John Muir in Alaska—Wrangell Island and Its Picturesque Attractions," *San Francisco Daily Evening Bulletin* (September 6, 1879), p. 1.

18. Melham, *John Muir's Wild America*, p. 136.

19. S. Hall Young, *Alaska Days with John Muir* (New York: Fleming H. Revell, 1915), pp. 29–30.

20. Muir, *Travels in Alaska*, p. 21.

21. Kollin, *Nature's State*, p. 29.

22. Peter A. Coates, *The Trans-Alaska Pipeline Controversy: Technology, Conservation, and the Frontier* (Anchorage: University of Alaska Press, 1993), p. 40.

23. D. K. Hall, C. S. Benson, and W. O. Field, "Changes of Glaciers in Glacier Bay, Alaska," in *Physical Geography* (Elsevier/Geo Abstracts, 1992), pp. 27–41.

24. Muir, *Travels in Alaska*, p. 13.

25. Michael F. Turek, "John Muir, Glacier Bay, and the Tlingit Indians: Rapid Landscape Change and Human Response in the Arctic and Sub-Arctic," June 15–18, 2005. (ICSU Dark Nature Project, Whitehorse, Yukon, Canada.)

26. Melham, *John Muir's Wild America*, p. 19.

27. Young, *Alaska Days with John Muir*, p. 99.

28. Wolfe, *John of the Mountains*, pp. 272–273. Also Muir, *Travels in Alaska*, pp. 142–146.

29. Kim Heacox, *Alaska's Inside Passage* (Portland, OR: Graphic Arts Center Public Library, 1997), p. 79.

30. Katherine Hocker, *Alaska's Glaciers: Frozen in Motion* (Anchorage: Alaska Natural History Association, 2006), p. 8.

31. Donald Worster, *A Passion for Nature: The Life of John Muir* (New York: Oxford University Press, 2008), p. 251.

32. Muir, *Travels in Alaska*, p. 156. Also John Muir, "Fort Wrangell, October 16, 1879," *San Francisco Daily Evening Bulletin*, November 8, 1879, p. 1.

33. Muir, *Travels in Alaska*, p. 145.

34. Young, *Alaska Days with John Muir*, pp. 108–112.

35. Molnia, *Glaciers of Alaska*, p. 8.

36. Ibid., pp. 97–111.

37. Ibid., p. 126.

38. Worster, *A Passion for Nature*, pp. 256–257.

39. Young, *Alaska Days with John Muir*, p. 71.

40. John Muir, "An Adventure with a Dog and a Glacier," *Century*, Vol. 54 (August 1897), p. 771.

41. John Muir, *Stickeen* (Boston, MA: Houghton Mifflin, 1910).

42. Worster, *A Passion for Nature*, pp. 260–261.

43. Muir, *Travels in Alaska*, p. 263.

44. Marcus Baker, *Geographic Dictionary of Alaska* (Washington, DC: Government Printing Office, 1906).

45. Young, *Alaska Days with John Muir*, p. 32.

46. Rod Miller, *John Muir: Magnificent Tramp* (New York: Forge, 2005), p. 112.

47. Author interview with the archivist Michael Wurtz, July 1, 2010.

48. John Muir, Alaska Glacier Drawings, University of the Pacific, Stockton, CA. (Unpublished inventory.)

49. Muir, *Travels in Alaska*, p. v.

50. Ibid.

51. Young, *Alaska Days with John Muir*, pp. 202–206.

52. John Muir, *The Writings of John Muir: The Cruise of the Corwin* (New York: Houghton Mifflin Company, 1917), p. xxvi.

53. Worster, *A Passion for Nature*, p. 204.

54. Muir, *The Writings of John Muir: The Cruise of the Corwin*, p. 24.

55. Ibid., p. 91.

56. Dan Flores, *Visions of the Big Sky: Painting and Photographing the Northern Rocky Mountain West* (Norman: University of Oklahoma Press, 2010), p. 113.

57. John Burroughs, "Narrative of the Expedition," in *Harriman Alaska Expedition*, 13 vols. (New York: Doubleday, Page, 1902), Vol. 1, pp. 18–80.

58. *Glaciers in Alaska* (Anchorage: Alaska Geographic Society, 2003), p. 72.

59. Laurie Lawlor, *Shadow Catcher: The Life and Work of Edward S. Curtis* (Lincoln: University of Nebraska Press, 2005), p. 37.

60. William H. Goetzmann and Kay Sloan, *Looking Far North: The Harriman Expedition to Alaska, 1899* (Princeton, NJ: Princeton University Press, 1982), p. 113.

61. Ibid., pp. 200–206.

62. Nancy Lord, *Green Alaska: Dreams from the Far Coast* (Washington, DC: Counterpoint, 1999), p. xix.

63. Quoted in Molnia, *Glaciers of Alaska*, p. 76.

64. Quoted in "Celebrating Wild Alaska: Twenty Years of the Alaska Lands Act" (Washington, DC: Alaska Wilderness League, December 2000). (Pamphlet.)

65. John Muir to Harry F. Reid, February 26, 1891, Muir Papers, University of the Pacific.

66. Muir, *Travels in Alaska*, p. 145.

67. Engberg and Merrell, *John Muir: Letters from Alaska*, pp. xx–xxvi.

68. Knut Hamsun, *The Cultural Life of Modern America* (Cambridge, MA: Harvard University Press, 1969), p. 78.

69. "The Philosophy of John Muir," in John Muir, Edwin Way Teale, and Henry Bugbee Kane, *The Wilderness World of John Muir* (New York: Houghton Mifflin Harcourt, 2001), p. 315.

1. Odyssey of the Snowy Owl

1. Charles Emmerson, *The Future History of the Arctic* (New York: Public Affairs, 2010), p. xiii.

2. Stephen Brown (ed.), *Arctic Wings: Birds of the Arctic National Wildlife Refuge* (Seattle, WA: Mountaineers, 2006), p. 115.

3. Paul Schullery, *American Bears: Selections from the Writings of Theodore Roosevelt* (Boulder, CO: Robert Rinehart, 1998), p. 77.

4. Laurie Lawlor, *Shadow Catcher: The Life and Work of Edward S. Curtis* (Lincoln: University of Nebraska Press, 2005), p. 37.

5. Richard Ellis, *On Thin Ice: The Changing World of the Polar Bear* (New York: Knopf, 2009), pp. 13–68.

6. Theodore Roosevelt, *Hunting Trips of a Ranchman: Sketches of Sport on the Northern Cattle Plains* (New York: Putnam, 1885).

7. Theodore Roosevelt, *The Wilderness Hunter* (New York: Putnam, 1893), p. 271. Also Paul Russell Cutright, *Theodore Roosevelt: The Making of a Conservationist* (Champaign: University of Illinois Press, 1985), p. 172.

8. Walter R. Borneman, *Alaska: Saga of a Bold Land* (New York: HarperCollins, 2003), pp. 18–19.

9. Nancy Lord, *Green Alaska: Dreams from the Far Coast* (Washington, DC: Counterpoint, 1999), p. 99.

10. Andromeda Romano-Lax, *Chugach National Forest: Legacy of Land, Sea, and Sky* (Anchorage: Alaska Natural History Association, 2007), p. 51.

11. Ibid., pp. 20–21.

12. Barry Lopez, *Arctic Dreams: Imagination and Desire in a Northern Landscape* (New York: Vintage, 2001), p. 339. (Originally published New York: Scribner, 1986.) Also Georg Wilhelm Steller, *Steller's History of Kamchatka* (Fairbanks: University of Alaska Press, 2003).

13. Stephen Haycox, *Alaska: An American Colony* (Seattle: University of Washington Press, 2002), pp. 50–51.

14. Frank Dufresne, "Foreword," in Corey Ford, *Where the Sea Breaks Its Back: The Epic Story of Early Naturalist George Steller and the Russian Exploration of Alaska* (Anchorage: Alaska Northwest, 1966), p. x.

15. Harry Ritter, *Alaska's History: The People, Land, and Events of the North Country* (Anchorage: Alaska Northwest, 1993), pp. 92–93.

16. John Muir, *The Writings of John Muir: The Cruise of the Corwin* (New York: Houghton Mifflin, 1917), p. 91.

17. Terry Gifford, *John Muir: His Life and Letters and Other Writings* (Seattle: Mountaineers, 1996), p. 804.

18. Lord, *Green Alaska*, pp. 134–135.

19. Dave Smith, *Alaska's Mammals* (Anchorage: Alaska Northwest, 2007), pp. 7–82.

20. John Muir, *Travels in Alaska* (Boston, MA, and New York: Houghton Mifflin, 1915), p. 56.

21. Lawlor, *Shadow Catcher*, pp. 37–38.

22. Glenn Holder, *Talking Totem Poles* (New York: Dodd, Mead, 1973), p. 44.

23. Stacey Bredhoff, *America's Originals* (Seattle: University of Washington Press, 2001), p. 58.

24. "President Talks to Alaskans," *Seattle Sunday Times*, May 24, 1903.

25. Peggy Wayburn, "The Last True Wilderness," in Mike Miller and Peggy Wayburn, *Alaska: The Great Land* (San Francisco, CA: Sierra Club, 1974), p. 127.

26. Ritter, *Alaska's History*, p. 47.

27. Joan M. Antonson and William S. Hanable, *Alaska's Heritage*, Unit 4, *Human History: 1867 to Present* (Anchorage: Alaska Heritage Society, 1985), pp. 228–229.

28. *The Works of Rudyard Kipling* (Wordsworth Editions, 2001), p. 123.

29. Wayburn, "The Last True Wilderness," p. 127. Also *The Alaskans* (New York: Time-Life Books, 1977), p. 180.

30. Borneman, *Alaska*, pp. 102–153.

31. Morgan B. Sherwood, *Exploration of Alaska: 1865–1900* (New Haven, CT: Yale University Press, 1965), pp. 36–56.

32. Frank Graham Jr., *Man's Dominion: The Story of Conservation in America* (New York: M. Evans, 1971), p. 185.

33. Rex Beach, *The Winds of Change* (New York: Putnam, 1945), p. 121.

34. William R. Hunt, *North of 53: The Wild Days of the Alaska-Yukon Mining Frontier, 1870–1914* (Fairbanks: University of Alaska Press, 1974), p. xv.

35. Miller and Wayburn, *Alaska: The Great Land*, p. 113.

36. Paul Brooks, *The Pursuit of Wilderness* (Boston, MA: Houghton Mifflin, 1971), p. 59.

37. Claus-M. Naske and Herman E. Slotnick, *Alaska: A History of the 49th State* (Norman: University of Oklahoma Press, 1994), p. 140.

38. Roderick Frazier Nash, *Wilderness and the American Mind* (New Haven, CT: Yale University Press, 2001), p. 154.

39. Frank M. Chapman to Theodore Roosevelt, June 10, 1911, Ornithology Department Archive, American Museum of Natural History, New York.

2. Theodore Roosevelt's Conservation Doctrine

1. John Burroughs, "Narrative of the Expedition," in *Harriman Alaska Expedition* (New York: Doubleday, Page, 1902), Vol. 1, pp. 18–80. Also William H. Dall, *Alaska and Its Resources* (Boston, MA: Lee and Shepard, 1870).

2. George Bird Grinnell, "What We May Learn from the Indian," *Forest and Stream*, Vol. 86 (March 1916), p. 846.

3. Linnie M. Wolfe, *John of the Mountains: The Unpublished Journals of John Muir* (Madison: University of Wisconsin Press, 1979), p. 400.

4. William H. Goetzmann and Kay Sloan, *Looking Far North: The Harriman Expedition, 1899* (Princeton, NJ: Princeton University Press, 1982), pp. 116–128.

5. George Kennan, *E. H. Harriman: A Biography* (Boston, MA: Houghton Mifflin, 1922).

6. Claus-M. Naske and Herman E. Slotnick, *Alaska: A History of the 49th State* (Norman: Oklahoma University Press, 1987), p. 3.

7. "Ex-Gov. John G. Brady Dies," *New York Times*, December 19, 1918.

8. *Harriman Alaska Expedition* (New York: Doubleday, Page, 1902), Vol. 2, p. 138.

9. Stephen Haycox and Alexandra J. McClanahan, *Alaska's Scrapbook: Moments in Alaska History 1816–1998* (Portland, OR: Graphic Arts Center, 2008), pp. 119–120.

10. Ibid., pp. 29–30.

11. Corinne Roosevelt Robinson, "My Brother, Theodore Roosevelt," *Scribner's*, Vol. 69 (1921), p. 132.

12. Goetzmann and Sloan, *Looking Far North*, p. 90.

13. Laurie Lawlor, *Shadow Catcher: The Life and Work of Edward S. Curtis* (Lincoln: University of Nebraska Press, 2005), p. 5.

14. Goetzmann and Sloan, *Looking Far North*, pp. 181–192.

15. Theodore Roosevelt, "Foreword," in Edward S. Curtis, *The North American Indian: Being a Series of Volumes Picturing and Describing the Indians of the United States and Alaska* (Author, 1907). (Foreword is dated October 1, 1906.)

16. Andromeda Romano-Lax, *Chugach National Forest: Legacy of Land, Sea, and Sky* (Anchorage: Alaska Natural History Association, 2007), p. 38.

17. Lawrence Martin, "Glacial Scenery in Alaska," *Bulletin of the American Geographical Society*, Vol. 47, No. 3 (1915), p. 173.

18. Ed Marston, "The Genesis of the West," *High Country News*, January 11, 2010.

19. Theodore Roosevelt to Serena E. Pratt, March 3, 1906, L. Dennis Shapiro Private Collection, Chestnut Hill, Massachusetts.

20. Ken Spotwood, "History of the Arctic Brotherhood," *Klondike Sun* (Archive of the Arctic Brotherhood, Seattle, WA).

21. "President Talks to Alaskans," *Seattle Sunday Times*, May 24, 1903.

22. Pinchot, quoted in Lawrence W. Rakestraw, *A History of the United States Forest Service in Alaska* (Anchorage: Cooperative Publication of Alaska Historical Commission, Department of Education, State of Alaska, and Alaska Region USDA Forest Service, 1981–2002). (Electronic version available courtesy of Forest History Society.)

23. T. J. Jackson Lears, *No Place of Grace: Antimodernism and the Transformation of American Culture, 1880–1920* (New York: Pantheon, 1981).

24. John Muir, *Travels in Alaska* (Boston, MA, and New York: Houghton Mifflin, 1915), p. 13.

25. Jonathan Raban, *Passage to Juneau: A Sea and Its Meaning* (New York: Random House, 1999), p. 332.

26. Quoted in "The Conservation of Wild Life," *Outlook*, Vol. 109 (January 20, 1915).

27. Naske and Slotnick, *Alaska*, pp. 101–102.

28. Polly Miller and Leon Miller, *Lost Heritage of Alaska: The Adventure and Art of the Alaskan Coastal Indians* (New York: Bonanza, 1967), pp. 243–252.

29. David E. Conrad, "Creating the Nation's Largest Forest Reserve: Roosevelt, Emmons, and the Tongass National Forest," *Pacific Historical Review*, Vol. 46, No. 1 (February 1977), pp. 65–83.

30. Barry Lopez, *Arctic Dreams: Imagination and Desire in a Northern Landscape* (New York: Vintage, 2001), p. 210.

31. George T. Emmons, "The Woodlands of Alaska," Tongass National Forest Archive, Ketchikan, AK.

32. Conrad, "Creating the Nation's Largest Forest Reserve."

33. Ibid.

34. Lawrence W. Rakestraw, *A History of the United States Forest Service in Alaska* (Anchorage: Alaska Historical Commission, 1981), Foreword.

35. Walter R. Borneman, *Alaska: Saga of a Bold Land* (New York: HarperCollins, 2003), pp. 4–15.

36. Romano-Lax, *Chugach National Forest*, p. 75.

37. John Burroughs, *Alaska: The Harriman Expedition, 1899* (New York: Doubleday, Page, 1902), p. 69.

38. Nancy Lord, *Rock, Water, Wild: An Alaskan Life* (Lincoln: University of Nebraska Press, 2009), p. 73.

39. Haycox and McClanahan, *Alaska's Scrapbook*, pp. 29–30.

40. Kathie Durbin, *Tongass: Pulp Politics and the Fight for the Alaska Rain Forest* (Corvallis: Oregon State University Press, 1999), p. 12.

41. Borneman, *Alaska: Saga of a Bold Land*, p. 239.

42. Lynn Readicker-Henderson and Ed Readicker-Henderson, *Inside Passage and Coastal Alaska*, 4th ed. (Edison, NJ: Hunter, 2002), pp. 55–57.

43. Durbin, *Tongass*, p. 11.

44. Mike Miller, "Discovery and Development," in Mike Miller and Peggy Wayburn, *Alaska: The Great Land* (San Francisco, CA: Sierra Club, 1974), p. 17.

45. Amy Gulick, *Salmon in the Trees* (Seattle, WA: Braided River, 2010), p. 13.

46. Lawrence Rakestraw (ed.), "A Mazama Heads North: Letters of William A. Langille," *Oregon Historical Quarterly* (June 1975), p. 1010; W. A. Langille, "Proposed Forest Reserve on the Kenai Peninsula, Alaska," in U.S. Senate, *Construction of Railroads in Alaska*, hearing before the Committee of Territories on 5.48 and 9.133 (63 Congress 1 session, GPO, 1913), pp. 681–699.

47. Peter A. Coates, *The Trans-Alaska Pipeline Controversy: Technology, Conservation, and the Frontier* (Bethlehem, PA: Lehigh University Press, 1991), p. 45.

48. Ira N. Gabrielson and Frederick C. Lincoln, *Birds of Alaska* (Harrisburg, PA: Stackpole, 1959), p. 10.

49. Elaine Rhode, *National Wildlife Refuges of Alaska* (Anchorage: Alaska Natural History Association, 2003), p. 53.

50. Bruce Woods, *Alaska's National Wildlife Refuges* (Anchorage: Alaska Geographic, 2003), p. 70.

51. Theodore Roosevelt, Executive Order No. 1039, U.S. Fish and Wildlife Archive, Anchorage, AK.

52. Timothy Egan, *The Big Burn: Teddy Roosevelt and the Fire That Saved America* (Boston, MA: Houghton Mifflin Harcourt, 2009), p. 78.

53. William H. Dall, "Geographical Notes in Alaska," *Bulletin of the American Geographical Society*, Vol. 28, No. 1 (1896), pp. 1–20.

54. Gabrielson and Lincoln, *Birds of Alaska*, pp. 14–15.

55. Aldo Leopold, *Game Management* (Madison: University of Wisconsin Press, 1933), p. 17.

56. Ibid., pp. 17–18.

57. Aldo Leopold, *A Sand County Almanac* (New York: Ballantine, 1970).

3. The Pinchot-Ballinger Feud

1. Theodore Roosevelt, "Introduction," in Robert E. Peary, *The North Pole* (New York: Cooper Square, 2001), p. xxxvii.

2. Ibid.

3. Theodore Roosevelt to William Robert Foran (September 12, 1909).

4. Barry Lopez, *Arctic Dreams: Imagination and Desire in a Northern Landscape* (New York: Vintage, 2001), pp. 377–386.

5. Roosevelt, "Introduction," *The North Pole*.

6. Alan Anderson, *After the Ice: Life, Death, and Geopolitics in the New Arctic* (New York: Smithsonian Books, 2009), p. 12.

7. Charles Emmerson, *The Future History of the Arctic* (New York: Public Affairs, 2010), p. 982.

8. Robert E. Peary, "Roosevelt—the Friend of Man," *Natural History*, Vol. 19, No. 1 (January 1919), p. 11.

9. Bill Streever, *Cold: Adventures in the World's Frozen Places* (New York: Little, Brown, 2009), p. 179.

10. Peter Matthiessen, *Oomingmak: The Expedition to the Musk Ox Island in the Bering Sea* (New York: Hastings House, 1967).

11. Theodore Roosevelt, *The Wilderness Hunter* (New York: Putnam, 1893), p. 271.

12. Theodore Roosevelt, "Is Polar Exploration Worth While?" *Outlook* (March 1, 1913).

13. Theodore Roosevelt, *A Book-Lover's Holidays in the Open* (New York: Scribner, 1916), pp. 336–337.

14. Hamlin Garland, *The Trail of the Goldseekers: A Record of Travel in Prose and Verse* (Norwood, MA: Norwood, 1899), p. 1.

15. Timothy Egan, *The Big Burn: Teddy Roosevelt and the Fire That Saved America* (New York: Houghton Mifflin-Harcourt, 2009), p. 81.

16. Archie Butt, *Taft and Roosevelt* (New York: Doubleday Doran, 1930), Vol. 1, pp. 244–257.

17. Roosevelt, quoted in Kathleen Dalton, *Theodore Roosevelt: A Strenuous Life* (Vintage Books, 2004), p. 357.

18. Theodore Roosevelt to Gifford Pinchot, June 17, 1910, quoted in *The Selected Letters of Theodore Roosevelt* (New York: Cooper Square, 2001), p. 529.

19. Dyan Zaslowsky and T. H. Watkins, *These American Lands: Parks, Wilderness, and the Public Lands* (Washington, DC: Island, 1994), pp. 287–288.

20. Nathan Miller, *Theodore Roosevelt: A Life* (New York: HarperCollins, 1993), p. 503.

21. Char Miller, *Gifford Pinchot and the Making of Modern Environmentalism* (Washington, DC: Island, 2001), p. 231.

22. Theodore Roosevelt to Gifford Pinchot, March 1, 1910, ibid., p. 231.

23. Edward J. Renehan Jr., *The Lion's Pride: Theodore Roosevelt and His Family in Peace and War* (New York: Oxford University Press, 1998), pp. 105–106.

24. Dalton, *Theodore Roosevelt: A Strenuous Life*, p. 358.

25. Douglas Brinkley, *The Wilderness Warrior: Theodore Roosevelt and the Crusade*

for America (New York: HarperCollins, 2009).

26. Elizabeth A. Tower, *Icebound Empire: Industry and Politics on the Last Frontier, 1898–1938* (Anchorage, AK: Publication Consultants, 1996).

27. Egan, *The Big Burn*, p. 79.

28. Patricia O'Toole, *When Trumpets Call: Theodore Roosevelt After the White House* (New York: Simon and Schuster, 2006), pp. 82–83.

29. Gifford Pinchot to R. E. Prouty, February 14, 1930, in Martin Nelson McGeary, *Gifford Pinchot: Forester-Politician* (Princeton, NJ: Princeton University Press, 1960), p. 116.

30. Katherine Hocker, *Alaska's Glaciers: Frozen in Motion* (Anchorage: Alaska Natural History Association, 2006), p. 11.

31. Egan, *The Big Burn*, p. 79.

32. Lawrence W. Rakestraw, *A History of the United States Forest Service in Alaska* (Anchorage: Alaska Historical Commission, 1981), chap. 3, "The Chugach National Forest Through 1910."

33. Edmund Morris, *Colonel Roosevelt* (New York: Random House, 2010), p. 44.

34. Henry Pringle, *The Life and Times of William Howard Taft* (New York: Farrar and Rinehart, 1939), p. 480.

35. Egan, *The Big Burn*, p. 81.

36. Richard Ballinger to William Hutchinson Cowles, December 9, 1909, Richard Ballinger Papers (microfilm), University of Washington, Seattle.

37. Charles Richard Van Hise, *The Conservation of Natural Resources in the United States* (New York: Macmillan, 1910), p. 12.

38. Tower, *Icebound Empire*, p. xi.

39. Peter A. Coates, *The Trans-Alaska Pipeline Controversy: Technology, Conservation, and the Frontier* (Bethlehem, PA: Lehigh University Press, 1991), p. 45.

40. McGeary, *Gifford Pinchot*, p. 134.

41. James Wickersham, *Old Yukon: Tales, Trails, and Trials* (Washington, DC: Law Book, 1938).

42. Coates, *The Trans-Alaska Pipeline Controversy*, p. 45.

43. McGeary, *Gifford Pinchot*, p. 133.

44. Stephen Haycox and Alexandra McClanahan, *Alaska Scrapbook: Moments in Alaska History: 1816–1998* (Portland, OR: Graphic Arts Center, 2008), pp. 43–44.

45. Tower, *Icebound Empire*, p. 153.

46. Miller, *Theodore Roosevelt*, p. 503.

47. McGeary, *Gifford Pinchot*, p. 130.

48. Egan, *The Big Burn*, p. 86.

49. Major-General A. W. Greely, *Handbook of Alaska: Its Resources, Products, and Attractions* (New York: Scribner, 1909), pp. 54–55.

50. Egan, *The Big Burn*, p. 86.

51. Ibid., p. 98.

52. "Are the Guggenheims in Charge of the Department of Interior?" *Collier's* (November 13, 1909).

53. George Edwin Mowry, *Theodore Roosevelt and the Progressive Movement* (Madison: University of Wisconsin, 1938), p. 86.

54. Gifford Pinchot, *Breaking New Ground* (Washington, DC: Island, 1998), pp. 498–500.

55. Rakestraw, *A History of the United States Forest Service in Alaska*, chap. 4.

56. Theodore Roosevelt to Gifford Pinchot, March 1, 1910, in Martin L. Fausold, *Gifford Pinchot, Bull Moose Progressive* (Syracuse, NY: Syracuse University Press, 1961), p. 36.

57. Theodore Roosevelt to Henry Cabot Lodge, March 4, 1910, in Miller, *Gifford Pinchot*, p. 176.

58. Elting E. Morison (ed.), *The Letters of Theodore Roosevelt*, Vol. VII (Cambridge, MA: Harvard University Press, 1954), p. 52.

59. Dalton, *Theodore Roosevelt*, p. 357.

60. McGeary, *Gifford Pinchot*, p. 176.

61. O'Toole, *When Trumpets Call*, p. 84.

62. Gifford Pinchot Diary, April 11, 1910, Library of Congress, Washington, DC.

63. Pinchot, *Breaking New Ground*, p. 502.

64. Gifford Pinchot, *The Fight for Conservation* (New York: Doubleday, Page, 1910), p. 6.

65. Miller, *Gifford Pinchot and the Making of Modern Environmentalism*, pp. 228–230.

66. Pinchot, *The Fight for Conservation*, first page of Introduction.

67. Ibid., p. 146.

68. Dalton, *Theodore Roosevelt*, p. 360.

69. Miller, *Theodore Roosevelt*, p. 508.

70. Joseph Bucklin Bishop, *Theodore Roosevelt and His Time: Shown in His Own Letters* (New York: Scribner, 1920), p. 122.

71. Theodore Roosevelt to David Grey, October 5, 1911, in Morison, *The Letters of Theodore Roosevelt*, Vol. VII, p. 407.

72. Viscount Grey of Fallodon, *Recreation* (Boston, MA: Houghton Mifflin, 1920), p. 32.

73. Jerome Jackson, William Davis Jr., and John Tautin (eds.), *Bird Banding in North America: The First Hundred Years* (Cambridge, MA: Harvard University Press, 2008), p. 3.

74. "History of Bird Banding," *Auk*, Vol. 38, No. 1 (January 1921), p. 220.

75. Curt Meine, *Aldo Leopold: His Life and Work* (Madison: University of Wisconsin

Press, 1988), p. 148. Also H. W. Henshaw, *Report of the Chief of the Bureau of Biological Survey* (Washington, DC: Government Printing Office, 1910), p. 11.

76. John F. Reiger, *American Sportsmen and the Origins of Conservation*, 3rd, rev. expanded ed. (Corvallis: Oregon State University Press, 2001), pp. 186–187.

77. Ibid.

78. Rakestraw, *A History of the United States Forest Service in Alaska*, chap. 4.

79. Samuel Trask Dana, *Forest and Range Policy: Its Development in the United States* (New York: McGraw-Hill, 1956), pp. 178–197; E. A. Sherman, "The Supreme Court of the United States and Conservation Policies," *Journal of Forestry* (December 1921), pp. 928–930.

80. Hamlin Garland, *Cavanaugh, Forest Ranger: A Romance of the Mountain West* (New York: Harper, 1910), p. 29.

81. Keith Newlin and Joseph B. McCullough (eds.), *Selected Letters of Hamlin Garland* (Lincoln: University of Nebraska Press, 1998), p. xvi.

82. Gifford Pinchot to Hamlin Garland, March 14, 1910, Pinchot Papers, Library of Congress, Washington, DC.

83. Susan Kollin, *Nature's State: Imagining Alaska as the Last Frontier* (Chapel Hill: University of North Carolina Press, 2001), pp. 62–63.

84. John Helper, "Michigan's Forgotten Son: James Oliver Curwood," *Midwestern Miscellany*, Vol. 7 (1979).

85. James Oliver Curwood, *The Alaskan* (Allison Park, PA: A Rose, 2008), pp. 5–14.

86. Ibid., pp. 13–44. Also see G. Edward White, *The Eastern Establishment and the Western Experience: The West of Frederic Remington, Theodore Roosevelt, and Owen Wister* (New Haven, CT: Yale University Press, 1968).

87. James Wilson, letter to Gifford Pinchot, February 1, 1905, Pinchot Papers, Library of Congress, Washington, DC.

88. Gifford Pinchot, testimony before the U.S. House Committee on the Public Lands, House of Representatives, 63rd Congress, 1913.

89. Robert W. Righter, *The Battle over Hetch Hetchy: America's Most Controversial Dam and the Birth of Modern Environmentalism* (New Haven, CT: Yale University Press, 2005), pp. 1–28.

90. Quoted in Roderick Nash, *Wilderness and the American Mind* (New Haven, CT: Yale University Press, 1967), p. 168.

91. John Muir, *The Yosemite* (New York: Century, 1912), p. 262.

4. Bull Moose Crusade

1. "Roosevelt Puts in a Strenuous Day," *New York Times*, June 23, 1910.

2. "Jungle Barks at Camp Fire Dinner," *New York Times*, January 10, 1914.

3. "Camp Fire Dinner for Buffalo Jones," *New York Times*, December 5, 1909.

4. Aldo Leopold, *Game Management* (New York: Scribner, 1933), p. 18.

5. "Roosevelt Puts in a Strenuous Day."

6. "Pinchot to Inspect Adirondack Forests," *New York Times*, July 21, 1911.

7. Briton Cooper Busch, *The War Against the Seals: A History of the North American Seal Fishery* (Kingston, ON: McGill-Queen's University Press, 1987), p. 96.

8. Ibid.

9. Gary Murphy, "'Mr. Roosevelt Is Guilty': Theodore Roosevelt and the Crusade for Constitutionalism, 1910–1912," *Journal of American Studies*, Vol. 36, No. 3 (December 2002), Part 1: "Looking Backward, Looking Forward: From the Gilded Age to the 1930s," p. 444.

10. *Theodore Roosevelt Cyclopedia* (New York: Roosevelt Memorial Association, 1941), p. 102.

11. David C. Scott and Brendan Murphy, *The Scouting Party: Pioneering and Preservation, Progressivism and Preparedness in the Making of the Boy Scouts of America* (Dallas, TX: Penland, 2010), p. 7.

12. Quoted in Paul Russell Cutright, *Theodore Roosevelt: The Making of a Conservationist* (Urbana: University of Illinois Press, 1985), p. 238.

13. Kathleen Dalton, *Theodore Roosevelt: A Strenuous Life* (New York: Random House, 2002), p. 352.

14. Theodore Roosevelt, "Foreword," in *African Game Trails* (New York: Scribner, 1910).

15. Neil Edward Stubbs, "Theodore Roosevelt and Ernest Hemingway," *Theodore Roosevelt Association Journal*, Vol. 25, No. 2 (2002), pp. 9–14.

16. Sean Hemingway, "Introduction," in *Hemingway on Hunting* (Guilford, CT: Globe Pequot, 2001), pp. xxv–xxvi.

17. "Millions in the Toy Trade," *New York Evening Post*, December 19, 1909, p. M2.

18. Dalton, *Theodore Roosevelt*, p. 356.

19. Ibid., pp. 352–358.

20. Roderick Frazier Nash, *Wilderness and the American Mind* (New Haven, CT: Yale University Press, 2001), p. 143.

21. Dan Beard, "The Boy Scouts," *Outlook* (July 23, 1916), p. 696.

22. Theodore Roosevelt to James Edward West, February 10, 1911, in Elting Elmore Morison (ed.), *The Letters of Theodore Roosevelt*, Vol. VII, *The Days of Armageddon, 1909–1914* (Cambridge, MA: Harvard University Press, 1954), p. 306.

23. "Boy Scout Leaders Dine Baden-Powell," *New York Times*, September 24, 1910, p. 8.

24. David C. Scott to Douglas Brinkley, February 18, 2010.

25. Nash, *Wilderness and the American Mind*, p. 148.

26. "Use Golden Rule in Play: Roosevelt to Boy Scouts," *Washington Post*, August 8, 1911, p. 5.

27. Morison, *The Letters of Theodore Roosevelt*, Vol. VII, p. 95.

28. Theodore Roosevelt to Gifford Pinchot, June 28, 1910.

29. Adolphus Washington Greely, *Handbook of Alaska: Its Resources, Products, and Attractions* (New York: Scribner, 1909), p. 62.

30. *Philadelphia Inquirer*, April 17, 1867.

31. John Muir, *Our National Parks* (Boston, MA, and New York: Houghton Mifflin, 1901), p. 11.

32. John Muir, quoted in Alfred Runte, *National Parks: The American Experience* (Lincoln: University of Nebraska Press, 1997), p. 48.

33. Peter A. Coates, *The Trans-Alaska Pipeline Controversy: Technology, Conservation, and the Frontier* (Bethlehem, PA: Lehigh University Press, 1991), p. 28.

34. Theodore Roosevelt to Theodore Roosevelt Jr., August 23, 1910.

35. Paul Brooks, *The Pursuit of Wilderness* (Boston, MA: Houghton Mifflin, 1971), p. 60.

36. Theodore Roosevelt to Theodore Roosevelt Jr., September 21, 1910.

37. Theodore Roosevelt to Willis Stanley Blatchley, December 9, 1910.

38. Char Miller, *Gifford Pinchot and the Making of Modern Environmentalism* (Washington, DC: Island, 2001), p. 357.

39. Theodore Roosevelt to Abraham Walter Lafferty, December 20, 1910.

40. David Harmon, Francis P. McManamon, and Dwight T. Pitcaithley (eds.), *The Antiquities Act: A Century of American Archaeology, Historic Preservation, and Nature Conservation* (Tucson: University of Arizona Press, 2006), p. 289.

41. "Sitka National Historical Park" (Sitka, AK: National Park Service Archive).

42. Theodore Roosevelt to Edmund Heller, February 10, 1911, in Morison, *The Letters of Theodore Roosevelt*, Vol. VII, p. 230.

43. William J. Long, "The Bull Moose," *Independent*, July 11, 1912, pp. 85–87.

44. Bruce Woods, *Alaska's National Wildlife Refuges* (Anchorage: Alaska Geographic Society, 2003), p. 16.

45. William T. Hornaday, *Wild Life Conservation in Theory and Practice: Lectures Delivered Before the Forest School of Yale University* (New Haven, CT: Yale University Press, 1914), p. 89.

46. Walter B. Borneman, *Alaska: Saga of a Bold Land* (New York: HarperCollins, 2003), p. 241.

47. Miller, *Gifford Pinchot and the Making of Modern Environmentalism*, p. 206.

48. Martin Nelson McGeary, *Gifford Pinchot: Forester-Politician* (Princeton, NJ: Princeton University Press, 1960), p. 208.

49. Ernest Gruening, *The State of Alaska: A Definitive History of America's Northernmost Frontier* (New York: Random House, 1954), pp. 130–135.

50. Gifford Pinchot to W. H. Downing, August 6, 1931, in McGeary, *Gifford Pinchot*, p. 449.

51. *Cleveland Press*, October 25, 1911, ibid., p. 209.

52. Theodore Roosevelt to Henry Fairfield Osborn, May 8, 1911, in Morison, *The Letters of Theodore Roosevelt*, Vol. VII, p. 264.

53. Theodore Roosevelt to William Kent, September 19, 1911, ibid., p. 343.

54. "The Roosevelt Letters," in Charles James Longman (ed.), *The Days of My Life:*

An *Autobiography by Sir. H. Rider Haggard* (London and New York: Longmans, Green, 1926), p. 182.

55. Theodore Roosevelt to Henry Rider Haggard, August 22, 1911, in Morison, *The Letters of Theodore Roosevelt*, Vol. VII, p. 329.

56. Charles Sheldon, *The Wilderness of the Upper Yukon: A Hunter's Explorations for Wild Sheep in the Sub-Arctic Mountains* (New York: Scribner, 1911).

57. Theodore Roosevelt to Anna Roosevelt Cowles, January 27, 1916, in Morison, *The Letters of Theodore Roosevelt*, Vol. VII, p. 43.

58. Francis Hobart Herrick, *Audubon the Naturalist: A History of His Life and Time* (New York: D. Appleton, 1917).

59. Theodore Roosevelt to Francis Hobart Herrick, January 15, 1912, in Morison, *The Letters of Theodore Roosevelt*, Vol. VII, p. 478.

60. Ibid.

61. Ibid.

62. Robert Griggs, "After the Eruption of Katmai, Alaska: The Story of the Effect on Cultivated and Native Vegetation," *Natural History*, Vol. 20 (1920), p. 390.

63. Robert F. Griggs, "The Valley of Ten Thousand Smokes: National Geographic Society Explorations in the Katmai District of Alaska," *National Geographic Magazine*, Vol. 31 (January–June 1917), p. 64.

64. Katmai National Park and Preserve, "History" (Katmai, AK: National Park Service Archive).

65. George Wuerthner, *Beautiful America's Alaska* (Portland, OR: Beautiful America, 1995), p. 66.

66. R. Craig Sautter and Edward M. Burke, *Inside the Wigwam: Chicago Presidential Conventions, 1860–1996* (Chicago, IL: Wild Onion, 1996), p. 121.

67. Kent Garber, "Teddy Roosevelt, on the Bull Moose Party Ticket, Battles Incumbent William Howard Taft," *U.S. News and World Report*, January 17, 2008.

68. Quoted in Cutright, *Theodore Roosevelt*, p. 238.

69. Quoted in Sidney M. Milkis, *Theodore Roosevelt, the Progressive Party, and the Transformation of American Democracy* (Lawrence: University Press of Kansas, 2009), p. 164.

70. Richard Cooley, *Politics and Conservation: The Decline of the Alaska Salmon* (New York: Harper and Row, 1963), pp. 96–98.

71. Edmund Morris, *Colonel Roosevelt* (New York: Random House, 2010), p. 234.

72. Daniel Ruddy, *Theodore Roosevelt's History of the United States: In His Own Words* (New York: HarperCollins, 2010), p. xv.

73. Milkis, *Theodore Roosevelt, the Progressive Party, and the Transformation of American Democracy*, p. 215.

74. Theodore Roosevelt, *The Works of Theodore Roosevelt*, Vol. XIX (New York: Scribner, 1926), p. 42.

75. T. H. Watkins, *Righteous Pilgrim: The Life and Times of Harold L. Ickes, 1874–1952* (New York: Holt, 1990), pp. 9–62.

76. NBC News Address, March 3, 1934, Speeches and Writings, Container 272, Secretary of the Interior File, Harold L. Ickes Papers, Library of Congress, Washington, DC.

77. Watkins, *Righteous Pilgrim*, p. 135.

78. Theodore Roosevelt to Kermit Roosevelt, September 27, 1912, Theodore Roosevelt Papers, Library of Congress, Washington, DC.

79. Patricia O'Toole, *When Trumpets Call: Theodore Roosevelt After the White House* (New York: Simon and Schuster, 2006) p. 218; H. W. Brands, *T.R.: The Last*

Romance (New York: Basic Books, 1997), p. 721.

80. Gifford Pinchot to Theodore Roosevelt, October 25, 1912, McGeary, *Gifford Pinchot*, p. 231.

81. Theodore Roosevelt to Gifford Pinchot, October 29, 1912, ibid.

82. James Chace, *1912: Wilson, Roosevelt, Taft, and Debs—The Election That Changed the Country* (New York: Simon and Schuster, 2004), p. 237.

83. Harold L. Ickes, *Autobiography of a Curmudgeon* (New York: Reynal and Hitchcock, 1943), p. 164.

84. Theodore Roosevelt to Kermit Roosevelt, November 5, 1912, Theodore Roosevelt Papers, Box 3, Library of Congress, Washington, DC.

5. Charles Sheldon's Fierce Fight

1. Kris Capps, *A Wildlife Guide: Denali National Park and Preserve, Alaska* (Santa Barbara, CA: ARA Leisure Services, 1994), p. 6.

2. Tom Walker, *McKinley Station: The People of the Pioneer Park That Became Denali* (Missoula, MT: Pictoral Histories, 2009), p. ix.

3. Charles Sheldon, *The Wilderness of the North Pacific Coast Islands: A Hunter's Experiences While Searching for Wapiti, Bears, and Caribou on the Larger Coast Islands of British Columbia and Alaska* (New York: Scribner, 1912), p. 3.

4. R. O. Polziehn, J. Hamr, F. F. Mallory, and C. Strobeck, "Phylogenetic Status of North American Wapiti (*Cervus elaphus*) Subspecies," *Canadian Journal of Zoology*, Vol. 76 (1998), pp. 998–1010.

5. Maria Pasitschniak-Arts, "Ursus arctos," *Mammalian Species Report*, American Society of Mammalogists (April 23, 1993).

6. Thomas McNamee, *The Grizzly Bear* (New York: McGraw-Hill, 1984), p. 248.

7. Sheldon, *The Wilderness of the North Pacific Coast Islands*, p. 178.

8. Roderick Frazier Nash, *Wilderness and the American Mind*, 4th ed. (New Haven, CT: Yale University Press, 2001), p. 285.

9. Neil B. Carmony and David E. Brown (eds.), *The Wilderness of the Southwest: Charles Sheldon's Quest for Desert Bighorn Sheep and Adventures with the Havasupai and Seri Indians* (Salt Lake City: University of Utah Press, 1979), pp. xiv–xv.

10. Ibid., p. xxiii.

11. Ibid., p. 204.

12. Theodore Roosevelt, "The American Hunter-Naturalist," *Outlook* (December 9, 1911), pp. 854–856.

13. Douglas Brinkley, *The Wilderness Warrior: Theodore Roosevelt and the Crusade for America* (New York: HarperCollins, 2009), pp. 585–630.

14. James B. Trefethen, *An American Crusade for Wildlife* (New York: Winchester, 1975), p. 192.

15. Theodore Roosevelt to Charles Sheldon, March 13, 1917, Charles Sheldon Papers, University of Alaska-Fairbanks.

16. Theodore Roosevelt to Charles Sheldon, May 5, 1910, Box 3, Folder 10, Roosevelt Correspondence, 1910–1917, University of Alaska-Fairbanks.

17. Jenks Cameron, *The Bureau of Biological Survey* (Baltimore, MD: Johns Hopkins University Press, 1929), p. 121.

18. Catherine Cassidy and Gary Titus, *Alaska's No. 1 Guide: The History and Journals of Andrew Berg 1869–1939* (Soldotna, AK: Spruce Tree, 2003), p. 314.

19. Charles Sheldon, *The Wilderness of the Upper Yukon: A Hunter's Explorations for Wild Sheep in Sub-Arctic Mountains* (New York: Scribner, 1911), p. 4.

20. James Gore King, *Attending Alaska's Birds: A Wildlife Pilot's Story* (Victoria, BC: Trafford, 2008), p. 166.

21. Margaret E. Murie, *Two in the Far North* (Anchorage: Alaska Northwest, 1962), p. 274.

22. D. S. Hik, S. J. Hannon, and K. Martin, "Northern Harrier Predation on Willow Ptarmigan," *Wilson Bulletin*, Vol. 98, No. 4 (1986), pp. 597–600.

23. William O. Douglas, *My Wilderness: The Pacific West* (Garden City, NY: Doubleday, 1960), pp. 11–12.

24. Sheldon, *The Wilderness of the North Pacific Coast Islands*, p. 104.

25. Charles Sheldon, "List of Birds Observed in the Upper Toklat River Near Mount McKinley, Alaska, 1907–1908," *Auk*, Vol. 26, No. 1 (January 1909).

26. Ira N. Gabrielson and Frederick L. Lincoln, *Birds of Alaska* (Harrisburg, PA: Stockpole, 1959), p. 114.

27. Tom Murphy, *The Comfort of Autumn: The Seasons of Yellowstone* (Livingston, MT: Crystal Creek, 2005), p. 100.

28. Sheldon, *The Wilderness of the North Pacific Coast Islands*, p. 294.

29. Hudson Stuck, *Ten Thousand Miles with a Dog Sled* (New York: Scribner, 1914).

30. Bill Sherwonit, Andromeda Romano-Lax, and Ellen Bielawski, *Travelers' Tales Alaska* (San Francisco, CA: Traveler's Tales, 2003), p. 14.

31. Hudson Stuck, *The Ascent of Denali* (New York: Scribner, 1918), p. xi.

32. C. Hart Merriam, "Introduction," in Charles Sheldon, *The Wilderness of Denali: Explorations of a Hunter-Naturalist in Northern Alaska* (New York: Scribner, 1930).

33. C. Hart Merriam, "Preliminary Synopsis of the American Bears," *Proceedings of the Biological Society of Washington*, Vol. 10 (1896), pp. 65–83. For more of Merriam's work on bears, see C. Hart Merriam, *The Mammals of the Adirondack Region, Northeastern New York* (New York: Henry Holt, 1886).

34. Sheldon, *The Wilderness of the North Pacific Coast Islands*, p. 48.

35. E. R. Hall, *The Mammals of North America*, 2nd ed. (New York: Wiley, 1981).

36. Charles Sheldon to Dr. C. Hart Merriam, February 20, 1911, Boone and Crockett Club Archives, Missoula, MT.

37. Theodore Roosevelt to Charles Sheldon, January 29, 1917, Sheldon Papers, University of Alaska-Fairbanks.

38. Charles Sheldon, Alaska Diary, March 27–29, 1908, in *The Wilderness of Denali* (New York: Scribner, 1960).

39. Henry P. "Harry" Karstens, Diary, January 12, 1908, Karstens Papers, University of Alaska-Fairbanks.

40. Nash, *Wilderness and the American Mind*, p. 286.

41. Sheldon, *The Wilderness of Denali*, p. 405.

42. Ibid., pp. 15–16.

43. James B. Trefethen, *Crusade for Wildlife* (New York: Stackpole, 1961), p. 179; John Isle, *Our National Park Policy: A Critical History* (Baltimore, MD: Johns Hopkins University Press, 1961), p. 226.

44. Capps, *A Wildlife Guide*, p. 8.

45. Harry Ritter, *Alaska's History: The People, Land, and Events of the North Country* (Anchorage: Alaska Northwest, 1993), pp. 74–75.

46. Madison Grant, "Establishment of Mount McKinley National Park," in William G. Sheldon (ed.), "A History of the Boone and Crockett Club" (unpublished).

47. Quoted in Peter A. Coates, *The Trans-Alaska Pipeline Controversy: Technology, Conservation, and the Frontier* (Bethle-

hem, PA: Lehigh University Press, 1991), p. 33.

48. Richard Slotkin, *The Fatal Environment: The Myth of the Frontier, 1776–1890* (New York: Atheneum, 1985), p. 20.

49. Sheldon, *The Wilderness of the North Pacific Coast Islands*, pp. 217–218.

50. Belmore Browne, *The Conquest of Mount McKinley* (Boston, MA: Houghton Mifflin, 1956).

51. Charles Sheldon to E. W. Nelson, October 10, 1915, Boone and Crockett Club Archives, Missoula, MT.

52. *Reports of the Secretary of the Interior for the Fiscal Year Ended June 30, 1918*, Vol. 1 (Washington, DC: U.S. Government Printing Office, 1919).

53. Frank Norris, *Crown Jewel of the North: An Administrative History of Denali National Park and Preserve* (Anchorage: Alaska Regional Office–National Park Service), p. 37.

54. Nash, *Wilderness and the American Mind*, p. 286.

55. Norris, *Crown Jewel of the North*, pp. 37–39.

56. Alfred Runte, *National Parks: The American Experience* (Lanham, MD: Taylor Trade, 2010), pp. 104–105.

57. Nash, *Wilderness and the American Mind*, p. 154.

58. Stewart Edward White, *The Forest* (New York: Phillips, 1903), p. 5.

59. "Park for Camp Fire Club," *New York Times*, July 22, 1917.

60. Author interview with Leonard Vallender, July 5, 2010.

61. Linnie Marsh Wolfe (ed.), *John of the Mountains: The Unpublished Journals of John Muir* (Madison: University of Wisconsin Press, 1979), p. 399.

62. T. S. Palmer to Frederick K. Vreeland, March 21, 1912; Vreeland to Palmer, March 22, 1912, Frederick Vreeland Papers, Box 5, General Correspondence, 1902–1931, Entry 138, Record Group 22, Native Archives, Washington, DC.

63. Frederick K. Vreeland, testimony before the Subcommittee on Public Lands of the Committee on Public Lands for the establishment of Mount McKinley National Park, House of Representatives, Washington, DC, May 4, 1916.

64. Wilfred Osgood, *A Biological Reconnaissance of the Base of the Alaska Peninsula, North America Fauna*, No. 24 (Washington, DC: Government Printing Office, 1904), pp. 25–26.

65. "Visit Alaska: Interest of Outdoor Life," *Anchorage*, August 1, 1912.

66. Frederick K. Vreeland to Dr. C. Hart Merriam, November 17, 1921, C. Hart Merriam Papers, Bancroft Library, University of California, Berkeley.

6. Our Vanishing Wildlife

1. William T. Hornaday, *Our Vanishing Wild Life: Its Extermination and Preservation* (New York: New York Zoological Society, 1913), p. 15. (Reprint, New York: Arno, 1970.)

2. V. B. Scheffer, "The Weight of the Steller Sea Cows," *Journal of Mammalogy*, Vol. 53, No. 4 (1972), pp. 912–914.

3. William G. Sheldon, "A History of the Boone and Crockett Club: Milestones in Wildlife Conservation," Boone and Crockett Club Archives, Missoula, MT.

4. Hornaday, *Our Vanishing Wild Life*, pp. ix–x.

5. Quinn Hornaday and Aline G. Hornaday (eds), *The Hornadays, Root and Branch* (Los Angeles, CA: Stockton Trade, 1979), pp. 77–89.

6. Stephen Fox, *The American Conservation Movement: John Muir and His Legacy*

(Madison: University of Wisconsin Press, 1981), pp. 157–158.

7. Hornaday and Hornaday, *The Horna-days, Root and Branch*, p. 86.

8. Frank Graham, *Man's Dominion: The Story of Conservation in America* (New York: M. Evans; distributed in association with Philadelphia, PA: Lippincott, 1971), p. 188.

9. "No More Slaughtering of Seals for Five Years," *New York Times*, September 1, 1912.

10. Hornaday, *Our Vanishing Wild Life*, p. 156.

11. Ibid., p. x.

12. Theodore Roosevelt, "Our Vanishing Wild Life," *Outlook* (January 25, 1913).

13. Fran Mauer to Douglas Brinkley, September 27, 2010.

14. Roosevelt, "Our Vanishing Wild Life." Also Arthur K. Willyoung, "Roosevelt the Great Outdoor Man," *Outing*, Vol. 74, No. 5 (August, 1919). For the history of the American Game Protection Association, see Fox, *The American Conservation Movement*, pp. 151–157.

15. Hornaday, *Our Vanishing Wild Life*, p. 269.

16. Nathan Miller, *Theodore Roosevelt: A Life* (New York: Morrow, 1992), p. 532.

17. Theodore Roosevelt, "The Conservation of Womanhood and Childhood," *Outlook* (December 23, 1911), p. 13.

18. Madison Grant to Dr. E. Lester Jones, January 11, 1915, Boone and Crockett Club Archive, Missoula, MT.

19. Roosevelt, "Our Vanishing Wild Life," p. 161.

20. Jennifer Price, "Hats Off to Audubon," *Audubon* (November–December 2004), p. 50.

21. Roosevelt, "Our Vanishing Wild Life," p. 161.

22. Ira N. Gabrielson, *Wildlife Refuges* (New York: Macmillan, 1943), p. 56.

23. Hornaday, *Our Vanishing Wild Life*, p. 150.

24. Ibid., p. 63.

25. David R. Klein and Robert G. White, "Parameters of Caribou Population Ecology in Alaska," *Biological Papers of the University of Alaska*, No. 3 (Fairbanks: University of Alaska, 1978).

26. Ibid., pp. 330–345.

27. Gabrielson, *Wildlife Refuges*, p. 67.

28. Hornaday, *Our Vanishing Wild Life*, p. 64.

29. Ibid., pp. 178–179.

30. Ibid., p. 269.

31. Curt Meine, *Aldo Leopold: His Life and Work* (Madison: University of Wisconsin Press, 1988), p. 287.

32. Charles Sheldon to George Bird Grinnell, December 23, 1918, Boone and Crockett Club Archives, Missoula, MT.

33. Peter A. Coates, *The Trans-Alaska Pipeline Controversy: Technology, Conservation, and the Frontier* (Bethlehem, PA: Lehigh University Press, 1991), p. 43.

34. William N. Wilson, *Railroad in the Clouds: The Alaskan Railroad in the Age of Steam, 1914–1945* (Boulder, CO: Pruett, 1977), pp. 7–11.

7. The Lake Clark Pact

1. Curt Meine, *Aldo Leopold: His Life and Work* (Madison: University of Wisconsin Press, 1988), p. 128.

2. Aldo Leopold, *Game Management* (Madison: University of Wisconsin Press, 1986), p. 19.

3. Neil B. Carmony and David E. Brown (eds.), *The Wilderness of the Southwest: Charles Sheldon's Quest for Desert Bighorn Sheep and Adventures with the Havasupai*

and Seri Indians (Salt Lake City: University of Utah Press, 1979), p. 12.

4. "Roosevelt to Cross Grand Canyon," *New York Times*, July 15, 1913, p. 2.

5. Theodore Roosevelt, *A Book-Lover's Holidays in the Open* (New York: Scribner, 1916), p. 7.

6. Ibid.

7. Ibid., pp. 1–10.

8. Lawrence W. Rakestraw, *A History of the United States Forest Service in Alaska* (Anchorage: Alaska Historical Commission, 1981), chap. 4.

9. Meine, *Aldo Leopold*, pp. 148–149.

10. Ibid.

11. The inscribed book is at the Leopold Library, University of California, Berkeley.

12. John Muir, *The Yosemite* (New York: Century, 1912), p. 257.

13. Robert W. Righter, *The Battle over Hetch Hetchy: America's Most Controversial Dam and the Birth of Modern Environmentalism* (New York: Oxford University Press, 2005), p. 133.

14. John Muir, *John Muir: The Eight Wilderness Discovery Books* (Seattle, WA: Mountaineers, 1992), p. 714.

15. John Muir, *Steep Trails* (Berkeley: University of California Press, 1974), p. 74.

16. Donald Worster, *A Passion for Nature: The Life of John Muir* (New York: Oxford University Press, 2008), p. 464.

17. Theodore Roosevelt, "John Muir: An Appreciation," *Outlook* (January 6, 1915), p. 27.

18. Roderick Frazier Nash, *Wilderness and the American Mind*, 4th ed. (New Haven, CT: Yale University Press, 2001), p. 358.

19. John Muir, *Travels in Alaska* (New York: Modern Library, 2002), p. 4.

20. Ibid., p. 13.

21. Ibid., p. 193.

22. Samuel Hall Young, *Alaska Days with John Muir* (Grand Rapids, MI: Fleming H. Revell, 1915), pp. 224–226.

23. Meine, *Aldo Leopold*, pp. 147–156.

24. Theodore Roosevelt to Aldo Leopold, June 18, 1917, Leopold Papers, University of Wisconsin, Madison.

25. Gene Fowler, William "Buffalo Bill" Cody's obituary, *Post* (January 1917).

26. Roosevelt, *Book-Lover's Holidays in the Open*, p. vii.

27. Bruce Woods, *Alaska's National Wildlife Refuges* (Anchorage: Alaska Geographic Society, 2003), p. 13.

28. Quoted in Paul S. Sutter, *Driven Wild: How the Fight Against Automobiles Launched the Modern Environmental Movement* (Seattle: University of Washington Press, 2002), p. 43.

29. Alfred Runte, *National Parks: The American Experience* (Lincoln: University of Nebraska Press, 1997), pp. 208–209.

30. Dian Olson Belanger and Adrian Kinnane, *Managing American Wildlife* (Rockville, MD: Montrose, 2002).

31. Frank Graham Jr., *Man's Dominion: The Story of Conservation in America* (New York: M. Evans, 1971), p. 201; James B. Trefethen, *An American Crusade for Wildlife* (New York: Winchester and Boone and Crockett Club, 1975), pp. 206–208.

32. Tanana Valley Association History (Fairbanks, AK: Tanana Valley Association Archive, April 23, 2009).

33. Theodore Roosevelt, "Is Polar Exploration Worth While?" *Outlook* (March 1, 1913).

34. Elaine Rhode, *National Wildlife Refuges of Alaska* (Anchorage: Alaska Natural History Association, 2003), pp. 10–12.

35. Roosevelt, "Is Polar Exploration Worth While?"

36. Madison Grant, "The Conditions of Wildlife in Alaska," in George Bird Grinnell (ed.), *Hunting at High Altitudes* (New York: Harper, 1913), p. 375.

37. Robert A. Jones, "Alaska Parks: Battle Lines Form Around Last Frontier," *Los Angeles Times*, September 5, 1977, p. 5.

38. Aldo Leopold, *A Sand County Almanac* (New York: Ballantine, 1970), p. 278.

39. Grant, "The Conditions of Wildlife in Alaska," p. 375.

40. John Muir, *The Writings of John Muir: The Cruise of the Corwin* (New York: Houghton Mifflin, 1917), p. 258.

41. John Branson, historian of Lake Clark–Iliamna National Park, is writing a biography of John W. Clark. He greatly helped me understand the history of this beautiful part of Alaska.

42. John B. Branson, *The Canneries, Cabins, and Caches of Bristol Bay, Alaska* (Anchorage, AK: U.S. Department of the Interior, 2007), pp. 1–10.

43. Brian Fagan, *"Where We Found a Whale": A History of Lake Clark National Park and Preserve* (Washington, DC: U.S. Department of the Interior, 2008), pp. 118–122.

44. Frederick K. Vreeland to E. A. Preble, November 29, 1921, Smithsonian Institution, Archives Record Unit 7176, U.S. Fish and Wildlife Service.

45. John Branson (ed.), *Lake Clark–Iliamna, Alaska 1921: Travel Diary of Colonel A. J. Macnab* (Anchorage: Alaska Natural History Association, 1996). (Booklet reprint.)

46. Ibid., p. 8.

47. Ibid., p. 27.

48. Colonel A. J. Macnab Diaries, August 18 and August 20, quoted in Branson, *Lake Clark–Iliamna, Alaska 1921*.

49. Branson, *Lake Clark–Iliamna, Alaska 1921*, p. 29. Special thanks to Branson for helping me get the expedition straight.

8. Resurrection Bay of Rockwell Kent

1. Rockwell Kent, *Wilderness: A Journal of Quiet Adventure in Alaska* (New York: Putnam-Knickerbocker, 1920), p. 191.

2. Ibid., p. 6.

3. Barry Lopez, *Arctic Dreams* (New York: Vintage, 2001), pp. 390–391.

4. Kent, *Wilderness*, p. 24.

5. Rockwell Kent, *Wilderness: A Journal of Quiet Adventure in Alaska* (Middletown, CT: Wesleyan University Press, 1996), p. xxxii. (From Los Angeles, CA: Wilderness, 1970.) Henceforth cited as Wesleyan University Press edition.

6. Kesler E. Woodward (ed.), *Painting in the North: Alaskan Art in the Anchorage Museum of History and Art* (Anchorage, AK: Anchorage Museum of History and Art, 1993). (From a plaque describing Sydney M. Laurence's *Mount McKinley*, oil on canvas, 1929.)

7. Linda Cook and Frank Norris, *A Stern and Rock-Bound Coast* (Anchorage, AK: National Park Service, Kenai Fjords National Park 1998); Doug Capra, letter to Douglas Brinkley, May 29, 2010.

8. Grace Glueck, "Celebrating an Artist's Spiritual Searches and Realist Findings," *New York Times*, August 26, 2005.

9. Judith H. Dabrzynski, "Adirondack Vistas in the Artist's Eye and the Visitor's," *New York Times*, July 23, 1999.

10. Rockwell Kent, *It's Me, O Lord: The Autobiography of Rockwell Kent* (New York: Dodd, Mead, 1955), p. 204.

11. Constance Martin, *Distant Shores: The Odyssey of Rockwell Kent* (Chesterfield, MA: Chameleon, 2000), p. 23.

12. Rockwell Kent, *N by E* (Middletown, CT: Wesleyan University Press, 1966), p. xvi.

13. Kent, *It's Me, O Lord*, p. 121.

14. Edward Hoagland, "Foreword," in Kent, *N by E*, pp. xvi–xvii.

15. "Rockwell Kent's Artistic Discovery of Alaska," *Current Opinion*, Vol. 67 (1919), p. 52.

16. Kent, *It's Me, O Lord*, p. 328.

17. Kent, *Wilderness*, p. 27.

18. Barry Lopez, *Winter Count* (New York: Vintage, 1999), p. 94.

19. "Historic Harrington Cabin" (Homer, AK: Homer Foundation, Pratt Museum, 2001). (Brochure.)

20. Kent, *Wilderness*, p. xi.

21. Bill Streever, *Cold: Adventures in the World's Frozen Places* (New York: Little, Brown, 2009).

22. Martin, *Distant Shores*, p. 26.

23. Kent, *Wilderness* (Wesleyan University Press edition), p. xx.

24. Rockwell Kent to Christian Brinton, winter 1919, Rockwell Kent Papers, Smithsonian Archives of American Art.

25. Kent, *Wilderness*, p. 161.

26. Jenks Cameron, *The Bureau of Biological Survey* (New York: Arno, 1974), pp. 124–127.

27. Doug Capra, "Introduction," in Kent, *Wilderness*, p. xi. (Wesleyan University Press edition.)

28. Martha Gruening, "The Freedom of Wilderness," *Freeman* (April 28, 1920), pp. 165–166.

29. Kent, *Wilderness*, p. 217.

30. Grace Glueck, "Cast into the Wilderness by Choice, He Found a Friend in the Landscape," *New York Times*, August 18, 2000.

31. Hoagland, "Foreword," in Kent, *N by E*, p. viii.

32. Garnett McCoy, "The Rockwell Kent Papers," *Archives of American Art Journal*, Vol. 12, No. 1 (January 1972), p. 6.

33. Scott R. Ferris, "Introduction," in Rockwell Kent, *Salamina* (Middletown, CT: Wesleyan University Press, 2003), p. xxi.

34. Kent, *It's Me, O Lord*, p. 328.

35. Gail Levin, *Twentieth-Century American Painting: The Thyssen-Bornemisza Collection* (New York: Sotheby's, 1987), p. 60.

36. Gary Snyder, "Raven's Beak River at the End," in *Mountains and Rivers Without End* (New York: Counterpoint, 1997), p. 65.

9. The New Wilderness Generation

1. Nathan Miller, *Theodore Roosevelt: A Life* (New York: HarperCollins, 1992), p. 560.

2. Theodore Roosevelt, speech to Colorado Livestock Association, Denver, August 29, 1910.

3. Edmund Morris, *Colonel Roosevelt* (New York: Random House, 2010), p. 576.

4. Patricia O'Toole, *When Trumpets Call: Theodore Roosevelt After the White House* (New York: Simon and Schuster, 2005), p. 401.

5. "Stop City Work in Colonel's Honor," *New York Times*, January 9, 1919, p. 4.

6. Aïda DiPace Donald, *Lion in the White House: A Life of Theodore Roosevelt* (New York: Basic Books, 2007), p. 265.

7. Edward Wagen Knecht, *The Seven Worlds of Theodore Roosevelt* (New York: Longmans, 1958), p. 20.

8. Morris, *Colonel Roosevelt*, p. 577.

9. Frederick S. Wood (ed.), *Roosevelt as We Knew Him* (Philadelphia, PA: Winston, 1927), p. 380.

10. "Wants Roosevelt Spirit Perpetuated," *New York Times*, February 19, 1919.

11. Hamlin Garland Diary, January 6, 1919, Huntington Library, San Marino, CA.

12. *Natural History: The Journal of the American Museum*, Vol. 19 (January 1919).

13. Michael J. Robinson, *Predatory Bureaucracy: The Exterminators of Wolves and the Transformation of the West* (Boulder: University Press of Colorado, 2005), p. 180.

14. "Want Park to Bear Name of Roosevelt," *New York Times*, January 14, 1919, p. 6.

15. Douglas Brinkley, *The Wilderness Warrior: Theodore Roosevelt and the Crusade for America* (New York: HarperCollins, 2009).

16. Gifford Pinchot, "Overturning Roosevelt's Work," *Christian Science Monitor*, February 24, 1919, p. 3.

17. John B. Branson, *The Canneries, Cabins, and Caches of Bristol Bay, Alaska* (Anchorage, AK: U.S. Department of the Interior, 2007), p. vi.

18. John Morton Blum, *The Republican Roosevelt* (New York: Atheneum, 1962), p. 146.

19. David Brower, "Foreword," in David Brower (ed.), *Wilderness: America's Living Heritage* (San Francisco, CA: Sierra Club, 1961), p. viii.

20. John Burroughs Journal, July 9, 1919, Berg Collection, New York Public Library.

21. Richard Lour, *Last Child in the Woods: Saving Our Children from Nature Deficit Disorder* (Chapel Hill, NC: Algonquin, 2005), pp. 1–11.

22. John Burroughs, *The Writings of John Burroughs* (Boston, MA: Houghton Mifflin, 1917), p. 16.

23. Edward Renehan, *John Burroughs: An American Naturalist* (Hensonville, NY: Black Dome, 1998), p. 313.

24. Jenks Cameron, *The Bureau of Biological Survey* (New York: Arno, 1974), pp. 118–119.

25. Carolyn Sheldon, "Vermont Jumping Mice of the Genus Zapus," *Journal of Mammalogy*, Vol. 19, No. 3 (August 1938), pp. 324–332.

26. Neil B. Carmony and David E. Brown (eds.), *The Wilderness of the Southwest: Charles Sheldon's Quest for Desert Bighorn Sheep and Adventures with the Havasupai and Seri Indians* (Salt Lake City: University of Utah Press, 1979), pp. xli–xlii.

27. William Sheldon, *The Book of the American Woodcock* (Amherst: University of Massachusetts Press, 1971).

28. Carmony and Brown, *The Wilderness of the Southwest*, pp. xl–xlii.

29. Charles Sheldon to George Bird Grinnell, February 28, 1920, Boone and Crockett Club Archives, University of Montana, Missoula.

30. Ben Casselman and Guy Chazan, "Disaster Plan Lacking at Deep Rigs," *Wall Street Journal*, May 18, 2010, p. 1.

31. Joan M. Antonson and William S. Hanable, *Alaska's Heritage*, Unit 4, *Human History: 1867–Present* (Anchorage: Alaska Historical Center, 1985), pp. 422–423.

32. Irvin Palmer Jr., "The History of Alaska Oil," *Alaska Business Monthly*, March 3, 2007.

33. Aldo Leopold, *A Sand County Almanac* (New York: Ballantine, 1970), p. 244.

34. Susan L. Flader and J. Baird Callicott, *The River of the Mother of God and Other Essays by Aldo Leopold* (Madison:

University of Wisconsin Press, 1991),
p. 52.

35. Ernest Walker, "Circular Letter
to Fur Wardens," April 1921, General
Correspondence, Bureau of Biological Survey, Record Group 22, National
Archives, Washington, DC.

36. Aldo Leopold, "Threatened Species,"
American Forests, Vol. 42, No. 3 (March
1936), pp. 116–119.

37. Stephen Fox, *The American Conservation Movement* (Madison: University of
Wisconsin Press, 1985), pp. 244–250.

38. Leopold, "Threatened Species."

39. Morgan Sherwood, *Big Game in
Alaska: A History of Wildlife and People*
(New Haven, CT: Yale University Press,
1981), p. 8.

40. Frank Dufresne, *Alaska's Animals
and Fishes* (Portland, OR: Metropolitan,
1946); Frank Dufresne, *My Way Was
North: An Alaskan Autobiography* (New
York: Holt, Rinehart and Winston,
1966); Frank Dufresne, *No Room for Bears*
(New York: Holt, Rinehart and Winston,
1965).

41. Frank Dufresne, "Alaska General
Correspondence," in Alaska Reports,
January 1924, General Bureau of the
Biological Survey, Record Group 22,
National Archives, Washington, DC.

42. Quoted in Sherwood, *Big Game in
Alaska*, p. 55.

43. Dufresne, *Alaska's Animals and
Fishes*, pp. 296–297.

10. Warren G. Harding: Backlash

1. James Garfield Diary, October 4,
1909, James R. Garfield Papers, Library
of Congress, Washington, DC.

2. Warren G. Harding, Executive Order
3421, U.S. Fish and Wildlife Service,
Anchorage, AK.

3. Hasia Diner, "Teapot Dome, 1924,"
in Arthur M. Schlesinger Jr. and Robert
Burns (eds.), *Congress Investigates: A
Documented History, 1792–1974* (New York:
Chelsea House, 1975).

4. Thomas Fleming, "History's Revenge," *New York Times*, February 23,
1998.

5. Warren Harding, Executive Order
No. 3797-A, February 27, 1923. Also
David L. Spencer, Claus-M. Naske,
and John Carnahan, *National Wildlife
Refuges in Alaska: A Historical Perspective*
(Anchorage, AK: U.S. Fish and Wildlife
Service, 1979), p. 102.

6. Morgan Sherwood, *Big Game in
Alaska: A History of Wildlife and People*
(New Haven, CT: Yale University Press,
1981), p. 73. See also Stephen Haycox and
Alexandra J. McClanahan, *Alaska Scrapbook* (Anchorage, AK: CIRI Foundation,
2007), p. 95.

7. Stephen Haycox, *Alaska: An American
Colony* (Seattle: University of Washington Press, 2002), p. 235.

8. Peter A. Coates, *The Trans-Alaska
Pipeline Controversy: Technology, Conserva-
tion, and the Frontier* (Bethlehem, PA:
Lehigh University Press, 1991), p. 53.

9. June Allen, "What Did Kill Warren
G. Harding," *Stories in the News* (Ketchikan, AK), July 23, 2003.

10. Actually a misquotation, according
to the Calvin Coolidge Memorial Foundation.

11. Bill Mares, *Fishing with the Presidents*
(Mechanicsburg, PA: Stackpole, 1999),
pp. 66–70.

12. Judith St. George, *The Mount Rushmore Story*, Part 2 (New York: Putnam,
1985), p. 128.

13. T. H. Watkins, *Righteous Pilgrim: The
Life and Times of Harold L. Ickes, 1874–1952*
(Holt, 1992), pp. 318–319.

14. William Skinner Cooper, "The Recent Ecological History of Glacier Bay, Alaska," *Ecology*, Vol. 4 (1923), pp. 93–128.

15. Charles Sheldon, *The Wilderness of Denali* (New York: Scribner, 1930).

16. Quoted in Kendrick A. Clements, *Hoover, Conservation, and Consumerism: Engineering the Good Life* (Lawrence: University Press of Kansas, 2000), p. 69.

17. Benjamin Franklin to Sarah Bache, January 26, 1784. Library of Congress, Manuscript Division, Washington, DC.

18. Peter Matthiessen, *Wildlife in America* (New York: Viking, 1959), p. 170.

19. Aldo Leopold to Karl T. Frederick, December 20, 1935, Leopold Papers, University of Wisconsin, Madison.

20. Susan L. Flader and J. Baird Callicott (eds.), *The River of the Mother of God and Other Essays by Aldo Leopold* (Madison: University of Wisconsin Press, 1991), p. 77.

21. Robert Lewis Taylor, "Oh, Hawk of Mercy!" *New Yorker*, April 17, 1948.

22. Keith L. Bildstein, *Migrating Raptors of the World: Their Ecology and Conservation* (Ithaca, NY: Cornell University Press, 2006), p. ix.

23. Dyana Z. Furmansky and Rosalie Edge, *Hawk of Mercy: The Activist Who Saved Nature from the Conservationists* (Athens: University of Georgia Press, 2009), p. 130.

24. Ibid., pp. 164–165.

25. Barry Lopez, *Arctic Dreams: Imagination and Desire in a Northern Landscape* (New York: Vintage, 2001), p. 390.

26. Kim Heacox, *Alaska's Inside Passage* (Portland, OR: Graphic Arts Center, 1997), p. 99.

27. Scott R. Ferris, "Introduction," in Rockwell Kent, *Salamina* (Middletown, CT: Wesleyan University Press, 2003), p. xiii.

28. Sherwood, *Big Game in Alaska*, pp. 92–93.

29. Heacox, *Alaska's Inside Passage*, p. 99.

30. Sherwood, *Big Game in Alaska*.

31. Matthiessen, *Wildlife in America*, p. 170.

11. Bob Marshall and the Gates of the Arctic

1. Roderick Nash, "The Strenuous Life of Bob Marshall," *Forest History* (October 1966), p. 19. See also Roger Kaye, *Last Great Wilderness* (Fairbanks: University of Alaska Press, 2000). Kaye's extraordinary book has informed this chapter. It explains fully how Marshall promoted the idea of wilderness in the 1930s.

2. Paul Schaefer, "Bob Marshall, Mount Marcy, and—the Wilderness," *Living Wilderness* (Summer 1966), pp. 12–16.

3. Charles Reznikoff (ed.), *Louis Marshall, Champion of Liberty: Selected Papers and Addresses*, Vol. 2 (Philadelphia: Jewish Publication Society of America, 1957), p. 1174.

4. Morton Rosenstock, *Louis Marshall, Defender of Jewish Rights* (Detroit, MI: Wayne State University, 1965), p. 19.

5. Edmund Morris, *Colonel Roosevelt* (New York: Random House, 2010), p. 739.

6. Quoted in James M. Glover, *A Wilderness Original: The Life of Bob Marshall* (Seattle, WA: Mountaineers, 1986), p. 13.

7. Robert Marshall, "The Problem of the Wilderness," *Scientific Monthly* (February 1930), pp. 141–148.

8. Terrence Cole, "Preface," in Robert Marshall, *Arctic Village* (Anchorage: University of Alaska Press, 2000), p. xiii. (Reprint.)

9. George Marshall, "Adirondacks to Alaska: A Biographical Sketch of Robert

Marshall," *Ad-i-ron-dac* (March–June 1951), p. 44.

10. Robert Marshall, "Why I Want to Become a Forester in the Future," April 17, 1918, Robert Marshall Papers, Bancroft Library, University of California, Berkeley. (Typescript.)

11. David A. Bernstein, "Bob Marshall: Wilderness Advocate," *Western Studies Jewish Historical Quarterly*, Vol. 13 (October 1980), p. 29.

12. Glover, *A Wilderness Original*, p. 2.

13. "The Alumnae," in *The Harvard Forest 1907–1934* (Cornwall, NY: Cornwall, 1935), p. 8.

14. Author interview, James N. Levitt, director of the Program on Conservation Innovation at Harvard Forest, Harvard University.

15. Robert Marshall, "Mountain Ablaze," *Nature* (June–July 1953).

16. Glover, *A Wilderness Original*, p. 73.

17. Robert Marshall, "Forest Devastation Must Stop," *Nation* (August 1929); Robert Marshall, "A Proposed Remedy for Our Forestry Illness," *Journal of Forestry* (March 28, 1930).

18. Robert Marshall, *Arctic Wilderness* (Berkeley: University of California Press, 1956).

19. John M. Kauffmann, *Alaska's Brooks Range* (Seattle, WA: Mountaineers, 1992), pp. 16–22.

20. Martin Wilmking and Jens Ibendorf, "An Early Tree-Line Experiment by a Wilderness Advocate: Bob Marshall's Legacy in the Brooks Range, Alaska," *Arctic*, Vol. 57, No. 1 (March 2004), pp. 106–109.

21. Robert Marshall, *Alaska Wilderness: Exploring the Central Brooks Range* (Berkeley: University of California Press, 1956, 1970, 2005).

22. Glover, *A Wilderness Original*, p. 122.

23. "A Letter to Foresters," February 7, 1930, Robert Marshall Papers, Bancroft Library, University of California, Berkeley.

24. Robert Marshall to Gerry and Lily Kempff, March 3, 1930, in Glover, *A Wilderness Original*, p. 114.

25. Robert Marshall to family and others, October 16, 1930, Robert Marshall Papers, State University of New York College of Environmental Science and Forestry. (Mimeographed letter.)

26. Robert Marshall, *Arctic Village* (New York: Literary Guild, 1933), pp. 57–58.

27. Marshall, *Alaska Wilderness*, p. 103.

28. Rick Bass, "Foreword to the Third Edition," ibid., p. xiii.

29. Roger Kaye, *The Last Great Wilderness: The Campaign to Establish the Arctic National Wildlife Refuge* (Fairbanks: University of Alaska Press, 2006).

30. Roderick Nash, *Wilderness and the American Mind* (New Haven, CT: Yale University Press, 1967), p. 274.

31. Frank Graham Jr., *The Adirondack Park: A Political History* (New York: Knopf, 1978), pp. 195–196.

32. Glover, *A Wilderness Original*, pp. 141–145.

33. Dayton Duncan and Ken Burns, *The National Parks: America's Best Idea* (New York: Knopf, 2009), p. 281.

34. Robert Marshall to E. Flint, February 21, 1933, Marshall Papers, Franklin Delano Roosevelt Library.

35. Gifford Pinchot to Franklin D. Roosevelt, January 20, 1933; in Glover, *A Wilderness Original*, p. 150; Edgar B. Nixon (ed.), *Franklin D. Roosevelt and Conservation, 1911–1945*, Vol. 1 (Hyde Park, NY: General Services Administration, 1957), pp. 129–132.

36. Address at the Laying of the Cornerstone of the Franklin D. Roosevelt

Library, Hyde Park, New York, November 19, 1939, in *The Presidential Papers of Franklin D. Roosevelt*, Vol. 8, 1939 (New York: Macmillan, 1941), p. 580.

37. John F. Sears, "Grassroots Democracy: F.D.R. and the Land," in David B. Woolner and Henry L. Henderson (eds.), *F.D.R. and the Environment* (New York: Palgrave Macmillan, 2005), chap. 1, p. 15.

38. U.S. Forest Service, Tongass National Forest, "Forest Facts: Ranger Boats," http://www.fs.fed.us/r10/tongass/forest_facts/resources/heritage/rangerboats.html, accessed March 13, 2010.

39. Curt Meine, *Aldo Leopold: His Life and Work* (Madison: University of Wisconsin Press, 1988), p. 342.

40. Quoted in Robert Sterling Yard to Aldo Leopold [n.d.], Leopold Papers, University of Wisconsin, Madison.

41. Bruce Woods, *Alaska's National Wildlife Refuges* (Anchorage: Alaska Geographic Society, 2003), p. 16.

42. Ira N. Gabrielson to Corey Ford, March 15, 1941, General Correspondence Relating to Wildlife Management—Kenai, 1932–1943, Record Group 22, National Archives.

43. John Leshy, "F.D.R.'s Expansion of Our National Patrimony: A Model for Leadership," in David B. Woolner and Henry L. Henderson (eds.), *F.D.R. and the Environment* (New York: Palgrave Macmillan, 2005), p. 178.

44. Glover, *A Wilderness Original*, p. 162.

45. Robert Marshall to Mardy Murie, July 31, 1933, Marshall Papers, Bancroft Library, University of California, Berkeley.

46. Robert Marshall, *The People's Forests* (New York: Smith and Haas, 1933), p. 219.

47. Glover, *A Wilderness Original*, p. 163.

48. Franklin Reed, "The People's Forest," *Journal of Forestry*, Vol. 32 (January 1934), pp. 104–107. (Review.)

49. Quoted in Richard N. L. Andrews, "Recovering F.D.R.'s Environmental Legacy," in David B. Woolner and Henry L. Henderson (eds.), *F.D.R. and the Environment* (New York: Palgrave Macmillan, 2005), p. 226.

50. Duncan and Burns, *The National Parks*, p. 290.

51. T. H. Watkins, "The Terrible Tempered Mr. Ickes," *Audubon* (March 1994), pp. 93–111.

52. William Cronon, "Foreword: Why Worry About Roads?" in Paul S. Sutter, *Driven Wild: How the Fight Against Automobiles Launched the Modern Wilderness Movement* (Seattle: University of Washington Press, 2002), p. xii.

53. B. Mackaye to H. A. Slattery, October 22, 1934, Robert Marshall Papers, Bancroft Library, University of California, Berkeley.

54. Ansel Adams, "Give Nature Time," Commencement Address, Occidental College, Los Angeles, CA, June 11, 1967.

55. Anthony B. Wolbarst, *Solutions for an Environment in Peril* (Baltimore, MD: Johns Hopkins University Press, 2001), p. 79.

56. Meine, *Aldo Leopold*.

57. Glover, *A Wilderness Original*, p. 177.

58. "The Wilderness Society," January 20, 1935, Wilderness Society Archive, Washington, DC. (Founding document.)

59. Stephen Fox, "We Want No Straddlers," *Wilderness*, Vol. 48, No. 167 (1984), pp. 5–19.

60. Ibid.

61. Joel H. Hildebrand, "Maintenance of Recreation Values in the High Sierra," *Sierra Club Bulletin*, Vol. 23 (1938), pp. 85–96.

62. Dyan Zaslowsky and T. H. Watkins, *These American Lands: Parks, Wilderness, and the Public Lands* (Washington, D.C.: Island, 1994), p. 292.

63. Robert Marshall, *Alaska Wilderness*, p. 123.

64. Robert Marshall, "Comments on the Report of Alaska's Recreational Resources Committee," *Alaska—Its Resources and Development*, U.S. Congress, House Doc. 485, 75th Congress, 3rd Session, Appendix B, p. 213.

65. Robert Marshall and Douglas K. Midgett (contributing author), *The People's Forests* (Iowa City: University of Iowa Press, 2002), pp. 64–65.

12. Those Amazing Muries

1. Verlyn Klinkenborg, "Margaret Murie's Vision," *New York Times*, October 24, 2003.

2. John Muir, *Travels in Alaska* (New York: Modern Library, 2002), pp. 277–278.

3. Harry Ritter, *Alaska's History: The People, Land, and Events of the North Country* (Anchorage: Alaska Northwest Books, 1993), p. 98.

4. Charles Craighead and Bonnie Kreps, *Arctic Dance: The Mardy Murie Story* (Portland, OR: Graphic Arts Center, 2006), pp. 12–22.

5. William Henry Smith, *The Life and Speeches of Hon. Charles Warren Fairbanks: Republican Candidate for Vice President* (Indianapolis, IN: W.B. Burford), p. 199.

6. Dermot Cole, *Historic Fairbanks: An Illustrated History* (San Antonio, TX: Historical Publishing Network, 2006), p. 7.

7. Syun-Ichi Akasofu, *The Northern Lights: Secrets of the Aurora Borealis* (Anchorage: Alaska Northwest, 2009), p. 8.

8. Ibid., p. 16.

9. Craighead and Kreps, *Arctic Dance*.

10. Jeff Schultz, *Dogs of the Iditarod* (Seattle, WA: Sasquatch, 2003), pp. 12–15.

11. Peter A. Coates, *The Trans-Alaska Pipeline Controversy* (Bethlehem, PA: Lehigh University Press, 1991), pp. 47–48.

12. Craighead and Kreps, *Arctic Dance*, p. 29.

13. Ibid., p. 31.

14. Stephen R. Fox, *The American Conservation Movement: John Muir and His Legacy* (Madison: University of Wisconsin Press, 1981), p. 267.

15. John F. Kauffmann, *Alaska's Brooks Range: The Ultimate Mountains* (Seattle, WA: Mountaineers, 2005), p. 83.

16. Olaus J. Murie, *Journeys to the Far North* (Palo Alto, CA: Wilderness Society/American West, 1973), pp. 104–106.

17. Olaus J. Murie to Mardy Murie, December 23, 1922.

18. Craighead and Kreps, *Arctic Dance*, p. 51.

19. Jenks Cameron, *The Bureau of Biological Survey* (New York: Arno, 1974), pp. 118–121.

20. Tom Walker, *Caribou: Wanderer of the Tundra* (Portland, OR: Graphic Arts Center, 2008), p. 22.

21. Margaret E. Murie, *Two in the Far North* (Anchorage: Alaska Northwest, 1962), p. 217.

22. There is a photo of the carved motto in Craighead and Kreps, *Arctic Dance*, p. 113.

23. Stephen Haycox and Alexandra J. McClanahan, *Alaska Scrapbook* (Anchorage, AK: CIRI Foundation, 2008), pp. 109–110.

24. Melody Webb, *The Last Frontier* (Albuquerque: University of New Mexico Press, 1985), p. 264.

25. Frank Dufresne, *Alaska's Animals and Fishes* (New York: A.S. Barnes, 1946), pp. x–xv.

26. Fox, *The American Conservation Movement*, pp. 267–268.

27. Margaret and Olaus Murie, *Wapiti Wilderness* (New York: Knopf, 1966), p. 7.

28. John Bowlby, *Charles Darwin: A New Life* (New York: Norton, 1992), p. 174.

29. Adolph Murie, *Ecology of the Coyote in Yellowstone*, Fauna of the National Parks of the U.S. Bulletin No. 4 (Washington, DC: U.S. Department of the Interior, 1940).

30. Debbie S. Miller, *Midnight Wilderness: Journeys in Alaska's Arctic National Wildlife Refuge* (Portland, OR: Alaska Northwestern, 2000), pp. 163–164.

31. Dyan Zaslowsky and T. H. Watkins, *These American Lands: Parks, Wilderness, and the Public Lands* (Washington, DC: Island, 1994), p. 293.

32. *Wolves, Bears, and Their Prey in Alaska: Biological and Social Challenges in Wildlife Management* (Washington, D.C.: National Academy, 1997), p. 55.

33. Aldo Leopold to Frederic Walcott, January 10, 1932, Leopold Papers, University of Wisconsin, Madison.

34. Edward Abbey, *The Journey Home: Some Words in Defense of the American West* (New York: Plume, 1991), p. 223.

35. William O. Douglas, *Go East, Young Man* (New York: Random House, 1974), p. 467.

36. William O. Douglas, "America's Vanishing Wilderness," *Ladies' Home Journal*, Vol. 81 (July 1964), pp. 37–41.

37. Douglas, *Go East, Young Man*, p. 143.

38. Victor B. Scheffer, *Adventures of a Zoologist* (New York: Scribner, 1980), p. 15.

13. Will the Wolf Survive?

1. David L. Mech, *The Wolf: The Ecology and Behavior of an Endangered Species* (Minneapolis: University of Minnesota Press, 1970), p. 348.

2. Adolph Murie, *Ecology of the Coyote in Yellowstone*, National Park Service Fauna Series, No. 4 (Washington, DC: U.S. Department of the Interior, 1940).

3. Adolph Murie, *A Naturalist in Alaska* (Tucson: University of Arizona Press, 1961), p. 208.

4. Michael J. Robinson, *Predatory Bureaucracy: The Extermination of Wolves and the Transformation of the West* (Boulder: University of Colorado Press, 2005), p. 1.

5. John McPhee, *Coming into the Country* (New York: Farrar, Straus and Giroux, 1976), p. 242.

6. Linda S. Franklin, "Adolph Murie: Denali's Wilderness Conscience," MA thesis, University of Alaska, Fairbanks (May 2004), p. 5.

7. Jim Rearden, *Alaska's Wolf Man: The 1915–1955 Wilderness Adventures of Frank Glaser* (Missoula, MT: Pictorial Histories, 1998), p. xii.

8. Charles Sheldon, *The Wilderness of Denali: Explorations of a Hunter-Naturalist in Northern Alaska* (New York: Scribner, 1960), p. 161.

9. Belmore Browne, *The Conquest of Mount McKinley* (Boston, MA: Houghton Mifflin, 1956), p. 210.

10. Robinson, *Predatory Bureaucracy*, p. 184.

11. Biological Survey, "Report of Chief of Bureau of Biological Survey, 1924," Record Unit 717, Box 24, Smithsonian Institute Archives, Washington, DC.

12. Aldo Leopold, *Game Management* (New York: Scribner, 1933), p. 19.

13. Olaus Murie to Aldo Leopold, October 30, 1931, Aldo Leopold Papers, University of Wisconsin, Madison.

14. Barry Lopez, *Of Wolves and Men* (New York: Simon and Schuster, 2004), p. 3.

15. Concordia College diploma, granting Adolph Murie a bachelor of science degree (June 1925), Adolph Murie Collection, Box 3, Alaska and Polar Regions Collections, Elmer E. Rasmuson Library, University of Alaska, Fairbanks.

16. F. C. Evans, "Lee Raymond Dice," *Journal of Mammalogy*, Vol. 59 (August 1978), pp. 635–644.

17. Robinson, *Predatory Bureaucracy*, p. 184.

18. Franklin, "Adolph Murie: Denali's Wilderness Conscience," p. 30.

19. David L. Mech, *The Wolves of Isle Royale*, Fauna Series No. 7 (Washington, DC: U.S. Government Printing Office, 1966), p. 22.

20. Adolph Murie, *Mammals from Guatemala and British Honduras*, Museum of Zoology Miscellaneous Publications No. 26 (Ann Arbor: University of Michigan Press, 1935).

21. Adolph Murie, *Following Fox Trails*, University of Michigan Museum of Zoology Miscellaneous Publications, No. 32 (Ann Arbor: University of Michigan Press, 1936), p. 44.

22. Louise Murie Macleod, "Adolph Murie 1899–1974" [n.d.], Adolph Murie Reference File, Denali National Park and Preserve Museum.

23. Victor H. Cahalane, "The Evolution of Predator Control Policy in the National Parks," *Journal of Wildlife Management*, Vol. 3 (1939), p. 235.

24. Murie, *A Naturalist in Alaska*, p. 220.

25. Adolph Murie, *The Wolves of Mount McKinley*, U.S. National Park Service Fauna Series 5 (Washington, DC: U.S. National Park Service, 1944), p. 3.

26. Stanley P. Young and Edward A. Goldman, *The Wolves of North America*, Vol. 1 (New York: Dover, 1944).

27. Joan M. Antonson and William S. Hanable, *Alaska's Heritage: 1867 to Present* (Anchorage: Alaska Historical Society, 1985), p. 286.

28. Terrence Cole, "Foreword," in Brian Garfield, *The Thousand-Mile War: World War II in Alaska and the Aleutians* (Fairbanks: University of Alaska Press, 1995), p. xi.

29. Samuel Eliot Morison, "Aleutians, Gilberts, and Marshalls, June 1942–April 1944," in *History of United States Naval Operations in World War II*, No. 1.7 (Boston, MA: Brown, 1951), pp. 3–4.

30. Herman E. Slotnick, *Alaska: A History of the 49th State* (Norman: University of Oklahoma Press, 1987), p. 131.

31. Douglas Brinkley, "Introduction," in Phillip J. Merrell (ed.), *The World War II Black Regiments That Built the Alaska Military Highway* (Jackson: University of Mississippi Press, 2002), pp. 5–12.

32. Harold L. Ickes to Harold W. Snell, September 25, 1944, Ickes papers.

33. Aldo Leopold, *A Sand County Almanac with Essays on Conservation from Round River* (New York: Ballantine, 1970), p. 138.

34. Ibid., pp. 138–139.

35. Thomas R. Dunlap, *Saving America's Wildlife: Ecology and the American Mind, 1850–1990* (Princeton, NJ: Princeton University Press, 1988), p. 75.

36. Murie, *A Naturalist in Alaska*, p. 79.

37. Ibid.

38. Sale, *A Complete Guide to Arctic Wildlife*, p. 396.

39. Murie, *A Naturalist in Alaska*, pp. 79–80.

40. Douglas H. Chadwick, "Wolf Wars," *National Geographic*, Vol. 217, No. 3 (March 2010), pp. 34–55.

41. William Brown, *Symbol of the Alaskan Wild: An Illustrated History of the Denali–Mount McKinley Region, Alaska* (Denali National Park: Alaska Natural History Association, 1993).

42. Franklin, "Adolph Murie: Denali's Wilderness Conscience," p. 50.

43. Brown, *Symbol of the Alaskan Wild*, p. 184.

44. Adolph Murie to Robert Sterling Yard, January 15, 1935, Adolph Murie Collection, Box 4, American Heritage Center, University of Wyoming, Laramie.

45. Olaus Murie to Adolph Murie, March 21, 1951, Martin Murie (personal papers), University of Alaska, Fairbanks.

46. Karsten Heuer, *Being Caribou: Five Months on Foot with an Arctic Herd* (Seattle, WA: Mountaineers, 2005).

47. Dunlap, *Saving America's Wildlife*, p. 105.

48. Lopez, *Of Wolves and Men*, pp. 159–160.

49. "History of Wolf Control in Alaska" (Washington, DC: Defenders of Wildlife, March 2010).

14. William O. Douglas and New Deal Conservation

1. William O. Douglas, *The Autobiography of William O. Douglas: The Court Years, 1939–1975* (New York: Random House, 1980), p. 371; William O. Douglas, *Go East, Young Man* (New York, Random House, 1974), p. 206.

2. William O. Douglas, *My Wilderness: The Pacific West* (Garden City, NY: Doubleday, 1968), p. 10.

3. Henry David Thoreau, *Walden or Life in the Woods* (New York: New American Library, 1960), p. 10.

4. Douglas, *My Wilderness*, p. 199.

5. Ibid.

6. Bruce Allen Murphy, *Wild Bill: The Legend and Life of William O. Douglas* (New York: Random House, 2003), p. 454.

7. Reprinted in Christopher Stone, *Should Trees Have Standing? Toward Legal Rights for Natural Objects* (Palo Alto, CA: Tioga, 1988).

8. *Sierra Club v. Morton*, 405 U.S. 727 (1972).

9. "Mr. Justice Douglas, Dissenting," *Living Wilderness* (Summer 1972), pp. 19–29.

10. Murphy, *Wild Bill*, pp. 454–457.

11. Author interview, June 14, 2010.

12. James O'Fallon, *Nature's Justice: Writings of William O. Douglas* (Corvallis: Oregon State University Press, 2000), pp. 293–294.

13. Douglas, *My Wilderness*, p. 160.

14. Adam W. Sowards, *The Environmental Justice: William O. Douglas and American Conservation* (Corvallis: Oregon State University Press, 2009), pp. 10–13.

15. William Douglas, *Of Men and Mountains: The Classic Memoir of Wilderness Adventure* (Guilford, CT: Globe Pequot, 2001), pp. 33–34.

16. Ibid., p. 168.

17. O'Fallon, *Nature's Justice*, p. 70.

18. Mark Wyman, *Hoboes: Bindlestiffs, Fruit Tramps, and the Harvesting of the West* (New York: Hill and Wang, 2010), p. 273.

19. *Current Biography*, Vol. 17 (New York: H.W. Wilson, 1942), pp. 233–235.

20. Douglas, *Of Men and Mountains*, p. 11.

21. O'Fallon, *Nature's Justice*, p. 38.

22. Douglas, *Of Men and Mountains*, p. 15.

23. O'Fallon, *Nature's Justice*, p. 5.

24. James F. Simon, *Independent Journey: The Life of William O. Douglas* (New York: Harper and Row, 1980), p. 72.

25. Ibid., p. 73.

26. Ibid., p. 75.

27. Douglas, *Go East, Young Man*, p. 137.

28. Ibid., p. 151.

29. O'Fallon, *Nature's Justice*, pp. 8–12.

30. Douglas, *Go East, Young Man*, p. 309.

31. Ibid., pp. 310–311.

32. Stephen Fox, *The American Conservation Movement* (Madison: University of Wisconsin Press, 1981), p. 212.

33. Stewart L. Udall, *The Quiet Crisis* (New York: Holt, Rinehart and Winston, 1963), p. 155.

34. William O. Douglas, *A Wilderness Bill of Rights* (Boston, MA: Little, Brown, 1965), pp. 178–179.

35. Kesler E. Woodward (ed.), *Painting in the North: Alaskan Art in the Anchorage Museum of History and Art* (Anchorage, AK: Anchorage Museum of History and Art, 1993).

36. Claus-M. Naske and Herman E. Slotnick, *Alaska: A History of the 49th State* (Norman: University of Oklahoma Press, 1994).

37. Glenn Holder, *Talking Totem Poles* (New York: Dodd, Mead, 1973), pp. 66–67.

38. "Glacier Bay National Monument," National Park Service History Report, Gustavus, AK.

39. William S. Cooper, "A Contribution to the History of the Glacier Bay National Monument" (Gustavus, AK: Glacier Bay National Park and Preserve Archive, March 1954).

40. Stephen Haycox and Alexandra McClanahan, *Alaska Scrapbook: Moments in Alaska History* (Portland, OR: Graphic Arts Center, 2008), pp. 119–120.

41. David L. Lendt, *Ding: The Life of Jay Norwood Darling* (Iowa City, IA: Maecenas, 2000).

42. Eric Jay Dolin and Bob Dumaine, *The Duck Stamp Story* (Iola, WI: Krause, 2000), p. 49.

43. Roosevelt to Henry L. Stimson, November 28, 1941, in Edgar B. Nixon (ed.), *Franklin D. Roosevelt and Conservation, 1911–1945*, 2 vols. (Hyde Park, NY: General Services Administration, 1957), Vol. 2, pp. 540–541.

44. Fox, *The American Conservation Movement*, pp. 220–223.

45. Raymond Blaine Fosdick, *John D. Rockefeller, Jr.: A Portrait* (New York: Harper, 1956), p. 129.

46. Tom H. Watkins, *Righteous Pilgrim: The Life and Times of Harold L. Ickes, 1874–1952* (New York: Holt, 1990), p. 829.

47. Douglas, *A Wilderness Bill of Rights*, p. 134.

48. Alden Whitman, "Vigorous Defender of Rights," *New York Times*, January 20, 1980, p. 28.

49. George Bookman, "Wonderful World of Walking," *Living Wilderness*, Vol. 20, No. 52 (Spring–Summer 1955), p. 1.

50. William O. Douglas, "The C&O Canal . . . 1959," *Living Wilderness*, Vol. 24, No. 68 (Spring 1959), p. 2.

51. Thoreau, *Walden*, p. 40.

52. Timothy Egan, *The Big Burn: Teddy Roosevelt and the Fire That Saved America* (Boston, MA: Houghton Mifflin Harcourt, 2009), pp. 271–272.

53. Michael J. Robinson, *Predatory Bureaucracy: The Extermination of Wolves and the Transformation of the West* (Boulder:

University Press of Colorado, 2005), p. 292.

54. David Brower, *Wilderness: America's Living Heritage* (San Francisco, CA: Sierra Club, 1961), pp. 102–103.

55. Murphy, *Wild Bill*, p. 455.

56. Ibid., pp. 454–457.

15. Ansel Adams, Wonder Lake, and the Lady Bush Pilots

1. Ansel Adams, *An Autobiography* (New York: Little, Brown, 1996), p. 236.

2. Ibid.

3. Kristin G. Congdon and Kara Kelley Hallmark, *Twentieth Century United States Photographers* (Westport, CT: Greenwood, 2008), p. 10.

4. Jonathan Spaulding, *Ansel Adams and the American Landscape: A Biography* (Berkeley: University of California Press, 1995), p. 236.

5. Mary Street Alinder and Andrea Gray Stillman (eds.), *Ansel Adams: Letters 1916–1984* (New York: Bulfinch, 2001), p. 402.

6. Ibid., p. 33.

7. Robert Turnage, "Ansel Adams: The Role of the Artist in the Environmental Movement," *Living Wilderness* (March 1980).

8. Dayton Duncan and Ken Burns, *The National Parks* (New York: Random House, 2009), p. 303.

9. Richard J. Orsi, Alfred Runte, and Marlene Smith-Baranzini, *Yosemite and Sequoia: A Century of California National Parks* (Berkeley and Los Angeles: University of California Press, 1993); see also Mike White, *Kings Canyon National Park: A Complete Hiker's Guide* (Berkeley, CA: Wilderness, 2004).

10. Duncan and Burns, *The National Parks*, pp. 304–305.

11. Mary Street Alinder, *Ansel Adams: A Biography* (New York: Holt, 1996), pp. 213–214.

12. Author interview with Michael Adams, June 8, 2010 (Carmel, CA).

13. Douglas Brinkley and Patricia Nelson Limerick (eds.), *The Western Paradox: A Bernard De Voto Conservation Reader* (New Haven, CT: Yale University Press, 2007), pp. 196–197.

14. Ansel Adams to Ted Spencer, February 8, 1947, in Alinder and Stillman, *Ansel Adams*, p. 190.

15. Adams, *An Autobiography*, p. 236.

16. Ibid.

17. Ibid., p. 238.

18. Alinder and Stillman, *Ansel Adams*, p. 217.

19. Adams, *An Autobiography*.

20. Ibid.

21. Spaulding, *Ansel Adams and the American Landscape*, pp. 235–236.

22. Author interview with Michael Adams, June 2, 2010 (Carmel, CA).

23. Robert Hirsch, *Seizing the Light: A History of Photography* (New York: McGraw-Hill, 2000), pp. 246–248.

24. Ibid.

25. Ansel Adams, *Examples: The Making of 40 Photographs* (New York: Little, Brown, 1983).

26. Susanne Lomatch, *Black and White Luminaries: Insights into Adams and Garrett* (Boise, ID: Idaho Photographic Workshop, February 21, 2010).

27. Spaulding, *Ansel Adams and the American Landscape*, p. 236.

28. Author interview with Michael Adams, June 2, 2010 (Carmel, CA).

29. Adams, *An Autobiography*, p. 241.

30. Julie Dunlap and Kerry Maguire, *Eye on the Wild: A Story About Ansel Adams*

(Minneapolis, MN: Millbrook, 1995), p. 50.

31. Ansel Adams to Beaumont Newhall, July 11, 1949, in Mary Street Alinder and Andrea Gray Stillman (eds.), *Ansel Adams: Letters and Images 1916–1984* (Boston, MA: Little, Brown, 1988), p. 209.

32. Ibid., pp. 208–209.

33. Roderick Frazier Nash, *Wilderness and the American Mind* (New Haven, CT: Yale University Press, 2001), p. 291.

34. Author interview with Virginia Wood, June 2010.

35. Ibid.

36. Virginia Wood to Mom, February 19, 1943, Wood Personal Papers, Fairbanks, AK.

37. Virginia Wood scrapbooks, private collection, Fairbanks, AK.

38. Author interview with Virginia Wood.

39. Ibid.

40. Ibid.

41. Christine Barnes, *Great Lodges of the National Parks*, Vol. 2 (Portland, OR: Graphic Arts, 2008), p. 150.

42. Dayton Duncan, *The National Parks: America's Best Idea* (New York: Knopf, 2010), p. 307.

16. Pribilof Seals, Walt Disney, and the Arctic Wolves of Lois Crisler

1. Neal Gabler, *Walt Disney: The Triumph of the American Imagination* (New York: Random House, 2006), p. 388.

2. Robert A. Henning (ed.), *Island of the Seals: The Pribilofs* (Anchorage: Alaska Geographic Society, 1982), p. 55.

3. Dave Ostlund, interview with Rachel Sibley, Kodiak Military History Museum at Miller Point, Fort Abercrombie, Kodiak, July 13, 2010.

4. Francis E. Caldwell, *Beyond the Trails: With Herb and Lois Crisler in Olympic National Park* (Port Angeles, CA: Anchor, 1998), pp. 152–189.

5. Richard Schickel, *The Disney Version: The Life, Times, Art, and Commerce of Walt Disney* (Chicago, IL: Ivan R. Dee, 1997), p. 278.

6. David Starr Jordan, Leonhard Hess Stejneger, Frederic A. Lucas, et al., *The Fur Seals and Fur-Seal Islands of the North Pacific Ocean: Special Papers Relating to the Fur Seal and to the Natural History of the Pribilof Islands* (Washington, DC: Government Printing Office, 1899).

7. Ira N. Gabrielson, *Wildlife Refuges* (New York: Macmillan, 1943), p. 77.

8. Schickel, *The Disney Version*, pp. 270–280.

9. Craig A. Hansen, "Seals and Sealing," in *Islands of the Seals: The Pribilofs* (Anchorage: Alaska Geographic Society, 1982), p. 55.

10. Gabler, *Walt Disney*, p. 446.

11. Ibid.

12. Ibid., p. 447.

13. Schickel, *The Disney Version*, p. 290.

14. Ralph H. Lutts, "The Trouble with Bambi: Walt Disney's Bambi and the American Vision of Nature," *Forest and Conservation History*, Vol. 36 (October 1992), pp. 160–171.

15. Susan Killon, *Nature's State: Imagining Alaska as the Last Frontier* (Chapel Hill: University of North Carolina Press, 2001), p. 96.

16. "Forest and Conservation History," *Forest History Society*, Vols. 36–37 (1992), p. 162.

17. Lutts, "The Trouble with Bambi."

18. Ibid.

19. Lois Crisler, "Santayana's Definition of Beauty," MA thesis (1925), University

of Washington, Seattle, Manuscripts and University Archives, University of Washington Libraries, Seattle.

20. Caldwell, *Beyond the Trails*, p. 191.

21. Ibid., p. 189.

22. Lois Crisler Papers, University of Washington Archives, Seattle.

23. Lois Crisler, "The True Mountaineer," *Natural History* (November 1950), pp. 422–428.

24. Sally Patrick Johnson, *Everyman's Ark: A Collection of True First-Person Accounts of Relationships Between Animals and Men* (New York: Harper, 1962), p. 178.

25. William O. Douglas, jacket copy for Lois Crisler, *Arctic Wild* (New York: Lyons, 1999).

26. Caldwell, *Beyond the Trails*, p. 207.

27. David Mech, "Foreword," in Crisler, *Arctic Wild*, p. x.

28. Rachel Carson to Lois Crisler, March 4, 1959, Rachel Carson Papers, Yale Collection of American Literature, Beinecke Rare Book and Manuscript Library, Yale University, New Haven, CT.

29. Crisler, *Arctic Wild*, p. 22.

30. Ibid., p. 290.

31. Jon T. Coleman, *Vicious: Wolves and Men in America* (New Haven, CT: Yale University Press, 2004), p. 160.

32. Lois Crisler, *Captive Wild* (New York: Lyons Press, 2000).

33. William O. Douglas, *Go East, Young Man* (New York: Random House, 1974), p. 207.

34. Jim Rearden, *Alaska's Wolf Man: The 1915–1955 Wilderness Adventures of Frank Glaser* (Missoula, MT: Pictorial Histories, 1998), pp. 323–324.

35. William O. Douglas, "For Every Man and Woman Who Loves the Wilderness," *Living Wilderness*, No. 58 (Fall and Winter 1956–1957).

36. Ibid.

37. J. Louis Giddings, *Ancient Men of the Arctic Wild* (New York: Knopf, 1967).

38. Gabler, *Walt Disney*, pp. 611–612.

17. The Arctic Range and Aldo Leopold

1. Aldo Leopold, *A Sand County Almanac* (New York: Oxford University Press, 1966), p. 70.

2. Ibid., p. 111.

3. Stephen Brown (ed.), *Arctic Wings: Birds of the Arctic National Wildlife Refuge* (Seattle, WA: Mountaineers, 2006), p. 74.

4. Margaret E. Murie, *Two in the Far North* (New York: Knopf, 1962), p. i.

5. Ibid., p. 254.

6. Peter A. Coates, *The Trans-Alaska Pipeline Controversy* (Bethlehem, PA: Lehigh University Press, 1991), p. 185.

7. George L. Collins to Louis Giddings Jr., Appendix B, "Genesis of the Arctic International Wildlife Range Idea, 1952," in George L. Collins, *The Art of Politics and of Park Planning and Preservation, 1920–1929* (Berkeley: University of California Press, 1980), p. 345.

8. Benton MacKaye, "Dam Site vs. Norm Site," *Scientific Monthly* (October 1950), pp. 241–247.

9. Hank Lentfer and Carolyn Servid, *Arctic Refuge: A Circle of Testimony* (Minneapolis, MN: Milkwood, 2001), p. 1.

10. Roger Kaye, *Last Great Wilderness: The Campaign to Establish the Arctic National Wildlife Refuge* (Fairbanks: University of Alaska Press, 2006), pp. 36–39.

11. Speaking to the U.S. Armed Forces Committee, 1952. *Bartlett's Familiar Quotations*, 15th ed. (Boston, MA: Little, Brown, 1980), p. 817.

12. Bosley Crowther, "The Legend of Lobo," *New York Times*, June 8, 2010.

13. Kaye, *Last Great Wilderness*, pp. 14–17. I couldn't have written this chapter without this pioneering work; it is far and away the most comprehensive book on the history of the Arctic Refuge.

14. Author interview with Virginia Wood, June 18, 2010.

15. Martha Sonntag Bradley, "Glen Canyon Dam Controversy" *Utah History to Go*, State of Utah online database, May 2010.

16. Ibid.

17. David Brower, *For Earth's Sake: The Life and Times of David Brower* (Salt Lake City, UT: Peregrine Smith, 1990), p. 347; Kevin Wehr, *America's Fight over Water: The Environmental and Political Effects of Large-Scale Water Surplus* (New York: Routledge, 2004), p. 212.

18. Brower, *For Earth's Sake*, p. 369.

19. "George Leroy Collins for 1959," Biological File, Arctic National Wildlife Refuge, Fairbanks, AK. See also A. Frank Willis, *Do Things Right the First Time: Administrative History of the National Park Service and the Alaska National Interest Lands Conservation Act of 1980* (Washington, DC: National Park Service, 1985).

20. John M. Kauffmann, *Alaska's Brooks Range: The Ultimate Mountains* (Mountaineers Books: 1992), p. 97.

21. George Collins and Lowell Sumner, "Background Information for Use in Connection with a Proposal for an Arctic International Wildlife Refuge," *University of British Columbia Law Review*, Vol. 6, No. 1 (June 1971), pp. 3–11.

22. Roderick Nash, quoted in Coates, *The Trans-Alaska Pipeline Controversy*, p. 34.

23. Kauffmann, *Alaska's Brooks Range*, pp. 100–101.

24. Kaye, *The Last Great Wilderness*, p. 24.

25. George L. Collins Diaries, April 20 and 24, 1952, U.S. Department of the Interior, National Park Service, George L. Collins Papers, File No. 207–10, Alaska and Polar Regions Department, University of Alaska, Fairbanks.

26. Barry H. Lopez, *Of Wolves and Men* (New York: Simon and Schuster, 1978), p. 144.

27. Larry Meyers, "He Wrestled a Wolf," *Alaska Sportsman*, No. 6 (June 1952), pp. 14–17, 40–45.

28. "Protect the Sacred Place Where Life Begins" (Fairbanks, AK: Gwich'in Steering Committee, 2010).

29. Kauffmann, *Alaska's Brooks Range*, p. 103.

30. Ibid., p. 81.

31. Charles Craighead and Bonnie Kreps, *Arctic Dance: The Mardy Murie Story* (Portland, OR: Graphic Arts Center, 2006).

18. The Sheenjek Expedition of 1956

1. Olaus Murie, "Alaska with O. J. Murie," *Living Wilderness*, No. 58 (Winter 1956–1957), pp. 28–30.

2. "National Wildlife Refuges in Region 6," U.S. Fish and Wildlife Report (Shepherdstown, WV: U.S. Fish and Wildlife Service, 1960).

3. Margaret Murie Diary, June 3, 1956, in Margaret E. Murie, *Two in the Far North* (Anchorage, AK: Northwest, 1962), p. 272.

4. Robert Krear, "The Olaus Murie Brooks Range Expedition," Roger Kaye Papers, Fairbanks, AK. (Unpublished manuscript.) Quoted in Roger Kaye, *Last Great Wilderness: The Campaign to Establish the Arctic National Wildlife Refuge* (Chicago, IL: University of Chicago Press, 2006).

5. Charles Craighead and Bonnie Kreps, *Arctic Dance: The Mardy Murie Story* (Portland, OR: Graphic Arts Center, 2006), p. 6.

6. William H. Rodgers Jr., "The Fox and the Chickens: Mr. Justice Douglas and Environment Law," in *He Shall Not Pass This Way Again: The Legacy of William O. Douglas* (Pittsburgh, PA: University of Pittsburgh Press, 1990), pp. 215–223. Also William O. Douglas, "The C&O Canal . . . 1959," *Living Wilderness*, Vol. 24, No. 68 (Spring 1959), pp. 1–2.

7. Author interview, George McGovern, June 16, 2010.

8. Murie, *Two in the Far North*.

9. William O. Douglas, "The Black Silence of Fear," *New York Times Magazine*, January 13, 1952, sec. 6, p. 7.

10. William O. Douglas, "People vs. Trout: A Majority Opinion," *New York Times Magazine*, April 2, 1950.

11. John F. Simon, *Independent Journey: The Life of William O. Douglas* (New York: Harper and Row, 1980).

12. William O. Douglas, *My Wilderness: The Pacific West* (Garden City, NY: Doubleday, 1960), p. 18.

13. James O'Fallon, *Nature's Justice: Writings of William O. Douglas* (Corvallis: Oregon State University Press, 2000), pp. 290–292.

14. Douglas, *My Wilderness*, pp. 65–74.

15. William O. Douglas, *Go East, Young Man* (New York: Random House, 1974), p. 207.

16. Ibid., pp. 206–207.

17. Bruce Allen Murphy, *Wild Bill: The Legend and Life of William O. Douglas* (New York: Random House, 2003).

18. Douglas, *Go East, Young Man*, pp. 469–470.

19. Douglas, *My Wilderness*, p. 23.

20. Murie, *Two in the Far North*, p. 335.

21. Douglas, *My Wilderness*, p. 15.

22. Stephen Fox, *The American Conservation Movement: John Muir and His Legacy* (Madison: University of Wisconsin Press, 1981), p. 244.

23. "Wilderness System Bill Urged," *Living Wilderness*, No. 58 (Winter 1956–1957), p. 30.

24. David Brower, quoted in *Sierra Club Bulletin* (June 1954).

25. John M. Kauffmann, *Alaska's Brooks Range: The Ultimate Mountains* (Seattle, WA: Mountaineers, 1992), p. 33.

26. Olaus J. Murie, "Wilderness Philosophy," quoted in Roger Kaye, *Last Great Wilderness: The Campaign to Establish the Arctic National Wildlife Refuge* (Fairbanks: University of Alaska Press, 2006), chap. 4.

27. Kaye, *Last Great Wilderness*, p. 84.

28. Murie, "Alaska with O. J. Murie," pp. 28–30.

29. Douglas, *My Wilderness*, p. 30.

30. George L. Collins, *The Art and Politics of Park Planning and Preservation, 1920–1979* (Berkeley: University of California Press, 1980), p. 190.

31. Douglas, *My Wilderness*, p. 9.

19. Dharma Wilderness

1. James I. McClintock, "Gary Snyder's Poetry and Ecological Science," *American Biology Teacher*, Vol. 54, No. 2 (February 1992), pp. 80–84.

2. John Halper (ed.), *Gary Snyder: Dimensions of a Life* (San Francisco, CA: Sierra Club, 1991), p. 340.

3. Jerry Crandall, "Mountaineers Are Always Free," in John Halper, *Snyder: Dimensions of a Life* (San Francisco, CA: Sierra Club, 1991), p. 4.

4. John Suiter, *Poets on the Peaks: Gary Snyder, Phillip Whalen, and Jack Kerouac* (Washington, DC: Counterpoint, 2007), p. 54.

5. J. Michael Mahar, "Scenes from the Sidelines," in John Halper, *Gary Snyder: Dimensions of a Life* (San Francisco, CA: Sierra Club, 1991), p. 9.

6. Author interview with Gary Snyder, April 17, 2010.

7. Lauren Danner, "Ice Peaks National Park," *Columbia* (Fall 2009).

8. Roger Tory Peterson, *A Field Guide to Western Birds* (New York: Houghton Mifflin, 1961).

9. Mahan, "Scenes from the Sidelines," p. 11.

10. Suiter, *Poets on the Peaks*, p. 34.

11. Ibid., p. 3.

12. Rod Phillips, *"Forest Beatnicks" and "Urban Thoreaus": Gary Snyder, Jack Kerouac, Lew Welch, and Michael McClure* (New York: Peter Lang, 2000), p. 14.

13. Gary Snyder, *A Place in Space* (Washington, DC: Counterpoint, 1996), p. 57.

14. Suiter, *Poets on the Peaks*, pp. 12–13.

15. Gary Snyder, *Earth House Hold: Technical Notes and Queries to Fellow Dharma Revolutionaries* (New York: New Directions, 1969), p. 12.

16. Ibid.

17. Gary Snyder, "A Berry Feast," in *The Back Country* (New York: New Directions, 1971), p. 3.

18. Suiter, *Poets on the Peaks*, p. 14.

19. Jeanne Abbot, "Gary Snyder," *Anchorage Daily News*, October 7, 1976.

20. Suiter, *Poets on the Peaks*, pp. 15–38.

21. Crandall, "Mountaineers Are Always Free," p. 3.

22. Daisetz Teitaro Suzuki, *Essays in Zen Buddhism* (New York: Grove, 1961), p. 262.

23. Jeremy Anderson, "My First Encounter with a Real Poet," in John Halper (ed.), *Snyder: Dimensions of a Life* (San Francisco, CA: Sierra Club, 1991), p. 30.

24. Suiter, *Poets on the Peaks*, p. 45.

25. John Suiter, "Rolling Toward the Mountain: Jack Kerouac's Last Great Adventure," *Sierra* (March–April 1958).

26. Ed Zahniser (ed.), *Where Wilderness Preservation Began: Adirondack Writings of Howard Zahniser* (Utica, NY: North Country, 1992), p. 1.

27. Han Shan, *Cold Mountain Poems* (Boston, MA: Shambhala, 2009), p. 30.

28. Quoted in Carol Baker, "1414 SE Lambert Street," in John Halper (ed.), *Snyder: Dimensions of a Life* (San Francisco, CA: Sierra Club, 1991).

29. Edward Abbey (ed.), *The Best of Edward Abbey* (San Francisco, CA: Sierra Club, 1984), p. 243.

30. Jack Kerouac, *The Dharma Bums* (New York: Penguin, 1976), p. 12.

31. Suiter, *Poets on the Peaks*, p. 66.

32. William Blake, *The Marriage of Heaven and Hell* (New York: Trianon, 1975), p. xxii.

33. Suiter, *Poets on the Peaks*, p. 71. It is impossible to write about Snyder at Sourdough without drawing on Suiter's book. All of my understanding of Snyder in the North Cascades emanates from *Poets on the Peaks*.

34. Travis Nicholas, " 'How Do You Like Your World?' The Zen of Philip Whalen," Poetry Foundation, March 25, 2008. (At Web site.)

35. Henry David Thoreau, *Walden* (London: J.M. Dent, 1955), p. viii.

36. Howard Zahniser, "The Need for Wilderness Areas," *Living Wilderness*, No. 59 (Winter–Spring 1956–1957), pp. 37–43.

37. Roger Kaye, *Last Great Wilderness: The Campaign to Establish the Arctic National Wildlife Refuge* (Fairbanks: University of Alaska Press, 2006), p. 93.

38. Margaret E. Murie, *Two in the Far North* (New York: Knopf, 1962), p. 371.

39. Trevor Carolan, "The Wild Mind of Gary Snyder," *Shambhala Sun Online* (April 29, 2010).

40. Allen Ginsberg, *Howl and Other Poems* (San Francisco, CA: City Lights, 1956), p. 17.

41. Rod Phillips, *"Forest Beatnicks" and "Urban Thoreaus,"* pp. 1–2.

42. Allen Ginsberg to John Allen Ryan (mid-September 1955), quoted in Bill Morgan and Nancy J. Peters (eds.), *Howl on Trial: The Battle for Free Expression* (San Francisco, CA: City Lights, 2006), p. 36.

43. Author interview with Michael McClure, July 7, 2010.

44. Michael McClure, *Humans to St. Geryon and Other Poems* (San Francisco, CA: Auerhahn, 1959), pp. 7–8.

45. Jonah Raskin, *American Scream: Allen Ginsberg, Howl, and the Making of the Beat Generation* (Berkeley: University of California Press, 2004), pp. 18–21.

46. Aldo Leopold, *A Sand County Almanac: With Essays on Conservation from Round River* (New York: Ballantine, Random House, Oxford University Press, 1970), p. 137.

47. Ginsberg, *Howl and Other Poems*.

48. Kerouac, *The Dharma Bums*, p. 14.

49. Michael McClure, *Scratching the Beat Surface: Essays on New Vision from Blake to Kerouac* (New York: Penguin, 1994), p. 13.

50. Norman Chance, "Project Chariot: The Nuclear Legacy of Cape Thompson, Alaska," *Arctic Circle*, Resource Database. Accessed July 7, 2010.

51. Roderick Frazier Nash, *Wilderness and the American Mind* (New Haven, CT: Yale University Press, 2001), p. 291.

52. Paul Brooks, *The Pursuit of Wilderness* (Boston, MA: Houghton Mifflin, 1971), p. 67.

53. Dan O'Neill, *The Firecracker Boys: H-Bombs, Eskimos, and the Roots of the Environmental Movement* (New York: Basic Books, 1994), pp. 36–37.

54. Author interview with Virginia Wood, June 18, 2010.

55. Chance, "Project Chariot."

56. Ibid.

57. Stephen E. Ambrose, *Eisenhower: The President*, Vol. 2 (New York: Simon and Schuster, 1984), pp. 479–480.

58. Gary Snyder, *The Back Country* (New York: New Directions, 1971), pp. 4–6.

59. Lawrence Ferlinghetti, " 'Howl' at the Frontiers," in Bill Morgan and Nancy J. Peters (eds.), *Howl on Trial* (San Francisco, CA: City Lights, 2006), p. xiv.

60. Snyder, *The Back Country*, p. 7.

61. Author interview with Ed Sanders, May 6, 2010.

62. Grace Glueck, "Cast into the Wilderness by Choice, He Found a Friend in the Landscape," *New York Times*, August 18, 2000.

63. Author interview with Lawrence Ferlinghetti, July 4, 2010.

64. Crick is quoted in Tom Montag (ed.), *Margins: A Review of Little Magazines and Small Press Books*, Issues 16–17 (1975), p. 24.

65. Gary Snyder, *The Back Country* (New York: New Directions, 1968), p. 20.

66. Gary Snyder, *Turtle Island* (New York: New Directions, 1974), p. 77.

20. Of Hoboes, Barefooters, and the Open Road

1. Allen Ginsberg to Robert LaVigne, August 3, 1956, in Bill Morgan (ed.), *The Letters of Allen Ginsberg* (New York: Da Capo, 2008), p. 139.

2. Stephen Haycox and Alexandra McClanahan, *Alaska Scrapbook: Moments in Alaska History* (Portland, OR: Graphic Arts Center, 2008), pp. 31–32.

3. Ginsberg to LaVigne, August 3, 1956.

4. Allen Ginsberg to Rebecca Ginsberg, August 11, 1956, in Morgan, *The Letters of Allen Ginsberg*, pp. 140–141.

5. Allen Ginsberg to Jack Kerouac, August 12–18, 1956, ibid., pp. 327–328.

6. Jack Kerouac, *The Scripture of the Golden Eternity* (New York: Corinth, 1960).

7. Allen Ginsberg to Carolyn Kizer, September 10, 1956, in Bill Morgan, *The Letters of Allen Ginsberg* (New York: Da Capo, 2008), pp. 141–143.

8. Jack Kerouac, *Lonesome Traveler* (New York: Grove, 1989), p. 182.

9. Steven Watson, *The Birth of the Beat Generation: Visionaries, Rebels, and Hipsters 1944–1960* (New York: Pantheon, 1995), p. 214.

10. Lindianne "Lady Greensleeves" Sarno-Glasgow (writer) and Mike "Spoonguy" Glasgow (proposal co-drafter), "Proposal to Sourdough Express" (Homer, AK: 2010). (Unpublished.)

11. Linnie Marsh Wolfe, *John of the Mountains: The Unpublished Journals of John Muir* (Madison: University of Wisconsin Press, 1979), p. 67.

12. Dolly Garza, *Common Edible Seaweeds in the Gulf of Alaska* (Fairbanks: Alaska Sea Grant College Program, 2005), pp. 3–4.

13. Janet R. Klein, *The Homer Spit: Coal, Gold, and Con Men* (Homer, AK: Kachemak Country, 1996), p. 55.

14. John Burroughs, *Far and Near* (Cambridge, MA: Houghton Mifflin, 1904), p. 80.

15. Thomas Locker, *John Muir: America's Naturalist* (Golden, CO: Fulcrum, 2003), p. 12.

16. Steve Turner, *Jack Kerouac: Angel-Headed Hipster* (New York: Viking, 1996), p. 161.

17. Martha Ellen Anderson, *Brother Asaiah* (Parker, CO: Thornton, 2006), p. 63.

18. Ibid.

19. Ibid., p. 322.

20. Author interview with Martha Ellen Anderson, July 3, 2010 (Homer, AK).

21. Jay Hammond, "Preface," in Anderson, *Brother Asaiah*, pp. 4–5.

22. Frederic Laugrand and Jarich Oosten, *The Sea Woman: Sedna in Inuit Shamanism and Art in the Eastern Arctic* (Anchorage: University of Alaska Press, 2009), pp. 34–108.

23. Gerald Nicosia, *Memory Babe: A Critical Biography of Jack Kerouac* (New York: Grove, 1983), p. 563.

24. Jack Kerouac, *The Dharma Bums* (New York: Penguin, 1976), p. 154.

25. Michael McClure, *Lighting the Corners: On Art, Nature, and the Visionary* (Albuquerque, NM: An American Poetry Book, 1993), p. 320.

26. Kerouac, *Lonesome Traveler*, p. 183.

27. Norman Podhoretz, "The Know Nothing Bohemians," *Partisan Review*, Vol. 25, No. 2 (Spring 1958).

28. Kerouac, *The Dharma Bums*, p. 77.

29. Ibid., p. 225.

30. Jack Kerouac, *Big Sur* (New York: Penguin, 1992), p. 45.

31. Kerouac, *Lonesome Traveler*, p. 175.

32. Ibid., p. 173.

21. Sea Otter Jones and Musk-Ox Matthiessen

1. Peter Matthiessen, *Wildlife in America* (New York: Viking, 1959), pp. 104–105.

2. Matthew J. Dufala, "Piece of History Finds Its Way to the 910th," *Airstream*, Vol. 17, Issue 8 (August 2001), p. 2.

3. Author interview with Dorothy Jones, August 26, 2010.

4. Richard P. Emanuel, "Robert 'Sea Otter' Jones, *Alaska Geographic*, Vol. 22, No. 2 (1995), p. 38.

5. Ibid., p. 32.

6. John Muir, *The Cruise of the Corwin* (New York: Houghton Mifflin, 1917), pp. 6–7.

7. L. J. Campbell, Penny Rennick, and Alaska Geographic Society, *The Aleutian Islands*, Vol. 22 (Ann Arbor: University of Michigan Press, 1995), p. 41.

8. Kenneth F. Wilson and Jeff Richardson, *The Aleutian Islands of Alaska: Living on the Edge* (Fairbanks: University of Alaska Press, 2008), p. 130.

9. James B. Trefethen, *An American Crusade for Wildlife* (Missoula, MT: Boone and Crockett Club, 1975), p. 335.

10. Robert Jones Reports, U.S. Fish and Wildlife Alaska Maritime National Wildlife Federation (Homer, AK, 1959).

11. Emanuel, "Robert 'Sea Otter' Jones," p. 42.

12. C. M. Mobly interview, Robert D. Jones, February 7, 1998, U.S. Fish and Wildlife Archive, Shepherdstown, WV.

13. Author interview with Dorothy Jones, August 26, 2010.

14. Alan Anderson, *After the Ice: Life, Death, and Geopolitics in the New Arctic* (New York: Smithsonian Books, 2009), p. 4.

15. Roderick Frazier Nash, *Wilderness and the American Mind* (New Haven, CT: Yale University Press, 2001), p. 291.

16. Author interview with Peter Matthiessen, June 17, 2010.

17. "A Conversation with Author Peter Matthiessen," *Charlie Rose* (PBS), May 27, 2008.

18. Author interview with Peter Matthiessen, June 18, 2010.

19. Ibid.

20. Matthiessen, *Wildlife in America*, p. 77.

21. Quoted in Diane Stupay, "Peter Matthiessen" (February 7, 2006). (Unpublished essay.)

22. Author interview with Peter Matthiessen, June 18, 2010.

23. Matthiessen, *Wildlife in America*, p. 233.

24. Author interview with Peter Matthiessen, June 18, 2010.

25. Peter Matthiessen, "In the Great Country," in Subhankar Banerjee (ed.), *Arctic National Wildlife Refuge: Seasons of Life and Land* (Seattle, WA: Mountaineers, 2003), pp. 40–57.

26. Matthiessen, *Wildlife in America*, p. 249.

27. Author interview with Peter Matthiessen, June 18, 2010.

28. Peter Matthiessen, *Oomingmak: The Expedition of the Musk-Ox in the Bering Sea* (New York: Hastings House, 1967), p. 28.

29. Author interview with Peter Matthiessen, November 15, 2010.

30. Peter Matthiessen, "Inside the Endangered Arctic Refuge," *New York Review of Books* (October 19, 2006); "Alaska: Big Oil and the Inupiat-Americans," *New York Review of Books* (November 22, 2007).

31. Matthiessen, "In the Great Country," p. 57.

22. Rachel Carson's Alarm

1. Robert C. Harris, *Johnny Appleseed: Source Book* (privately published, 1945), pp. 17–18.

2. Neal Gabler, *Walt Disney: The Triumph of the American Imagination* (New York: Random House, 2006), pp. 514–516.

3. *Science News Letter* (November 8, 1958).

4. Gregg Mitman, *Reel Nature: America's Romance with Wildlife on Film* (Cambridge, MA: Harvard University Press, 1999), p. 115.

5. "Cruelty to Animals in the Entertainment Business," Canada Broadcasting Corporation, April 1982.

6. Lois Crisler, *Arctic Wild* (New York: Lyons, 1999), p. 151.

7. David Mech, "Introduction," in Lois Crisler, *Arctic Wild* (New York: Lyons, 1999), pp. ix–xi. (Reprint.)

8. Walt Disney to Olaus Murie, December 4, 1953, Box 264, Olaus Murie Papers, Denver Public Library, Denver, Colorado.

9. Olaus Murie to Walt Disney, September 18, 1950, ibid.

10. Olaus Murie, "Wilderness Is for Those Who Appreciate," *Living Wilderness*, Vol. 5 (1940), p. 5.

11. Mitman, *Reel Nature*, p. 123.

12. Frank Graham Jr., *Since Silent Spring* (Boston, MA: Houghton Mifflin, 1970), p. 198.

13. Rachel Carson to Lois Crisler, August 19, 1961, Rachel Carson Papers, Yale Collection of American Literature, Beinecke Rare Book and Manuscript Library, Yale University, New Haven, CT.

14. Author interview with Cathy Stone (Douglas's fourth wife), August 19, 2010.

15. Adam W. Sowards, *The Environmental Justice: William O. Douglas and American Conservation* (Corvallis: Oregon State University Press, 2009), p. 119.

16. William O. Douglas, *A Wilderness Bill of Rights* (Boston, MA: Little, Brown, 1965), p. 166.

17. Edwin O. Wilson, "Afterword," in Rachel Carson, *Silent Spring* (New York: Houghton Mifflin, 2002), p. 357.

18. Sally Ann Grumaer Rannery, "Heroines and Hierarchies: Female Leadership in the Conservation Movement," in Donald Snow (ed.), *Voices from the Environmental Movement* (Washington, DC: Island, 1992), p. 115.

19. Isabella Bird, *A Lady's Life in the Rocky Mountains* (New York: Putnam, 1900), p. 167.

20. Grumaer Rannery, "Heroines and Hierarchies: Female Leadership in the Conservation Movement," pp. 116–119.

21. Edna Ferber, *Ice Palace: A Novel for Alaska Statehood* (Garden City, NY: Doubleday, 1958), p. 18.

22. Mitman, *Reel Nature*, p. 116.

23. Vera Norwood, *Made from This Earth* (Chapel Hill: University of North Carolina Press, 1993), p. 162.

24. *Audubon* (November–December 2004), pp. 52–53.

25. Paul Brooks, *The House of Life: Rachel Carson at Work* (Boston, MA: Houghton Mifflin, 1989), p. 253.

23. Selling the Arctic Refuge

1. Margaret E. Murie, "Foreword," in Debbie S. Miller, *Midnight Wilderness: Journeys in Alaska's National Wildlife Refuge* (Portland, OR: Alaska Northwest, 2000), p. x.

2. T. H. Watkins, *Vanishing Arctic: Alaska's National Wildlife Refuge* (Washington, DC: Aperture Foundation, 1988), p. 47.

3. Roger Kaye, *Last Great Wilderness* (Anchorage: University of Alaska Press, 2006), pp. 106–107.

4. Olaus J. Murie to George L. Collins, November 29, 1956, Margaret Murie Papers, Box 2, Folder 18, American Heritage Center, University of Wyoming, Laramie.

5. Gary Snyder, *Look Out: A Selection of Writings* (New York: New Direction, 2002), pp. 115–116.

6. Rick Bass, *Caribou Rising: Defending the Porcupine Herd, Gwich-'in Culture, and the Arctic National Wildlife Refuge* (San Francisco, CA: Sierra Club, 2004), p. 109.

7. Kaye, *Last Great Wilderness*, p. 107.

8. Daniel Nelson, *Northern Landscapes: The Struggle for Wilderness Alaska* (Washington, DC: Resources for the Future, 2004), p. 45.

9. Mark Harvey, *Wilderness Forever: Howard Zahniser and the Path to the Wilderness Act* (Seattle: University of Washington Press, 2005), p. xii.

10. Olaus Murie, in *Living Wilderness*, Vol. 58 (1956), p. 30.

11. Murie to Collins, November 29, 1956.

12. Lois Crisler to Olaus Murie, October 24, 1956, Margaret Murie Papers, Box 2, Folder 18, American Heritage Center, University of Wyoming, Laramie.

13. Olaus J. Murie to Fairfield Osborn, November 4, Margaret Murie Papers, Box 2, Folder 18.

14. Lois Crisler, "Where Wilderness Is Complete," *Living Wilderness*, Vol. 60, (Spring 1957), p. 4.

15. Ibid., pp. 1–4.

16. Tom Walker, *Caribou: Wanderer of the Tundra* (Portland, OR: Graphic Arts Center, 2005), p. 24.

17. Crisler, "Where Wilderness Is Complete."

18. Bass, *Caribou Rising*, p. 109.

19. Ernest Thompson Seton, *The Arctic Prairies* (Ann Arbor: University of Michigan, 1911), p. 209.

20. Kaye, *Last Great Wilderness*, p. 114.

21. Olaus Murie to Fairfield Osborn, February 18, 1957.

22. U.S. Congress, Senate, Committee on Interstate and Foreign Commerce, *Hearing Before the Merchant Marine and Fisheries Subcommittee on S. 1899*, 86th Congress, 1st Session, June 30, 1959, p. 55.

23. Quoted in Kaye, *Last Great Wilderness*, p. 119.

24. David Backes, *The Life of Sigurd F. Olson* (Minneapolis: University of Minnesota Press, 1999), p. 254.

25. Ibid., p. 271.

26. "Fred Seaton," Kansas Historical Society, Topeka, Kansas. (Biography.)

27. "Fred A. Seaton, Interior Chief Under Eisenhower, Dies at 64," *New York Times*, January 18, 1974.

28. Ibid.

29. Dermot Cole, "The Road to Statehood," in *Alaska 50: Celebrating Alaska's 50th Anniversary of Statehood 1959–2009* (Tampa, FL: Faircount Media Group, 2008), p. 23.

30. F. Seaton, speech at the Eisenhower Library.

31. Herb and Lois Crisler to Secretary of the Interior Fred Seaton, May 14, 1958.

32. Kaye, *Last Great Wilderness*, p. 246.

33. "Wildlife Range—Boon to State," *Fairbanks Daily News-Miner*, May 21, 1959. (Editorial.)

34. Debbie S. Miller, *Midnight Wilderness: Journeys in Alaska's National Wildlife Refuge* (Portland, OR: Alaska Northwest, 2000), p. 172.

35. Kaye, *Last Great Wilderness*, pp. 119–121.

36. Peter A. Coates, *The Trans-Alaska Pipeline Controversy* (Anchorage: University of Alaska Press, 1993), p. 92.

37. Miller, *Midnight Wilderness*, p. 158.

38. Ibid., p. 174.

39. Clarence J. Rhode to Olaus Murie, May 21, 1957, Margaret Murie Papers, Box 2, Folder 18.

40. Kaye, *Last Great Wilderness*, p. 83.

41. Margaret Murie, *Two in the Far North* (Anchorage: Alaska Northwest, 1962), p. 357.

42. Miller, *Midnight Wilderness*, p. 170.

43. John M. Kauffmann, *Alaska's Brooks Range: The Ultimate Mountains* (Seattle, WA: Mountaineers, 2005), p. 102. See also Kaye, *Last Great Wilderness*, pp. 121–125.

44. Elaine Rhode, *National Wildlife Refuges of Alaska* (Anchorage: Alaska Natural History Association, 2003), p. 20.

45. Neil M. Maher, *Nature's New Deal: The Civilian Conservation Corps and the Roots of the American Environmental Movement* (New York: Oxford University Press, 2008), p. 5.

46. Miller, *Midnight Wilderness*, p. 172.

47. Kaye, *Last Great Wilderness*, p. 132.

48. William O. Douglas, *My Wilderness: The Pacific West* (Garden City, NY: Doubleday, 1968), pp. 16–17.

49. Gerald E. Bowke, *Reaching for a Star: The Final Campaign for Alaskan Statehood* (Fairbanks, AK: Epicenter, 1989).

50. Kaye, *Last Great Wilderness*, p. 138.

51. Dermot Cole, "The Road to Alaska's Statehood," in *Alaska 50: Celebrating Alaska's 50th Anniversary of Statehood 1959–2009* (Tampa, FL: Faircount Media Group, 2008), p. 19.

52. Ross Coen, "Eisenhower Was Reluctant Supporter of Alaska Statehood," *Anchorage Daily News*, July 6, 2008.

53. Claus-M. Naske and Herman E. Slotnick, *Alaska: A History of the 49th State* (Norman: University of Oklahoma Press, 1979), p. 174.

54. Miller, *Midnight Wilderness*, p. 155.

55. Kaye, *Last Great Wilderness*, p. 151.

56. John A. Murray, *The Mountain Reader* (Old Saybrook, CT: Globe Pequot, 2000), p. 110.

57. Ansel Adams to J. F. Carithers, December 19, 1959. Personal papers of Douglas Carithers, Tucson, AZ.

Epilogue: Arctic Forever

1. William O. Douglas, "Foreword," in *Farewell to Texas: A Vanishing Wilderness* (New York: McGraw-Hill, 1967).

2. William O. Douglas, *My Wilderness: The Pacific West* (Garden City, NY: Doubleday, 1968), p. 94.

3. Ibid., pp. 30–31.

4. William O. Douglas, *A Wilderness Bill of Rights* (New York: Little, Brown, 1965), p. 86.

5. James F. Simon, *Independent Journey* (London: Penguin, 1981), p. 328. (Originally published New York: Harper and Row, 1980.)

6. Author interview with Ethel Kennedy, June 20, 2010.

7. Terry Tempest Williams, *The Open Space of Democracy* (Barrington, MA: Orion Society, 2004).

8. William H. Rodgers Jr., "The Fox and the Chickens: Mr. Justice Douglas and Environmental Law," in *He Shall Not Pass This Way Again: The Legacy of William O. Douglas* (Pittsburgh, PA: University of Pittsburgh Press, 1990), pp. 48–219.

9. Celia Hunter, "Statement: Before the Committee on Interstate and Foreign Commerce," October 20, 1954, Arctic Wildlife Range–Alaska.

10. Author interview with Virginia Wood, June 18, 2010.

11. Ibid.

12. *Alaska Conservation Society Newsletter*, No. 1 (March 1960).

13. Virginia Wood testifying on behalf of Alaskan wilderness preservation, quoted in Roger Kaye, *Last Great Wilderness* (Anchorage: University of Alaska Press, 2006), p. 196.

14. David Backes, *The Life of Sigurd F. Olson* (Minneapolis: University of Minnesota Press, 1997), pp. 298–299.

15. Sigurd F. Olson, "Alaska's Land and Scenic Grandeur," *Living Wilderness* (Winter 1971–1972).

16. Backes, *The Life of Sigurd F. Olson*, p. 298.

17. Author interview with Virginia Wood, June 16, 2010.

18. Kaye, *Last Great Wilderness*, p. 202.

19. "Governor's Office—News Release," September 26, 1960. Alaska Conservation Society Papers, Box 57, Folder 623, Fairbanks, AK.

20. David L. Spencer, Claus-M. Naske, and John Carnahan, *National Wildlife Refuges of Alaska* (January 1979), p. 109.

21. Sigurd F. Olson to George L. Collins, December 2, 1970, Sigurd Olson Papers, Box 80, Arctic Range Folder, Minnesota Historical Society, St. Paul.

22. Douglas, *My Wilderness*, pp. 30–31.

23. Author interview with Stewart Udall, March 16, 2009.

24. Margaret Murie, *Two in the Far North* (Anchorage: Alaska Northwest, 1962).

25. "Secretary Seaton Establishes New Arctic National Wildlife Range," Department of the Interior, Washington, DC, December 7, 1960.

26. David Petersen, *Elkheart, A Personal Tribute to Wapiti and Their World* (Black Earth, WI: Big Earth, 1998), p. 100.

27. Kaye, *Last Great Wilderness*, p. 206.

28. Author interview with Stewart Udall, September 6, 2009.

29. Ibid.

30. Douglas Brinkley, "Eisenhower: His Farewell Speech as President Inaugurated the Spirit of the 1960s," *American Heritage*, Vol. 52 (September 2001).

31. Author interview with Carl Rowan, March 19, 1997.

32. William Schwarz (ed.), *Voices for the Wilderness* (New York: Ballantine, 1969), pp. 109–121.

33. Geoffrey L. Haskett, "Background: ANWR," *Federal Register*, Vol. 75, No. 66 (April 7, 2010). (Notice 17764.)

34. Curt Meine, *Correction Lines: Essays on Land, Leopold, and Conservation* (Washington, DC: Island, 2009), p. 108.

35. Mardy Murie to Fairfield Osborn, January 7, 1961, Margaret Murie Papers, American Heritage Center, University of Wyoming, Laramie.

36. Author interview with Cathy Stone, August 18, 2010.

37. William O. Douglas, *Muir of the Mountains* (San Francisco, CA: Sierra Club Books for Children, 1961), p. 57.

38. Adam W. Sowards, *The Environmental Justice: William O. Douglas and American Conservation* (Corvallis: Oregon State University Press, 2009), pp. 2–3.

39. William O. Douglas, *My Wilderness: East to Katahdin* (Garden City, NY: Doubleday, 1966), pp. 31–33.

40. Douglas, *Muir of the Mountains*, p. 101.

INDEX

BOOKS BY DOUGLAS BRINKLEY

THE WILDERNESS WARRIOR
Theodore Roosevelt and the Crusade for America

ISBN 978-0-06-056531-2 (paperback)

A historical narrative and eye-opening look at the pioneering environmental policies of President Theodore Roosevelt.

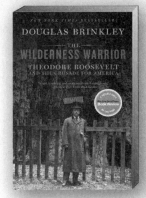

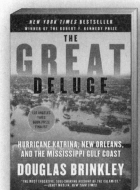

THE GREAT DELUGE
Hurricane Katrina, New Orleans, and the Mississippi Gulf Coast

ISBN 978-0-06-114849-1 (paperback)

The complete tale of the terrible storm, offering a piercing analysis of the ongoing crisis, its historical roots, and its repercussions for America.

THE BOYS OF POINTE DU HOC
Ronald Reagan, D-Day, and the U.S. Army 2nd Ranger Battalion

ISBN 978-0-06-056530-5 (paperback)

A chronicle of the men who conquered Pointe du Hoc in 1944, as well as the Presidential speech made forty years later in tribute to their duty, honor, and courage.

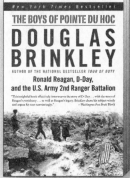

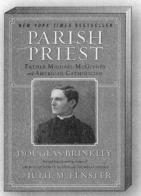

PARISH PRIEST
Father Michael McGivney and American Catholicism

ISBN 978-0-06-077685-5 (paperback)

An in-depth biography of the Roman Catholic priest who stood up to anti-Papal prejudice in America and founded the Knights of Columbus.

TOUR OF DUTY
John Kerry and the Vietnam War

ISBN 978-0-06-056529-9 (paperback)

Douglas Brinkley explores Senator John Kerry's odyssey from highly-decorated war veteran to outspoken antiwar activist.

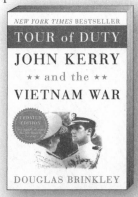